AQUINAS ON METAPHYSICS

AQUINAS ON METAPHYSICS

A HISTORICO-DOCTRINAL STUDY OF THE *COMMENTARY ON THE METAPHYSICS*

by

JAMES C. DOIG
Clayton Junior College

MARTINUS NIJHOFF / THE HAGUE / 1972

ISBN 90 247 5045 8

PRINTED IN THE NETHERLANDS

TABLE OF CONTENTS

PART THREE

INTRODUCTION

Thomas Aquinas' *Commentary on the Metaphysics* has long been considered by many as one of the most interesting, most rewarding of all his works. Yet strangely enough, there has been no extensive study of this work, at least none that has ever reached print. It is in the hope of partially filling this gap in medieval research that the present study of the metaphysical system of the *Commentary* was conceived.

However, the discussion of the *Commentary's* metaphysics must simultaneously be an investigation into the reasons which motivated Aquinas in the composition of his work. Did he wish to expose only the theories of Aristotle, or did he simultaneously intend to present his own metaphysical views? Obviously, we must learn the answer to this before we can proceed to disentangle the metaphysical system, or systems, operative in Aquinas' *Commentary*.

Up to the present day this problem, the nature of Aquinas' exposition has not been answered in a manner acceptable to all. Generally speaking, three theories have been advanced. A first one would see the *Commentary* as an objective exposition of Aristotle.[1] A second opinion views Aquinas' exposition as an attempt to express his own personal theories on metaphysics.[2] And finally, the third view divides within the *Commentary* paragraphs containing Aquinas' personal thought

[1] For the discussion of this view, see M. Grabmann, "Die Aristoteleskommentare des heiligen Thomas von Aquin," *Mittelalterliches Geistesleben. Abhandlungen zur Geschichte der Scholastik und Mystik*, Band I, Hueber, München, 1926, pp. 297–98. See also: W. Turner, "St. Thomas' Exposition of Aristotle: A Rejoinder," *The New Scholasticism*, XXXV, 1961, p. 213. For an author who holds this view see: E. Gilson, *Elements of Christian Philosophy*, Doubleday, New York, 1960, p. 282, footnote 6.

[2] Grabmann, *op. cit.*, p. 300; Turner, *op. cit.*, p. 213. For recent articles accepting this view of the *Commentary*, see: J. Isaac, O.P., "Saint Thomas interprète des oeuvres d'Aristote," *Scholastica ratione historico-critica instaurando*, Acta cong. schol. internat. Romae anno sancto MCML celebrati, Rome, Pont. Athen. Antonianum, 1951, pp. 353–63; in the same volume, B. Nardi, "L'aristotelismo della scolastica e i Francescani," pp. 607–26; T. Heath, O.P., "Saint Thomas and the Aristotelian *Metaphysics:* Some Observations," *The New Scholasticism*, XXXIV, 1960, pp. 483–60.

from those containing Aquinas' exposition of Aristotle; thus the work is basically objective, but on occasion personal.[1] Yet even though these theories have been seriously advanced, we have been given no study of the *Commentary* made on a scale sufficiently large to substantiate any of the three views.

On first glance, any one of the three theories appears to exclude the other two. Yet such is not the case, for all three are true to some extent.

It is my belief that, if one asks "how" Aquinas wrote his *Commentary*, then ultimately the third theory would be the correct one to hold. To question "how" Aquinas wrote is to question his method, his procedure. As the investigations which follow will show, on some occasions he seeks to expose the words of Aristotle, on other occasions the intention of Aristotle, and on yet further occasions, a meaning not found in Aristotle, a meaning which transforms Aristotelianism. Now there is no doubt that Aquinas wrote in this fashion, and hence one can legitimately view the *Commentary* as containing paragraphs where Aquinas presents his own theories, as well as paragraphs where he gives the doctrine of Aristotle.

In so far as this third opinion on the nature of the *Commentary* pretends to speak solely of Aquinas' procedure or of "how" he wrote, then it is quite evidently correct. Yet unfortunately it is possible to begin with this "how" of the *Commentary* and imperceptibly to transform the theory into an explanation of "why" Aquinas wrote. A recent article exemplifies this passage from "how" to "why." As one historian sees it, *In IV Meta.*, 2, 554–58 contains passages where Aquinas presents the sense of the Aristotelian text (par. 554–55) and others where Aquinas gives his own theory (par. 556 and 558). Viewed in the light of "how" Aquinas wrote, such a division of these paragraphs is legitimate, for Aristotle never gave the doctrine of *esse* found in paragraphs 556 and 558, and he did quite clearly teach the theories of paragraphs 554–55. However, from these facts one can not conclude (as the author in question does) anything about "why" Aquinas wrote.[2] Simply because paragraphs 554–55 contain Aristotle's thought, whereas 556 and 558 present Aquinas', one can not say that Aquinas wanted to give Aristotle's views in some paragraphs and his own doctrine in others.

Accordingly, we must keep separate the "how" and the "why" of Aquinas' book. The fact that the *Commentary* is basically an objective

[1] GRABMANN, *op. cit.*, pp. 298–99; TURNER, *op. cit.*, p. 213. This is the opinion accepted by Turner as well as by most present day historians.

[2] J. OWENS, C.Ss.R., "The Accidental and Essential Characteristics of Being in the Doctrine of St. Thomas Aquinas," *Mediaeval Studies*, XX, 1958, pp. 2–4; 6–8.

exegesis of Aristotle, although containing an occasional Thomist passage, this fact does not permit us to say anything about "why" Aquinas wrote.

The third view of the *Commentary* is then correct if one views it as an expression of "how" Aquinas wrote. On the other hand, the first and second views would both be correct explanations of "why" Aquinas wrote. The first view mentioned above understands the *Commentary* as an objective exposition of Aristotle. As I believe the present study will show, Aquinas wrote because he desired to oppose his view of Aristotle's metaphysics to the interpretations of Aristotle given by Avicenna, Averroes, and Albert the Great. Hence if we speak of the "why" of the *Commentary,* we would be correct in maintaining that it is an objective exegesis of Aristotle, for it was Aquinas' intention to explain Aristotle's thought.

However, our investigations will also show that with some probability Aquinas accepted as his own the metaphysics he considered to be Aristotle's. Hence the "why" of the *Commentary* is simultaneously Aquinas' wish to explain the metaphysics proper to Aristotle (the first view) and the metaphysics he accepted as his own (the second view).

To discover that the *Commentary* was indeed written for this purpose, it is necessary to compare Aquinas' work on the one hand with Aristotle's *Metaphysics,* and on the other, with the *Commentaries* on the *Metaphysics* written by Avicenna, Averroes, and Albert. If one lays open side by side the *Metaphysics* of Aristotle and the four *Commentaries,* one is amazed to realize that Aquinas must have had the three works of his predecessors constantly before him. The evidence is abundant. The most important point of Avicenna's work – the formal object of metaphysics is "here-and-now existence" – is categorically rejected by Aquinas. The very structure of Averroes' understanding of Aristotelian metaphysics is just as strongly opposed; Averroes had insisted that there is no proof of a first cause of being – this theory, together with its repercussions on the method and on the development of metaphysics, is strongly opposed by Aquinas on every occasion Averroes refers to it. And much the same is true of Aquinas' relation to Albert's work.

It is extremely important to underline this opposition evidenced by a comparative study of these four *Commentaries.* The very points on which Aquinas rejected the views of his predecessors are issues central to any conception of metaphysics. The views given by Aquinas on these points are thus the very same views which enable one to obtain a glimpse of the thread running through each and every discussion of the many topics taken up in Aquinas' *Commentary*. Hence the very presence of the opposition of the latter's work to that of the three predecessors founds the belief that the *Commentary* of Aquinas is to be considered as a unit. One can not distinguish within it sections where he wanted only to expose Aristotle's thought. Here, it is the "why" of Aquinas' work that is at issue, for we are dealing with Aquinas' intention: to correct the unfaithful interpretations given by Avicenna, Averroes, and Albert.

At this point the value of a further comparison of Aquinas and Aristotle is immediately evident. The comparison of Aquinas and the other three commentators reveals that Aquinas has but one aim: to counteract the influence of previous expositions of the *Metaphysics*. If we turn now to a comparison of Aristotle and Aquinas perhaps we shall be forced to modify that conclusion slightly; perhaps we shall discover that Aquinas, although writing against his predecessors, is presenting his metaphysics and not that of Aristotle. In other words, if the synthesis of the metaphysics operative in Aquinas' *Commentary* does not reveal a system that a medieval could have legitimately found in the Aristotelian *Metaphysics*, or if Aquinas' manner of exposing Aristotle shows that he does not attribute the *Commentary*'s metaphysics to the latter, then obviously we must conclude that Aquinas did not intend to present pure Aristotelianism. We should be forced to conclude that the *Commentary* opposes Aquinas' thought, not Aristotle's, to the theories given by Avicenna, Averroes, and Albert.

However although one does discover a system not identical to that of Aristotle, it is not at all impossible that Aquinas believed it to be truly the system of the *Metaphysics*. One discovers a metaphysics cast in the general framework outlined by Aristotle: a study of all substances as such – the universal science – which suddenly realizes the need of an ultimate cause of being, and so becomes the first science. Such a metaphysical synthesis is operative in Aquinas' *Commentary*, and as Aquinas appears to inform us, it is the system he attributed to Aristotle.

Thus this comparative study of the five philosophers leads one to conclude that Aquinas believed he was exposing Aristotle's thought. However the relation of Aquinas' work to that of Aristotle is a complicated one, for *de facto* the system Aquinas presents is transformed by the doctrine of *esse*, the co-principle of substance. Yet there are no indications that Aquinas believed the doctrine of *esse* to be anything other than the link between universal and first science. The metaphysics operative in Aquinas' *Commentary*, while not that present in

Aristotle's work, nevertheless represents what Aquinas apparently believed to be Aristotelian philosophy. Accordingly, the "why" of the *Commentary* is to present Aristotle's system.

It is such a comparative study that I have chosen to give here. On the weight of the evidence brought to light by this study we shall be permitted to conclude several important facts regarding the *Commentary*. First, it is a unit, written in the light of one conception of metaphysics; thus the doctrine of each and every paragraph follows from one philosophy of being. Second, it was directed against the interpretations of Avicenna, Averroes, and Albert. Third, Aquinas apparently believed this metaphysics to be that of Aristotle. And finally, despite this belief, the metaphysical synthesis of the *Commentary* can not be attributed to Aristotle, for it is centered around a theory of *esse* conceived as a limited sharing in Actuality.

The question naturally arises: is the philosophy of being implied in Aquinas' writing not only Aquinas' understanding of Aristotle, but Aquinas' own personal thought as well? It would require a long and detailed study to answer this question fully, definitively. One would need to delve deeply into the other writings of Aquinas. Since the present comparative study of the five philosophers is of necessity quite long, it has been thought best to limit severely the investigation into the question of the Thomist character of the *Commentary*. This is not to say that we shall not be able to conclude with some probability that the *Commentary* presents Aquinas' own theories. Thus the "why" of the *Commentary* shall be seen to be two-fold: to present Aristotle's metaphysics, but also to present Aquinas', for in Aquinas' eyes the two systems are identical.

In the exposition to follow a three-fold division has been adopted. Part One will consider several preliminary issues. In Chapter I, the chronology of the composition of the *Commentary*, and the identity of the Latin versions of the *Metaphysics* used by Aquinas are discussed. A synthetic view of the interpretations given by Avicenna, Averroes, and Albert follows in Chapter II. Chapter III discusses the metaphysics contained in the *Prooemium* to Aquinas' *Commentary*. Since the *Prooemium* expresses Aquinas' own views on metaphysics, the *Commentary*'s doctrines can be styled "Thomist" only if they agree with those views.

In Part Two the *Commentary* proper is studied; it is on the basis of the comparative studies of the three chapters of this division that one is permitted to conclude to the unity of the metaphysics operative in

Aquinas' exposition, as well as to the fact that Aquinas apparently viewed this metaphysics as that of Aristotle. However, because of the striking similarity between the *Commentaries'* metaphysics and that of the *Prooemium*, one can conclude as well that Aquinas' goal in writing was the exposition of the system proper both to himself and to Aristotle.

Finally, Part Three presents a synthetic view of the metaphysics operative in Aquinas' *Commentary*, as well as further evidence that Aquinas accepted that metaphysics as his own.

Concerning the mechanics of the present study, it is best to add a word about the footnotes. Quite often, in the midst of a discussion of one of Aquinas' expositions in the *Commentary*, I refer the reader's attention to a footnote containing an examination of some view attributed to Aquinas on the basis of a recent study of his work. Quite often I explain why I do not agree with a given interpretation of Aquinas' thought, and my view is based on the doctrine of the *Commentary*. For example, M. Gilson's view that *esse* is known in a judgment may be criticized in the light of the doctrine of the *Commentary*. Ideally, it would have been more proper to discuss such problems after the personal or Thomist character of the metaphysics of the *Commentary* had been established. However practical considerations have led me to judge it best to treat such matters at the moment I deal with the similar theory in the *Commentary*. When such matters are discussed in footnotes, the reader should not be mislead into thinking that I believe already proved the fact that the *Commentary* presents Aquinas' own metaphysical system.[1]

It is difficult to grasp clearly the depth of one's indebtedness. Yet one thing is certain: this book would never have been written were it not for the interest, encouragement, and counsel of numerous friends. To them this book is dedicated.

April, 1967

[1] Unless otherwise indicated, all English translations of Aristotle's works are taken from the Oxford translation. All quotations of Albert the Great's *Commentary* are taken from the 1960–64 edition of B. Geyer. In quoting Averroes' *Commentary* I have used the Junctas edition of 1562; in the case of Avicenna's *Metaphysica* only the Junctas 1508 edition has been quoted. Finally, Aquinas' *Commentary* has been cited according to the paragraph numbers of the Spiazzi-Cathala edition.

PART ONE

PRELIMINARIES

Aquinas' *Commentary on the Metaphysics*, as any other philosophical creation of man, has numerous relationships, some historical, some literary, some more properly philosophical in nature. Aquinas wrote his *Commentary* at a certain point in time, in answer to a certain need, to attain a certain goal. As a philosophical work, it undoubtedly contains ideas current before its composition, and perhaps some never before expressed; moreover, as an exposition of Aristotle's *Metaphysics*, it must have been in some respects moulded by the Latin translations upon which it was based. These relationships – philosophical, historical, literary – must be taken into account before we can ascertain the nature of the work itself. Hence, in Chapter I the questions of the chronology of the composition, and of the Latin versions of Aristotle's *Metaphysics*, will be reviewed. Chapter II will trace briefly the interpretations of Aristotelian metaphysics found in the *Commentaries* of Avicenna, Averroes, and Albert the Great.

As is true of many a philosophical work, Aquinas' *Commentary* contains a preface, the *Prooemium*. There is no doubt that this preface contains its author's personal ideas on metaphysics. Hence it is imperative that we attempt to grasp them; if the *Commentary* does contain Aquinas' own theory of metaphysics, that theory will have to agree with the one presented in the *Prooemium*. To a study of that latter conception Chapter III is devoted.

CHAPTER I

LITERARY AND CHRONOLOGICAL ASPECTS OF THE *COMMENTARY*

1. MEDIEVAL LATIN VERSIONS OF THE *METAPHYSICS*

Since we are to read the *Commentary* in relation to Aristotle, we must discover *what* Aristotle we are supposed to read, for one dare not take just any edition of Aristotle to read in conjunction with Thomas Aquinas' work. For Aquinas possessed a very special Aristotle, not the Aristotle of the Greek, but rather the Aristotle of medieval Latin; in fact, it appears that he knew at least five Aristotle's, each slightly different, and a detailed study of his *Commentary* reveals the use of each of these versions of the Aristotelian *Metaphysics*.

The number and the identity of the medieval Latin versions of Aristotle's *Metaphysics* and their use by Aquinas have long been of great interest to scholars. Even today, however, the question remains to some extent an open one. It is likely to remain so until we have been given critical editions of all the Latin versions. But there are numerous facts which have come to light and which have been accorded some form of credence. It is to a listing of such facts that this section is devoted.

Metaphysica Vetustissima. This is the oldest of translations made from the Greek[1] and is found conserved in two manuscripts of the second half of the 12th century.[2] In the form in which we possess it today, the *Vetustissima* contains Books I–IV, 4, 1007a 32–33.[3] After much discussion, this version is commonly believed to be the work of Giacomo di Venezia.[4]

[1] M. GRABMANN, *Guglielmo di Moerbeke, O.P., il traduttore delle opere di Aristotele*, Miscellanea Historiae Pontificiae, Vol. XI; Pont. Univ. Gregoriana, Roma, 1964, p. 96.

[2] L. MINIO-PALUELLO, "Note sull' Aristotele latino medievale," *Rivista de Filosofia Neo-Scolastica*, XXXXII, 1950, p. 222.

[3] *Aristoteles Latinus, Pars Prior*, Corpus Philosophorum Medii Aevi, Roma, 1939, p. 62.

[4] MINIO-PALUELLO, "Note sull' Aristotele...," pp. 222–24.

Did the version in its original form have more than four books? If we agree with Minio-Paluello that the *Vetustissima* is the *Litera Boethii* cited by Thomas, then the *Vetustissima* might have been complete. He argues that since the *Litera* was cited according to Books V and XII as well as I and III, the *Vetustissima* possessed these books.[1] On the other hand, if one accepts Geyer's thesis and identifies the *Litera Boethii* with the *Vetus*, then one has no means of knowing how many books were originally contained in the *Vetustissima*.[2]

The most important question for us is whether Thomas knew and used this version. We cannot answer with any certainty. If it is this version to which he referred as "*Litera Boethii*," then neboth knew and used it and in this event, a critical edition of the *Vetustissima* would have to be consulted in a definitive study of the Thomist *Commentarium*. If, on the other hand, the *Vetus* is the *Litera Boethii*, then as far as we know, Thomas did not cite the *Vetustissima*.

Metaphysica Vetus. The *Vetus*, as the *Vetustissima*, was translated from a Greek text.[3] In manuscript form, we have today only Books I–IV, 4, 1007a 32, or one line less than the *Vetustissima*.[4]

As recent publications indicate, the nature of the *Vetus* is not yet the subject of anything such as universal approval.[5] There is, however, agreement on the existence of the *Vetus* (containing at least four books) before the career of Aquinas began. Aquinas cited this version in fact in the *Commentarium in Sententiarum* and in the *Commentarium in*

[1] *Ibid.*, pp. 222; 224–26. cf. A. Dondaine, O.P., "Bulletin d'histoire," *Revue des Sciences Philosophiques et Théologiques*, XXV, 1936, pp. 714–15.

[2] B. Geyer, "Die Übersetzungen der aristotelischen Metaphysik bei Albertus Magnus und Thomas von Aquin," *Philosophisches Jahrbuch*, 1917, p. 392 sqq. This thesis is adopted by Grabmann, *Guglielmo de Moerbeke...*, p. 101 and by D. Salman, O.P., "Saint Thomas et les traductions latines des Métaphysiques d'Aristote," *Archives d'histoire doctrinale et littéraire du Moyen Age*, VII, 1932, p. 85. An argument to show the *Vetustissima* did not have Books XI, XIII, and XIV is given by Salman in the same article. Quoting *De unitate intellectus*, Salman maintains that here Thomas refers to these books as inexistent in Latin; hence these books were not contained in the three versions had by Aquinas at that time (*Vetustissima*, *Vetus*, *Arabica*). Cf. Salman, *op. cit.*, pp. 87–89. Salman's theory concerning the references of *De unitate intellectus* to Books XI, XIII, and XIV has since, we believe, been proved false: cf. A. Dondaine, O.P., "Notes et Commentaires. Saint Thomas et les traductions latines des Métaphysiques d'Aristote," *Bulletin Thomiste*, I, 1931–33, pp. 205*–210*.

[3] F. Pelster, S.J., "Die Übersetzungen der aristotelischen Metaphysik in den Werken des hl. Thomas von Aquin," *Gregorianum*, XVI, 1935, pp. 326–27.

[4] *Aristoteles Latinus, Pars Prior*, p. 63.

[5] Cf. F. Pelster, S.J., "Die griechisch-lateinischen Metaphysikübersetzungen des Mittelalters," *Abhandlungen zur Geschichte der Philosophie des Mittelalters*, Münster, 1923, pp. 91–96; 101–103. By the same author: "Neuere Forschungen über die Aristotelesübersetzungen des 12. und 13. Jahrhunderts. Eine kritische Übersicht," *Gregorianum*, XXX, 1949, pp. 52–53. Grabmann, *Guglielmo de Moerbeke...*, p. 97. Minio-Paluello, "Note sull' Aristotele...," p. 222.

Metaphysicorum.[1] In the metaphysical work moreover, it played an important role. At the beginning of his *Commentary*, Aquinas took the *Vetus* reading rather than the *Media*; even more, the divisions of the Aristotelian text in Books I–IV are given according to the *Vetus.*[2] Although other versions (*Media* and *Moerbecana*) were also used in these first four books, gradually the *Vetus* assumed more importance. This lasted until Book IV, *lectio* 6 when the *Vetus* suddenly disappeared.[3]

The fact that Aquinas knew and used the *Vetus* version of the Aristotelian *Metaphysics* requires, then, that in a definitive study of Aquinas' *Commentary*, one make every effort to ascertain the role played by this version in forming his exposition. Fortunately, a critical edition of the *Vetus* is available, an edition showing that this version is a "remarkably accurate" translation of Aristotle's Greek.[4] However, the text as edited does not always agree with the phrases Aquinas quotes from the *Vetus.*[5] Nevertheless, this version will be of great use, especially in Chapter IV when we examine Aquinas' exposition formed on the basis of the *Vetus.*

Metaphysica Media. There is no more agreement on the nature of the *Media*, nor on its date, than there is on the nature and date of the *Vetus.*[6] According to *Aristoteles Latinus* and Minio-Paluello, the *Media* must be prior to 1230 since the *Vetus* (dependent on the *Media*, they say) was already in use at that time.[7] For Pelster, the *Media* is at least anterior to 1261–64 as it was used by Albert the Great in his *Metaphysics* (1262–68), and by Aquinas in the *Summa Theologiae, Pars Prima* (1266), in the *Contra Gentiles* (1261–64); perhaps, it was even used in *De veritate* (1256–59).[8] Considering the general state of disagree-

[1] D. SALMAN, O.P., "Versions latines et commentaires d'Aristote," *Bulletin Thomiste*, XIV, 1937, p. 100.

[2] V. DE COUESNONGLE, O.P., "La causalité du *maximum*. II. Pourquoi Saint Thomas a-t-il mal cité Aristote?" *Revue des Sciences Philosophiques et Théologiques*, XXXVIII, 1954, p. 674. Pelster, "Die Übersetzungen...," pp. 553–58.

[3] PELSTER, "Die Übersetzungen...," pp. 553–58.

[4] The *Vetus* is found in: *Opera Hactenus inedita Rogeri Boconi*, Fasc. XI: *Questiones altere supra libros prime philosophie Aristotelis. Questiones supra de plantis. Metaphysica Vetus Aristotelis e cod. Vetustissimis*, ed. R. Steele, Clarendon Press, Oxford, 1932. (pp. 255–312). For the accuracy of the translation, see: p. xxiv.

[5] *Ibid.*, p. xxiii.

[6] *Aristoteles Latinus, Pars Prior*, p. 62. PELSTER, "Die griechisch-lateinischen Metaphysikübersetzungen...," pp. 103–105.

[7] *Aristoteles Latinus, Pars Prior*, p. 62. MINIO-PALUELLO, "Note sull' Aristotele...," p. 226.

[8] PELSTER, "Die griechisch-lateinischen Metaphysikübersetzungen...," pp. 105–106. Salman held the version of the *Metaphysics* in Albert the Great and in Aquinas could not, in these cases, be the *Media;* he based his argument of the reference in the *De unitate intellectus* of Aquinas. Cf. SALMAN, "Saint Thomas et les traductions...," pp. 87–89. For an attack on Salman's positions, an attack we believe conclusive, see DONDAINE, "Notes et Commentaires. Saint Thomas et les traductions...," pp. 205*–210*. A recent discovery indicates the necessity of a complete revision of the chronology of Albert's extensive work. Concerning this discovery, Fr. D. A. Callus, O.P., remarks that in this newly discovered writing, around 1271

ment centered around the *Media*, it is to be expected that the translator's identity is also a matter of discussion.[1]

Though the issues mentioned be uncertain, there is no doubt as to the importance of this version for the *Commentary* of Aquinas. At the beginning of the first book (*lectio* 1) we find the *Media* used as a secondary text. From this point until Book IV, *lectio* 6 the *Media* retains its position as *alia litera*. Hence, one can easily conclude that although it was not being used as the base of the exposition (the *Vetus* was the basic text througout these books), nevertheless Aquinas did have the *Media* version before him as he wrote. After Book IV, *lectio* 6, however, the *Media* becomes the primary text (and the *Vetus* the *alia litera*). This situation remains stable until Book V, *lectio* 20, par. 1068, when the *Media* becomes once again an *alia litera*. After Book VI a reference to the *Media* is exceptional.[2]

As was true of the *Vetus*, so too the *Media* must be taken into account before the final view of the *Commentary* can be had. We are fortunate, thus, that an edition of the *Media* has recently been published.[3] In our work, it will only be the famous *lectio* 9 of Book V that must be interpreted in the light of the *Media*, however.

Metaphysica Moerbecana.[4] The *Moerbecana* is so called because it was translated by William of Moerbeke for Aquinas' use.[5] Pelster has demonstrated this version to be a revision of the *Media* with the addition of Book XI from a Greek text; Book XI was thus for the first time translated into Latin.[6] The date of the translation is difficult to as-

Albert stated that he was growing blind due to his advanced age; thus, concludes Fr. Callus, Albert's work had been completed by that date. *The Commentary on the Metaphysics* dates, thus, from before 1262–63. Cf. D. A. CALLUS, O.P., "Une oeuvre récemment découverte d'Albert le Grand: *De XLIII problematibus ad Magistrum Ordinis* (1271)," *Revue des sciences philosophiques et théologiques*, pp. 259–60.

[1] MINIO-PALUELLO, "Note sull' Aristotele...," pp. 227–31. *Aristoteles Latinus, Pars Prior*, p. 62.

[2] PELSTER, "Die Übersetzungen...," pp. 547–58.

[3] This edition is found in: ALBERTI MAGNI, *Metaphysica*, in *Opera Omnia*, Tomus XVI, pars. I. and II, edidit B. Geyer, Aschendorf, Münster, 1960–64. The *Media* is printed at the bottom of the pages, below Albert's expositions.

[4] This version is often called the *novae translationis*; the term "Moerbecana" will be used here, however. Besides being simpler, the latter name avoids any possible confusion of this version with the *Arabica* which is sometimes called "nova."

[5] GRABMANN, *Guglielmo de Moerbeke*...," p. 99.

[6] PELSTER, "Die griechisch-lateinischen Metaphysikübersetzungen...," pp. 107–110. In recent years Pelster has discovered a note on folio 140r of cod. Vat. lat. 2081 written between 1240–60 which confirms anew the fact that Book XI was unknown until it appeared in the *Moerbecana*. Cf. PELSTER, "Neuere Forschungen...," pp. 52–53. Dondaine, some thirty years ago, advanced the hypothesis of a double redaction of the *Moerbecana*; the first draft contained no Book XI, while in the second this book appeared for the first time in Latin. Cf. DONDAINE, "Notes et Commentaires. Saint Thomas et les traductions...," p. 199*. Recent work on the manuscript Napoli, Bibl. Naz. VIII F. 16 has revealed the real nature of the

certain with the precision one would wish. From the facts of Moerbeke's career, it would seem to have been as early as 1260.[1] However the first use of it appears to be in the *Summa Theologiae* I, 17, 2 where Book IV is quoted. Accordingly, Books I–IV must have been finished by 1266, the date of the *Prima Pars*.[2] It is strange then to note that Aquinas did not possess Book XI until as late as the end of 1270.[3]

The *Moerbecana*, as the *Vetus* and the *Media*, plays an important role in Aquinas' *Commentary*. In the first *lectio* of Book I, it appears as an *alia litera* and until Book IV, 6 rare uses of it continue to occur; from Book IV, 6 on, it becomes more important, until finally in Book V, 20, 1068 it becomes the basis for the *Commentary*. Of the last six books, however, only Book XI is based solely on it.[4]

Contrary to what was once held, the Latin version printed with the exposition of Thomas is not the *Moerbecana*; rather, it is a combination of the *Media* and the *Moerbecana* and actually is closer to the *Media*.[5] Nor do we possess a critical edition of the *Moerbecana*. Accordingly, for this reason an absolutely definitive study of the *Commentarium*, at least in regard to its sources, is as yet impossible.

Metaphysica Arabica or Nova. The *Arabica*, as the name indicates, was translated from an Arabic version.[6] It is the version which is found together with the Latin translation of Averroes' *Commentary on the Metaphysics*.[7] Since the *Commentary* of Averroes was translated by

problem Dondaine's double redaction of the *Moerbecana* was called upon to solve, and has thus shown the lack of any need to postulate such a double redaction. Cf. J. Duin, "Nouvelles précisions sur la chronologie du "Commentum in Metaphysicam" de S. Thomas," *Revue Philosophique de Louvain*, LIII, 1955, pp. 511–34.

[1] Grabmann, *Guglielmo de Moerbeke...*," p. 99.

[2] Pelster, "Die Übersetzungen...," p. 333. When he wrote this article, the author also maintained that Book VIII was cited in *De unitate intellectus* in 1270; this in vol. XVI of *Gregorianum*, 1935. The following year, when he wrote the third article in this series, he changed his opinion of Book VIII and decided it was not cited in *De unitate intellectus*. Cf. "Die Übersetzungen...," *Gregorianum*, XVII, 1936, pp. 404–405. It has also been claimed that indications in some of Roger Bacon's writings revealed the existence of the *Moerbecana* in 1267; Msgr. Mansion believes the text of Bacon too vague to be accepted as undoubtedly referring to the version in question. Cf. A. Mansion, "Quelques travaux récents sur les versions latines des Éthiques et d'autres ouvrages d'Aristote," *Revue Philosophique de Louvain*, XXXIV, 1936, p. 91.

[3] Salman, "Saint Thomas et les traductions...," pp. 98–103. The author argues that the use of the expression "Book XI" in citing Book XII reveals ignorance of the existence of the true Book XI.

[4] Pelster, "Die Übersetzungen...," *Gregorianum*, XVI, 1935, pp. 547–61.

[5] Dondaine, "Notes et Commentaires, Saint Thomas et les traductions...," p. 200*. Even though admitting this conclusion, J. P. Rowan appears to maintain that the text of the *Metaphysics* printed in the Cathala-Spiazzi edition of the *Commentary* is the self-same version used by Aquinas. Cf. J. P. Rowan, "Introduction," in St. Thomas Aquinas, *Commentary on the Metaphysics of Aristotle*, Vol. I, pp. vii and xx.

[6] *Aristoteles Latinus, Pars Prior*, p. 64.

[7] De Couesnongle, "La causalité du *maximum*. IV. Pourquoi St. Thomas...," p. 661.

Michael Scot, the *Arabica* is also thought to be from his pen.[1] It is said to have been known at Paris by 1230 and so composed around 1220.[2] It was thus earlier than the *Media*.[3] In the form in which it is found in manuscripts it has the following order: Book II, Book I (5sqq.), Books III–X, Book XII.[4]

The *Arabica* was no stranger to Aquinas. Quoted in the *Commentary on the Sentences, De ente et essentia, Quodlibetum VII, De veritate,* it also played a role in the *Commentary on the Metaphysics*.[5] In fact, this version was most likely at hand during the entire composition of the latter work.[6]

Although we do not have a critical edition of the *Arabica*, we do have a printed text which is more than we have of the *Moerbecana*.[7] Thus it is possible to determine rather easily the role of this version in the Thomist *Commentary*.

Litera Boethii. The *Litera* is the most mysterious of the seven medieval Latin versions of the *Metaphysics*.[8] In fact, we know of it only through seven references to it by Aquinas, through one by Albert the Great,[9] and, so Pelster maintains, through the use of it in the *De motu*

[1] PELSTER, "Die Übersetzungen...," *Gregorianum*, XVI, 1935, pp. 346–48.

[2] GRABMANN, "*Guglielmo de Moerbeke*...," p. 97. Salman, however, prefers to place the date of composition nearer to 1230. Cf. SALMAN, "Versions latines...," pp. 104–105.

[3] PELSTER, "Neuere Forschungen...," pp. 52–53.

[4] *Aristoteles Latinus, Pars Prior*, p. 64.

[5] PELSTER, "Die Übersetzungen...," *Gregorianum*, XVI, 1935, pp. 344–46. De Couesnongle has shown the importance of the *Arabica* in Book II, *lectio* 2. Cf. DE COUESNONGLE, "La causalité du *maximum*. I. L'utilisation par saint Thomas d'un passage d'Aristote. II. Pourquoi Saint Thomas...," pp. 433–44; 658–80.

[6] Several studies of Thomas' *Commentary* have disclosed the fact that Aquinas had Averroes' *Commentary* before him as he wrote. Since the latter work is found joined to the *Arabica*, Thomas probably referred to this version throughout the composition of his own work. For the studies illustrating this fact, see the work of DE COUESNONGLE, mentioned in the preceding note, as well as the following articles: G. DUCOIN, S.J., "Saint Thomas Commentateur d'Aristote. Étude sur le commentaire thomiste du livre Λ des *Métaphysiques* d'Aristote," *Archives de philosophie*, XX, 1937, pp. 78–117, 240–71, 392–445. A. FESTUGIERE, "Notes sur les sources du commentaire de S. Thomas au livre XII de *Métaphysiques*," *Revue de sciences philosophiques et théologiques*, XVIII, 1929, pp. 282–90, 657–63. A. MANSION, "'Universalis dubitatio de veritate.' S. Thomas in Metaphy., Lib. III, lect. I," *Revue philosophique de Louvain*, LVII, 1959, pp. 513–42.

[7] This text is given in *Aristotelis Opera Omnia*, Vol. VIII, *Metaphysicorum Libri XIIII*, Venetiis, 1562. The *Arabica* is given in italics.

[8] One of these, the *Fragmentum Vaticanum*, was not known to Thomas and so does not interest us here. Cf. PELSTER, "Die griechisch-lateinischen Metaphysikübersetzungen...," p. 93. An eighth version, the *Secunda*, was proposed in 1932 by Salman. The problem that *Secunda* was postulated to answer has since, I believe, been solved less radically. Cf. SALMAN, "Saint Thomas et les traductions...," pp. 87–91. For the solution of Salman's difficulty, cf. DONDAINE, "Notes et Commentaires. Saint Thomas et les traductions...," pp. 205*–210*.

[9] SALMAN, "Versions latines...," p. 103. DONDAINE, "Notes et Commentaires. Saint Thomas et les traductions...," p. 212*.

cordis of Alfredo Anglico (before 1215).[1] We do not, then, possess it in manuscript form. Some however have seen a resemblance between the *Litera* and the *Vetus* and so identified them;[2] others prefer to identify the *Litera* and the *Vetustissima,* but would maintain Aquinas possessed this version only in the form of marginal notes on the manuscript of another version.[3]

Quite obviously, the unknown *Litera Boethii* presents another stumbling block for the study of the *Commentarium.* Adding to the general confusion is the presence of fourteen other references made by Thomas; are these to the *Litera* or to some other, as yet unknown, version?[4]

St. Thomas, it appears, knew and used the *Vetus, Media, Moerbecana, Arabica,* and *Litera Boethii.*[5] The *Arabica* and the *Vetus* were the first to appear in his works: the *Arabica* in *De ente* (1253–55) and the *Vetus* in *Commentarium in Sententiarum* (1252–57); the *Media* and *Moerbecana* followed in *Summa Theologiae, Prima Pars* (1267), while the *Litera* was the last to appear, in *Commentarium in Metaphysicorum* (1268–72).[6] All these versions, moreover, appeared in the *Commentarium in Metaphysicorum.* The following list will enable us to see the complexity of the question.[7]

Book	*Basic Text*	*Alia Litera*
I–IV, 6	Vetus	Media
	(Arabica, II, 1)	Moerbecana (rarely)
		Litera Boethii (I, 4, 72; five times in III)
IV, 6–V, 20	Media	Moerbecana
	(Moerbecana, V, 7)	Vetus (IV, 6)
		Litera Boethii (V, 21, 1109)

[1] Pelster, "Die Übersetzungen...," *Gregorianum,* XVI, 1935, pp. 343–44.

[2] Geyer, "Die Übersetzungen...," pp. 392–415. Pelster maintains a dependence of *Litera* on *Vetus.* Cf. Pelster, "Die Übersetzungen...," *Gregorianum,* XVI, 1935, pp. 342–43.

[3] Dondaine has found a *Moerbecana* with marginal notes identifiable as the *Litera.* Cf. Dondaine, "Bulletin d'histoire," pp. 714–15. Minio-Paluello has found an instance where the work of the translator of the *Vetustissima* is given as marginal notes under the title "translatio Boethii." Cf. Minio-Paluello, "Note sull' Aristotele...," pp. 222–25.

[4] Salman, "Versions latines...," p. 104.

[5] The *Vetustissima* alone was not used.

[6] For the chronology of the works of Aquinas, see: Walz, *Saint Thomas d'Aquin* pp. 221–26. I have consulted as well: Grabmann, *Die Werke des hl. Thomas von Aquin. Eine literarhistorische Untersuchung und Einführung.* Beiträge zur Geschichte der Philosophie und Theologie des Mittelalters; Aschendorf, Munster, 1949, 3. Auflage; p. 287 (*Com. in Sent.*); p. 295 (*Sum. Theol.*); p. 343 (*De ente*); pp. 282–83 (*Com. in Meta.*).

[7] This list does not pretend to be a complete expression of the use of the various translations; it only serves to summarize the results of recent studies on this problem. It is based on: Pelster, "Die Übersetzungen...," *Gregorianum,* XVI, 1935, pp. 547–48; 553–54; 558–59. Duin, "Nouvelles précisions...," p. 523.

V, 20–V, end	Moerbecana	Media
VI–X, end	Moerbecana	Media (rarely)
XI	Moerbecana	
XII	Moerbecana	Media (rarely)

2. THE CHRONOLOGY OF THE COMPOSITION OF THE *COMMENTARIUM*

The last forty years have seen great progress in determining with precision the date of the composition of the Thomist *Commentary*. An attempt is made here to summarize the results of this work.

Great interest has been shown in a manuscript of the *Commentarium*, Napoli, Bibl. Naz. VIII F. 16. To some this manuscript appears to be, in part, the autograph of St. Thomas – not in the sense that it was written by his hand, but that it was dictated by him and thus represents his final mind regarding the *Metaphysics* of Aristotle.[1] The key to the interest generated by this document lies in the fact that in it several different handwritings have been distinguished. These are as follows:[2]

Books	*Folio Number*	*Scribe or Hand*
Prooem. – I	lra–14vb	1 (1ra–3vb) 2 (4ra–11vb) 1 (12ra–14vb)
II–III	15ra–31vb	3
IV–V, 7, 855	32ra–42v	2
V, 7, 856–VII, 16, 1647	43ra–69ra	2B
VII, 17	69ra–70rb	4
VIII	70va–74vb	1 (70va–73vb) 5 (74ra–vb)
IX	76ra–85ra	6
X	86ra–93vb	7
XI–XII	94ra–117vb	6

Some parts of the text of this copy of the *Commentary* are distinguished by a large number of corrections. Words or parts of sentences are canceled out and replaced immediately by other formulas. It is as if someone were dictating; after the scribe had written down the dictated words, he was asked to efface them and to put others in their place.

[1] Cf. T. Käppeli, "Mitteilungen über Thomashandschriften in der Biblioteca Nazionale in Neapel. II. Ein Autograph des Metaphysikkommentars des hl. Thomas?", *Angelicum*, X, 1933, pp. 116–25. A. Mansion, "Saint Thomas et le "Liber de causis." A propos d'une édition récente de son Commentaire," *Revue Philosophique de Louvain*, LIII, 1955, pp. 63–4. Duin, "Nouvelles précisions...." p. 512. This manuscript presents roughly the same text as our published editions.

[2] Käppeli, "Mitteilungen über Thomashandschriften...," p. 117. Duin discovered a new hand (2B) overlooked by Käppeli. Cf. Duin, "Nouvelles précisions...," pp. 511–12.

This peculiar situation led Käppeli to conclude that this text was dictated by the author, by Thomas himself.[1] Books II–III, the work of hand 3, and Books V, 7, 856–VII, 16, 1647, the work of hand 2B, exhibit the unusual type of corrections mentioned. This permits one to conclude that scribes 3 and 2B were working as secretaries of Thomas, writing while he dictated.[2] The remaining books, however, possess only corrections explainable by the simple fact of copying.[3]

It is interesting to note that Books IV–V, 7, 855, the work of scribe 2, seem to have been intended as a completion of the folios containing the books dictated by Thomas. Not only did hand 2 stop at the point where hand 2B began, but, in addition, scribe 2 took care to cut away the rest of the folio, thus ensuring that nothing would be added between the point where he stopped and the point where the dictated portion began.[4] This indicates, of course, that the work of 2B (Books IV–V, 7, 856–VII, 16, 1647) was written by Thomas earlier than the work of scribe 2 was added. It does not, however, indicate whether the former books were composed in this form prior to the latter, or vice versa.

Other investigations have been carried out during the past years. The results of this work, if added to the conclusions of the study of the Naples manuscript, will enable us to progress toward determining the date of composition of the *Commentary*. The other work to which reference is made concerns the citation of Book Λ (Book XII) of Aristotle's *Metaphysics* as well as the use of Simplicius' *Commentary on De Caelo*.

Book XI (K) of the *Metaphysics* is contained solely in the *Moerbecana*: for the first time, when this version appeared, Book XI was available to medieval students in Latin. As Salman has pointed out Aquinas never refers to a text which is not available to his readers. Accordingly, if it can be shown that Thomas does not use Book XI, and that Book XII (Λ) is referred to as "XI," then one has conclusive evidence that Book XI (K) was not available to him nor to his readers at that time. This is precisely what Salman has accomplished. We can never find Book XI (K) cited, he noted, although this is not too surprising as it contained nothing that was not contained in a better ex-

[1] Käppeli, "Mitteilungen über Thomashandschriften...," pp. 118–19.

[2] Duin, "Nouvelles précisions...," p. 512.

[3] *Ibid.*

[4] *Ibid.*, p. 511. As we will note later, the work of scribe 2 overlaps slightly that of scribe 2B, paragraphs 856–58 being common to both. Since the two versions of paragraphs 856–58 are different from one another, scribe 2B didn't realize these were the opening paragraphs of the work he intended to complete.

position in other parts of the Aristotelian *corpus*. On the other hand, we do find book XII (Λ) referred to as "Book XI." In the *Contra Gentiles*, for example, there are 11 references to Book XII (Λ), in all cases as "XI." In the *Commentary on the Physics*, there are three references to Book XII (Λ) under the form of "Book XI," while two refer to it as "Book XII." The latter references (to "Book XII") concern the well known theory of the prime mover, a doctrine immediately known as belonging to Book XII (Λ). These, then, could have been corrected, Salman argues, once it was known there was another Book XI, this time the true one – K, and that Λ was actually Book XII. The other three references to Book XII (Λ) as "Book XI" refer to less well known theories and hence were most likely left uncorrected by scribes. The same types of references are found in other books, in *De potentia, De malo, Commentarium in Perihermeneias, Q. D. de anima,* and *De unitate intellectus*. In all of these Book XII (Λ) is called "Book XI," although editors have changed some of the references to read "XII" in the course of the centuries since they were written. All of these works, Salman concludes, were written before the translation of Book XI (K) in the *Moerbecana*. Other works referring constantly to Book XII (Λ) as "Book XII" were most certainly written after Book XI (K) appeared in Moerbeke's version, for example, *Commentarium in de caelo, Commentarium in de generatione*.[1]

The last of the works which are thus seen as prior to the translation of Book XI (K) was the *De unitate intellectus*, written late in the year 1270.[2] On the other hand, the commentaries on *De caelo* and on *De generatione* are certainly later than 1270. The former is probably from the year 1272, since it uses the commentary of Simplicius on *De caelo*, the translation of which was completed by Moerbeke on June 15, 1271.[3] The *Commentary on de generatione* followed very closely, probably in 1272–73.[4] Hence one can date the Moerbecana translation of Book XI (K) between late 1270 and late 1271.

If, with this new chronological information, we return to the Naples manuscript, we can make some interesting discoveries. Book VI, 2, 1188 refers to Book XII (Λ) as "Book XI"; Book VII, 1, 1245 and 1269 refer to Book XII (Λ) as "Book XII." Since all three of these paragraphs are found within the section ascribed to hand 2B, and thus dictated by

[1] SALMAN, "Saint Thomas et les traductions...," pp. 98–103.

[2] For the date, cf. A. WALZ, *Saint Thomas d'Aquin*, p. 225; GRABMANN, *Die Werke des hl. Thomas von Aquin...*, p. 327.

[3] WALZ, *op. cit.*, p. 222; GRABMANN, *op. cit.*, p. 276.

[4] WALZ, *op. cit.*, p. 222; GRABMANN, *op. cit.*, p. 277.

St. Thomas, we are permitted to conclude that Book XI (K) was translated by Moerbeke between the time of dictation of the first two of these three paragraphs. Moreover, since the entire work of scribe 2B continues through several folios without interruption, we can exclude any long interruption in the course of the composition of this text. Hence, the work of scribe 2B (Book V, 7, 856–VII, 16, 1647) would both precede and follow the appearance of Book XI (K). It could thus date from the beginning of the school year of 1270 –71,that is from late 1270, and continue into late 1271.[1]

In Book I, 12, 192 and 17, 267, we have again references to Book XII (Λ), but now as "Book XII". Other manuscripts, however, have "Book XI." Therefore, these "Book XII" references of the Naples manuscript were not due to Aquinas, but to the scribe who copied this section. The composition of Book I, then, must be located before the translation of Book XI (K), that is before the last month of 1270.[2]

In Books II–III, there are the unusual type of corrections we have previously noted. This section was written by scribe 3. In these books, Book XII (Λ) is always referred to as "XII." Hence, this version of Book II–III must date after the translation of Book XI (K) by Moerbeke. It must also then be posterior to the version of Books V, 7, 856–VII, 16, 1647. Books II–III would date from 1271 at the very earliest.[3]

We noted previously that the work of scribe 2 (Books IV–V, 7, 858) was included in this manuscript after the work of hand 2B (Books V, 7, 856–VII, 16, 1647). In Book IV, 2, 563, Book XII (Λ) is called "XI." This section, too, must have been composed prior to the translation of *Moerbecana* Book XI (K), therefore before the end of 1270.[4]

The other books, Book VII, 17, 1648 – XII, all refer to Book XII (Λ) as "Book XII." Hence, all these must be posterior to the period running from late 1270–early 1271.[5]

The dates allotted some of these books are confirmed, and in some cases made more precise by other research. In Book III, 11, 468, one finds a citation from the commentary of Simplicius on *De caelo*. The translation of this work was completed by Moerbeke on June 15, 1271.[6]

[1] Duin, "Nouvelles précisions...," p. 516.
[2] *Ibid.*, p. 517.
[3] *Ibid.*, p. 516.
[4] *Ibid.*
[5] *Ibid.*, p. 518.
[6] Msgr. Mansion has proved that the translation of Simplicius used by Aquinas was that of Moerbeke and not that of Grossetest as was once suggested. Cf. A. Mansion, "Date de quelques commentaires de Saint Thomas sur Aristote. (*De anima, Metaphysica*)," in *Studia Mediaevalia in honorem admodum Rev. Patris R. J. Martin*, Bruges, 1948, pp. 284–86.

Hence, the composition of Book III must be after this date. Book II, the work of the same scribe who wrote Book III, and like Book III, the result of Aquinas' dictation, could not have been very much earlier than June, 1271.[1]

Book XII also refers to the same commentary of Simplicius; this time, however, it is to certain historical conceptions of the Greek. The thought of Simplicius has not been thoroughly mastered by Aquinas in the sections where reference is made to it (Book XII, 12, 2537, 2578, 2582). Later, in his *Commentarium in de caelo*, Aquinas had assimilated Simplicius' work. Hence, we can assume Thomas composed Book XII during the second half of 1271 or in the early months of 1272, shortly after Moerbeke's translation of Simplicius, and before beginning the *Commentarium in de caelo* which was left unfinished at his death.[2]

A negative criterion – the failure of Thomas to use in Book I any of the historical data available through Simplicius' *Commentary on the Heavens* – has prompted some to speculate that this might be additional proof that Book I dates from before June 15, 1271.[3]

Additional research, doctrinal in this case, confirms the date we have assigned to Book IX, that is, after the first months of 1271. In the *Commentarium in de anima*, III, 11–12, Thomas treats the problem: "is human intelligence capable of knowing purely spiritual substances?" The problem is treated again in *De unitate intellectus*. The latter work obviously contains a later position of Aquinas. Yet in Book IX, 11, 1916, Thomas appears to have progressed even further. Whereas in *In III de anima*, 11–12 he spoke of books "which we do not have," and in the *De unitate intellectus*, he said "Aristotle's position on this cannot be known by us," in *In IX Meta.*, 11, 1916 he noted: "From this it is evident that according to Aristotle's opinion, the human intellect can reach an understanding of simple substances. In the third Book of *De anima* he seemed to have left this in doubt." From a com-

[1] Duin, "Nouvelles précisions...," p. 516. Grabmann first noticed the importance of this citation of Simplicius. He mistook it, however, for a citation of the *Commentary on the Categories* and thus set the date of Aquinas' work as after March, 1266. Cf. M. Grabmann, *Die echten Schriften des hl. Thomas von Aquin. Auf Grund der alten Kataloge und der handschriftlichen Überlieferung festgestellt.* Beiträge zur Geschichte der Philosophie des Mittelalters, Band XXII, Aschendorf, Munster, 1920, p. 60.

[2] Duin, "Nouvelles précisions...," p. 520. It was Msgr. Mansion who was the first to notice that the citations of Simplicius proved Book XII dated from after June, 1271. Cf. A. Mansion, "Pour l'histoire du commentaire de saint Thomas sur la Métaphysique d'Aristote," *Revue Néo-Scolastique*, XXVI, 1925, pp. 276–78.

[3] M. Deman, "Remarques critiques de saint Thomas sur Aristote interprète de Platon," *Les Sciences Philosophiques et Théologiques*, 1941–42, pp. 140–41.

parison of these three texts, it seems that Book IX was composed after the *De unitate intellectus*, after late 1270.[1]

Because of the results obtained through these investigations, the chronological order of the composition of the various books of the *Commentary* can be stated as follows:

Book	*Date of Composition*
Prooem.–I	before the last months of 1270
II	probably immediately before Book III
III	shortly after June 15, 1271; after Books V, 7, 856–VII, 16, 1647
IV–V, 7, 858	before June 15, 1271
V, 7, 856–VII, 16, 1647	begun after late 1270; finished before late 1271
VII, 17–XII	begun shortly after early 1271; finished in early 1272; after Books V, 7, 856–VII, 16, 1647

There is, however, still other evidence, both internal and external to the *Commentary*, which necessitates the introduction of nuances into this chronological order. First of all, within the *Commentary* itself, there is the matter of the divisions Thomas gives at the beginnings of Books VII, XI, and XII. In Book VII, 1, 1245–46, Thomas divides the remainder of the *Metaphysics* in this fashion:

I. de ente per se
  A. de ente
    1. diviso per decem praedicamenta
      a. ostendit Aristoteles quod oportet determinare de sola substantia (Book VII, 1)
      b. incipit Aristoteles de substantia determinare (Book VII, 2)
    2. diviso per potentiam et actum (Book IX)
  B. de uno et de his quae consequuntur ad unum (Book X)
II. de primis principiis entis (Book XII)

Yet in Book XI, 1, 2146, he wrote:

I. de communibus quae sequuntur ens commune (Books VII–X)
II. de substantiis separatis
  A. recolligit Aristoteles utilia ad cognitionem substantiarum separatarum (Books XI–XII, 4)
    1. recolligit ea quae praecedunt considerationem substantiae (Book XI)
    2. recolligit ea quae ad considerationem substantiae pertinent (Book XII, 1–4)
  B. inquirit de substantiis separatis (Book XIII, 4-end)

[1] For the relative chronology of the *Commentarium in De anima*, and the *De unitate intellectus*, cf. G. VERBEKE, "Notes sur la date du commentaire de saint Thomas au De anima d'Aristote," *Revue Philosophique de Louvain*, L, 1952, pp. 60–62. For the introduction of Book IX of the *Commentary on the Metaphysics* into the discussion, cf. DUIN, "Nouvelles précisions...," pp. 518–19.

Finally, in Book XII, 1, 2416; 2, 2424 and 2428:

I. recolligit Aristoteles ea quae dicta erant de entibus imperfectis (Book XI)
II. recolligit quae dicta sunt de ente simpliciter, idest de substantia
    A. ostendit quod ad istam scientiam pertinet considerare de substantia (Book XII, 1)
    B. determinat de substantia
        1. dividit substantiam (Book XII, 2, 2427)
        2. determinat de substantia
            a. de substantia sensibili (Book XII, 2–4)
            b. de substantia immobili (Book XII, 5-end)

These divisions cause a bit of confusion at first sight. If we consider the division given in Book VII, we notice that no place has been allotted Book XI.[1] Yet if Thomas had not mentioned Book XII explicitly, "In secunda de primis principiis entis, in duodecimo libro, ibi, De substantia quidem etc.", then it would not have been at all difficult to integrate this division with that of Book XI: the division of Book VII states we will treat separate substances while that of Book XI says this treatment has two parts (Book XI–XII, 4, and Book XII, 5–end). Moreover, there is no doubt that the reference "in duodecimo libro" is correct; in recent years Msgr. Mansion has discovered in Napoli, Bibl. Naz. VIII F. 16, an almost obliterated reference to Book XII in the passage of Book VII we have discussed (paragraph 1245).[2] How, then, is one to interpret this lack of coherence?

As has been noted, this passage from Book VII is contained in the text written by scribe 2B at Aquinas' direction after the translation of Book XI (K). On the one hand then, we have a division which, if there were no reference to Book XII, would cause no difficulty whatsoever; on the other hand, we have the reference to "Book XII" which, while indicating Thomas knew the existence of Book XI, leaves no place for it. Does this not suggest the possibility of at least two drafts of Book VII of the *Commentary*, the first draft gave the division of the books and refers for the treatment of separate substances to Book XII (Λ) as "Book XI"; the second draft, dictated by Aquinas to scribe 2B, saw the inclusion of the reference "Book XII"? In this case, of course, the second draft was not always as thorough as it should have been. In the case of paragraph 1245 of Book VII, 1, for example, if Aquinas had referred to Book XI, rather than to Book XII, the division of

[1] MANSION, "Pour l'histoire du commentaire de saint Thomas sur la Métaphysique...," pp. 281–83.

[2] For the publication of this discovery, cf. DUIN, "Nouvelles précisions...," pp. 519–20, footnote 16.

Book XI, 1, 2146 would fit into the proposed study of separate substances announced in Book VII. What probably happened is that, wherever he had referred to Book XII (Λ) as "Book XI" in the first draft, in the second draft he referred to it as "Book XII" when he dictated it to his scribe. This, it appears, would account for the lack of coherence among the divisions given in Books VII and XI.[1]

When one compares the division given in Book XII with that of Book VII, no new difficulties arise. Nor does a comparison of the divisions of Book XII and Book XI raise doubts. Book VII forecasts a treatment of the first principles of being; Book XII supplies a treatment of the first immobile substance as well as of the first causes of motion. Book XI promises a summary of all the principles useful in rising to knowledge of separate substances and then, a treatment of these substances; Book XII seems to complete the first project and carry out the second as well.[2]

The foregoing study of these divisions seemed to indicate the possibility of two redactions of the *Commentary*. A second indication, this time the historical writings of Ptolemaeus Lucencis, seems also to point to such a double draft. Of Thomas' stay at Rome in the years 1265–67, Ptolemaeus notes: "Quasi totam philosophiam sive moralem sive naturalem exposuit et in scriptum seu commentum redegit, sed praecipue Ethicam et Metaphysicam."[3] Since we have already proved that parts of the *Commentary* were written in 1270–72, does not this testimony point to the necessity of two drafts, one in 1265–67, the other ending in 1270–72?[4]

[1] All that Msgr. Mansion had concluded from the comparison of these divisions was that Thomas didn't know Book XI (K) when he wrote Book VII, and that Book XI (K) was translated into Latin after the composition of Book VII. Cf. MANSION, "Pour l'histoire du commentaire de saint Thomas sur la Métaphysique...," p. 284. It is not surprising that this is the only conclusion drawn from a comparison of these divisions. Taken alone they will not bear the weight of further hypothesis. It is only because of further evidence (that is: Book VII was dictated by Thomas after the translation of Book XI (K)), that one can bring forth the theory of a double draft.

[2] Msgr. Mansion feels that Book XII, although compatible with Book XI, could have been written first, and later, after the translation of Book XI (K), altered slightly to be more in harmony with it. Cf. MANSION, "Pour l'histoire du commentaire de saint Thomas sur la Métaphysique...," p. 284. Duin considers Msgr. Mansion's conclusion to be too strong if the latter wishes to maintain an actual anteriority for the bulk of Book XII; Duin feels that if Book XII were written twice, the second draft saw substantial changes, e.g. the introduction of the information gleaned from Simplicius' *Commentary on the heavens*. Cf. DUIN, Nouvelles précisions...," p. 520.

[3] PTOLEMAEUS, *Historia ecclesiastica*, Lib. XXII, C. XXIV, quoted in P. MANDONNET, *Des écrits authentiques de S. Thomas d'Aquin*, 2e édit., Fribourg, 1910, p. 60.

[4] In Ch. IV–VI, it will be maintained that Aquinas' exposition is in opposition to Albert's *Commentary*. Even if Aquinas wrote first in 1265–67, he could still have had Albert's work in mind. Cf. footnote 8, p. 5.

Let us note, first of all, that Ptolemaeus has been called a witness whose testimony merits serious consideration.[1] Moreover, Ptolemaeus speaks of himself as pupil, as confidant, as well as confessor of St. Thomas; in the years 1272–73 he was with the saint in the convent of San Domenico Maggiore in Naples.[2] Does not this clothe his testimony with some degree of credibility? For after all, he wrote at the beginning of the 14th century, certainly before 1318.[3]

There are other considerations, however, which incline one to reject the witness of Ptolemaeus. First of all, he places Thomas' stay at Rome (1265–67) during the pontificate of Urban IV; yet Urban IV reigned from 1261–64.[4] Secondly, Thomas is said to have written both the commentaries on the *Ethics* and on the *Metaphysics* during the years 1265–67; it has been shown, however, that the *Commentarium in Ethicorum* was composed around 1260.[5] Finally, on several other facts Ptolemaeus has been proved wrong.[6]

In favor, however, of Ptolemaeus' witness are three items which seem to point to a double draft of the *Commentarium*, one of which could have been composed during the period 1265–67. These items are: first, Thomas is said to have left Italy in 1268, arriving at Paris in January, 1269, where he remained until April 1272; then he returned to Italy, arriving at Naples in the Autumn of 1272.[7] Second, in the year 1274 there was no copy of the commentary of Simplicius on the *De caelo* to be found in Paris; thus it does not seem possible that there would have been a copy there two or three years earlier during Aquinas' stay; hence, Thomas could not have written his *Commentarium in Metaphysicorum* between 1269–April, 1272.[8] And third, Siger of Bra-

[1] E. Janssens, "Les premiers historiens de la vie de S. Thomas," *Revue Néo-Scolastique de Philosophie*, XVIII, 1924, pp. 205–206.

[2] See Walz, *Saint Thomas d'Aquin*, pp. 176–79; 181; 186. Grabmann, *Die Werke des hl. Thomas von Aquin*, p. 102.

[3] Grabmann, *op. cit.*, p. 102.

[4] Cf. Mandonnet, *Des écrits authentiques...*," p. 60.

[5] G. Verbeke, "Authenticité et chronologie des écrits de saint Thomas d'Aquin," *Revue Philosophique de Louvain*, XLVIII, 1950, pp. 261–62. For the date of the composition of the *Commentary on the Ethics*, cf. G. Verbeke, "La date du commentaire de S. Thomas sur l'Éthique à Nicomaque," *Revue Philosophique de Louvain*, XLVII, 1949, pp. 203–20.

[6] J. Isaac, O.P., "Études critiques. No. 142. A. Mansion, "Date de quelques...," *Bulletin Thomiste*, XXIV–XXIX, 1947–52, p. 176. Msgr. Mansion's discussion of the divisions of Books VII, XI, and XII, plus the citations of Simplicius found in Aquinas' *Commentary*, had led Isaac to reject Ptolemaeus' testimony completely.

[7] A. Walz, "Chronotaxis vitae et operum S. Thomas," *Angelicum*, XVI, 1939, p. 477. For a detailed account of Aquinas' activity during these years, see the same author's *Saint Thomas d'Aquin*, pp. 149–78.

[8] This information concerning Simplicius' *Commentary* is contained in a letter sent from the faculty of arts at Paris to the chapter of Lyons. Cf. Pelster, "Die Übersetzungen...," *Gregorianum*, XVII, 1936, pp. 389–90.

bant, in his *Quaestiones in Metaphysicam* knew and used the *Commentary* of Aquinas; since Siger's work cannot be much after 1272, it is possible that he knew a first draft of Thomas' *Commentary*.[1]

In the light of this conflicting evidence we have three possibilities.

1) We can reject Ptolemaeus' statements as erroneous. Thomas, then, wrote the *Commentary* in Paris. Its composition might extend throughout his entire stay, beginning before the translations of Book XI (K) and of Simplicius' work and ending after these works had been received by Thomas. In this event, Moerbeke must have sent both translations from Italy to Paris, and Aquinas must have taken them, or at least, the work of Simplicius, back to Italy with him.

2) We can reject Ptolemaeus' statements. Thomas, we could say, wrote a first draft of the *Commentarium* in Paris, but redacted it anew in Italy late in the Autumn of 1272. During this second draft, he introduced elements from Book XI (K) and from Simplicius.

3) One can accept Ptolemaeus' statements as referring to a first draft. A second one could have been finished in Paris or in Italy.

Before we attempt to choose between these hypotheses, let us return to the Naples manuscript of the *Commentary*. There is much to be learned from it regarding the possibility of more than one redaction.

Folios 42v–42ra present an interesting phenomenon. The first folio is the last page written by scribe 2. The last part of this folio presents what appears to be an earlier version of paragraphs 856–858 of our present editions. When hand 2B begins on folio 43ra, however, one has again the same three paragraphs. On folio 42v, an *alia translatio* is mentioned. The text given is the *Moerbecana*, which is thus opposed to the basic text, the *Media*. On folio 43ra, however, the exposition is based entirely on the *Moerbecana*, and the reading of the *Media* is explicitly said to be false. The text given on folio 42v represents, then, an earlier version, while that on folio 43ra represents a second draft.[2] In addition the unusual character of the corrections in the text of scribe 2B – words or sentences written, then crossed out and immediately replaced – seems to imply a previously existing draft, which was being read and corrected simultaneously, by Thomas.[3]

An interesting aspect of these paragraphs 856–58 has seemingly passed unnoticed. No one has wondered, at least not in print, why

[1] For the date of Siger's *Quaestiones*, cf., VAN STEENBERGHEN, *Siger de Brabant d'après ses oeuvres inédites*, Vol. II: *Siger dans l'histoire de l'Aristotelisme*, Tome XIII Les Philosophes Belges, Louvain, p. 564.

[2] KÄPPELI, "Mitteilungen über Thomashandschriften...," pp. 121–22.

[3] DUIN, "Nouvelles précisions...," p. 523.

scribe 2 stopped after paragraph 858. Since he seems to have felt his work was completed when he finished that paragraph, something like the following hypothesis seems to be called for. Scribe 2 compared the manuscript from which he was copying, with the manuscript of Naples. Seeing that paragraph 859, as well as the next few paragraphs, were nearly identical in both manuscripts, he concluded that his work of completing the Naples manuscript was finished. This similarity of the two manuscripts, then, is proof that Thomas was dictating from an older copy and that he did not completely change his text in the new draft.

Books II–III, the work of scribe 3 in the Naples codex, also presents evidence favoring the theory of a double draft. First of all, we must note that these books were composed after June 15, 1271 since they cite the commentary of Simplicius on *De caelo*. Secondly, they constantly cite Book XII (Λ) as "XII," thus indicating composition posterior to that of Books V, 7, 856–VII, 16, 1647, the work of scribe 2B. In the third place, after Book V, 20–22, Thomas opts definitively for the *Moerbecana* as the basic text.[1] Yet in Books II–III, although composed after Book V, 20–22, it is still the *Vetus* upon which Thomas comments; the *Moerbecana*, on the other hand, is hardly used. If there were only one draft of Books II–III, and that after the translation of Simplicius and of Book XI (K), why didn't Thomas use the *Moerbecana* as the basis of his work? The version of Books II–III in the Naples manuscript must, then, be a second draft. It was not complete, however, as it did not see a complete elimination of the former basic text, the *Vetus*, nor the *alia literae*, the *Litera Boethii* and the *Media*.[2]

Other than what we have already discussed concerning Books II–III, V, 7, 856–VII, 16, 1647, there is no additional evidence contained in the Naples manuscript which can shed light on the possibility of a second draft of the *Commentary*. The most we can say of the remaining books, is that Books VII, 17–XII, at least in the form in which they are found in this manuscript and in our editions, are posterior to the second draft of books II–III, and V, 7, 856–VII, 16, 1647.[3]

What, then, are the results of present day research into the chronology of the *Commentarium*? First, a double draft of Books II–III and of Books V, 7, 856–VII, 16, 1647 seems proved. Secondly, it does not

[1] The choice of the *Moerbecana* over the *Media* in Book V, 7, 856–58 must be considered an exception. These paragraphs are one of the few places where the second draft was a complete one.

[2] DUIN, "Nouvelles précisions...," pp. 523–24.

[3] *Ibid.*, p. 524.

seem necessary to choose any one of the three possible hypotheses indicated on page 19 to the exclusion of the other two. Perhaps, Thomas began and finished the *Commentary* in Paris between 1269 and April, 1272; or, perhaps he wrote an early draft in Italy (1265–67) and redacted it in Paris (1269–72) or in Italy (after 1272); or maybe the first draft dates from the 1269–72 Paris period, the second after the return to Italy. The only motive for eliminating any of these hypotheses would be the prodigious philosophical and theological output of the period from 1269 to 1273. Salman has indeed judged the quantity of Aquinas' work during this period to be so great, that it is impossible to add to the list of his writings, the *Commentary*.[1] One would, however, be justified in refusing to lend credence to this argument, for what difference would the addition of the 150 *lectiones* of the *Commentary* make? If Aquinas could find time to write as much as he is said to have done during these years, he could have written the *Commentary* too without being overworked.

Perhaps the best manner of concluding this section is to list the data generally accepted.

1) Book XI (K) was not known in *De unitate intellectus* (late 1270) and was known in *Commentary on the heavens* (after 1272).
2) Concerning the manuscript Napoli, Bibl. Naz. VIII F 16 we can say:
   a. Book XI (K) was not known when Book I was composed.
   b. Book XI (K) was known when scribe 3 wrote, when Thomas revised, Books II–III.
   c. Book XI (K) was not known when Books IV–V, 7, 858 were written.
   d. Book XI (K) was not known in the early passages dictated by Aquinas to scribe 2B (Book VI, 2, 1188); it was known by the time the scribe reached Book VII, 1, 1245.
   e. Book XI (K) was known when Books VII, 7–XII were written.
3) The translation of Simplicius' *Commentarium in de caelo* was completed on June 15, 1271. Hence:
   a. Book I was probably written before June, 1271.
   b. Books II–III and XII were written after June, 1271.
4) On the basis of the handwriting of the Naples manuscript, Books II–III, V, 7, 856–VII, 16, 1647 are a revised text.
5) The chronological order of composition is as follows:

Book	*Date of Composition*
Prooem.–I	before the last months of 1270
II–III	revised, and in part written after June 1271; written after Books V, 7, 856–VII, 16, 1647
IV–V, 7, 858	before June, 1271
V, 7, 856–VII, 16, 1647	revised after late 1270; finished before late 1271; not always a complete revision
VII, 17–XII	begun shortly after early 1271; finished in early 1272; after Books V, 7, 856–VII, 16, 1647

[1] SALMAN, "Saint Thomas et les traductions...," pp. 111–13.

A chapter containing as many unrelated elements as this one deserves some sort of concluding word. This chapter had but one aim, to serve as an introduction to certain important elements of the historical, literary, and chronological orders which must be kept in mind as we pursue our study of Aquinas' *Commentary*.

It was noted in this chapter, first, that Thomas used five different versions of the *Metaphysics* in composing his *Commentary*. Secondly, we noted the different periods of the composition of Thomas' book. These facts must be taken into account in our reading of the *Commentary*. Each passage of Thomas' work must be traced back if possible to the version used as the base. In addition, we must be on our guard for possible doctrinal or methodological differences in the *Commentary* which result from the different periods of composition. With these few rules of method in mind, we can now turn to the three interpretations which confronted Aquinas when he wrote – the works of Avicenna, Averroes, and Albert the Great.

CHAPTER II

THE METAPHYSICAL VIEWS OF AVICENNA, AVERROES, AND ALBERT

A comparative study of the *Commentaries* on Aristotle's *Metaphysics* written by Aquinas, Avicenna, Averroes, and Albert the Great yields abundant evidence that Aquinas referred continually to the expositions of his predecessors. In fact, one realizes that Aquinas must have had the three earlier commentaries constantly before him as he wrote. Because of this it appears best to present a brief, yet inclusive view of the three metaphysical syntheses that faced Aquinas as he worked.[1]

1. AVICENNA'S *FIRST PHILOSOPHY*

In the opening chapter of his *Philosophia prima sive scientia divina* Avicenna raises the question of the subject to be studied in this new science. He is very careful to point out that it can not be God, for in the initial stage of metaphysics one is totally ignorant of the existence of such a Necessary Being, one is ignorant of the existence of a Creator. God is one of those things which are "separated from matter" and it is quite evident that only the divine science studies such things;[2] hence, it is no wonder that His existence is still unknown.

This notion of divine science as the study of whatever is separated from matter recurs constantly in the introductory chapters of Avicenna's work (T. I, cc. 1–9). For example, divine science is said to limit its investigations to "things separated from matter both in existence and in definition."[3] As one would suspect, the meaning of the name "divine science" is determined by reference to such "separated things"; yet at the same time, Avicenna leaves room for this discipline to study material things, provided they are studied in so far as "their existence

[1] The basic ideas exposed in this chapter will be treated anew in Chapters IV-VI, ordinarily in greater detail.

[2] T. I, c. 1, fol. 70r, B–C ("B–C" signifies the marginal letters; folios not lettered will be cited according to the column, e.g. "73v, col. A").

[3] *Ibid.*, fol. 70r, A.

does not require matter."[1] Even something like motion and rest can be examined in so far as its existence is not dependent on matter![2]

Avicenna's view on the subject of first philosophy or divine science becomes apparently more intelligible if one reflects on those passages where he explains that this science, as any other science, must study both a subject or genus and the essential attributes of that genus.[3] For example, chapter 2 of Tractatus 1 contains an enumeration of the items not discussed in the sciences studied prior to first philosophy; among the objects listed are these: substance as *ens*, body as substance, measure and number as having existence, immaterial or formal entities. All of these, Avicenna concludes, should be studied in first philosophy in so far as they can be grouped around a common subject; that common subject, it is stated, is nothing other than existence – *esse*.[4] For Avicenna, then, the genus to be studied by divine science is *esse*, or existence; all the various items studied in this science – God, number, measure, motion, rest, etc. – are to be considered the accidents or essential attributes of existence as such.

However, the theory is not quite so simple, for Avicenna uses interchangeably *esse inquantum est esse*, *ens inquantum est ens*, and the study of *res omnino separata a materia*. The causes, for example, are said to be considered "in so far as they have existence" and so the subject of the study is said to be *ens inquantum est ens* – concrete being as such, let us say.[5] In the face of this identification of *esse* (existence) and *ens*, one's initial inclination is to view *esse* as merely an alternate way of expressing "concrete being" or *ens*. But that such is not what is meant by Avicenna's use of *esse* becomes evident from a reading of these introductory chapters. If, for example, to study the *esse* of motion

[1] "Cum autem inquiritur in hac scientia de eo quod non praecedit materiam, non inquiritur in ea nisi secundum hoc quod eius esse non eget materia." *Ibid.*, c. 2, fol. 71r, F.

[2] "Eorum autem quae inquiruntur in ea [scientia] quatuor sunt: ... et quedam sunt res materiales, sicut motus et quies, sed de his non inquiritur in hac scientia secundum quod sunt in materia, sed secundum quod esse quod habent. Cum igitur haec pars divisionis accepta fuerit cum aliis partibus divisionibus, tunc omnes communicant in hoc quod inquisitio de his non est, nisi secundum modum quod esse eorum non est existens per materiam." *Ibid.*

[3] "... subiectum omnis scientie est res, que conceditur esse, et ipsa scientia non inquirit, nisi dispositiones illius subiecti...." *Ibid.*, c. 1, fol. 70r, B. Cf. the discussion of *ens* as the subject matter in *Ibid.*, c. 6, fol. 73v, col. B *in fine*; also the discussion of the three speculative sciences in *Ibid.*, c. 1; fol. 70r, A–B.

[4] After an enumeration of these objects, Avicenna writes: "Sed non potest poni eius subiectum commune, ut illorum omnium sint dispositiones et accidentalia communia nisi esse." *Ibid.*, c. 2, fol. 70v, A–B.

[5] "Sed [consideratio de causis] vel in hac scientia, et tunc conveniens esset eas esse subiectum eius, vel in alia. Sed in alia scientia esse non potest. Nulla enim alia inquirit de causis ultimis nisi ista scientia; si autem consideratio de causis fuerit inquantum habent esse, et de omni eo, quod accidit eis secundum hunc modum, oportebit tunc ut ens inquantum est ens sit subiectum, quod est convenientius." *Ibid.*; c. 1, fol. 70v, E.

is to study the concrete reality of motion, how can one speak of the divine science as investigating motion, "not in so far as it is in matter, but according to the *esse* it has?"[1] On the positive side, in favor of understanding *esse* as "existence" rather than as "concrete being," one can point to the fact that only divine science can ask *an sit*, the traditional Aristotelian statement of the question which yields knowledge "that a thing exists."[2] In the context of the explanation of why divine science does not study God as its subject, Avicenna makes this point: when we begin the divine science, it has not yet been established whether there is a God, for the prior sciences could not ask *an sit Deus?* – only the divine science can ask such a question. That this question is proper to the divine science is to be shown in a later passage, Avicenna remarks.[3] When he turns to the proof of this point, Avicenna explains that no science prior to the divine science has ever proved the *existence* of its subject; no science has studied substance as *ens*, body as substance, number and measure as having *esse*, and so on. But the science that does study these, the science that proves the existence of the subjects of the other sciences, this science will be considering *esse* as such, for all these things are the "dispositions and accidents" of *esse*.[4] The only conclusion that can be drawn is, I believe, the following: the divine science asks *an sit Deus?* because it alone studies *esse*, it alone establishes the *existence* of things.

But if this is so, Avicenna is stating that the divine science is the study of "concrete being as such" (*ens qua ens*) to the extent that it studies "existence as such" (*esse qua esse*). However, this is not the only problem, for the study of the existence (*esse*) of motion and rest is the study of what is "totally separated from matter." The key to these ambiguities lies in Avicenna's notion of *esse*, and in his understanding of what it means to study *esse inquantum esse*.

If we accept the distinction of existence and essence, which, as is commonly admitted, Avicenna did, we would not ordinarily say that

[1] Cf. quotation in note 2, p. 24.

[2] Cf. Aristotle, *Posterior Analytics*, II, c. 1. Note that metaphysics alone asks "is it?" "because it belongs to the same kind of thinking to show what it is and that it is." *Metaphysics*, E 1, 1025b16–18. Also *Meta.*, Z, c. 17.

[3] "... quomodo potest concedi esse eius [idest, Dei]. Restat ergo, ut ipsum inquirere non sit, nisi huius scientie. De eo autem inquisitio sit duobus modis. Unus est quo inquiritur an sit. Alius est quo inquiruntur eius proprietates. Postquam autem inquiritur in hac scientia an sit. Tunc non potest esse subiectum huius scientie, nulla enim scientiarum debet stabilire esse suum subiectum. In proximo etiam ostendam quod an sit non potest quaeri, nisi in hac scientia." *Meta.*, T. I, c. 1, fol. 70r, C.

[4] *Ibid.*, c. 2, fol. 70v, A–B. E.g. "Sed non potest poni eis subiectum commune, ut illorum omnium sint dispositiones et accidentalia communia nisi esse."

a concrete being is existence: we would not say that *ens est esse*. Nor would Avicenna make such a statement about a concrete being.[1] Rather, he says that "a concrete thing is an essence which is, or which exists" – *ens est certitudo seu essentia quae est*.[2] In all beings one has to distinguish the *esse* of a thing and the essence or certitude by which a thing is what it is. It is precisely the *fact of having* a certitude or an essence that Avicenna refers to as *esse proprium* – proper existence.[3] In attempting to explain exactly what is meant by *esse*, Avicenna clearly indicates that he is referring to the *de facto* presence, here and now, of a quiddity. *Esse* signifies what is often referred to by contemporary philosophers as "brute facticity."[4] Accordingly, one can say that the study of *esse qua esse* is the study of the "fact of having a quiddity, considered precisely as such a fact."

There is but one more element to note before Avicenna's theory is meaningful. In the passage to which we have been referring, Avicenna writes:

> Everything has a certitude by which it is what it is... And this is what perhaps we call proper *esse*; nor do we mean by that expression anything but the meaning [or concept] of affirmative *esse*, because the word *ens* signifies many meanings, [or concepts], among which is the certitude by which each thing is; and this is like the proper *esse* of a thing.[5]

By "proper *esse*" then is meant the "affirmative *esse*"; *ens* is said to mean many things, among which is essence – these doctrines appear

[1] Avicenna does not of course state that "ens, is not esse" yet this is evidently implied by the thought of *Ibid.*, c. 6, fol. 72r, A–73r, col. A.

[2] "Sed dum dixeris quod certitudo huius vel certitudo illius est certitudo, erit superflua enunciatio et inutilis. Si autem diceres quod certitudo huius est res, erit etiam haec enunciatio inutilis ad id quod ignorabamus. Quod igitur utilis est dicere hoc est, scilicet ut dicas quod certitudo est res, sed haec res intelligitur ens, sicut si diceres quod certitudo huius est certitudo quae est." *Ibid.*, fol. 72v, C.

That *certitudo* is a synonym for "essence" in this text, see the discussion in A.-M. Goichon, *La distinction de l'essence et de l'existence d'après Ibn Sīnā (Avicenna)*, Desclée, Paris, 1937, p. 48. As Mlle. Goichon explains, *certitudo* can have three meanings: 1) the ontological truth of a thing: a thing realized conformably to its essence; 2) the truth in our intellects: the exact conception of a thing; and 3) quiddity. In the present case, the *certitudo quae est* would appear to be the third meaning: "the quiddity or essence which exists."

[3] "Unaquaquae enim res habet certitudinem qua est id quod est ... et hoc est quod fortasse appellamus esse proprium; nec intendimus per illud nisi intentionem esse affirmativi, quia verbum ens signat etiam multas intentiones, ex quibus est certitudo qua est unaqueque [sic] res; et est sicut esse proprium rei." *Meta.*, T. I, c. 6, fol. 72v, C.

[4] *Esse* as *de facto* presence or as brute facticity seems to be implied in statements such as the following: "Redeamus igitur et dicamus quod de his quae manifesta sunt est hoc quod unaquaque res habet certitudinem propriam quae est eius quiditas. Et notum est quod certitudo cuiusque rei quae est propria ei est praeter esse quod multivocum est cum aliquid. Quoniam cum dixeris quod certitudo talis rei est in singularibus vel in anima vel absolute, ita ut communicet utrisque, erit tunc haec intentio apprehensa et intellecta." (This text continues with the citation given in note 2). *Ibid.*

[5] Cf. the text of note 3.

to be references to the Aristotelian theory of *ens* as meaning either "predicaments of substance and accidents" or "truth". In speaking of *ens* in the first sense, one refers to "what a thing is"; in the later, to "that a thing is." To know "that a thing is," to give an affirmative answer to the question *an est?*, is to posit the "affirmative *esse*" – the *Est!*[1] It would appear, thus, that by the proper *esse* of a thing, Avicenna is thinking of the knowledge we have "that a thing is" – *Est!* Not that Avicenna confuses our *knowledge* with the *existence* of a quiddity. Rather, in the passage quoted above, he merely wishes to pin-point the act in which we grasp the existence proper to a thing. It is in answering the question "Is it?" that we grasp the "fact of having a quiddity."

Avicenna claims that the divine science studies the "concrete being as such" (*ens qua ens*), which appears to be identified with "existence as such" (*esse qua esse*). That he can legitimately identify *ens* and *esse* stems from the meaning attributed to *esse* – the "fact of having a *quiddity*." What is the difference between studying a concrete being as *concrete* here and now, and studying the fact of having quiddity considered precisely as that fact! In other words, the concrete thing is the essence which is, the essence as factually present.

But Avicenna also claims that divine science is the study of what is completely separated from matter in *esse* and definition: motion and rest, for example, are studied "not in so far as they are in matter, but in so far as their *esse* does not require matter"! The solution to this enigmatic statement rests in the reference to "affirmative *esse*." If one emphasizes the grasp of existence, if one studies actual instances of motion and rest as affirmed, then one is not concerned with matter. One asks: *An est motus?*, and the reply comes: *Est.* It is in that *est* that one grasps the fact of motion, a grasp which is not a grasp of matter, but of an *esse*, of an *est.* And if one focuses one's attention on that grasp of existence precisely as a grasp of existence, one can study motion "not in so far as it is in matter, but in so far as its existence does not require matter."

That this is indeed Avicenna's thought appears as soon as one realizes the connection between these first chapters of his *Commentary* and the puzzling first chapter of Book E of Aristotle's *Metaphysics*. For his part, Aristotle explained the division of speculative philosophy into natural, mathematical, and theological science. Each of these

[1] Cf. note 10 for the Aristotelian doctrine on the questions "is it?" and "what is it?." For the divisions of "being", see ARISTOTLE, *Meta.*, Δ, c. 7.

divisions is specified by the separation from matter, or lack of it, found in the objects studied in each science.[1] These Aristotelian ideas are reproduced in the opening paragraphs of Avicenna's exposition, e.g. "the divine sciences investigate only those things separated from matter in their existence and their definition."[2] In Aristotle's work, one discovers two other doctrines which would appear to be present also in Avicenna's work: theology, or the divine science, is simultaneously the universal study of being as such; and "it belongs to the same kind of thinking [the theological, universal science] to show what it is and that it is," while the other sciences accept this knowledge from the first or theological, universal science.[3] That these thoughts are at the basis of these Avicennian doctrines can scarcely be denied: "...*is it?* can be sought only in this science," that is, in divine science; the explanation that the divine science must establish the existence of the subject of the science of mathematics, as well as the existence of the subject of natural philosophy.[4]

Avicenna has, then, a view of metaphysics in which *esse*, considered as the fact of having a quiddity, assumes the role of formal object: it is the metaphysician's task to analyze the "factual presence of a quiddity considered precisely as a fact." Avicenna, moreover, has no difficulty explaining why man should ever begin to think metaphysically. His explanation is grounded in his theory of "being" as man's first idea: *ens* can be explained by nothing else, and is itself presupposed by everything else.[5] As one might already suspect from what has been said about "existence," this idea of "being," signifying *esse*-essence composition, is close to a formal category; rather than expressing the essence and *esse* of all things, it expresses only what it means to have here and now an essence.[6] Yet because all men understand this "to be here and now a quiddity," and because moreover, "to be here and now" does not necessarily involve matter, every one realizes that a study of the being of things must be carried through. One need not, in other words, discover the existence of an immaterial prime mover before understanding that material beings do not exhaust the range of being.[7]

[1] *Ibid.*, E 1, 1026a6–16.

[2] AVICENNA, *Meta.*, T. I, c. 1, fol. 70r, A–B.

[3] *Meta.*, E 1, 1025b3–18.

[4] Cf. the text of note 3, p. 25.

[5] I will not attempt to cite Avicenna's statements here; this will be amply discussed in Ch. IV, section 4, A.

[6] Cf. Ch. V, section 4, B. On the idea of being as the poorest concept of all, see GOICHON, *La distinction de l'essence et de l'existence...*, pp. 2–5.

[7] Here Avicenna parts company with Aristotle, and for this will be criticized by Averroes. For Averroes' thought, cf. notes 3, p. 35 and 1–4, p. 36.

Even if we admit Avicenna's claim that all men understand "being," it does not seem that we must also admit any universal recognition that *esse* is not the same as essence. Avicenna's procedure would seem to concede this point, for he undertakes to prove this distinction to his readers. It is at this point that his metaphysics appears to begin.

In his study of the distinction between essence and existence, Avicenna adopts what Averroes and Aquinas will call "a logical investigation – an investigation of predication," that is to say, an examination of our use of language, which use reveals our understanding of the reality of things. To paraphrase Avicenna's exposition of the distinction, we can note: If we explain to Socrates the definition of "horse," Socrates will learn nothing provided he already knows what a horse is; but if we tell him: "That is a horse," it is possible that we will tell him something that he does not yet know.[1] That statement "That is a horse" is an expression of the knowledge had in a judgment, while to understand what "horse" means involves the first operation of the intellect. Thus, while the mere repetition of a definition of "horse" is of no value to one who knows what a horse is, the judgment that a horse is found here and now can on occasion be new to some one. It seems evident that Avicenna implies here an investigation into predication: when we understand an object we do not know its existence; we discover the separation or distinction of *esse* and essence if we realize that it is in the judgment of the here and now presence of a quiddity ("horse") that we have expressed something not expressed in the concept of that quiddity.[2]

The same type of examination is carried through as well in the discussions of the meaning of *res*, of *aliquid*, of *unum*, of substance.[3] However, one must not overlook the important fact that, for Avicenna, the key to this type of analysis of predication is the second operation of the intellect, the judgment of the here and now presence of a quiddity. It is this second operation which is at the base of the spirit of Avicennian metaphysics, for what this science studies is nothing other than the "fact of the here and now presence of a quiddity."[4]

[1] *Meta.*, T. I, c. 6, fol. 72r, B–v, C.

[2] Cf. Ch. IV, section 4, A.

[3] It is impossible to explain *ens* through any other word, and so Avicenna was reduced to dividing and distinguishing *ens*, in the hope that by so doing we would grasp the intelligibility of *ens*; cf. GOICHON, *La philosophie d'Avicenne et son influence en Europe médiévale*, 2e édit., Adrieu-Maisonneuve, Paris, 1951, pp. 21–24; by the same author: *La distinction de l'essence et de l'existence...*, pp. 7–8.

[4] Although Mlle. Goichon has often noted how the distinction of *esse* and essence touches

The first step within Avicenna's system coincides with the realization that *esse* and essence are distinct, one would suppose. The second step concerns the opposition of the "possible" and the "necessary." In this second step, emphasis is placed again on the judgment of *esse*: Whatever falls under the intellectual knowledge of *esse* can be divided into two camps, writes Avicenna – 1) possible being, whose *esse* is neither necessary nor impossible; and 2) necessary being, whose *esse* is necessary.[1] This distinction is, in turn, followed by that of "substance" and "accidents"; this latter distinction too is made according to the uses of "is." *Esse*, writes Avicenna, belongs to a thing either *per essentiam* or *per accidens*. An example of *esse* which belongs to a thing essentially is "to be a man"; this *esse* belongs essentially to any given man. An example of the accidental *esse* is the "to be white" of a wall; because the wall could just as easily be blue as white, this "to be white" is only *per accidens* the *esse* of this wall.[2] (This *per accidens esse* is the "happens to be" of Aristotle and Aquinas.)[3] The *per essentiam* type of *esse* must be further divided, explains Avicenna. The first division is that of "substance": "what *is* not in a subject." Opposed to "substance" are the "accidents": "what *is* in a subject."[4]

In all these investigations Avicenna emphasizes the judgment of existence. This same emphasis remains when he studies the distinction of matter and form. In the introductory chapters of his *Metaphysica* Avicenna affirms that the first philosophy alone can study material bodies as composed of *yle* and form.[5] In fact metaphysics is to prove that these two principles of material essence are distinct. However,

every other doctrine in Avicenna's work, no one has ever explained, as I am here, that the formal object of metaphysics is the "here and now existence". Cf. GOICHON, *La distinction de l'essence et de l'existence*...

[1] "Dicemus igitur quod ea quae sunt esse possunt in intellectu dividi in duo. Quorum unum est quod cum consideratum fuerit per se eius esse non est necessarium; et palam est etiam quod eius esse non est impossibile, alioquin non cadet sub esse, et hoc est in termino possibilitatis. Alterum est, quod cum consideratum fuerit per se eius esse erit necesse." *Meta.*, T. I, c. 7, fol. 73r, col. A. Cf. GOICHON, *La distinction de l'essence et de l'existence*..., pp. 162–64. M.-D. ROLAND-GOSSELIN, O.P., *Le "De ente et essentia" de St. Thomas d'Aquin*, Vrin, Paris, 1948, 151–53.

[2] "Dicemus igitur quod esse vel est rei per essentiam sicut homini esse hominem, vel est ei per accidens sicut Petro esse album." *Meta.*, T. II, c. 1, fol. 74v, A.

[3] ARISTOTLE, *Meta.*, Δ 7, 1017a13–14. AQUINAS, *In V Meta.*, 9, 887. For a discussion of this doctrine, see below Ch. VI, section 1, C.

[4] "Accingamur ergo nunc et tractemus de esse secundum quod est esse per essentiam. Quod autem prius est ex omnibus divisionibus eorum quae sunt per essentiam substantia est. Quoniam esse duobus modis est. Unum est id quod cum sit in aliquo eius existentia et species acquisita est in seipsa; non est sicut pars eius, nec potest esse sine eo. Et hoc est quod est in subiecto. Aliud est quod est, sed non est in aliquo hoc modo, quoniam nullatenus est in subiecto, et hoc est substantia." *Meta.*, T. II, c. 1, fol. 74v, A.

[5] *Ibid.*, T. I, c. 2, fol. 70v, A.

Avicenna makes no reference to the method used to discover this distinction. Undoubtedly though, Avicenna examines predication; moreover, it appears rather certain that the judgment of *esse* is essential to this proof. The distinction of matter and form involves our knowledge of essence, to be sure, but the reason for introducing this proof into metaphysics is precisely that the judgment of *esse* – the determining factor in the formal object of metaphysics – underlies the entire proof.

The proof of the distinction of matter and form is quite long as Avicenna gives it. Suffice it to note that he speaks of the following steps. 1) Some say a body is a substance having length, breadth, and width.[1] 2) However, this is not universally true; a sphere, for example, has none of these dimensions.[2] 3) We know the existence of a body in so far as we know the existence of corporality: "Iam enim certificatum est ipsum esse corpus per id quo certificatur corporeitas."[3] 4) Thus, we must understand a body to be a substance whose form is that by which it is what it is.[4] 5) And so, corporeity is the form which underlies dimensions; and such a form can not be considered by mathematics, for mathematics considers bodies which can be measured.[5] 6) A "natural" body, that is, one studied by physics, can change, can alter; however, body as such can not alter.[6] Hence physics does not study body as body, Avicenna seems to imply. 7) Although it is true that we can divide the bodies we see around us, corporeity as such is not divisible. Rather, corporeity is what can receive division; corporeity is a form existing in something. The very ability to divide objects without depriving ourselves of bodies requires that dimensions be existing in a subject.[7] 8) Body as having forms, notes Avicenna, is an effect; but

[1] *Ibid.*, T. II, c. 2, fol. 75r, A.

[2] *Ibid.*

[3] *Ibid.*

[4] "Sic igitur oportet intelligi corpus quod ipsum est substantia cuius haec est forma qua est id quod est". *Ibid.*

[5] "Corporeitas igitur vera est forma continuitatis recipiens id quod diximus de positione trium dimensionum; et haec intentio est extra mensuram et extra corporeitatem disciplinalem. Hoc enim corpus secundum hanc formam non differt ab alio corpore sive sit maius, sive sit minus, nec comparatur ei sive fit equale sive sit numeratum per illud sive communicans ei sive incommunicans. Hoc enim non est ei nisi inquantum est mensuratum, et inquantum aliqua pars eius numerat illud; et haec omnia considerantur in eo, absque consideratione corporeitatis quam assignavimus." *Ibid.*

[6] "... est possibile, ut cum unum corpus rarificatur et densatur calefactione et infrigidatione permutetur eius quantitas. Sed corporeitas eius quam diximus, non permutetur nec alteretur. Igitur corpus naturale est substantia secundum hunc modum." *Ibid.*

[7] "Quod autem nobis necessarium est hic [,] hoc est scilicet ut natura corporeitatis non prohibeat hoc inquantum est natura corporeitatis. Prima igitur dicam nos certificasse quod corporeitas inquantum est corporeitas non est nisi receptibilis divisionis. Igitur in natura corporeitatis est recipere divisionem; igitur manifestus est ex hoc quod forma corporis et dimensiones sunt existentes in aliquo.... Id autem quod est ipsa continuatio vel continuum in

in so far as it can have a certain dimension, a body is in potency. This fact demands that the potential aspect of a body is not identical to its act.[1] 9) Thus the form of the body must be joined to something else: in so far as the body is in act, we have an effect of form. In so far as the body is in potency, we have something else – matter or *yle*.[2]

On first sight, this proof does not appear to involve a judgment of *esse*; we are simply reasoning with the data we see around us. But if we recall Avicenna's explanation of the formal object of metaphysics, we recognize the role of *esse* in this discussion. As Avicenna notes, the sole reason why it is possible to consider body, *ens*, measure, etc. in metaphysics is that the *intentio essentis* is the common meaning of all these. The only thing they have in common is *esse*.[3] It is only if he considers a body as "here and now realized" that Avicenna can distinguish between the physical, the metaphysical, and the mathematical manners of considering body. The physician will study the changes taking place in bodies (point 6 above); the mathematician will study the dimensions of a body (point 5 above). It is only if he emphasizes the here-and-now of a body, does a metaphysician have an opportunity to treat bodies: as "here and now" a material object is considered neither as white nor green, and any change in its color will make no difference to the here-and-now of the body; as "here and now" a material body is considered neither as big nor as little, and any statement about its size will have absolutely no interest for the metaphysician. Thus to consider a material object as "here and now" is to abstract from these physical and material aspects; it is to consider, probably even, to assume, that a body as such is somehow something else than changeable (natural) or measurable (mathematical). To consider a body as "here and now" is to assume there is something underlying these natural and mathematical aspects; it is to assume there is form (corporeity)

se impossibile est ut remaneat ipsum continuatione destructa. Omnia enim continuatio est dimensio quae cum separatur destruitur illa dimensio et acquiruntur aliae duae dimensiones.. Igitur in corporibus est aliquid quod est subiectum continuationi et discontinuationi propter id continuationis quod accidit mensuris terminatis." *Ibid.*, fol. 75v, B.

[1] "Corpus enim inquantum est corpus habens formas corporeas, est quodam in effectu. In quantum vero aptum est ad quamlibet mensuram est in potentia. Res autem secundum quod est in potentia est aliqua res et ipsa est alia res, secundum quod est in effectu. Est igitur potentia corpori, sed non inquantum est sibi effectus." *Ibid.*

[2] "Igitur forma corporis coniuncta est alii inquantum est et forma. Corpus igitur est substantia composita ex quodam per quod habet potentiam, et ex quodam per quod habet effectum. Id autem per quod habet effectum est forma eius; per quod vero habet potentiam est materia eius, et hoc est yle." *Ibid.*, fol. 75v, B–C. As with the case of Avicenna's formal object, so too in regard to the distinction of matter and form, we have found no treatment of it. E.g. there is no mention of this in ROLAND-GOSSELIN, *Le "De ente et essentia"* ..., pp. 61–65, where he treats the principle of individuation in Avicenna's thought.

[3] *Meta.*, T. I, c. 2, fol. 70v, B.

underlying these other aspects (points 4–5). But if one thus considers a body to be constituted as body in so far as there *is* corporeity, then one must explain the body's ability to change naturally and mathematically by some principles other than corporeity. Matter or *yle* will do fine. – In this entire discussion, Avicenna is considering bodies as "separated from matter"; he considers them as being here and now.

Turning now to the proof for the existence of God, we run into a thorny problem: does Avicenna prove the existence of God, or does he suppose from the beginning that the world is created? His words on the subject of metaphysics, namely that it is not God for he is one of the things sought in this science, would imply that he must prove the existence of God.[1] But does he?

As Avicenna remarks, one can not demonstrate God after the fashion of a *propter quid* demonstration. We do not know the "cause" of God, for He has none; that is, we do not deduce His existence as if it were necessarily contained within the knowledge we have of some formal cause. Rather, we simply *point* to Him, a *demonstratio quia*.[2] Be that as it may, the problem is to find the knowledge which permits Avicenna to affirm the existence of God. Averroes will accuse Avicenna of presupposing the God of faith; it was because of this presupposition, according to his critic, that Avicenna considered the sensible universe to consist of possible beings (necessary *ex alio*), and hence to conclude the existence of a necessary cause of being.[3] What is more, some modern studies on Avicenna have decided that Averroes' accusation

[1] Cf. the texts given in note 3, p. 25. Mlle. Goichon has noted that Avicenna on one occasion, tried to prove the existence of God from the internal evidence of the idea of being; however, the only book in which Avicenna explained this doctrine was the Isārāt, a book never translated into Latin. Cf. Goichon, *La distinction de l'essence et de l'existence...*, pp. 335–37.

[2] "Nec fit demonstratio de eo, quia ipse non habet causam. Similiter non quaeritur de eo quare. Cum enim scies posteaquam eius actio non habet quare." *Meta.*, T. VIII, c. 4, fol. 99r, B. "Inquiramus ergo quid sit subiectum huius scientiae: et consideremus an subiectum huius scientiae sit ipse deus excelsus; sed non est; imo ipse est unum de hiis, quae quaeruntur in hac scientia." *Ibid.*, T. I, c. 1, fol. 70r, B. "Non est autem manifestum per se..." *Ibid.*, fol. 70r, C. "... de eo autem inquisitio fit duobus modis; unus est quo inquiritur an sit, alius est quo inquiruntur eius proprietates." *Ibid.*

[3] In the context of an argument against Alexander's theories of the creation of all heavenly bodies, Averroes remarks that these theories, plus an unfortunate Aristotelian remark on the finitude of all heavenly bodies, had led Avicenna to his theory of the possible, or necessary from another (creatures), and the necessary from itself (God). Cf. *In VIII Phys.*, c. 79, fol. 426v, L–M. Aquinas knew this passage; cf. *In VIII Phys.*, 21, 3, where he shows he is reading both Avicenna and Averroes. In the *Commentary on the Metaphysics*, Averroes accuses Avicenna of listening to the "Loquentes in nostra lege, cum quorum sermonibus admiscuit ipse suam scientiam Divinam." *In IV Meta.*, c. 3, fol. 67r, B. In this case, Averroes is objecting to Avicenna's reasons for positing existence as something added to essence. Ultimately, therefore, Averroes is attacking the foundation of Avicenna's proof for God.

is fundamentally correct.[1] However, one modern exegete of Avicennian thought has been able to reconstruct a strictly philosophical proof on the basis of Avicenna's writings. Such a proof would proceed as follows. Not only has Avicenna remarked that "being" is our first idea, but that "thing" and "necessary" enjoy the same title of absolute priority.[2] However, in the world around us we soon realize that beings, although necessary, nevertheless begin to be; thus they are not absolutely necessary, but only in some secondary way. ("Necessary" in this discussion means, of course, "contingent.") Thus the beings of our experience, while being necessary, are not necessary by or through themselves; hence they are necessary through another. – This reconstruction follows in general that of Mlle. Goichon; it is most probably a valid reading of Avicenna.[3]

This then is Avicenna's proof for God. The remainder of his metaphysical speculation centers around the questions of the divine attributes, emanation, and so on. These questions do not interest us here for they have little or no relation to Aquinas' *Commentary*.[4]

The un-Aristotelian character of this synthesis is not difficult to grasp. Fundamentally, Avicenna parts company with Aristotle in assigning to metaphysics the view-point of *esse*. All other important differences between the two philosophers (discounting the Neo-Platonistic tendencies of Avicenna's treatment of the divine attributes,

[1] D. SALIBA, *Étude sur la métaphysique d'Avicenne*, Presses Univ. de France, Paris, 1926, pp. 101–106; 54–55. L. GARDET, *La pensée religieuse d'Avicenne (Ibn Sīnā)*, Vrin, Paris, 1951, pp. 38–42.

[2] "... ens et res et necesse talia sunt quod statim imprimuntur in anima prima impressione quae non acquiritur ex aliis notioribus se...." *Meta.*, T. I, c. 6, fol. 72r, A.

[3] Cf. GOICHON, *La philosophie d'Avicenne*..., pp. 22–23. ROLAND-GOSSELIN, *Le "De ente et essentia"*..., pp. 153–55. As far as can be determined, there is one element of this proof that Avicenna does not give: "when we look around us we realize that beings are not necessary through themselves". However, I am inclined to accept Mlle. Goichon's exposition as it appears to be the only way to understand texts such as these: "... ea quae cadunt sub esse possunt in intellectu dividi in due. Quorum unum est quod cum consideratum fuerit per se eius esse non est necessarium; et palam est etiam quod eius esse non est impossibile, alioquin non cadet sub esse: et hoc est in termino possibilitatis. Alterum est, quod cum consideratum fuerit per se eius esse erit necesse. Dicemus igitur quod necesse esse per se non habet causam et quod possibile esse per se habet causam." *Meta.*, T. I, c. 7, fol. 73, Col. A. "Quicquid autem est cuius esse per aliquid cum consideratum fuerit per se non habebit esse necessarium. Quicquid autem consideratum per se sine alio non habet esse necessarium non est necesse esse per se... Quicquid enim est ad cuius esse agit aliud eius esse non est necessarium in se... Si autem acquiratur [esse] ab alio a se tunc illud aliud est causa. Si vero entia [sic.] acquiratur ab alio a se manifestum est autem quod quicquid potest non esse habet esse iam appropriatum est per aliquid quod sibi abvenit ab alio a se... Igitur manifestum est quod quicquid possibile est esse non habet esse nisi cum necessarium est respectu suae causae"... *Ibid.*, Col. B.

[4] For a discussion of these Neo-Platonic aspects of his natural theology, cf. GOICHON, *La philosophie d'Avicenne*..., pp. 33–53. GARDET, *La pensée religieuse d'Avicenne*, pp. 62–66; 51–56.

emanation, etc.) are reduceable to the opposition between Avicennian *esse* and Aristotelian "being." More about this in later chapters.

Aquinas' chief objection to Avicenna will be noted as well in the discussions which follow; fundamentally, Aquinas felt that the *esse* of Avicennian philosophy is not a fit subject for metaphysical speculation. Twice Aquinas attacked his predecessor – and by name – each time clearly indicating the irreducible opposition of the Thomist and Avicennian systems.[1] This very attack, by emphasizing that the subject of metaphysics is understood and expressed in a concept, will mark Thomas' adherence to the general outlines of what he understood to be Aristotle's system. However, as will be explained in the pages to come, this adherence to Aristotle is simultaneously a transformation from within through the introduction of participated *esse*.

2. THE *COMMENTARY* OF AVERROES

Just as Avicenna openly declares that metaphysics alone can prove the existence of God, so too Averroes openly proclaims that Avicenna's view is completely false. Avicenna followed Alexander on this point, remarks Averroes.[2] In Alexander's mind, the physician could not prove the existence of the principles of natural beings; the physician could but assume the principles of sensible beings that are posited by the metaphysician. This is wrong, Averroes believes, for the eternal substance is proved to be the first principle of natural beings in the final book of the *Physics*.[3] Even more, notes Averroes, the physician's proof

[1] Cf. Ch. IV, section 4, B. Aquinas says nothing about Avicenna's theory on the impossibility of proving God in physics. But since Averroes shows the un-Aristotelian character of this theory, one can also put forth the hypothesis: Thomas did not bother to refute this theory, since his contemporaries had already read the refutation in Averroes (as well as the discussion in Thomas' *In VIII Phys., lectiones* 12–13).

[2] Aquinas did not directly know Alexander's work, but only in so far as Averroes referred to it. For the manner in which Thomas at times adopted a suggestion of Alexander, cf. DUCOIN, "Saint Thomas commentateur d'Aristote...," 78–117; 240–71; 392–445.

[3] "Et Alexander dixit, demonstrare enim principia entium, quae sint, est primi Philosophi: et quod istis utitur Naturalis, est quia non declarat ea, sed quia ponit ea... Cum enim dixit, quod demonstrare principia entium est primi Philosophi non naturalis: et quod Naturalis ponitur ea positione, etc., si intendebat, quod principia entium, de quibus Philosophus considerat, et quae Naturalis concedit esse, sunt principia substantiae sensibilis aeternae, quae sunt substantia abstracta, et quod principia, de quibus considerat Naturalis, declarando ea esse, sunt principia generabilis et corruptibilis, erit sermo non rectus. Substantia enim aeterna declarata est in naturalibus in fine Octavi Physici, sicut declarata sunt principia substantiae generabilis et corruptibilis in Primo istius libri. Quomodo igitur dicitur, quod naturalis ponit hanc positionem, cum impossibile sit ut declaretur nisi in naturalibus? et quomodo est rectum dicere, quod Naturalis non declarat nisi principia substantiae generabilis et corruptibilis, cum non solummodo consideret de substantia generabili et corruptibili, sed etiam de non generabili et non corruptibili? considerat enim de ente mobili sive generabili sive non. Secundum hunc intellectum iste sermo est falsus." *In XII Meta.*, c. 5, fol. 292v, K-fol. 293r, B.

through motion is the sole valid proof.[1] Unfortunately Avicenna thought Alexander was correct. Hence Avicenna gave as his own this same false doctrine concerning physics and metaphysics.[2]

The true relation between physics and metaphysics is as follows, concludes Averroes. The first philosopher must study the principles of substance as such; thus we will be led to state that immaterial substance is the principle of natural substance.[3] However, one must underline the fact that the first philosopher accepts from physics the intrinsic substantial principles of matter and form, as well as the extrinsic principle of motion – the immaterial first mover.[4] After he accepts these doctrines, the first philosopher explains how his first mover is the principle of natural substances.[5]

Thus Averroes would have metaphysics begin only after the physician has proved the existence of the prime mover. Naturally enough however, metaphysics is not a mere continuation of physics; rather, the metaphysician is a student of the first causes and principles of substance precisely as causes and principles of substances.[6] So it is, that although the first philosopher accepts the existence of the principles of substance (matter and form; the prime mover), he studies

[1] No science can prove its own first principles, Averroes admits. But, he explains, this means one can not deduce one's principles from higher ones. This does not mean that physics does not prove the existence of God "... facultas inferior non habet considerare principia sui generis secundum simplicem demonstrationem: secundum vero demonstrationes, quae vocantur signa, possibile est. Et ideo principia subiecti Naturalis non demonstrantur nisi per res posteriores in scientia Naturali. Et ideo impossibile est declarare aliquid abstractum esse, nisi ex motu. Et omnes viae, quae reputantur esse ducentes ad primum motorem esse praeter viam motus aequaliter, sunt insufficientes." *Ibid.*, fol. 293 r, C.

[2] "Et tamen Avicenna reputavit hunc sermonem esse Verum, scilicet quod nulla facultas habet probare sua principia. Et Avicenna hoc absolute dixit, quod primus Philosophus habet declarare prima principia sensibilis substantiae sive aeternae, sive non: et Naturalis ponit positione naturam esse; et quod primus Philosophus demonstrat ea esse, et non distinxit inter duas substantias... Divinus igitur habet probare principia naturalis subiecti: et Naturalis ponere ea positione." *Ibid.*, fol. 293r, D–E.

[3] "Et dicemus nos quidem, quod Philosophus inquirit quae sunt principia substantiae secundum quod substantiae, et declarat, quod substantia abstracta est principium substantiae naturalis." *Ibid.*, fol. 293r, E.

[4] "... sed hoc ponendo accepit pro constanti, hoc quod declaratum est in naturalibus de principiis substantiae generabilis et corruptibilis, scilicet quod declaratum est in primo Physi, scilicet quod est compositum ex materia et forma... et quod declaratum est in Octavo, scilicet quod movens aeternam substantiam est abstractum a materia." *Ibid.*, 293r, E–F. Also *In II Meta.*, c. 4, fol. 30r, C.

[5] Averroes writes: "(Aristotle gives here) ... quod declaratum est in naturalibus... (and then Aristotle begins to consider these) ... consideratione propria huic scientiae, verbi gratia quod est substantia et prima forma, et quod est principium et finis." *In XII Meta.*, c. 5, fol. 293v, G–H.

[6] "... si igitur prius decem entium sit substantia, ut post declarabitur: et Philosophus est ille, qui scit primas causas omnium, necesse est ut cognitio principiorum substantiae sit accepta in definitione Philosophi. Dicitur enim, quod Philosophus est ille, qui scit principia et causas substantiae." *In IV Meta.*, c. 2, fol. 66r, C. (In this context the *Philosophus* is the metaphysician).

these in a new way.[1] Whereas the physicist merely studies matter, form, and the first mover as the principles of motion, the metaphysician is interested in these as principles of substance.[2] Ultimately, the metaphysician differs from the physicist in that the former eventually learns to know the first form and the last end of substance.[3]

Averroes' metaphysics is a study of being as such in so far as it is a study of the first form and ultimate end of all beings. It is in this spirit that Averroes notes that metaphysics, even though it considers material beings, considers them as if they were not material; it studies natural beings as "potential" or "actual," or as "one" or "many", or as "substance" or "accident."[4] In this same spirit Averroes interprets Aristotle's first aporie as concluding that only the first formal and final causes are to be studied by the metaphysician.[5] Finally, when he reaches Book E, Averroes returns again to his view of metaphysics as the study of the first form and first end. This science, he writes, will investigate the "simple" causes of being, that is, the causes of being as such.[6] Moreover this science studies only what exists independently of matter; it studies the eternal causes of the heavenly bodies.[7] Thus metaphysics is said to study whatever can be defined in terms of "God".[8]

Thus in the Averroistic conception of metaphysics, one finds identified:

[1] "... de principiis autem substantiae sensibilis generabilis et corruptibilis locutus est in... Naturalibus, sed alio modo. Consideratio enim de principiis eius, in eo quod est substantia, alia est a consideratione de eis, inquantum sunt causae transmutationis..." *In XII Meta.*, c. 1; fol. 290v, K–L. Also *Ibid.*, fol. 293v, G.

[2] "Sic igitur intelligendo est communicatio istarum duarum scientiarum in consideratione de principiis substantiae, scilicet quod Naturalis declarat ea esse secundum quod sunt principia substantiae mobilis. Philosophus autem considerat de eis secundum quod sunt principia substantiae in eo quod est substantia, non substantia mobilis. Et etiam Philosophus declarat abstractum esse, quod est principium substantiae sensibilis..." *Ibid.*, fol. 293v, H–I. Also: *In VII Meta.*, c. 39, fol. 192r, B–D.

[3] "Et ista quaestio differt ab illa, quae est in scientia Naturali. Quoniam ista quaestio inducit ad sciendum primam formam omnium entium, et ultimum finem: et prima quaestio, quam incoepit in scientia Naturali, inducit ad sciendum primam materiam, et formas naturales, et primum motorem." *In VII Meta.*, c. 5, fol. 156r, B–C. Cf. Festugière, "Sources du Commentaire de S. Thomas...," pp. 660–62.

[4] Cf. *In II Meta.*, c. 16, fol. 35v, L.

[5] "... declaratum est igitur ex hoc, quod cognitio uniuscuiusque generum istarum causarum habet scientiam propriam: et quod est haec una scientia quae dicitur Philosophia, quae propriae considerat de prima forma, et primo fine." *In III Meta.*, c. 3, fol. 41v, M. Also *In IV Meta.*, c. 1, fol. 64r, C–D.

[6] *In VI Meta.*, c. 1, fol. 144r, F.

[7] "... cum res abstractae sint priores in esse non abstractis, necesse est ut prima scientia in esse sit scientia rerum abstractarum. Deinde dixit et quae non moventur, etc. et intendit, quod res immobiles necesse est ut sint aeternae magis quam aeternae mobiles divinae, scilicet corpora coelestia. Sunt enim causae istorum, scilicet quod substantia abstracta est causa corporum coelestium." *Ibid.*, c. 2, fol. 146v, G.

[8] "... manifestum est igitur ex hoc sermone, quod modi Philosophiae speculative sunt tres, scilicet rerum mathematicarum, et naturalium, et divinarum, scilicet substantiarum, in quarum definitione accipitur Deus." *Ibid.*, fol. 146v, G–H.

1) the study of being as such, 2) the study of matter, form, and the first mover, as principles of substance, 3) the study of the ultimate final and formal causes of all beings, and 4) the study of what is defined in terms of God. The method by which the metaphysician manages to identify these seemingly diverse aspects is basically one of logic.

The actual construction of Aristotelian metaphysics commences with Book Z, according to Averroes. The first six books are only preparatory.[1] Nevertheless, this does not mean that the consideration of the use of words, such as it found in Book Δ, is foreign to metaphysics. Quite the contrary, the metaphysician must discover by a logical method, what intentions one is to study.[2] Apropos of this logical study, Averroes explains that there are two ways of using logic; one way is to use logic as an instrument – such use is found in all sciences; a second way is to accept logical statements about being, for example, statements giving definitions or descriptions, and to use these to discover being as such.[3] Thus it is that the metaphysician must discover what intentions he is to study; this discovery moreover involves the acceptance of logical statements. Averroes summarizes his method in a few words as follows:

> Considerare enim de definitionibus est commune Logico et Philosopho: sed duobus modis diversis. Logicus enim considerat de definitionibus, secundum quod est instrumentum, quod inducit intellectum ad intelligere quidditates rerum; Philosophus autem, secundum quod significat naturas rerum...[4]

However clear Averroes' words may seem, his method is nevertheless best understood by noting how it is used in practice.

The divisions of *entia per se*, Averroes writes, are the divisions in the types of predicates; just as there are only a limited number of types of predicates, so too are there the same limited number of types of being.[5] Moreover, the nine types of accidental predicates all refer to a predicate of a substantial type.[6]

What Averroes has done in this investigation of being is not difficult to grasp: he has carried through an examination of our use of predicates. His first point was to note that there are ten types of pre-

[1] Cf. *In VII Meta., introductio*, fol. 152v, L–M.

[2] "... expositio istorum nominum est pars istius scientiae per se, non sicut consuetudo est in scientia Naturali... quoniam distinctio nominum fuit illic solummodo ad distinguendum intentionem quaesitam ab alia intentione: hic autem est ad numerandum ea, de quibus considerat ista scientia." *In V Meta.*, c. 1, fol. 100v, I–K.

[3] In *VII Meta.*, c. 2, fol. 153v, K–L. When he speaks of the logical statements used by the sciences, he notes: "Et propositiones dialecticae sunt *entis simpliciter*, sicut definitiones et descriptiones, et alia dicta in eis." *Ibid.*, Italics added.

[4] *Ibid.*, c. 42, fol. 194r, A–B.

[5] *In V Meta.*, c. 14, fol. 116v, M – fol. 117r, A.

[6] *Ibid.*, fol. 117r, A–B.

dicates; then he noted a reference of the use of some predicates to the use of other types: the nine types of predicates known as "accidents," when predicated, refer to the tenth type of predicate, "substance." Now it is certainly not at all difficult to see why he says that logic is involved in this study: Averroes' procedure consisted in examining the way concepts or predicates are used; the discovery of the various types of predicates revealed the types of being or reality.[1]

Following this investigation of predicates which resulted in the knowledge of the ten types of being, Averroes builds on this result by beginning his examination of the tenth type of being, substance. He turns thus to the use of the word "substance"; by examining this use, he should be able to discover the reality of substance.[2] The first use of "substance" is to predicate it of individual material bodies or substances, he explains.[3] Secondly the forms of individual substances are called "substance."[4] In the third use, "substance" is said of the material aspects of substances; that is, the subject of a form is called "substance."[5] Fourth, one sometimes calls definitions "substance."[6] Fifth but improperly, we call the Platonic ideas by the name "substance."[7] Finally, the answer to the question: "What is it?" is the "substance" of a thing.[8] It is this last use of the name "substance" that is of interest to us, for it is this substantial use of *ens* that is the principal one.[9]

Such a study has been logical, Averroes remarks. However, this should not be surprising because many demonstrations of metaphysics are logical. That is to say, the metaphysician accepts many logical statements about *ens simpliciter* and uses them in his work.[10] We have certainly caught Averroes in the act of examining predication. Averroes calls such investigations "logical," for they involve the use of statements made about *ens simpliciter* by the logician. The question naturally arises: Why does Averroes use the word *ens simpliciter* instead of *ens rationis*? Surely Averroes does not mean that the logician has the same subject matter as the metaphysician: *ens simpliciter*! Or does he?

1 Cf. *In III Meta.*, c. 1, fol. 36v, M – fol. 37r, A. *In VII Meta.*, c. 11, fol. 161r, D. *Ibid.* c. 42, fol. 194r, A–B.
2 *In V Meta.*, c. 15, fol. 118r, C.
3 *Ibid.*, fol. 118r, C–D.
4 *Ibid.*, fol. 118r, D–E.
5 *Ibid.*, fol. 118r, E.
6 *Ibid.*
7 *Ibid.*, fol. 118r, E–F.
8 *Ibid.*, fol. 118r, F – fol. 118v, G.
9 *In VII Meta.*, c. 1, fol. 153r, B–C; *Ibid.*, c. 2, fol. 153v, G.
10 *Ibid.*, fol. 153v, K–L.

If we turn to Averroes' discussion of the relation of sophistry, dialectics, and metaphysics as given in his Book IV, we gain little in the way of solving this problem. Averroes says simply that the three disciplines have the same subject matter, *ens simpliciter*. The difference between them is reduced to their different methods of treating their common subject: the philosopher demonstrates something about reality, while the logician arrives at probable knowledge of being; the sophist merely imitates the philosopher.[1]

Such a treatment is of little value; it serves only to highlight the problem: what does the logician say about *ens simpliciter*? That the logician would treat "being of reason" is not surprising, but how could he study real being – *ens* pure and simple? A survey of the investigation Averroes has thus far carried through, as well as of those which follow, reveal that this problem is only illusory – the sole problem is Averroes' language, not his theories.

Because "being" is predicated of all the objects falling under the ten predicaments, the metaphysician must search for the first being which is the cause of all objects called "being."[2] How does he search for this first being? He attacks the problem of the uses of the name "substance"; as already noted, there are several different meanings of this word, but the principal one is "the answer to the question: 'What,' when asked of an individual substance." When the metaphysician understands completely this meaning of "substance," he will understand the first cause of all beings, the prime mover.[3] Naturally enough, the metaphysician must begin with a study of sensible substances.[4] He must understand matter and form, and the first motor *qua* principles of substance.[5] The study of matter and form will lead the metaphysician

[1] *In IV Meta.*, c. 5, fol. 70v, I–K.

[2] "Et intendit quod, quia hoc nomen ens dicitur de omnibus, oportet artificem huius scientiae considerare de omnibus, et attribuere illa ad primum ens, quod est causa in omnibus entibus." *In VI Meta.*, c. 4, fol. 147v, L–M. Note also *In X Meta.*, c. 7, fol. 256v, G – fol. 257r, B, where Averroes speaks of the need of seeking the *unum* in virtue of which all substances are *unum*. However, "seek" in Averroes' philosophy means take the prime mover of the *Physics* and then say: "... quod hoc non solummodo est principium tamquam motor, sed tanquam forma et finis...," and so one can say: "... quod illud est unum, de quo declaratum fuit hoc, quod est principium substantiae." *Ibid.*, fol. 257r, A–B.

[3] "... quaerendum est de substantia illud, quid semper fuit quaesitum usque ad hoc tempus, in quo nullus dixit aliquid demonstrative, et est, quid est illud, quid est quidditas huius substantiae. Et augmentavit hanc quaestionem. Quoniam, cum fuerit scitum illud, quid est quidditas istius substantiae, tunc erit scita prima causa omnium entium." *In VII Meta.*, c. 5, fol. 156r, A–B.

[4] *Ibid.*, fol. 155v, M.

[5] "... illic, scilicet in Physico auditu fuit perscrutatus alio modo ab isto. Illic enim perscrutatum est de principiis corporis secundum quod est naturale, idest secundum quod quiescit, aut movetur, hic autem perscrutatus est de eis secundum quod est substantia tantum." *Ibid.*, fol. 156r, B.

to knowledge of the first form of all beings; the study of the first motor will result in knowledge of the ultimate end of all creatures.[1]

If a study of the matter and form of sensible substance is carried through from the point of view of "substance," what would result? That is, if the metaphysician studies matter and form to ascertain their role in the constitution of substance, what will he conclude? As we have just been told, Averroes feels he will reach knowledge of the first form of all beings. There is an element of ambiguity in Averroes' explanations, however. Metaphysics is to study only the formal and final causes, and yet Averroes would have us examine matter and form as the principles of substances. Actually what he means is not as confusing as it may seem: we are to begin an investigation of matter and form in sensible substances; once we see the role of each of these principles, we will realize that form alone is the "cause" of substance, and hence *qua* metaphysicians we can then view all things as substances, that is from the viewpoint of form. Thus when Averroes indicates that this science does not study the material cause, he does not wish to say that one does not even investigate it; rather he means only that the *completed science* of metaphysics will be capable of speaking metaphysically about all beings without ever mentioning even implicitly the materiality of things.[2]

The first step on the road to knowledge of the role of form in substance is made as soon as one recognizes that "form" is the cause of any given substance; no substance to which we can point would be a substance if it were not for its form.[3] Yet is this view really correct? Is it sufficient to say that form is the "cause" of substance? Although it does indeed seem that this is so at first glance, it is yet more obvious that matter is the cause of substance, and hence has a greater claim than form to the title "substance."[4] Yet if one examines matter's claims, one realizes finally that matter is outside the entire gamut of categories. One perceives as well that one does not understand a substance by grasping its matter; just the contrary, one understands an object through its forms. Hence, we were right in our initial statement:

[1] *Ibid.*, fol. 156r, B–C; cf. note 3, p. 37.

[2] Cf. Averroes' discussion of the first aporie of Book B: *In III Meta.*, c. 3, fol. 41v, G–K.

[3] "... oportet prius nos ponere principium considerationis in principio substantiae, quam omnes homines concedunt esse primum, et unum, et dicere quid sit... Omnes homines conveniunt in hoc, quod individua corporum existentium per se sunt substantiae, et quod in eis est principium... et ista substantia, de qua dicit se perscrutaturum, est illa, de qua declarabitur esse formam: et dixit ipsam esse primam, quia est causa substantiae demonstratae: et substantiae demonstratae non sunt substantiae nisi per illam..." *Ibid.*, fol. 156, E–F.

[4] *Ibid.*, c. 8, fol. 158v, I.

form is more worthy of being called "substance" than anything else is.[1] Because we see that substances around us are composed of matter and form, we did not have to examine the composite itself, for this is an effect of matter and form.[2] Rather we had to look to the causes of the composite, only look to matter and to form rather than to the effect, to the composite. Such a study reveals that matter is a substance in potency.[3] Form, on the other hand, is the true cause of the substance. Hence we can now turn our attention to the first motor as a form, for since form is the cause of substance, the first immaterial mover will be a substance.[4] In physics, one could have studied forms of sensible beings from the aspect of their action in the sensible world. However, they have no common action, and hence one could not have argued that the first mover, *qua* form, is a substance. In metaphysics, we note that the action of sensible form – in so far as it is the action of substantial forms – is to be the cause of substance. That is, the study of form in its influence in the constitution of substance reveals that form is the cause of substance. Hence we can speak now of the first form, of the first mover, and say of it: "It is a substance."[5]

The examination of sensible substance has led Averroes to say the first motor as form is a substance. As we know, Averroes would not accept this procedure as a proof for the existence of a first substance. We have already noted his remark: "Et ideo impossible est declarare aliquid abstractum esse, nisi ex motu. Et omnes viae, quae reputantur esse ducentes ad primum motorem esse praeter viam motus aequaliter, sunt insufficientes."[6] Thus the metaphysician does not prove the existence of a first substance; rather he proves only that the first or immaterial mover – a pure form – must be a substance, since form is the constitutive

[1] *Ibid.*, fol. 159r, E–F.

[2] *Ibid.*, c. 9, fol. 159v, I.

[3] *Ibid.*, fol. 159v, I–K.

[4] "Deinde dixit Et perscrutemur igitur de tertia, idest forma. Deinde dixit Et concessum est, etc. et quia manifestum est per se, quod individua substantiae sunt substantiae, oportet ut primo quaeramus de substantia, quae est prima substantia istarum substantiarum, et causa earum...." *Ibid.*, fol. 159v, K–L. There is the temptation to interpret the "prima substantia" of these paragraphs as "substance as such"; however, the immediate context of this quote reveals such a procedure to be incorrect. Note the following footnote, which immediately follows the passage just quoted.

[5] "... consideratio de formis naturalibus, secundum quod sunt naturales, non inducit ad primam formam. Consideratio enim de forma, inducens ad primam formam, est consideratio de ea secundum quod est substantia, et scientia Naturalis non considerat de rebus secundum quod sunt substantiae... Formae vero, quia sunt plures, et non habet eandem actionem communem, ideo non potuit declare de suis actionibus in scientia Naturali primam formam. Sed illud, quod fuit illic declaratum, est ex hoc, quod actionem primae formae, quae non est in materia, et est primus motor..." *Ibid.*, fol. 159v, L – fol. 160r, B. Note also the quote given in footnote 1, p. 43.

[6] *In XII Meta.*, c. 5, fol. 293r, C.

cause of substance. Thus metaphysics begins after the physical proof of an immaterial first mover; metaphysics, through its examination of sensible substance as substance, concludes that any form will be a substance. If the form is in matter, then one has a material substance; if there exists some being which is only form, then one has an immaterial substance.[1]

And so Averroes rises from sensible substances to a knowledge of the first form. Now such a doctrine on the nature of metaphysics clarifies the priority of metaphysics over physics; whereas physics knows the existence of a pure form prior to any metaphysical consideration, only further, metaphysical, study reveals this pure form to be a substance. Clear too is the reason behind the identity Averroes places between 1) the study of being as such (universal science), 2) the study of matter, of form, and of the first mover, as principles of substances, 3) the study of the first formal causes of all beings. Two aspects of Averroes' metaphysics are not yet properly integrated into this system: Why is metaphysics also the study of the ultimate final cause of all beings? Why is metaphysics the study of what is defined in terms of God?

It is not difficult to discover the explanation that underlies the conception of metaphysics as the study of objects definable in terms of "God." Ultimately, this explanation is rooted in the meaning Averroes gives to "universal science." To know all beings as such (universal science) is to know substance as such; to know substance as such is to know how form is the cause of substance. Yet to know form as the cause of substance is to know God, for He is the only pure form known in philosophy. Now this same explanation underlies Averroes' view of the divine science, the study of all things in terms of "God." As he explains, man must study form in sensible substances because these substances are better known to us than the first mover; he notes that it is proper to man to go from knowledge of singulars to knowledge of universals.[2] Or to say the same thing, man passes from knowledge of what is better known to us to knowledge of what is better known in

[1] "(In the *Physics*) ... non fuit declaratum de actione istius, utrum est substantia aut non. Et in hac scientia quaerit de principio primae substantiae. Et ideo posuit principium perscrutationis ex hoc de principiis substantiarum sensibilium. Et, cum fuerit ei hoc declaratum, quod principia substantiarum sensibilium, scilicet formae, sunt substantiae, et est declaratum in scientia Naturali hoc esse corpus sensibile, quod est causa in aliis substantiis sensibilis, et prius eis, declarabitur, quod forma istius corporis est principium ultimae substantiae, quae est prior aliis substantiis, et quod ipsum est illud, quid dat aliis substantias, et formas substantiales generabiles et corruptibiles." *In VII Meta.*, c. 9, fol. 160, B–C.

[2] "... rectum est igitur, ut ex cognitione substantiae sensibilis declaratur substantia quaesita, scilicet separata aeterna." *Ibid.*, c. 59, fol. 207r, A.

itself.[1] In the case of metaphysical knowledge of being, man begins with a very slight and imperfect awareness; this imperfect knowledge is had in so far as man knows sensible beings.[2] Thus, the story of metaphysics is a motion from this imperfect knowledge of being to the perfect knowledge of being; this perfect knowledge is had in so far as one knows the simple substance, the prime mover. To know the prime mover, writes Averroes, is to know how he contains all other substances.[3] Since, as Averroes explains, metaphysics' knowledge of the prime mover contains knowledge of all substances, one can understand how metaphysics will define all things it studies in terms of "God." That is to say, because the metaphysician studies the principles of substances, he is studying principally form. Moreover, when he realizes that form is the cause of substance, he knows the first mover, the pure form, to be a substance. Finally, metaphysical knowledge of being as such is primarily knowledge of substance as such. But to state the metaphysical knowledge of substance as such is to state how form is the cause of substance. And if one is speaking only of substance in relation to form (and not also in relation to matter), then one is referring solely to the prime form, to the prime mover, to the sole being which is form alone, to God. Thus everytime one speaks metaphysically of any substance, one speaks in terms of the pure form – one simply relates substance to form: one speaks of God.

Because of Averroes' insistence on speaking of metaphysics as defining all things in terms of "God," it is not surprising to discover that he holds man to be capable of understanding the quiddity of an immaterial being.[4] After all, to understand the "form as the cause of substance" is, in Averroes' mind, to understand "quiddity of immaterial being."[5]

[1] *Ibid.*, c. 10, fol. 160v, H–L.

[2] *Ibid.*, fol. 160v, L.

[3] "... necesse [est] scire ex eis [particularibus], quae sunt male cognitionis, quia habent cognitionem particularem ad illa, quae cognoscuntur per cognitionem universalem, transferendo de cognitione naturaliter istarum substantiarum sensibilium ad substantiam simplicem, quae cognoscitur modo universali, scilicet quae continet omnes substantias." *Ibid.*, fol. 160v, M.

[4] "Et, quia dispositio intellectus de re intelligibili est sicut dispositio sensus de re sensibili, assimilavit virtutem intellectus in comprehendo intellecta abstracta a materia modo debilissimo visui in sentiendo, scilicet vespertilionis, non comprehendo maximum sensibilium, scilicet Solem. Sed hoc non demonstrat res abstractas intelligere esse impossibile nobis: sicut inspicere Solem est impossibile vespertilioni; quoniam, si ita esset, tunc ociose egisset, quia fecit illud, quod est in se naturaliter intellectum, non intellectum ab alio: sicut si fecesset Solem non comprehensum ab aliquo visu." *In II Meta.*, c. 1, fol. 29r, B–C. Also *In IX Meta.*, c. 22, fol. 248v, K–L.

[5] It is difficult to see any difference between the view of Averroes on man's method of learning to know God, and that of Avempace, which Aquinas describes as "frivolous" in *Con. Gen. III*, c. 41. Note Aquinas' reference to Averroes when he treats the same problem:

Metaphysics is also said to be the study of the ultimate goal of all beings. The prime mover of the *Physics*, who has become the first substance of the *Metaphysics*, moves all things because he is their goal, Averroes explains.[1] In other words, because this first mover is completely immobile, he can only "draw" things to himself, he can only move things in so far as they desire him.[2]

This then is Averroes' version of Aristotle's metaphysical system. Although the treatment of the ultimate end of all beings seems to contain no "logical studies," the other sections of Averroes' *Commentary* are filled with investigations best characterized as "logical." This distinction between the logical part and the section containing no logical procedure parallels the distinction between the study of the principles of matter and form, and the study of the prime mover. In so far as one studied matter and form with a view to discovering their influence in the constitution of substance, one used a logical examination; but once one had realized that the presence of form in a being implies the presence of a substance, then one leaves behind the investigation of predication; one states then that the prime mover is the first substance, and that he draws all things to himself.

In the course of our exposition the question was raised why Averroes speaks of logical statements as "statements about *ens simpliciter*."[3] The answer is discovered if we deduce the meaning of that phrase from Averroes' use of such statements in Books VII–VIII. On this basis, one must conclude that Averroes is merely guilty of a slight impreciseness of language. For Averroes, an example of a proposition about *ens simpliciter* would be: "A definition signifies the quiddity of an object."[4] To someone more accustomed to the language of Aquinas, this is a proposition about a being of reason. To call an idea a "definition" is to relate it to the subject of which it can be predicated: an idea is a definition if it expresses the quiddity of the subject of which it is said. Now obviously Averroes would have no objection to this Aristotelian doctrine; only its language would appear uncomfortable. In other words, Averroes prefers to say an idea *qua* definition tells us something about *ens simpliciter*, rather than to call an idea *qua* definition an *ens rationis*. Averroes in short emphasizes the reality which an idea expres-

"Ex quo apparet falsum esse quod Averroes hic dicit in *Commento*, quod Philosophus non demonstrat hic, res abstractas intelligere esse impossibile nobis... Et ratio sua, quam inducit, est valde derisibilis." *In II Meta.*, 1, 286.

[1] *In XII Meta.*, c. 5, fol. 293v, H–G.

[2] *Ibid.*, c. 36, fol. 318r, E–D.

[3] E.g. *In IV Meta.*, c. 5, fol. 70v, I–K. *In VII Meta.*, c. 2, fol. 153v, K–L.

[4] *In VIII Meta.*, c. 2, fol. 210r, C.

ses. To call an idea a "definition" is to specify that aspect of reality expressed by the idea: quiddity. Thus, only Averroes' language is different, not his theory of logic.

Averroes' metaphysics is both "universal science" and "first science": universal science, because it speaks of all beings *qua* being when it explains that form is the cause of substance; first science, because in speaking of form as the cause of substance, it refers directly to the immaterial first mover. It is at the same time, and for the same reason the "divine science": it treats whatever is definable in terms of "God," that is, in terms of "form as the cause of substance." Finally, and quite obviously, it proceeds by using logical statements, by searching for the aspects of reality corresponding to logical propositions. Moreover, it is *meta-physica* because it presupposes the physical proofs both of the intrinsic principles of mobile being (matter and form) and of the extrinsic principle (the prime mover).

Hence, Averroes' work is truly a God-centered system. Quite obviously, his view is not that of Avicenna. Conveniently, Averroes has pointed out why he disagrees with his Arabian predecessor. First and foremost, Averroes could not accept the Avicennian (and Alexandrian) theory of metaphysics as proving, and then presenting to physics, the principles of mobile being: matter, form, and prime mover. Such a theory runs directly contrary to the heart of Averroes' own views. Secondly, there was the foolish Avicennian doctrine of *esse*. This theory, at the center of Avicenna's system, is in Averroes' eyes based on the grossest of mistakes – the identification of the *ens ut verum* and extramental reality.[1]

3. THE *COMMENTARY* OF ALBERT

Physics considers mobile beings in so far as they are mobile, and is thus interested in all four causes as the causes of motion, writes Albert. On the other hand, there is metaphysics; metaphysics too investigates all four causes, although from a new point of view. Physics, Albert explains, proves the existence of a first cause of motion – the prime

[1] "Sed deceperunt eum [Avicennam] duo, scilicet quia opinabatur quod unum... [etc.] Et etiam fuit deceptus, quia ignoravit differentiam inter hoc nomen ens, quod significat genus, et quod significat verum. Quod enim significat verum, est actus, et quod significat genus, significat unumquodque decem praedicamentorum multipliciter." *In X Meta.*, c. 8, fol. 257r, F – v, G. Cf. *In V Meta.*, c. 14, fol. 117, D–F. See also. ROLAND-GOSSELIN, *Le "De ente et essentia"*..., pp. 157–59; E. GILSON, *Being and Some Philosophers*, 2nd edit., Pont. Inst. of Med. Studies, Toronto, 1952, pp. 51–56. This point is treated at greater length below, Ch. IV, section 4, A and B.

mover. This doctrine metaphysics accepts and builds upon, for metaphysics considers this prime mover as the first form and the ultimate end of all beings; even more, metaphysics explains how the first mover "sends forth" all beings through the instrumentality of motion. In other words, the prime mover is considered as the efficient cause in so far as all beings flows from Him through motion. Thus, concludes Albert, I can not agree with those writers (e.g. Averroes) who maintain that physics considers the efficient and material causes, while wisdom studies the formal and final causes.[1]

What procedure does one follow in metaphysics? Not a logical one, answers Albert. For example, one incorrect method is exemplified by those who conclude that universals are more truly substances than particular objects are. These philosophers have made only a logical investigation of things.[2] Or again: we must avoid errors such as thinking of substance as "what stands under." Some philosophers have committed such an error because of an investigation of the relations of predicates to the subjects of which they are said.[3] Such statements might lead one to believe that Albert chooses a different procedure than the logical one noted in Averroes. This impression is increased by a reading of Albert's exposition of Book Γ's comparison of the dialectician, the sophist, and the philosopher. Under Albert's pen, the dialectician differs from the philosopher in so far as he works from the "signs which are found in many things."[4] Apparently, because he would deal with "signs" rather than with reality, the dialectician can not aid the metaphysician. Yet notwithstanding these points, the investigation of predication is central to Albert's work. Examples of this method are abundant in the treatments of *ens* and substance. Despite his warning regarding the evils of using a logical method in metaphysics,

[1] "Nec etiam est dicendum, sicut quidam alii dixerunt, quod physicus consideret efficientem et materiam, et metaphysicus formam et finem. Quia licet nos in III huius sapientiae libro iam ostenderimus, quod primus philosophus principaliter considerat formam et finem, per quae demonstrat secundum proprietatem suae sapientiae... tamen tam physicus quam metaphysicus considerant omnes quatuor causas. Sed physicus considerat eas prout sunt principia mobilis, primus autem philosophus reducit efficientem in formam primam et finem ultimum, et sic ipse est causa universi esse et forma et finis. Et si accipiatur per motum prima causa, hoc non est, ut sciatur, inquantum est movens talem motum, sed potius inquantum ipse ambit virtute et forma sua mobile et motum, quod est instrumentum fluxus totius entis ab ipso, et hoc modo non considerat ipsum physicus." *In XI Meta.*, T. I, c. 3, p. 462, l. 81–p. 463, l. 10. Among the writers who held the notion Albert opposes: Averroes: *In XII Meta.*, c. 6, fol. 294v, K–M; *Ibid.*, c. 2, fol. 290v, M. Aristotle's Book XI was unknown to Albert, and hence Book XII is called Book XI. Thus Albert's exposition in the text cited above is actually an exposition of Aristotle's Book XII.

[2] *In XI Meta.*, T. I, c. 2.

[3] *In V Meta.*, T. II, c. 5, p. 242, l. 78–93.

[4] *In IV Meta.*, T. I, c. 7, p. 170, l. 52–55. (Averroes: *In IV Meta.*, c. 5, fol. 70v, G–K).

Albert is one with Averroes in employing what can only be termed a logical investigation of predication. So much then for Albert's *statements* on method; let us note rather the method actually used in the construction of metaphysics.

Albert explains that the particular sciences such as mathematics and physics study particular types of beings, and from restricted points of view. Thus none of them investigates what it means "to be."[1] Accordingly, metaphysics has a legitimate claim to be heard.

The starting point of metaphysics must be the study of sensible beings, for we must begin with what is better known to us. Only after we understand sensible beings as being, only then can we examine separate substances.[2] The initial step in the effort to grasp the being of things around us consists in dividing the various uses of the name "being"; as well one must discover how we use names such as "principle," "cause," etc., as all these are involved in the discussion of the reality of objects. Once this examination of the use of names is completed, we shall see perfectly what must be investigated regarding our subject "being", as well as its principles and the parts of being.[3]

The first of the important names to be analyzed is, of course, "being". One can not explain this term "being" by using any other word; "being" is the concept man understands best. Even more, being is itself the most knowable of things. Hence, we must look for a different way of explaining this term.[4] The only possible method appears to be a division of different uses or ways of *saying* "being".[5] The more important use of "being" is found in essential predication. In this type, we predicate one thing of another in such a way that we express the "what" of the subject. Yet we can distinguish ten ways of stating what a thing is – ten ways of stating that something is a being. Thus, there are predicates expressing the "how" of a what, or the "how much" of a what, and so on. In each of these ten types of predicates, we must express at least secundarily the "what" of the subject.[6]

As a consequence of the discovery of this division of "being", we are led to the study of "substance", for to predicate "substance" is the

[1] *In IV Meta.*, T. I, c. 7, p. 169, l. 91 – p. 170, l. 39.
[2] *Ibid.*, T. I, c. 6, p. 168, l. 5–15.
[3] *In V Meta.*, T. I, c. 1, p. 207, l. 6–24. – Albert's explanations show great dependence on Averroes' *In V Meta.*, c. 1. fol. 100v, I–L.
[4] *In V Meta.*, T. I, c. 11, p. 233, l. 55–63. – Albert's work here is dependent on Avicenna's *Meta.*, T. I, c. 6, fol. 72r, A. (In turn, Albert aided Thomas: *In V Meta.*, 9, 889).
[5] *In V. Meta.*, T. I, c. 11, p. 233, l. 64–65.
[6] *Ibid.*, p. 234, l. 22–33.

principal example of calling something "being" in an essential way.[1] – Thus far Albert has carried through the type of investigation Averroes considered to be logical. At this point in his *Commentary* – after stating that we must now examine the use of "substance" – Albert interrupts his exposition of Aristotles' thought to introduce a discussion billed as "the exposition of substance and its true modes and principles".[2] The doctrine given in this disgression is but a summary of that presented, or better, discovered elsewhere through an investigation of predication.[3] Hence, these explanations can be omitted, and instead his exposition of the "common way to divide substance" must be noted. This "common way", to take Albert at his word, consists in noting what concepts are not predicated.[4] The inability to predicate certain concepts permits one to realize that the objects thus conceived are substances. The one exception, a concept of a substance which can be predicated, is the concept expressing the essential principles of an object; yet even this concept is not predicated as if it expressed something added to a completed substance; rather it expresses what the subject is.[5] By such an examination of predication, Albert arrives at a four-fold division of "substance": 1) "corporeal simples" such as earth, fire, etc.; 2)essential principle of substance, the substantial form; 3) "material and formal particles" which exist in bodies and which can not be destroyed without destroying the bodies; and 4) the quiddity of a thing.[6] The common meaning intended when these four types are named "substance" is "being, existing on its own."[7]

Hence Albert proceeds by examining the use of predicates. Any of his anti-logical affirmations must be understood in this light; he is not against using logic, but rather against mistaking the logical characteristics of being for the reality of things. Just as was true of Averroes, so too Albert begins his examination of substance by noting *logicae rationes* concerning the quiddity of sensible substance.[8]

[1] *Ibid.*, T. II, c. 1, p. 236, l. 8–16.

[2] *Ibid.*, T. II, c. 5, p. 241, l. 13–14.

[3] *Ibid.*, T. II, c. 1–4.

[4] *Ibid.*, T. II, c. 5, p. 241, l. 14–16.

[5] *Ibid.*, p. 241, l. 37–48.

[6] For division no. 1) see *Ibid.*, p. 241, l. 16–37; for division no. 2) see *Ibid.*, p. 241, 1. 49–77; for division no. 3) see *Ibid.*, p. 241, l. 78 – p. 242, l. 46; for division no. 4) see *Ibid.*, p. 242, l. 47–51.

[7] "... si acciapiatur ipsa natura generis substantiae, hoc non ab actu substandi dicitur, sed potius per oppositionem ad accidens. Et sic substantia est ens per se existens sive non in alio..." *Ibid.*, p. 242, l. 78–82.

[8] "Primum igitur sic procedentes a sensibilibus substantiis ad substantias alias, dicemus quaedam rationibus logicis de eo quod est quid est, ostendo primum, quid est in unoquoque sensibilium secundum se." *In VII Meta.*, T. I, c. 6, p. 326, l. 45–48.

It is important to know just what quiddity is to be examined, for one can legitimately distinguish between a quiddity considered in itself and as found in individual things.[1] As Albert decides the issue, it is quiddity in the latter sense that is to be considered, for metaphysics investigates the "substantial being of first substance."[2]

The individual results of all the investigations made are, naturally enough, those of Aristotle. However, it is not the results, but the application of these results to God that is of interest. As Albert explains, echoing Averroes, the study of quiddity will lead us to knowledge of the first cause in so far as it is the first form and the ultimate end of all beings.[3] The reasons behind this affirmation are as follows. The quiddity of first substance (Albert does not mean God by "first substance", but is speaking of the first creature of God) is prior to accidents and to matter. In the case of accidents, this is obvious.[4] That quiddity is prior to matter is a bit more difficult to see; but, as Albert remarks, "to be material" happens to a substance in so far as a substance is in matter.[5] Moreover, if we search for the reason why any given quiddity considered in itself is found to exist, we can only conclude that the quiddity is a reality because it is a "radius and a light" of the first form, that is of the divine intellect.[6] In other words, because a real quiddity is intelligible in itself, it must be impregnated, saturated with the "light of the intellect from which it comes."[7] Thus if we studied the quiddities around us, we shall be able to learn something about the first formal cause of all beings. This situation, this resemblance of beings to God, explains why men have always sought to know God: since He is the source of all forms, the first form, all try to learn something about Him.[8]

[1] *Ibid.*, T. I, c. 1, p. 316, l. 34–39. *In V Meta.*, T. VI, c. 5.

[2] *In VII Meta.*, T. I, c. 1, p. 316, l. 39–41. *Ibid.*, T. I, c. 2, p. 318, l. 4–43.

[3] "Inquisitio enim de ipsa [quidditate] conducit ad notitiam causae primae, secundum quod ipsa est prima forma et ultimus finis..." *Ibid.*, T. I, c. 4, p. 320, l. 13–15.

[4] That metaphysics studies created being: cf. *In I Meta.*, T. I, c. 1, p. 3, l. 1–4; *In II Meta.*, c. 2, p. 93, l. 5–29. That quiddity is prior to accidents: cf. In *VII Meta.*, T. I, c. 4, p. 320, l. 16–22.

[5] "... esse materiale accidit ei per hoc quod quidem est in materia, sed accipitur etiam secundum se, et sic habet esse immateriale et simplex." *Ibid.*, p. 320, l. 32–34. It is because "substance" does not demand matter, but rather can be material almost by accident, that Albert insists terms such as *ens mobile* and *ens quantum* are not subalternated to *ens metaphysicum*. Cf. *In I Meta.*, T. I, c. 2, p. 5, l. 25–33. *In XI Meta.*, T. I, c. 3, p. 462, l. 49–80.

[6] "Et si quaeritur origo huius esse quod forma secundum se accepta sic habet, non potest ad aliquid referri, nisi quod habet hoc, inquantum est radius quidam et lumen primae formae, quae est intellectus divinus." *In VII Meta.*, T. I, c. 4, p. 320, l. 34–38.

[7] "Cum igitur sit intelligibilis per seipsam, oportet, quod hoc habeat, inquantum immixtum est ei lumen intellectus primi, a quo exiit." *Ibid.*, p. 320, l. 41–43.

[8] "... quidditas rei sensibilis conducit ad notitiam causae primae formalis. His igitur de causis ipsa est, quod olim et nunc et semper quaesitum est et quaeritur et inquiritur, quia omnes homines natura scire desiderant hoc modo dictam formam primam, quae est fons

It is against the background of such theories that Albert rises from sensible beings to God. He knows the existence of a prime mover and so is content to look around the world of sense with the hope of finding out what the prime mover is. Thus, even though he must investigate sensible substances, it is the prime mover that is of primary interest.[1] For by knowing the prime mover, he knows how other things flow from Him.[2] Accordingly, there is no proof of the prime mover, or of a cause of being, to be found in Albert's work other than the famous physical argument from motion.[3]

When he turns to speak of the prime mover, Albert notes among other things, that God is the first substance.[4] He must also be immaterial. The only argument for God's immateriality is, Albert remarks, the fact that He can not be divided in place and in form. These possibilities of division follow from matter, not from form, and hence God has no matter.[5] Or again, because the first substance is not material, he is a pure intellect; and because he is universally the agent of motion, he bestows all perfections and is the highest good.[6] Thus, he is also the most beloved of all things.[7]

As is to be expected, a system such as Albert's can include Averroes' theory of metaphysical definitions as containing the notion of "God".[8]

formarum, et ultimum finem, qui sicut in duce exercitus universorum, quae sunt, est in motore primo..." *Ibid.*, p. 320, l. 44–51.

[1] Albert notes that he will show how the prime mover is the principle of all other beings: "Et in hoc erit huius operis quod sapientia vocatur, finis et complementum..." *In XI Meta.*, T. I, c. 3, p. 462, l. 37–39.

[2] Cf. *Ibid.*, p. 462, l. 32–37. Also: "Cum enim... constet tres esse substantias, quarum principia inquirenda sunt, quarum una est sensibilis mobilis et corruptibilis, cuius principia determinata sunt, et secunda est sensibilis et mobilis et incorruptibilis et tertia immobilis et insensibilis et incorruptibilis, et tertia quidem est principium secundae, nec potest sciri secunda nisi per tertiam, non enim potest determinari aliquid secundum proprietatem istius sapientiae de motis motu perpetuo nisi per scientiam motorum..." *Ibid.*, T. II, c. 1, p. 482, l. 9–18.

[3] Cf. *Ibid.*, the entire chapter. As Albert notes, however, one must not understand the proof as if it implies the eternity of motion and of time: *Ibid.*, p. 483, l. 45–55.

[4] *Ibid.*, T. II, c. 6, p. 490, l. 11–14. Cf. *Ibid.*, T. II, c. 1, p. 482, l. 38–59.

[5] "Quod autem penitus est indivisibilis per ubi et formam, est omnino carens materia, cum omnis materia per hoc sit materia, quia per ubi vel per formam est divisibilis et extensibilis. Et ista est vera demonstratio, quod substantia prima sit immaterialis... Et quidquid aliud ad hoc inducitur, non habet veritatem demonstrationis." *Ibid.*, c. 3, p. 485, l. 66–73.

[6] "Cum igitur probatum sit primum substantiam non esse materialem, constat, quod ipsa est intellectus purus... et cum sit universaliter agens, constat, quod largitur omnes perfectiones et est summum bonum..." *Ibid.*, c. 6, p. 490, l. 11–16.

[7] "Ipsa igitur est maxime amata et desiderata ab omnibus, et omne quod movetur, desiderat ipsam. Hoc igitur modo movet non-mota, eo quod ipsa nihil desiderat extra se. Desiderium autem est causa motus omnis. Movet autem sicut desiderabile..." *Ibid.*, p. 490, l. 19–23.

[8] "... in prima philosophia omnia dicuntur divina, eo quod in diffinitione eorum cadit deus..." *In VI Meta.*, T. I, c. 3, p. 305, l. 38–40.

This doctrine is not surprising, for metaphysics consists in discovering the substance of any being, and as Albert says, substance is saturated with the "light of the divine intellect".[1] Thus, the metaphysician naturally resolves all sensible beings back to the "light from which they came".[2]

If one is inclined to wonder how Albert would interpret St. Paul's reference to the possibility of knowing God "through those things which have been made", one need but look to the early part of the *Commentary*.[3] There Albert explains that we can not completely understand God, due to the imperfection of our intellects.[4] However, man is capable of rising from the physical world to knowledge of what God is, to knowledge of the source of the light which proceeds into the entity of things.[5] This must not be understood as implying the possibility of a metaphysical proof of God. The entire movement of metaphysics is the process of searching for the "light proceeding from God into the first entity and substance of things, before this entity and substance is contracted to imaginary and sensible being by the determination of quantity and contrariety."[6] The process is achieved when we have discovered the *esse substantiale substantiae primae*, for this is the first creature of God, this is the basic, common way of imitating God.[7] Metaphysics thus discovers what God is, not that He is.

Creatures (*esse*) serve then as the instrument leading us to a discovery of what God is; they are able to fulfill this function since they are expressions of God, since they are impregnated with His light. In the early part of his work, Albert says that *esse simplex*, the first creature

[1] "Adhuc autem forma substantialis per esse materiale non est intelligibilis, sed per seipsam et non per aliud sicut accidens. Cum igitur sit intelligibilis per seipsam, oportet, quod hoc habeat, inquantum immixtum est ei lumen intellectus primi, a quo exiit." *In VII Meta.*, T. I, c. 4, p. 320, l. 39–43. (The lines immediately following this text are given in note 8, p. 50).

[2] "... omnia exeunt a primis principiis divinis et in ipsis sunt artificiata in mente artificis. Et sicut artificiata resolvuntur ad lumen intellectus primi activi et per ipsum diffiniuntur, ita omnia resolvuntur ad lumen separatarum substantiarum, et ipsae separatae substantiae resolvuntur ad lumen intellectus dei, per quod subsistunt, et per ipsum sicut per primum principium diffiniuntur. Et haec est causa, quod divina et theologica dicitur haec sapientia." *In VI Meta.*, T. I, c. 3, p. 305, l. 41–49.

[3] If Albert seems at times a bit un-Christian, then one should remember his affirmation: "We do not give our own opinions, but those of Aristotelians; our readers must judge what is right." Cf. *In XI Meta.*, T. II, c. 1, p. 482, l. 23–29.

[4] *In II Meta.*, c. 2, p. 92, l. 60–68.

[5] "Intellectus humanus... incipit ab eo lumine quod est permixtum tenebris et per separationem apud se factam tandem venit in intelligibile sincerum... Venit enim ab intellectu obscuro ad lumen sincerum et a lumine sincero coadunato visu venit ad lumen perpurum et per grados ascendens tandem accipit ipsum in fonte luminis, sicut aquila contuetur lumen in rota solis." *Ibid.*, p. 93, l. 81 – p. 94, l. 2.

[6] *Ibid.*, p. 93, l. 20–23.

[7] *In VII Meta.*, T. I, c. 4, p. 320, l. 34–54. Cf. the texts given in footnotes 6–8, p. 50.

of God, is the object of metaphysics.[1] And in the subsequent parts of his work, he does not contradict this theory. However, in some texts it is difficult to distinguish between Averroes and Albert, although the difference is capital. Averroes' metaphysics is directly a study of God. Albert, as Averroes, accepts the physician's proof of the prime mover, and henceforth searches for means to know this mover. Yet whereas Averroes makes God, understood as the prime mover, the object of metaphysics, Albert does not; for Albert the first creature of God, being, is the object studied. Thus Albert starts in the universal science (the study of being as such); when he discovers what a substance is, then Albert speaks of God, then he enters first philosophy. Then too, for Averroes, to know form as the constitutive cause of substance is to understand the immaterial prime mover; for Albert, man is much too chained to earth to permit him any such claim to comprehend God, although man is capable of discovering something about what God is. Hence, Averroes and Albert are close, but not at all identical in their views on metaphysics.

These three commentators – Avicenna, Averroes, Albert the Great – pretend to expose the metaphysical theories proper to Aristotle. As we shall discover in the course of the investigations of Chapters IV–VI, Aquinas felt that none of the three presented a satisfactory exposition of the Aristotelian *Metaphysics*. Accordingly, Aquinas composed his own *Commentary*. In reading Chapters IV–VI it will be necessary to keep before us the salient points of Aristotelian metaphysics as understood by the three predecessors of Aquinas. In summary form, these points are as follows:

1) Material Object of Metaphysics:
   Avicenna: all realities
   Averroes: God – the prime mover
   Albert: created being – *esse simplex*
2) Formal Object of Metaphysics:
   Avicenna: *esse* or existence as affirmed
   Averroes: being or substance *qua* constituted by form
   Albert: being or substance *qua* constituted by form
3) Method Employed in the Construction of Metaphysics:
   Avicenna: investigation of affirmation of existence
   Averroes: logical investigation of predication
   Albert: logical investigation of predication

[1] *In I Meta.*, c. I, p. 3, l. 1–4. Also: *In II Meta.*, c. 2, p. 93, l. 5–29.

4) Movement of Metaphysics:

Avicenna: one begins in the universal science, and subsequently proves the existence of the first efficient cause of *esse*; consequently, one studies this cause (God) in first science.

Averroes: one recalls what was proved in physics (i.e., prime mover, and matter-form composition of material beings); next one engages in a logical investigation of predication to discover what is meant when we speak of anything as "being"; following this preparation, one begins metaphysics by studying the prime mover as being, that is, as substance constituted what it is by form; this latter study, and it alone, is metaphysics.

Albert: one begins in the universal science by carrying through a logical investigation of predication; thus one discovers what it means to be real; by knowing being as such in this way, one knows what God is as being, albeit imperfectly; subsequently one studies God more specifically in first science by discussing the consequences flowing from the being of God as it has been understood.

5) Proof of the Existence of God:

Avicenna: because of the distinction of essence and existence, there must be a first cause of existence who is pure existence.

Averroes: Avicenna's distinction of essence and existence is invalid; there is no metaphysical proof of God, only the physical proof of the prime mover.

Albert: there is no metaphysical proof of God, although the physical proof of the prime mover is valid.

CHAPTER III

THE *PROOEMIUM* TO AQUINAS' *COMMENTARY*

Before beginning the lengthy investigations into the intentions motivating Aquinas in the composition of his *Commentary on the Metaphysics*, it seems worthwhile to note that a Thomist conception of metaphysics is readily available to the reader of the *Commentary*. It is found in the *Prooemium*, the one part of the *Commentary* undeniably expressive of Aquinas' own thought. Accordingly, before we begin our investigation into the body of Thomas' exposition of the Aristotelian *Metaphysics*, we should seek the conception to be found in the *Prooemium*; that conception will serve as the framework into which must fit any doctrine of the *Commentary* which can be styled "Aquinas' personal thought." As we shall discover, the conception of metaphysics presented in the *Prooemium* is actually the barest outline of the conception at work in the body of the *Commentary*.

1. THE DOCTRINE OF THE *PROOEMIUM*

The *Prooemium* is a strange piece of philosophical writing. Due to his unusual emphasis on the intellect, one is somewhat at a loss to know whether Aquinas intends to give elements of a scientific introduction, or whether the affair is a rather loose, dialectical proof of the object of metaphysics. Whatever may be the case, for the present we must limit ourselves to grasping the meaning of Aquinas' words. Later we can with some certainty determine the nature of the "intellectualism" of the *Prooemium*.

Whenever several things are found ordered to the same end, Aquinas writes in beginning the *Prooemium*, in such a case one of these things must rule the others. Now the same must be true of all the intellectual disciplines which aid man to achieve his perfection: one of these must be the ruling discipline. Let us name such a discipline "wisdom," for the wise man is said to be he who rules others.[1]

The exact nature, the identity of this discipline – Aquinas calls it "science," no doubt because sciences are higher than other intellectual disciplines – the nature of this science will be known if we can discover the most intelligible object. The reason given is quite simple: intellectuals naturally rule other men, and so the ruling intellectual will be the most intellectual of all, he will be the person who studies the most intelligible objects.[2]

To discover the nature of the most intelligible objects, three paths are open. First, if one considers that the greatest certitude would be had from the most intelligible things, the intelligible objects studied in wisdom would be the first causes – for the knowledge of causes is the source of certitude.[3] A second path leading to the identity of the most intelligible objects involves the recognition that the intellect understands universals, whereas the senses grasp only particulars. Thus wisdom, as intellectual, would study those aspects of things which correspond to our most universal concepts, for these would be the most intellectual. These universals are *ens* and other things which in some way "come with" *ens*.[4] Thirdly, one can discover the most intelligible object if one reflects on the fact that an intellect is such because of its immateriality. Thus the more immaterial a thing is, the more intellectual and the more intelligible it will be. Accordingly, whatever is separated from sensible matter both in its existence and in its definition is the most intelligible object – e.g., God and other separated intelligences.[5]

[1] "...quando aliqua plura ordinantur ad unum, oportet unum eorum esse regulans, sive regens, et alia regulata, sive recta... Omnes autem scientiae et artes ordinantur in unum-scilicet ad hominis perfectionem, quae est eius beatitudo. Unde necesse est, quod una earum sit aliarum omnium rectrix, quae nomen sapientiae recte vindicat. Nam sapientis est alios ordinare."

[2] "Quae autem sit haec scientia, et circa qualia, considerari potest, si diligenter respiciatur quomodo est aliquis idoneus ad regendum... homines intellectu vigentes, naturaliter aliorum rectores et domini sunt... ita scientia debet esse naturaliter aliarum regulatrix, quae maxime intellectualis est. Haec autem est, quae circa maxime intelligibilia versatur."

[3] "Maxime autem intelligibilia tripliciter accipere possumus. Primo quidem ex ordine intelligendi. Nam ex quibus intellectus certitudinem accipit, videntur esse intelligibilia magis. Unde, cum certitudo scientiae per intellectum acquiratur ex causis, causarum cognitio maxime intellectualis esse videtur. Unde et illa scientia, quae primas causas considerat, videtur esse maxime aliarum regulatrix."

[4] "Secundo ex comparatione intellectus ad sensum. Nam, cum sensus sit cognitio particularium, intellectus per hoc ab ipso differre videtur, quod universalia comprehendit. Unde et illa scientia maxime est intellectualis, quae circa principia maxime universalia versatur. Quae quidem sunt ens, et ea quae consequuntur ens, ut unum et multa, potentia et actus."

[5] "Tertio ex ipsa cognitione intellectus. Nam cum unaquaeque res ex hoc ipso vim intellectivam habeat, quod est a materia immunis, oportet illa esse maxime intelligibilia, quae sunt

Thus, writes Aquinas, we know what the most intelligible objects are, and so we know what wisdom must study: first causes, being, and separated intellects. Yet we must recognize that these separated intellects are the first causes of being, which itself is a genus. Thus wisdom will take the genus of being for its subject and will study its first *causae essendi* (the separated intellects).[1]

Now in so far as this ruling science of wisdom studies the separate substances, it can be called "theology," a study of divine things; and in so far as it studies the first causes of things, we can call it "first philosophy."[2] This science has yet another name which is given it because of its subject matter. Being, or *ens commune,* and all the other common notions connected with being are the most common of all notions, the most universal notions. In fact, if one does not know them, one will be unable to know completely what is proper to any genus or species of beings;[3] that is, it seems to be implied, one could not understand completely any generic or specific term, nor the individuals to whom the term applies, without knowing *ens* and its retinue of notions. In so far as these notions are studied after one has studied less common notions, the science studying them can be called *transphysica* or "metaphysics."[4]

Aquinas makes two additional points which must be mentioned: (1) *ens commune,* the subject of metaphysics, is predicated of many things; and (2) metaphysics has a place in the *via resolutionis* in much the same way as a more common thing follows a less common thing.

First then, of what is *ens commune* predicated? On this point Aquinas is absolutely clear only if one knows beforehand what his doctrine is! *Ens commune,* he writes, is predicated totally of everything which

[sic: read "sunt"] maxime a materia separata. Intelligibile enim et intellectum oportet proportionata esse... Ea vero sunt maxime a materia separata, quae non tantum a signata materia abstrahunt,... sed omnino a materia sensibili. Et non solum secundum rationem, sicut mathematica, sed etiam secundum esse, sicut Deus et intelligentiae. Unde scientia, quae de istis rebus considerat, maxime videtur esse intellectualis, et aliarum princeps sive domina."

[1] "Eiusdem autem scientiae est considerare causas proprias alicuius generis et genus ipsum:... Unde oportet quod ad eamdem scientiam pertineat considerare substantias separatas, et ens commune, quod est genus, cuius sunt praedictae substantiae communes et universales causae."

[2] "Dicitur enim scientia divina sive theologia, inquantum praedictas substantias considerat... Dicitur autem prima philosophia, inquantum primas rerum causas considerat."

[3] "Huiusmodi [idest: 'ens et ea quae consequuntur ens'] autem non debent omnino indeterminata remanere, cum sine his completa cognitio de his, quae sunt propria alicui generi vel speciei, haberi non possit... cum his unumquodque genus entium ad sui cognitionem indigeat..."

[4] This science is called: "Metaphysica, in quantum [sic.] considerat ens et ea quae consequuntur ipsum. Haec enim transphysica inveniuntur in via resolutionis, sicut magis communia post minus communia."

does not exist in matter and which is defined without matter ("... dicitur tamen tota de his quae sunt separata a materia secundum esse et rationem"). So far his thought is straightforward. But when he wants to explain what things can receive the predication of "being", he writes:

Quia secundum esse et rationem separari dicuntur, non solum illa quae numquam in materia esse possunt, sicut Deus et intellectuales substantiae, sed etiam illa quae possunt sine materia esse, sicut ens commune. Hoc tamen non contingeret, si a materia secundum esse dependerent.

Thus Aquinas says there are two classes of things which are said to be separated from matter both according to *esse* and *ratio*; one supposes therefore that *ens commune* is predicated of both classes, and hence they are both studied by metaphysics in so far as they receive the predication of "being." These classes are: (1) all immaterial beings, those thing which can never exist in matter, for example, God; and (2) all those things ("illa") which can be without matter since they do not necessarily exist in it; *ens commune* is itself an example of this class. What are the other things denoted by "illa"? And is *ens commune* taken as a concept? Or does Aquinas mean: a being *qua* being does not necessarily need matter? Although one might wish to substitute for "illa" something such as "real beings *qua* being," there is nothing in these lines themselves which permits or which requires such a reading. On the other hand it seems that "illa" is supposed to be taken as standing for notions such as *ens commune*; thus "illa" would stand for "ens, et ea quae consequuntur ens, ut unum et multa, potentia et actus." Yet only a few sentences earlier Aquinas had mentioned that *ens commune* is caused by God; obviously there he meant that the real aspects of things which we express by the concept *ens commune* are caused. Hence, when he says here that metaphysics studies the immaterial substance and "illa," the "illa" would stand for those real aspects of things which we express as *ens, unum, multa, actus*, and so on.

Yet there is still a puzzle in these lines; Aquinas had said that God is studied as the cause of *ens commune* and yet here he says that God is called "common being," that this term *ens commune* is predicated of God. Is God both the cause and a member of the class of things falling under the concept *ens commune*? The solution of this problem is not to be found in the *Prooemium*. One wonders whether we are supposed to understand Aquinas' thought from the earlier writings, or whether we are expected to look for the solution in the body of the *Commentary* itself. For the moment we can only assume that we are expected to be familiar with this terminology from an acquaintance with Aquinas' earlier works.

There is a second doctrine in the *Prooemium* which is also not clear. Aquinas writes:"Haec enim transphysica invenitur in via resolutionis, sicut magis communia post minus communia." Earlier he said the *trans*- or meta-physics will study *ens* and its companion notions (or the reality corresponding to these notions). It is evident that these are the *magis communia.* This passage begs for comparison with the earlier one in which it was noted that every science needs to use these common notions in its work: unless one know *ens commune,* one can have no specific or generic knowledge of anything. On the basis of the implied proportion:"other sciences are reduced or resolved to metaphysics, as less common notions are reduced or resolved to more common ones," one is inclined to see metaphysics as giving us some content of knowledge which is included in all other knowledge. One suspects that by relating *resolutionis* and metaphysics, Aquinas is referring to something touching the essence of metaphysics; one suspects as well that we have not yet fully grasped Aquinas' doctrine.

Thus the *Prooemium* presents several problems: (1) What is the content of the idea *ens commune*? (2) How is it possible to predicate this concept of God when He is the cause of the reality corresponding to that idea? and (3) What is the exact meaning of *via resolutionis*? The solution of these three problems is a must if we are to understand the *Prooemium.* Hence let us attempt to discover what Aquinas' contemporaries, his disciples even, would have understood on the basis of their knowledge of his earlier works.

2. THE IDEA OF *ENS COMMUNE*

The intention of this section is not to give either a complete, nor an extremely detailed account of "common being." Rather we must be content with a brief exposition of what Aquinas understood by this term. There are two reasons necessitating such a limitation. First a practical one: because contemporary historians of Thomist metaphysics are not at all in agreement regarding some of the most important aspects of Aquinas' theory of "being," to give here more than a skeletal view of the doctrine found in his independent writings would in effect render suspect the present discussion of the *Prooemium.* Secondly, a speculative reason: Aquinas' *Prooemium* is extremely compact, very much a summary of metaphysical theories; hence, it would appear that he wished to recall to his readers only the skeletal meaning of *ens commune*; if one were to present here more than a minimal account of "common being," one would risk overturning the delicate balance of Aquinas' *Prooemium*: that is, one would raise further problems, problems not answered, nor even assumed as existent by the *Prooemium*. Consequently we must be content with a mere sketch of the meaning of *ens commune*, with an exposition only of what is essential to the understanding of the *Prooemium*.

Ens commune, in the first place, is not an expression found more than once or twice in each of Thomas Aquinas' works. Among its early occurrences we find for example, that it is mentioned in the *De veritate* as synonymous with *ens quod est primum per communitatem* and *prima conceptio nostri intellectus in statu viae*. By equating the concept of *ens commune* with the concept which is "first through community," Aquinas place emphasis on "being" as a concept which is contained within every other concept. To be conceived correctly however, "being" must not be thought a "core-meaning" in the sense of the minimum meaning central to all others, and from which all other meanings are constructed. "Common being" is not such a concept, not a concept with the least of all possible meanings. On the contrary, everything is being and thus "common being" expresses everything about everything: far from being the least meaningful of our ideas, "common being" is the most meaningful of all.[1]

But *ens commune* is also said to be the first concept man has. This is a rather common teaching of the *De veritate*, occurring at least three times. In affirming this characteristic of "common being," Aquinas affirms that "being,' is predicated of everything,[2] that "being" ex-

[1] This doctrine is contained in objection 10 and the answer to it of *De verit.*, 10, 11. Note that in the objection Thomas maintains that we would know the essence of God immediately in this life because of our idea of *ens*. Thomas' answer consists in denying that we have knowledge of God; despite this, we do have the type of idea of *ens* affirmed in the objection although we do not have it perfectly, only *sufficienter*. This objection reads: "Praeterea, sicut ens quod de omnibus praedicatur, est primum in communitate, ita ens a quo omnia causantur, est primum in causalitate, scilicet Deus. Sed ens quod est primum in communitate, est prima conceptio nostri intellectus in statu viae. Ergo et ens quod est primum in causalitate, statim per essentiam suam in statu viae cognoscere possumus." "Ad decimum dicendum, quod ens quod est primum per communitatem, cum sit idem per essentiam rei cuilibet, nullius proportionem excedit; et ideo in cognitione cuiuslibet rei ipsum cognoscitur. Sed ens quod est primum causalitate, excedit improportionaliter omnes alias res: unde per nullius alterius cognitionem sufficienter cognosci potest. Et ideo in statu viae, in quod per species a rebus abstractas intelligimus, cognoscimus ens commune sufficienter, non autem ens increatum."

[2] Cf. the objection 10 cited in the previous footnote, where "ens quod de omnibus praedicatur" is identified with "prima conceptio nostri intellectus in statu viae." The answer to this objection does not deny this doctrine, but rather uses it to show that this concept of *ens* is involved in our knowledge of any material being.

presses every aspect of every being, even though it expresses every aspect only implicitly.[1]

A later use of *ens commune* occurs in the *Pars Prima* of the *Summa Theologiae* in the context of an explanation of the difference between "similarity" and "image". According to Thomas, "similarity" is first, the genus of "image"; "similarity" thus expresses the properties common both to images and to other types of similarity.[2] Second, "similarity" is a perfection of a being which is *de facto* an image of something else.[3] It is in the midst of this discussion that Aquinas speaks of *ens commune*. The objection is raised: anything representing or reflecting the divine essence must be called an "image"; thus, the soul as immortal and indivisible is an image.[4] Aquinas' answer can be phrased as follows. The soul is properly, or in itself, of an intellectual nature, and hence it is an image of the divine essence; but in so far as it is indivisible the soul is not an image; *qua* indivisible the soul is only similar to God; that is, *qua* indivisible the soul is a particular instance or embodiment of *ens commune*. In short, *ens commune* is related to the divine essence and to the soul after the fashion in which "similarity," is related to "image": just as "image" is a species of "similarity," and each image is a particular good, so too in much the same way, the divine essence and the soul are species of *ens commune*, and each instance of the divine essence and of the soul would be particular examples embodying the perfections of *ens commune*.[5] Thus, *ens*

[1] "Illud autem quod primo intellectus concipit quasi notissimum, et in quo omnes conceptiones resolvit, est ens... Unde oportet quod omnes aliae conceptiones intellectus accipiantur ex additione ad ens. Sed enti non potest addi aliquid quasi extranea natura, ...quia quaelibet natura essentialiter est ens....", *De verit.*, 1, 1c. Cf. *De verit.*, 21, 1c.

[2] "Et similiter similitudo consideratur ut praeambulum ad imaginem, inquantum est communius quam imago,... Et sic similitudo hominis ad Deum attenditur secundum ea quae sunt communiora proprietatibus naturae intellectualis, secundum quas proprie attenditur imago." *Sum. Theol. I*, 93, 9c; cf. ad *Ium.*

[3] "Et similiter similitudo ...consideratur etiam ut subsequens ad imaginem, inquantum significat quandum imaginis perfectionem; dicimus enim imagines alicuius esse similem vel non similem ei cuius est imago, inquantum perfecte vel imperfecte repraesentat ipsum." *Ibid.*.

[4] "Praeterea, ratio imaginis attenditur non solum secundum representationem divinarum Personarum, sed etiam secundum repraesentationem divinae essentiae: ad quam repraesentationem pertinet immortalitas et indivisibilitas. Non ergo convenienter dicitur (Sent., 2, d. 16) quod similitudo est in essentia, quia est immortalis et indivisibilis; imago autem in aliis." *Ibid.*, objection 2.

[5] My phrasing of the answer is anything but a simple translation of Thomas' words; yet I feel it is the correct reading. "Ad secundum dicendum quod essentia animae pertinet ad imaginem, prout repraesentat divinam essentiam secundum ea quae sunt propria intellectualis naturae; non autem secundum conditiones consequentes ens in communi, ut est esse simplicem et indissolubilem." *Ibid.*, ad *2um.* Join this answer to these words: "Unum autem, cum sit de transcendentibus, et commune est omnibus, et ad singula potest aptari; sicut et bonum et verum. Unde, sicut bonum alicui rei particulari potest comparari ut praeambulum ad ipsam, et ut subsequens, prout designat aliquam perfectionem ipsius; ita etiam est de com-

commune appears to be an universal concept, much like a genus, which is to be considered as applicable to God and creatures, a concept which is particularized to form concepts of beings such as God and creatures.

One of the terms used as a synonym for *ens commune* in *De veritate, prima conceptio nostri intellectus in statu viae,* is rather common in Aquinas' writings. This term, "the first concept of our intellect in this life," or a variation of it, is found in the *Commentary on the Sentences,*[1] in *De ente et essentia,*[2] in *De veritate,*[3] in *In Boethii De Trinitate,*[4] in *Pars Prima*[5] and *Pars Prima Secundae* of the *Summa Theologiae,*[6] as well as the body of the *Commentary on the Metaphysics.*[7] Excepting from consideration the passages in the latter work, one discovers that Aquinas uses this expression to inculcate the following doctrines concerning "being": (1) "being" is a type of *a priori* on the side of the intellect to which all objects must conform; at the same time it is a quality of things which enables them to be objects of the intellect;[8] (2) "being" is apprehended at the core of all concepts in the sense already explained;[9] (3) "being" is the first of all ideas formed by the intellect.[10]

From this very sketchy view of the use of *ens commune* in the *De veritate* and the *Prima Pars,* and from the use of the expression *ens quod primo cadit in intellectu,* we note a certain continuity of meaning. In the first place, and most important of all, the meaning of "being" penetrates the meaning of all other concepts; it is predicated of everything. Moreover, "being" is our first idea.[11] In the *Prooemium,* then, by his reference to *ens commune* Aquinas expects us to understand a theory such as this. (Actually, the *Prooemium* points to this meaning when it

paratione similitudinis ad imaginem. Est enim bonum praeambulum ad hominem, secundum quod homo est quoddam particulare bonum: et rurus bonum subsequitur ad hominem, inquantum aliquem hominem specialiter dicimus esse bonum, propter perfectionem virtutis.", *Ibid., cor.* Much the same doctrine is found in the earlier work: *Sum con. Gen.,* I, 26.

[1] *In I Sent.,* d. 38, q. 1, *div. quaes..*

[2] *De ente et essentia, prooemium.*

[3] *De verit.,* 1, 1c; 10, 11, obj. 10; 21, 1c.

[4] *In Boeth. De Trin.,* q. 1, a. 3, obj. 3, p. 69, l. 11–12 (Decker edit.)

[5] *Sum. Theol. I,* 87, 1c; and 3, ad *Ium.*

[6] *Sum. Theol. I–II,* 94, 2c.

[7] *In I Meta.,* 2, 46; *In IV Meta.,* 6, 605; *In X Meta.,* 4, 1998; *In XI Meta.,* 4, 2210; *Ibid.,* 5, 2211.

[8] *In I Sent.,* d. 38, q. 1, *div. quaes..*

[9] *De verit.,* 1, 1c; 21, 1c; *Sum. Theol. I,* 87, 3, ad *Ium*; *Sum. Theol. I–II,* 94, 2c; this is implied in *De ente., prooem..*

[10] *De ente., prooem.*; *De verit.,* 1, 1c; *Ibid.,* 21, 1c; *In Boeth. De Trin.,* q. 1, a. 3, obj. 3; *Sum. Theol. I–II,* 94, 2c.

[11] No mention is made in this conclusion of the teaching of *In I Sent.,* d. 38, q. 1, *div. quaes.*: *ens* is a type of intellectual a priori. Emphasis on this aspect would carry us too far afield at the moment.

notes that to understand any genus of being, one needs a prior comprehension of "being.")

But how can "being" – *ens commune* – be predicated of God if He is the cause of the genus of "common being"? How can Aquinas say as he did in *Sum. theol. I*, 93, 9 that *ens commune* is something like a genus under which fit God and other beings? The most succinct answer is contained in *De veritate*. (The same answer is found in much more detail in the analyses of the human attempt to speak of God, e.g. *Sum. theol. I*, 13; *Sum. con. gent. I*, cc. 30–36; etc..) In the *De veritate* Aquinas notes that we do not know God's essence in this life because we are limited to knowledge of the material beings around us, and God's essence is far superior to all such beings. This doctrine raises a problem, however, for it seems to contradict the fact that *ens commune*, our first concept, is predicated of all things, including God, and at the same time expresses totally any aspect of the thing of which it is predicated. Thus if we wish to retain the doctrine that we do not know God's essence, it appears that we must say either (1) we do not know *ens commune*, or (2) *ens commune* does not express every aspect of the beings of which it can be predicated. Aquinas, as one would expect, escapes this dilemma by steering a middle course, and so neither totally denies nor totally affirms either alternative. His theory can be paraphrased as follows. *Ens commune* is included in every concept. If we knew all beings we would be able to form *perfectly* the concept of "common being." But we do not know all beings, and hence we do not possess in any perfect way that knowledge described as *ens commune*. Nevertheless, we do know a great deal about many beings and hence we can form a concept common to, and expressing all aspects of, those beings – material beings. Moreover, not only can we say that God is the cause of *ens commune* (that is, being which is referred to by our concept *ens commune* is created or caused by God) but we can predicate the content of *ens commune* of God as well. This is so, Aquinas would say, for although we do not know what God is, we do not know what He is not. That is to say, when we look at creatures we can realize that they are similar to their creator; we can know God by understanding that He is above all that we know here. What Aquinas implies is that in speaking of God we use our concept *ens commune* which is formed from contact with creatures, but we simultaneously tell ourselves that God is not *ens* as creatures are: He is being, but not after our fashion. Thus (without going into all the intricacies and difficulties of this theory) we can use the expression "God is being," and by "being" predicate only the positive content of *ens commune*. To be sure, the concept predicated of God is modified from without in such a way that it is not the same concept of "being" used of creatures; but that modification consists in a judgment that God is not being as we are, and thus such a judgment or modification involves no positive change in the content of the concept *ens commune*. So it is that the concept "common being" can be predicated both of God and of creatures, although it is still correct to speak of God as the cause of "common being."[1]

When Aquinas wrote his *Prooemium* then, he expected his readers to understand *ens commune* in the sense outlined above. *Ens commune* expresses everything about every material being, as has been noted; thus the reality corresponding to this concept has God as its cause. Moreover, *ens commune* is said to be a concept we use to speak of all beings, even of God; the legitimacy of predicating this concept of God lies in a negative aspect of its use – we remind ourselves that God is not being in the same sense in which we are.

Now there are many difficulties with this theory as it has been presented here, but it would not be of value to consider such issues at this time. For the moment we have a general idea of *ens commune*. That is sufficient for our purpose: the comprehension of the statements involving *ens commune* in the context of the discussion of metaphysics as presented in the *Prooemium*.

3. THE DOCTRINE OF THE *VIA RESOLUTIONIS*

A complete study of the *via resolutionis* or of the use of the *resolutio* in Aquinas' works would be too detailed and would lead us too far afield from our principal interest to be carried out here. Thus it

[1] Thomas' doctrine is contained in various parts of *De verit.*, 10, 11; Obj. 10: "Praeterea, sicut ens quod de omnibus praedicatur, est primum in communitate, ita ens a quo omnia causantur, est primum in causalitate, scilicet Deus. Sed ens quod est primum in communitate, est prima conceptio nostri intellectus in statu viae. Ergo et ens quod est primum in causalitate, statim per essentiam suam in statu viae cognoscere possumus." "Ad decimum dicendum, quod ens quod est primum per communitatem, cum sit idem per essentiam rei cuilibet, nullius proportionem excedit; et ideo in cognitione cuiuslibet rei ipsum cognoscitur. Sed ens quod est primum causalitate, excedit improportionaliter omnes alias res: unde per nullius alterius cognitionem sufficienter cognosci potest. Et ideo in statu viae, in quo per species a rebus abstractas intelligimus, cognoscimus ens commune sufficienter, non autem ens increatum." "Ad quartum dicendum, quod intellectuali visione in statu viae Deus cognoscitur, non ut sciatur de eo quid est, sed quid non est. Et quantum ad hoc eius essentiam cognoscimus, eam super omnia collectam intelligentes, quamvis talis cognitio per aliquas similitudines fiat." Cf. also ad. 5*um*. Ad. 6*um*: "Intellectus autem ex affectibus in causas procedens, tandem pervenit in ipsius Dei cognitionem aliqualem, cognoscendo de eo quid non est...." Of great importance in these texts is the *sufficienter* which qualifies our knowledge of *ens commune* (ad 10 *um*); the context of the objection seems to imply that although perfect knowledge of *ens commune* is not had, we still can predicate this concept of God in so far as we realize that it is an expression of what He is not.

is fortunate that several independent studies have already been made from which we can draw.

To achieve an adequate idea of what Aquinas understood by *resolutio* one must recall his teaching on *ratio* and *intellectus*. *Ratio*, in Aquinas' writings, expresses that type of intellectual operation by which man passes from the state of potentially knowing to that of actually knowing; *intellectus* expresses the process by which man elaborates sensible elements and forms concepts through which he says to himself a portion of the truth contained in an object.[1]

Ratio however is not an end in itself; rather it is ordained to an *intellectus*.[2] Through an *intellectus* man approaches the level of the pure spirits[3] in so far as he thus grasps an essence through a simple regard or contemplative act of looking.[4] The *intellectus* serves, however, not only as the goal of a *ratio*, but as well as the point of departure of a new *ratio*; the activity of *ratio* begins always from a simple grasp of a truth (that is, from an *intellectus*) just as after a process of reasoning, the mind arrives at an *intellectus* of a new truth.[5]

The question of the validity of a particular instance of movement from *intellectus* to *intellectus* by means of a *ratio* is sometimes answered by Aquinas in terms of the *resolutio*. That is to say, the human intellect, beginning from a firm grasp of first principles, can reduce or resolve all syllogistic conclusions to the first principles.[6] Yet for Aquinas *resolutio* did not merely express this reduction of reasoning to its first principles, but it expressed as well a method of apprehension and judgment. In every case, however, *resolutio* is connected with an *intellectus*, with a simple, contemplative grasp of truth. In apprehension, resolution can be described as the activity of the agent intellect in so far as that faculty recedes from the complexity of sense to produce a simple intelligibility in act.[7] *Resolutio* is found in immediate judgments whether these judgments deal with necessity or with contingency; the necessary immediate judgments can be resolved to the first judgments

[1] J. Peghaire, C.S.Sp., *Intellectus et ratio selon s. Thomas d'Aquin*, Publications de l'Institut d'Etudes Médiévales d'Ottawa. VI, Vrin, Paris, 1936; pp. 281–82.

[2] *Ibid.*, p. 282.

[3] *Ibid.*.

[4] *Ibid.*, pp. 247–49; 261–62.

[5] *Ibid.*, pp. 245; 261–69.

[6] *Ibid.*, pp. 269–72.

[7] L.-M. Régis, O.P., Analyse et synthèse dans l'oeuvre de saint Thomas," in *Studia Mediaevalia in honorem Reverendi Patris Raymundi J. Martin, O.P.*, Brugis Flandrorum, 1948; p. 317.

such as the principle of non-contradiction, whereas the contingent immediate judgments are reducible to sense data.[1]

As far as resolution can be styled a tool of the metaphysician, two different types of resolution are to be distinguished. In the *resolutio secundum rem* one begins with complex sensible data and seeks the ultimate extrinsic cause of the object sensed.[2] There is the second type, however, the *resolutio secundum rationem* (also called a *reductio*); in this resolution, one is interested in the intrinsic causes of the object sensed: in the particular, the universal forms are disengaged, and from these in turn are disengaged still more universal forms.[3] It is by these two resolutions that one arrives at the objects studied in metaphysics. By the *resolutio secundum rem* one arrives at knowledge of the First Cause; by the *resolutio secundum rationem* one reaches the most common universal of all – the notion of "being."[4]

Examples of these two resolutions are numerous in Aquinas' works. The proof of God according to the *prima via*, for instance. Becoming demands – for its intelligibility – the relation of potency to act; this passage from "becoming" to "potency-act" is a *resolutio secundum rationem*.[5] Once one knows the necessity of a potency-act relationship as a presupposition of the intelligibility of becoming, one uses the *resolutio secundum rem* to posit the existence of the Act, of God.[6]

Such is, in brief, the Thomist theory of *resolutio*. However, it is somewhat unfortunate that the three authors quoted in this discussion of resolution (Rabeau, Régis, and Peghaire) have relied heavily upon texts taken from the *Commentaries* on the *Physics*, on the *Metaphysics*, on the *Posterior Analytics*, and especially on the *De Trinitate*. Thus for our purposes their expositions are suspect unless we can discover essentially the same doctrine in Aquinas' independent, or non-commentary, writings. Such a discovery would require a lengthy examination of texts, and accordingly can not be carried through in a complete fashion here. Rather we must be content to note first, an interesting similarity between seven widely scattered texts of Aquinas, one of which is taken from the *Commentary on Boethius' De Trinitate*; second a similarity between the *Prooemium*, the exposition of the *De Trinitate*, and a text from the *Summa theologiae*. Because of these similarities, we shall be

[1] *Ibid.*, pp. 320–22.

[2] G. Rabeau, *Species. Verbum. L'activité intellectuelle élémentaire selon S. Thomas d'Aquin*, Bibl. Thomiste XII, Vrin, Paris, 1938; p. 175.

[3] *Ibid.*.

[4] *Ibid.*.

[5] *Ibid.*, p. 176.

[6] *Ibid.*, p. 175.

led to view the exposition Aquinas gives of the various types of *resolutiones* in his commentary on the *De Trinitate* as representative of his personal thought. Thus the doctrine expressed in the *Prooemium* under the guise of the theory of *via resolutionis* should become clear.

1) *De veritate, q. 10, a. 8, ad 10um*: written between 1256–59.[1]
In this passage Aquinas uses *resolvendo* in explaining that a doctrine taken from Dionysius really does not have anything to do with the issue discussed previously in article 8. The objection, taken from *De divinis nominibus*, concerns the circular motion of human knowledge: the soul of man goes out to external things, and then returns from them to itself. Aquinas, in his answer, explains that this idea of circular knowledge was Dionysius' manner of showing that the knowledge of the soul is less perfect than that of angels; there is a circle in knowledge, Aquinas notes, in so far as the soul reasons from principles to conclusions in the way of discovery (*via inventionis*) and then traces back (*resolvendo*) the conclusions to the principles in the way of judging (*via judicii*).[2]

The word *resolvendo* and the idea of resolving conclusions back to principles were nothing new to Aquinas' writings. In the *Commentary on the Sentences* one finds at least five examples of the same doctrine: to assure oneself that a particular act of knowledge is certain, one need only resolve the conclusions back to the first principles from which all knowledge flows.[3] In the *De veritate* too one finds other passages in which Aquinas mentions this same idea.[4] In none of these texts from the *Sentences* nor these just mentioned from *De veritate* is there any mention of the circularity of knowledge, nor of Dionysius.

2) *De veritate, q. 15, a. 2, ad 3um.* In this text, although there is no explicit reference to Dionysius nor to his idea of circularity, the conception of the resolution is the same as in the earlier text of *De veritate.* If there is any difference at all between the two texts, it is that this second one is more ample in its description of *resolutio*. In the passage of interest, Aquinas distinguishes the "scientific power" from the

[1] WALZ, *Saint Thomas d'Aquin*, p. 223. GRABMANN, *Die Werke des hl. Thomas von Aquin...*, p. 304.

[2] "Ad decimum dicendum, quod circulus quidam in cognitione animae attenditur, secundum quod ratiocinando inquirit existentium veritatem; unde hoc dicit Dionysius ut ostendat in quo animae cognitio deficiat a cognitione angeli. Haec autem circulatio attenditur in hoc quod ratio ex principiis secundum viam inveniendi in conclusiones pervenit, et conclusiones inventas in principia resolvendo examinat secundum viam judicandi. Unde hoc non est ad propositum." ad 10*um*. Objection 10 refers to *De Divinis Nominibus* C. VI, but does not give a text; actually the reference should be to Chapter VII.

[3] *In II Sent.*, d. 9, q. 1, a. 8, ad 1*um*; *Ibid.*, d. 7, q. 1, a. 1, *sol.*; *In III Sent.*, d. 23, q. 2, a. 2, *sol.* 1; *Ibid.*, d. 35, q. 1, a. 2, *sol.* 2; *Ibid.*, d.31, q. 1, a. 8, ad 1*um*.

[4] E.g.: *De verit.*, 11, 1, ad 13*um*.

"reasoning power." The reasoning power is the ability to pass from the principles of an essence to a conclusion; the scientific power is the ability to return from conclusions to principles, the ability to resolve (*resolvere*) the conclusions back into the principles.[1]

3) *In Boethium De Trinititate, q. 6, a. 1, ad 3am quaestionem, corpus*; written between 1257–58.[2] Here one finds mention both of Dionysius and of the *via resolutionis*. The text of Dionysius is quoted this time whereas in *De veritate* q. 10, a. 8 it was only referred to but not quoted. The text concerns the circularity of human knowledge: the soul is rational in so far as it "goes around" (*circumeunt*) or "circles things." Aquinas mentions this text as if it showed only that a human mind must consider many different things, and after such a consideration that it can form one simple conception. – On the other hand, there are angels who know all things by knowing one thing. Again a text of Dionysius is cited. – It would appear that these two texts were the ones referred to in the *De veritate* article mentioned above.

Immediately following the last of the two quotations of Dionysius, Aquinas begins a lengthy explanation in terms of *via resolutionis* and *via compositionis vel inventionis*, an explanation which is presented as if it were somehow implied by Dionysius' words. The salient points of this explanation are as follows. *Via resolutionis* is the process in which one reasons from one thing to another, and finally gathers all things together in one view, or in one act of knowledge. Now each science must resolve or trace everything back to the first cause which that science studies; the divine science must, in its turn, reduce the first causes studied in each individual science to the first causes studied in the divine science. Since there are both intrinsic and extrinsic causes in every scientific domain, the intrinsic ones must gradually be reduced to more and more universal notions until, in the divine science, one comes to the most universal notions of all, to those common to all beings – to the notions of *ens* and its *per se* accidents; in a parallel manner all extrinsic causes must be gradually traced back to the simple, supreme

[1] "Scientificum autem et ratiocinativum diversae quidem potentiae sunt, quia quantum ad ipsam rationem intelligibilis distinguuntur. Cum enim actus alicuius potentiae se non extendat ultra virtutem sui obiecti, omnis operatio quae non potest reduci in eamdem rationem obiecti, oportet quod sit alterius potentiae, quae habeat aliam obiecti rationem. Obiectum autem intellectus est quod quid est... et propter hoc, actio intellectus extenditur quantum potest extendi virtus eius ad quod quid est. Per hanc autem primo ipsa principia cognita fiunt, ex quibus cognitis ulterius ratiocinando in conclusionum notitiam: et hanc potentiam quae ipsas conclusiones in quod quid est nata est resolvere, Philosophus scientificium appellat." ad 3*um*.

[2] Walz, *Saint Thomas d'Aquin*, p. 222. Grabmann, *Die Werke des hl. Thomas von Aquin...*, pp. 359–60.

first cause – the separated substances. Thus divine science must be learned after the other sciences; in so far as the natural terminus of rational, discursive knowledge is the unitary, intellectual grasp of all things in one notion, this science is called "metaphysics" or *transphysica* "quia post physicam resolvendo occurrit." On the other hand, there is the *via inventionis*; metaphysics is called "first philosophy" in this way since it gives principles to all other sciences.[1]

In the first text noted above (*De veritate*, q. 10, a. 8) Dionysius' "circularity" and the resolution (*resolvendo*) were mentioned; there Aquinas spoke universally in reference to the process carried through in individual acts of demonstration. In the *De Trinitate* text, he joins again the Dionysian idea of circularity and the resolution; here however the context has been enlarged. Yet there is no contradiction at all between the two expositions of resolution – the first speaks only of the necessity of resolving every conclusion into its proper or proximate principles, while the second would have these same proper principles resolved back to the metaphysical first causes. Whereas in the *De veritate* Aquinas was treating the self-knowledge enjoyed by the soul, in the *De Trinitate* he is treating the science of metaphysics. Thus, it could easily be maintained that the narrower view of the *De veritate* is narrow, not because Aquinas held only a theory of resolution in that restricted sense, but because the contest demanded no broader view. Unfortunately however the *De veritate* text only makes us suspect the full import of resolution; it does not explicitly inform us what exactly that import may be. Since the exposition of resolution in the *De Trinitate* is an exposition of Boethius' writing – even though the idea of resolution is not explicitly in Boethius – one can not assume

[1] Differt autem ratio ab intellectu... Est enim rationis proprium circa multa diffundi et ex eis unam simplicem cognitionem colligere. Unde Dionysius dicit 7c. De divinis nominibus quod animae secundum hoc... Intellectus autem e converso per prius unam et simplicem veritatem considerat et in illa totius multitudinis cognitionem capit... Unde Dionysius ibidem dicit quod angelicae mentes... Sic ergo patet quod rationalis consideratio ad intellectualem terminatur secundum viam resolutionis, inquantum ratio ex multis colligit unam et simplicem veritatem... Tota autem consideratio rationis resolventis in omnibus scientiis, ad considerationem divinae scientiae terminatur. Ratio enim,... procedit quandoque de uno in aliud secundum rem,... quasi resolvendo, cum proceditur ab effectibus ad causas, eo quod causae sunt effectibus simpliciores et magis immobiliter et uniformiter permanentes... Ultimus igitur terminus resolutionis in hac vita est, cum pervenitur ad causas supremas maxime simplices, quae sunt substantiae separatae. Quandoque vero procedit de uno in aliud secundum rationem, ut quando est processus secundum causas intrinsecas, ...resolvendo autem quando e converso [idest: a particularibus in maxime universales formas], eo quod universalius est simplicius. Maxime autem universalia sunt, quae sunt communia omnibus. Et ideo terminus resolutionis in hac via ultimus est consideratio entis et eorum quae sunt entis inquantum huiusmodi. Haec autem sunt, de quibus divina scientia considerat,... substantiae scilicet separatae, et communia omnibus entibus." ad 3*am quaes. corpus*, p. 211, l. 1–p. 212, l. 18 (Decker edit.).

that that exposition of *resolutio* is Aquinas' personal theory (except of course in so far as such a theory would appear to be the explanation demanded by the *Prooemium* of the *Commentary on the Metaphysics* – a point to which we must return shortly.)

4) *In Dionysii De divinis nominibus, C. VII, l. 2, n. 713*; written between 1261–64.[1] This text presents Aquinas' commentary on the passage of Dionysius referred to in the first and the third texts already considered. Aquinas explains the circularity of knowledge in this fashion: man must understand the quiddity of a thing by observing its properties; then he argues from that essence – and here the circle is closed – back to its properties. Thus the circle consists in beginning from principles, going out to things and then back to the principle. (Also of note is the similarity between this passage and the second text studied above: *De veritate*, q. 15, a. 2.)[2]

5) *In Dionysii De divinis nominibus, C. IV, l. 7, nn. 375–76.* In this passage Aquinas discusses again the circular motion of the human soul. Since circularity expresses the essence of uniform motion, he writes, there must be something circular about the operations of the soul if these operations are in any sense uniform. But are these operations uniform? Certainly there is no uniformity in those activities by which the soul's attention is directed exclusively to the multitude of diverse material things exterior to itself; thus in so far as the soul receives its knowledge from these diverse things, it is not acting uniformally, and so there is no circular motion in its activity. However, in so far as the soul turns away from these external things and turns back into itself, there is an uniform, a circular motion; the same uniformity of operation is discernible when the soul turns from itself to angels, and from itself to God.[3] – Aquinas' statements are somewhat vague for they are too

[1] GRABMANN, *Die Werke des hl. Thomas von Aquin...*", p. 364. 1261–62 seem to be the years in which Thomas began; the date of completion is not very well determined.

[2] "Et ad exponendum secundum quod animae rationales dicuntur, subiungit quod circumeunt circa veritatem existentium, diffusive et circulo. Veritas enim existentium radicaliter consistit in apprehensione quidditatis rerum, quam quidditatem rationales animae non statim apprehendere possunt per seipsam, sed diffundunt se per proprietates et effectus qui circumstant rei essentiam... Haec autem circulo quodam efficiunt, dum ex proprietatibus et effectibus causas inveniunt et ex causis de effectibus iudicant... Inquisitio enim rationis ad simplicem intelligentiam veritatis terminatur, sicut incipit a simplici intelligentia veritatis quae consideratur in primis principiis; et ideo, in processu rationis est quaedam convolutio ut circulus, dum ratio, ab uno incipiens, per multa procedens, ad unum terminatur."

[3] "...cum enim motus Angeli et Animae sit operatio eius, circularitas autem motus rationem uniformitatis exprimat, necesse est eo modo circularem motus Angelo et Animae attribuere, inquantum competit eis uniformitas intellectualis operationis... Animae autem connaturale est quod intelligat accipiendo a rebus exterioribus quae sunt multiformes et divisae, unde in hac receptione non potest attendi circularitas motus eius, sed magis in hoc quod a rebus exterioribus revocatur: primo quidem, in seipsam conversa; secundo, elevata in considerationem angelicarum virtutum; tertio autem, usque ad ipsum Deum." n. 375.

general, too lacking in detail to be readily comprehensible. However the detail is added, and the key to the meaning of the passage given by the introduction of the theory of *resolutio*.

The soul, Aquinas goes on to note, grasps the first principles by a simple *intellectus*, by a simple act of understanding; there is no "discourse," no reasoning process needed. In so far as the soul considers these first principles, it can be said to have a circular or uniform motion, for the soul has gone out to things and then turned back to itself. Whenever the soul is engaged in reasoning, whenever it is moving from one external thing to another, there is no circularity, of course; in such cases the soul's operation can best be described as "discrete, rectilinear motion" (although Aquinas does not actually employ this expression). The soul, after such a process of reasoning, is able to turn back on itself, to itself; thus the soul judges the validity of its reasoning process by a *resolutio* to the first principles. By this resolution the soul is able to guard against the possibility of error, for these principles are the signposts according to which the soul can safely journey. – Such a *convolutio*, such a circular motion from self to things and then from things to self through a *resolutio*, is had in three ways; or better there are three distinct circles which the soul can make. First, after having known external, material objects, the soul returns to itself and discovers what are the necessary characteristics of the soul which are presupposed for knowledge of external material objects; by this circle, the soul is made similar to itself: it knows itself. Secondly, in so far as the soul thus knows its own nature, it can to some extent grasp the uniformity of angelic knowledge. Thirdly, by the same knowledge of its own nature, the soul is raised to knowledge of God.[1] – Setting aside the difficulties connected with man's knowledge of angels and of God, as well as the reasons for seeing that knowledge as circular, one can still see the core of Thomas' discussion: the soul has a circular activity in so

[1] "Hoc est ergo quod dicit, quod motus circularis animae est secundum quod ab exterioribus intrat ad seipsam et ibi uniformiter convolvitur, sicut in quodam circulo, secundum suas intellectuales virtutes; quae quidem convolutio dirigit virtutem animae, ut non erret: manifestum est enim quod Anima, discurrendo de uno ad aliud vel de contrario in contrarium, ratiocinatur multipliciter; sed omnis ista ratiocinatio diiudicatur per resolutionem in prima principia, in quibus non contingit errare, ex quibus Anima contra errorem defenditur, quia ipsa prima principia simplici intellectu absque discursu cognoscuntur et ideo eorum consideratio, propter sui uniformitatem, circularis convolutio nominatur. Per hanc ergo convolutionem, primo congregatur ad seipsam, considerans id quod in natura sua habet ut cognoscat; deinde, sic uniformis facta, unitur per huiusmodi convolutionem, unitis virtutibus, scilicet angelicis, inquantum per similitudinem huius uniformis apprehensionis, uniformitatem Angelorum aliquo modo considerat; et ulterius per istam convolutionem, manducitur ad pulchrum et bonum, idest Deum... et ideo circularitas motus animae, completur in hoc quod ad Deum manuducit." n. 376.

far as it goes out to external things and then returns by a *resolutio* to its first principles to judge the validity of its reasoning processes.[1]

6) *Summa theologiae I, q. 79, a. 8c*; written between 1267–68.[2] Here we have no direct quote from Dionysius, yet in the *corpus* reference to him is made on two occasions. In this article Aquinas wishes to show that *ratio* and *intellectus* are not separate potencies in man. He first suggests that to see this doctrine properly, one should consider the acts of *ratio* and *intellectus*, whereas to reason (*ratio*) is to proceed from one thing to another, *intellectus* is a simple apprehension of truth. Angels, moreover, are said to know simply, after the fashion of *intellectus*, although men must use *ratio* to attain finally an *intellectus*. During this exposition Dionysius is appealed to, after the fashion of an *auctoritas* ("ut Dionysius dicit"; "ut ibidem dicitur"). In addition Aquinas compares reasoning to understanding (*intelligere*) as "moving" is related to "resting"; the first is imperfect and the second perfect. Moreover, just as movement begins from an immobile state and terminates in rest, in this same way reasoning begins its *via inquisitionis vel inventionis* from something understood "simply" – or through an *intellectus* of the first principles – and later in the *via judicii* resolves (*resolvendo*) all conclusions back to the first principles.[3] It is interesting to note that the elements "motion and rest," "perfect and imperfect," "beginning and ending in an immobile state," are all parts of an explanation of an idea taken from Boethius' *De consolatione philosophiae*, C. IV, prosa 6. There is no mention of Boethius in the *corpus*, but in objection two his name, the *De consolatione*, and the following objection are all found: *intellectus* is to *ratio* as eternity to time; therefore *intellectus* and *ratio* are different potencies. This same comparison was given and identically

[1] In earlier passages of the exposition on Dionysius' work, there are references to the *via resolutionis* in the sense of an ontological return of material things to God; e.g.: *In Dion. De Div. Nom.*, C. 1, l. 2, n. 51.

[2] WALZ, *Saint Thomas d'Aquin*, p. 223. GRABMANN, *Die Werke des hl. Thomas von Aquin..*, p. 295.

[3] "Respondeo dicendum quod ratio et intellectus in homine non possunt esse diversae potentiae. Quod manifeste cognoscitur, si utriusque actus consideretur. Intelligere enim est simpliciter veritatem intelligibilem apprehendere. Ratiocinari autem est procedere de uno intellecto ad aliud, ad veritatem intelligibilem cognoscendam. Et ideo angeli, qui perfecte possident, secundum modum suae naturae, cognitionem intelligibilis veritatis, non habent necesse procedere de uno ad aliud; sed simpliciter et absque discursu veritatem rerum apprehendunt, ut Dionysius dicit, 7 cap. *De div. nom.*. Homines autem ad intelligibilem veritatem cognoscendam perveniunt, procedendo de uno ad aliud, ut ibidem dicitur: et ideo rationales dicuntur. Patet ergo quod ratiocinari comparatur ad intelligere sicut moveri ad quiescere, vel acquire ad habere: quorum unum est perfecti, aliud autem imperfecti. Et quia motus semper ab immobili procedit, et ad aliquid quietum terminatur; inde est quod ratiocinatio humana, secundum viam inquisitionis vel inventionis, procedit a quibusdam simpliciter intellectis, quae sunt prima principia;ia iudicii, resolvendo redi et rursus, in vt a prima principia, ad quae inventa examinat."

explained in the article already noted from the exposition of Boethius' *De Trinitate*; not only was it given, but the explanation was presented as an introduction to Dionysius' words on circularity, which in turn introduced the exposition of *via resolutionis* and its role in the science of metaphysics. In a certain sense then, Aquinas can be said to be copying here in the *Summa* what he wrote earlier in his exposition of the *De Trinitate*; the narrow contest of the topic discussed in the *Summa* would explain the limit placed on the scope of resolution in the latter work.[1]

7) *Summa theologiae II–II, q. 180, a. 6, ad 2um and 3um;* written between 1269–72.[2] Once again the Dionysian doctrine concerning the circularity of human knowledge is brought forth. The context is a discussion of the contemplative life. In the second objection the point is made that the contemplative life belongs to man's intellect in so far as man resembles an angel; yet since Dionysius in *De divinis nominibus*, C. IV assigns different types of circularity to angelic and human knowledge, it does not appear correct to speak of all contemplation in univocal terms such as "circularity." In his answer, Aquinas opts for "circularity" as a term useful in the explanation of contemplation. Because the angelic intellect is much higher, much more perfect than the human one, he writes, the motions by which men and angels arrive at a simple uniform view of God are different. The angel does not pass through any process of discussion, but immediately intuits God. Man, however, has a two-fold difficulty to overcome before he arrives at the contemplation of his Maker: first, he must relinquish the multiplicity of exterior things and return to himself; second, he must cease all activity of reasoning and must rest in a simple grasp of truth. This last operation – the simple grasp – is the same type man performs when he understands the first principles of all knowledge. It is interesting to note the closing words of this article: "Unde patet quod Dionysius multo sufficientius et subtilius motus contemplationis describit."[3]

[1] Compare the last sentences of footnote 3, p. 72, with these texts: "Praeterea, Boethius dicit, in libro *De consol.*, quod intellectus comparatur ad rationem sicut aeternitas ad tempus." *Sum. Theol. I*, 79, 8, obj. 2. (This same example appeared earlier when Thomas wrote:) "Differt autem ratio ab intellectu, sicut multitudo ab unitate. Unde dicit Boëtius in IV *de Consol.*, quod similiter se habet ratio ad intellectum, sicut tempus ad aeternitatem et circulus ad centrum. Est enim rationis proprium circa multa diffundi et ex eis unam simplicem cognitionem colligere. Unde dicit Dionysius... Sic igitur patet quod rationalis consideratio ad intellectualem terminatur secundum viam resolutionis...", *In Boeth. De Trin.*, q. 6, a. 1, ad 3*am quaes. cor.*

[2] WALZ, *Saint Thomas d'Aquin*, p. 223. GRABMANN, *Die Werke des hl. Thomas von Aquin..*, p. 295.

[3] "...Dionysius motum circularem in angelis assignat inquantum uniformiter et indesi-

– Thus in this article of the *Summa* Aquinas clearly adopts the position of Dionysius; that is, Aquinas adopts the Dionysian theory explained exactly as he, Aquinas, had exposed it years earlier in his *Commentary on the Divine Names* (with the single exception that the *word* "resolutio" does not appear in the *Summa.*)

What has been learned from this investigation of seven texts, written over a period of approximately 16 years, can be summarized as "an almost constant linking of a Dionysian theory of circularity to the idea of *resolutio.*" This theory and the resolution are explained in more or less universal senses in various passages; for example,

a) there is a circle in knowledge in so far as one argues from an essence to the properties of the essence (text no. 2);
b) there is also a circle in so far as one moves from knowledge of properties to knowledge of a quiddity and then resolves the quiddity back to the properties (text no. 4);
c) still more, there is a circle in so far as the soul moves from principles to conclusions and then resolves the conclusions back to the principles (text no. 1);
d) there is circularity in so far as the soul goes out to things and then resolves its knowledge back to the first principles (texts no. 3, 5, and 6);
e) there is circularity in so far as the soul goes out to things and then back to itself and from itself to God (texts no. 5, and 7);
f) there is a circle in so far as man through the individual sciences goes out to things, forms conclusions, and resolves these back to "being" and to God in metaphysics (text no. 3).

Thus, the chief difference between these seven texts appears to lie in the manner in which Aquinas applied this principle: "there is circularity in knowledge in so far as man goes from A to B and then resolves B back to A." In the *Commentary on Boethius' De Trinitate* the principle is applied to the entire spectrum of speculative knowledge; in the exposition of Dionysius' *On the Divine Names,* C. VII, and in *De veritate,* q. 15, Aquinas applied the principle to knowledge of essence and its properties; in the other texts some of the intermediate applications are illustrated.

nenter, absque principio et fine, intuentur Deum: sicut motus circularis... In anima vero antequam ad istam uniformitatem perveniatur, exigitur quod duplex eius difformitas amoveatur. Primo quidem, illa quae est ex diversitate exteriorum rerum, prout scilicet relinguit exteriora. Et hoc est quod primo ponit in motu circulari animae introitum ipsius ab exterioribus ad seipsam. – Secundo autem oportet quod removeatur secunda difformitas, quae est per discursum rationis. Et hoc idem contingit secundum quod omnes operationes animae reducuntur ad simplicem contemplationem intelligibilis veritatis. Et hoc est quod secundo dicit, quod necessaria et uniformis convolutio intellectualium virtutum ipsius: ut scilicet cessante discursu, figatur eius intuitus in contemplatione unius simplicis veritatis. Et in hac operatione animae non est error: sicut patet quod circa intellectum primorum principiorum non erratur, quae simplici intuitu cognoscimus.", ad *2um.* Compare this treatment with Thomas' commentary on Dionysius quoted in footnotes 3, p. 70, and 1, p. 71.

Thus it would appear that the treatment of *resolutio* given in the context of the exposition of Boethius' *De Trinitate* is actually an exposition of Aquinas' personal theory: this exposition differs from the others examined only in the degree to which Aquinas invoked the principle of the circularity of human knowledge. This conclusion would appear to be justified by a further comparison, for the *Prooemium*, as well as a text in the *Prima Pars* seem to demand that we understand *resolutio* after the fashion of the exposition of the *De Trinitate*.
From the *Prooemium*:

Haec enim transphysica inveniuntur in via resolutionis, sicut magis communia post minus communia.

From *In Boeth. De Trin.*:

Haec autem sunt, de quibus divina scientia considerat, ...substantiae scilicet separatae et communia omnibus entibus ...et nihilominus ipsa scientia addiscitur post physicam et ceteras scientias, in quantum consideratio intellectualis est terminus rationalis, propter quod dicitur metaphysica quasi trans physicam, quia post physicam resolvendo occurrit.[1]

In the *Prooemium* metaphysics is said to study *ens commune*, a notion which is in some way connected with all knowledge; without knowing *ens commune*, one can not know any genus of being perfectly. In addition, metaphysics is said to study God and the other separate substances who are the first causes of *ens commune*. When one turns to the exposition of the *De Trinitate*, one notes that metaphysics is described as last in the *via resolutionis* because it concerns itself with the most common notion of all – "being"; to that term are reduced all the intrinsic causes of any particular being. In addition, because it is last in the *via resolutionis*, metaphysics studies the separate substances, for all the extrinsic causes of any particular being are reduced to the first of all extrinsic causes, a separate substance. The similarity between the *Prooemium*'s metaphysics and that of the *De Trinitate* is striking, to say the least.

Another similarity is found by a comparison of the commentary on Boethius with the *Pars Prima*.

From *In Boeth. De Trin.*:

Ratio ...procedit quandoque de uno in aliud secundum rem, ...quasi resolvendo cum proceditur ab effectibus ad causas, eo quod causae sunt effectibus simpliciores, et magis immobiliter et uniformiter permanentes. Ultimus ergo terminus

[1] *In Boeth. De Trin.*, q. 6, a. 1, ad 3*am quaes. corpus*, p. 212, l. 16–25 (Decker edit.). For the lines immediately preceding these, cf. footnote 1, p. 69.

resolutionis in hac vita est, cum pervenitur ad causas supremas maxime simplices, quae sunt substantiae separatae. Quandoque vero procedit de uno in aliud secundum rationem, ut quando est processus secundum causas intrinsecas: componendo quidem, quando a formis maxime universalibus in magis particularia proceditur; resolvendo autem quando e converso, eo quod universalis est simplicius. Maxime autem universalia sunt, quae sunt communia omnibus entibus. Et ideo terminus resolutionis in hac via ultimus est consideratio entis et eorum quae sunt entis inquantum huiusmodi.[1]

From the *Pars Prima*:

Oportet enim universaliores effectus in universaliores et priores causas reducere. Inter omnes autem effectus, universalissimum est ipsum esse. Unde oportet quod sit proprius effectus primae et universalissimae causae, quae est Deus.[2]

In the *De Trinitate* Aquinas speaks of two resolutions; first one can resolve the concept of the particular intrinsic causes of a being to the concept of the intrinsic causes of every being; secondly one can resolve the extrinsic causes of a being to the extrinsic cause of every being. Thus one can resolve all causes to the idea of "being" and to God. In the *Pars Prima* the conception is not so broad, so general; there Aquinas speaks of the most universal of all effects which is *esse*, and the most universal of all causes which is God. Yet there is not a great deal of difference between the *De Trinitate* and the *Pars Prima*: to reduce all intrinsic causes to "being" is not much else than to reduce all effects to *esse*. Hence, just as the conception of *resolutio* in the *Prooemium* seems to demand something such as the conception of the *De Trinitate*, so too, the theory of the *Pars Prima* seems to demand a general theory such as that of the *De Trinitate*. One feels justified in maintaining that Aquinas' explanation of Boethius, at least in the one article studied, is an expression of Aquinas' personal thought and not only his understanding of Boethius' doctrine.

On the basis of these investigations into the doctrinal background against which Aquinas spoke of *ens commune* and *via resolutionis*, what is the final word to be said of the *Prooemium*? In the latter work, Aquinas speaks of metaphysics as the science which studies the reality expressed by the concept *ens commune*; in this concept the metaphysician has an expression of every aspect of material beings; he has a concept which is found to be included in every other concept. Moreover, the reality expressed by this "common being" is a reality caused by God. However, the concept "common being" can be predicated of God

[1] *Ibid.*, p. 212, l. 3–16 (Decker edit.).
[2] *Sum. Theol. I*, 45, 5c.

as long as one realizes that its positive content does not express God but rather expresses what He is not. The relationship of *ens commune* and all other concepts is expressed in terms of "resolution": because *ens commune* expresses every aspect of any other concept, and is contained within all other concepts, all concepts can be "resolved," "reduced," or "traced back" to *ens commune*. An analogical type of resolution holds in matters of extrinsic causality: God is the ultimate cause, and to Him can be "resolved" all other extrinsic causes. *Via resolutionis* expresses not only a type of static posteriority of the metaphysical concepts of "being" and of God, but expresses a dynamic aspect of metaphysics: metaphysics is the last science to be learned, and it appears that one learns it by tracing or by resolving concepts and extrinsic causes back to "being" and to God.

Such then is the conception of metaphysics presented by Aquinas as the introduction to his *Commentary* on Aristotle's metaphysical writings. Did Aquinas arrive at such a conception entirely on his own? Does its presentation represent Aquinas' participation in a dialogue with other philosophers or commentators? The present chapter can only touch on this issue. An examination of the introductions written by Albert, Avicenna, and Aristotle can underscore some important aspects of this problem, for it reveals that Aquinas wrote to counterbalance the thought of Albert and Avicenna, and as well that Aquinas' *Prooemium* is basically Aristotelian in spirit.

4. THE INTRODUCTION TO ALBERT'S *COMMENTARY*

Albert the Great devotes the first three chapters of his *Commentary* to introductory remarks concerning the science of metaphysics. Metaphysics, he writes, is the highest of the three speculative sciences.[1] The other two are, in fact, founded on metaphysics.[2] That is to say, metaphysics or wisdom studies *esse simplex*;[3] this *esse simplex*, or the "existing act of the first essence" as it is called, is the first creature of God;[4]

[1] *In I Meta.*, T. I, c. 1, p. 1, l. 10–15; p. 1, l. 57–p. 2, l. 30.

[2] "Haec autem speculatio est rerum altissimarum divinarum, quae sunt esse simplicis differentiae et passiones praeter conceptionem cum continuo et tempore, nihil accipientes principiorum essendi ab eis, eo quod priora illis sunt et causae eorum, et ideo ista stabiliunt in esse omnia continua et omnia temporalia. Quod esse stabilitum et fundatum supponitur et non quaeritur in eis in scientiis doctrinalibus et physicis, partes entis continui vel mobilis considerantibus. Sicut enim causa tertia in ordine fundatur in secundaria et secundaria fundatur in primaria et primaria non fundatur in aliquo, sed est fundamentum omnium consequentium, ita naturalia et doctrinalia fundantur in divinis, et divina non fundantur, sed fundant tam mathematica quam physica." *Ibid.*, p. 2, l. 16–30.

[3] *Ibid.*, p. 2, l. 31–p. 3, l. 5; especially p. 2, l. 67–70 and l. 88–92.

[4] "...esse, quod est simplex esse, actus existens primae essentiae,..." *Ibid.*, p. 2, l. 68–69. "Esse enim, quod haec scientia considerat, non accipitur contractum ad hoc vel illud, sed

it is that which is had by any substance logically prior to quantification or to motion.[1] Physics must suppose that there is a mobile body, and mathematics that there is quantified or discrete continuum; metaphysics precisely by studying the *esse simplex* of everything, gives to the other two sciences their objects and their principles.[2]

The principles of *esse simplex* are principles which transcend whatever is physical; because it studies these principles metaphysics is called *transphysica*.[3] It is also called "divine" because the principles of *esse simplex* are divine and the best, the first, and the complement of everything else.[4] These principles are divine things, in the conception of which time and the continuum are not included; as divine and first, these principles are the causes of all *continua* (or mathematical properties, let us say) and of all temporal beings.[5]

But how does metaphysics study both the *esse simplex* and its principles, divine things? It is precisely a question such as this that Albert categorically rejects. There were "quidam Latinorum logice persuasi", he notes, who maintained that God was the subject of metaphysics in so far as metaphysics studied what is most noble, most divine, and highest of all things. These same philosophers held that being (*ens*) was the subject of the science in so far as the subject is "quod communius subicitur", that is, in so far as the subject is that which is said of all things. And finally, these same thinkers placed a cause as the subject of metaphysics, in so far as metaphysics studied what is most certain.[6]

potius prout est prima effluxio dei et creatum primum, ante quod non est creatum aliud". *Ibid.*, p. 3, l. 1–4. Many of Albert's expositions in later books tempt one to believe that it is God, not being, which is the subject of metaphysics. Yet in the light of the text just quoted, it would be wrong to say that God is the subject. For an author who makes that mistake: M. SCHOOYANS, "La distinction entre philosophie et théologie d'après les commentaires aristotéliciens de saint Albert le Grand", *Revista da Universidade Catholica de São Paulo*, XVIII, 1959, pp. 274 sqq.

[1] *Ibid.*, p. 2, l. 47–75.

[2] *Ibid.*, p. 2, l. 75–92.

[3] "... transphysica vocatur, quoniam quod est natura quaedam determinata quantitate vel contrarietate, fundatur per principia esse simpliciter, quae transcendunt omne sic vocatum physicum." *Ibid.*, p. 2, l. 89–92.

[4] "Vocatur autem et divina, quia omnia talia sunt divina et optima et prima, omnibus aliis in esse praebentia complementum." *Ibid.*, p. 2, l. 92–p. 3, l. 1.

[5] *Ibid.*, p. 2, l. 16–21. Cf. footnote 2, p. 77.

[6] "Sunt autem quidam Latinorum logice persuasi, dicentes deum esse subiectum huius scientiae, eo quod nobilissimae scientiae dicunt debere esse nobilissimum subiectum et primae scientiae primum subiectum et divinae divinum et altissimae altissimum; et huiusmodi multa ponunt secundum logicas et communes convenientias, et hi more Latinorum, qui omnem distinctionem solutionem esse reputant, dicentes subiectum tribus modis dici in scientia, scilicet quod communius subicitur aut quod certius aut quod in scientia dignius est. Et primo modo dicunt ens in ista scientia et secundo causam et tertio modo deum, et hanc scientiam non a toto, sed a quadam sui parte dignissima vocari divinam." *In I Meta.*, T. I, c. 2, p. 5, l. 34–47.

Albert leaves us in no doubt as to his opinion of the distinctions made by these men: "Sed ego tales logicales convenientias in scientiis de rebus abhorreo, eo quod ad multos deducunt errores."[1]

Having made known his opinion of these ideas, Albert proceeds to make clear his own ideas. In effect, he says that metaphysics is not called "divine" because it studies beings which are divine, in the sense of things existing apart from sensible things after the fashion of God. Rather metaphysics is a "divine" science because it studies the most manifest aspects (*manifestissima*) of all things, and such manifest aspects are the most divine, the most noble, and prior to all other aspects.[2] It is in this sense that metaphysics can be said both to study the simple to-be (the existing act of the first essence) and the principles of the simple to-be: these most manifest aspects "ens et entis partes et principia."[3] Because metaphysics studies these most honorable, most wonderful, and most certain aspects in their universality ("per totum et non in quadam sui parte") it can be called "first philosophy".[4] Thus it is that Albert sees himself as a member of the Peripatetic school, casting his metaphysics as a study of being as such (*esse simplex* in its universality) and of all those things which follow, not on a particular being, but on being as such.[5] Because of its subject matter metaphysics is able to found all the other sciences; that is to say, all composites are ultimately *resolved* into the simple to-be, the existing act of the first essence, the act of being.[6]

This conception of metaphysics is not without its *prima facie* diffi-

[1] *Ibid.*, p. 5, l. 47–49.

[2] "Nec denominatio ideo fit, quod divina dicitur. Omnia enim apud naturam omnium rerum manifestissima sunt divinissima et nobilissima et priora omnibus, et haec sunt ens et entis partes et principia...", *Ibid.*, p. 5, l. 51–54. Cf. L. De Raeymaeker, "Albert le Grand, Philosophe. Les lignes fondamentales de son système métaphysique," *Revue Néo-scolastique*, XXXV, 1933, p. 13.

[3] *Ibid.* Cf. *Ibid.*, p. 5, l. 15–20.

[4] "Et ideo et honorabilissimorum et mirabilissimorum et certissimorum per totum et non in quadam sui parte est scientia ista, quam ideo merito primam vocamus philosophiam.", *Ibid.*, p. 5, l. 55–58.

[5] *Ibid.*, p. 4, l. 51–56.

[6] Speaking of the manner in which *esse simplex* founds all the determinations of material beings, Albert writes: "Et utrumque istorum [id est: the two opposed types of determinations] fundatur in esse, quod est simplex esse, actus existens primae essentiae, quae est, in qua stat omnis compositi resolutio ultima." *In I Meta.*, T. I, c. 1, p. 2, l. 67–70. Speaking of the objection that other sciences are useless if metaphysics studies being as such, Albert writes: "Causae autem et principia non probantur hic per ea quae sunt cuidam naturae propria et proxima, sed potius per entis principia stabiliuntur ea quae propria sunt, non in eo quod propria, sed resoluta et reducta ad ens vel partes entis, secundum quod est ens et non secundum quod est hoc ens. Et ideo ad sciendas res in propria natura, summe requiruntur scientiae particulares, quae tamen nihil penitus probare possent, nisi eo modo quo subiecta et principia sua omnia relinquunt per entis principia esse stabilita in scientia ista." *In I Meta.*, T. I, c. 2, p. 5, l. 15–25. Cf. also *Ibid.*, c. 1, p. 3, l. 18–25.

culties, for Albert appears to identify "simplex esse, actus existens primae essentiae", or "creatum primum" on the one hand, with *ens* on the other. *Esse simplex,* for Albert is opposed to *esse mobile* and to *esse continuum.* This is of course understandable. The simple to-be is not the same as a mobile to-be or a quantified to-be. But how is it possible to identify the simple to-be and *ens*? This constituted no problem for Albert. In fact, this identity of *esse,* to-be, and *ens* is not infrequent in his writings. Père Roland-Gosselin has noted this frequency, remarking that this identity of to-be and being permitted Albert to affirm that to-be is the first creature of God. Albert, obviously, did not understand *esse* and essence in the same fashion as this couple is discussed in the Thomistic school. *Esse* is taken in the sense of what is not non-being; it is in itself common and communicable as is any other form.[1] It is the form of the whole, as opposed to the form of the matter.[2] The essence, on the other hand, is what constitutes a being as possible; it is what makes a thing the type of thing it is. For example, the essence "man" makes Socrates a possible man; if God creates *ens* or *esse,* the form of the whole, Socrates is a real man.[3]

Not only does metaphysics study *esse simplex,* but all the following must be considered as well: "per accidens et per se esse, potentia et actus, unum et multum, idem et diversum, conveniens et contrarium, separatum et non-separatum."[4] These would all appear to be reduced or resolved to being as such, either as "parts of being", or as what follows on being as such.[5]

[1] ROLAND-GOSSELIN, *Le "De ente et essentia" de S. Thomas d'Aquin...*, pp. 175–79.

[2] *Ibid.*, p. 174.

[3] *Ibid.*, p. 178. – Albert underwent at varying moments the influence of Boethius, Avicenna, and William of Auvergne. This accounts for a certain variation in his approach to *esse* and essence (*Ibid.*, p. 172). In the beginning of his career, in the *Commentary on the Sentences,* he does not use the Boethian distinction of *id quod est* and *id quo est* to explain the composition of creatures (p. 173). When he treats the soul however, there is *id quod est,* (the hypostasis which makes the soul subsistent) and the *quo est* or *esse,* the essence in act in the subject, the form of the whole (p. 174). In the second book of the *Sentences,* however, the *quod est* in separated substances appears as what is possible, while the *esse* is what makes the possible real (p. 174–75). And in the third book of the work, the treatment of Christ shows the lack of anything such as a Thomist sense of *esse* and essence (p. 176). In the commentaries on Aristotle, Albert rarely mentions the problem of the composition of created being; however, in speaking of angels in the *Metaphysics,* Albert distinguishes between the angel's possibility, and its reality, given by *ens* or *esse* (pp. 177–78). It is only in books of a later period that there is anything like a distinction of *esse* and essence (pp. 178–81). Thus when Albert wrote his *Metaphysics* (?–1263), *esse* was not the metaphysical principle of Thomas; *esse simplex* or *esse* or *ens* must be considered as the form of the whole; the *prima essentia,* of which *esse* is the existing act, is the possibility of being. Also: J. PAULUS, *Henri de Gand. Essai sur les tendances de sa métaphysique,* Vrin, Paris, 1938, pp. 111–12.

[4] *In I Meta.*, T. I, c. 2, p. 3, l. 72–77.

[5] "Haec enim scientia una est, quoniam licet sit de multis, de omnibus tamen illis est,

In explaining how all things are considered in metaphysics, Albert brings in the notion "unity of analogy." All beings have a unity in so far as all are related to "one"; of course, the diverse beings are all related to "one" in diverse ways: "Et ideo ens in omnibus his unitatem habet analogiae, quae unitas... est multorum ad unum respicientium ...sic, quod illa diversa aliquo modo sunt unius, et ille modus quod sunt unius, est diversus in diversis."[1] One of these diverse modes of belonging, or of being, to the one (*unius*) is *ens simpliciter* or substance; quantity, quality, and so on, are other modes which are ways of belonging to the "one" which is substance. Thus the parts of being (quality, quantity, etc.) are united to being, to substance, the subject of metaphysics.[2]

There are still the *passiones* which must be brought into the unity of metaphysics. Being, the subject of the science, is the "immediatio substandi passionibus", that is, the immediate substrate of all "passions."[3] In so far as the metaphysician studies the genus of "immediate substrate of passions", he brings all the passions of being into his science. Why does he attempt such a thing is not an unfair question; and Albert's answer is ready: because the principles of the *quod quid est* are the principles of the passions of that quiddity.[4] So it is that because the subject of this science (being) is first, or prior to all other aspects of a thing, and because the passions are the first and the simple passions, that the metaphysician can determine what the other sciences must presuppose, namely that their subjects exist.[5]

Such is the introduction to metaphysics composed by Albert the Great. There are several points of comparison with the *Prooemium* of Aquinas that must be noted.

First, one is struck by the resemblance between Aquinas' ideas and those which Albert confessed were abhorrent to him. Aquinas investigates metaphysics from the view-point of the study of the most certain

prout reducta sunt in ens ut partes et prout sunt ens consequentia in eo quod est ens.", *In I Meta.*, T. I, c. 3, p. 5, l. 61–64.

[1] *Ibid.*, p. 5, l. 64–71.

[2] *Ibid.*, p. 5, l. 71–76.

[3] *Ibid.*, p. 5, l. 77–78.

[4] *Ibid.*, p. 5, l. 79–84.

[5] "...cum physicus supponit esse corpus mobile et cum mathematicus supponit esse continuum quantum vel discretum, ideo ponit esse, quia ex suis propriis principiis esse ipsum probare non potest, sed oportet, quod esse probetur ex principiis esse simpliciter. Et ideo ista scientia metaphysica stabilire habet et subiecta et principia omnium aliarium scientiarium. Non enim possunt stabiliri et fundari ab ipsis scientiis particularibus, in quibus "quia sunt" vel esse relinquuntur vel supponuntur." *In I Meta.*, T. I, c. 1, p. 2, l. 75–81. Cf. *Ibid.*, p. 2, l. 16–30; *Ibid.*, c. 2, p. 4, l. 57–68.

object (first causes), of the most universal object ("being" and its retinue of notions), of the most immaterial object (God and the separated substances). Of course, only being is called the subject, or the genus studied, while God and the separated substances are said to be the first causes of being. The logically minded philosophers to whom Albert refers attributed a three-fold subject to metaphysics: the most certain object (cause), the most common object (being), the most noble, divine, or high object (God). One feels obliged to see in these logically minded philosophers, as reported by Albert, a source of inspiration for Aquinas' thought.

Just as Aquinas' ideas resemble the abhorrent one of the "quidam Latinorum logice persuasi", so they differ from those of Albert. For the former, metaphysics is the "divine science" or theology because it considers the separated substances and God. For Albert, metaphysics is "divine" because "ens et entis partes et principia" are divine in the sense of the most honorable, most wonderful, and most certain of all aspects of reality.

In assigning the name "first philosophy" to metaphysics, Albert notes the following reason: the aspects of things which metaphysics studies are studied universally. In Aquinas' *Prooemium*, however, "first philosophy" is the study of the first extrinsic causes of being.

What of the conformity between Albert and Aquinas? Certainly for both *ens* is the subject of metaphysics. Moreover, both place metaphysics last in *resolutio*; but whereas Albert says all is resolved to the simple to-be, the form of the whole, for Aquinas all appears to be resolved to being, the basic reality of things. The difference between these two men is anything but slight however. For Albert, the concept of *esse simplex* or *ens* does not in any way contain such concepts as *esse mobile* or *esse quantum*.[1] In Thomas' mind, on the other hand, *ens simpliciter* or *ens commune* expresses implicitly every aspect expressed by any other concept. Thus it would seem that one must recognize some divergence between Albert's *ens*, the form of the whole, and Thomas' *ens*, the basic commodity of reality.

There is another point of comparison: it does not seem incorrect to view as similar the several reasons given for calling metaphysics *transphysica*. Aquinas' reason is that the science of being is the last science in the "way of resolution", and he compares this posteriority to that of

[1] "Nec etiam propriae et determinatae scientiae cuiusdam entis isti scientiae metaphysicae subalternantur, quia ad subalternationem non requiritur, quod subiectum sit sub subiecto tantum, sed quod medium, quod est causa propter quid, sub alterius scientiae medio concludatur vel contineatur.", *In I Meta.*, T. I, c. 3, p. 5, l. 25–30.

the more common notions in relation to the less common. Albert on the other hand sees this science as *transphysica* because the principles of *simplex esse* transcend the physical world, a transcendance viewed in terms of *resolutio*. (Aquinas however appears to view metaphysics as *trans-* or *meta-* physics for an additional reason as well: metaphysics is the last science to be learned; as we noted, the *resolutio* concerns the learning of sciences as well as the static hierarchical relation between them.)

For Albert's conception of metaphysics, the unity of analogy is important. Because of the diverse relations of the "parts" of being to *ens simpliciter*, metaphysics is able to study many different aspects of being. There was no such doctrine present in the *Prooemium* of Aquinas, but as we shall have reason to note in the chapter that follows, such a conception is present in Aquinas' work, to say nothing of Albert's predecessors – Aristotle, Avicenna, and Averroes.

Schematically, these comparisons can be summarized as follows.

1) Albert rejects the threefold division of the subject given by some logically minded Latin philosophers.	1) Thomas accepted the threefold division, but not as a division of the subject of metaphysics; rather, it is a division of the things studied in the science.
2) Albert calls metaphysics "divine" because it studies the most manifest or most divine aspect of things.	2) Aquinas calls metaphysics "divine' because it studies God, the cause of its subject.
3) Albert calls metaphysics "first philosophy" because it studies its object in a universal manner.	3) For Aquinas, this science is "first philosophy" because it studies the separate substances which are the first causes of being.
4) Albert places metaphysics last in the order of resolution because all composition is reduced to the *simplex esse*, the form of the whole.	4) Thomas places metaphysics last in the order of resolution for two reasons: i) all concepts are reduced to being, the espression of the basic reality of a thing; and ii) this science is the last to be learned.
5) Albert refers to this science as *transphysica* because the principles it studies transcend all physical reality; this transcendency is viewed in terms of *resolutio*.	5) "Metaphysics" or *transphysica* refers to the place this science occupies in the *via resolutionis*. But this place is to be understood in both of the ways mentioned above.

This comparison of Albert and Aquinas illustrates an important characteristic of the conception of metaphysics given in the *Prooemium*. Albert's exposition was finished before Aquinas began to write.[1] Thus

[1] For the date of Albert's work (?–1263) see the discussion of footnote 8, Ch. I, p. 5. For the date of Aquinas' work (1270–72), see the discussion of Ch. I, section 2, especially the summary on p. 21.

it appears necessary to view Aquinas' thought as formulated in relation to that of Albert. The latter had spoken of his personal metaphysics as "divine", as "first philosophy", as *transphysica,* as the study of *esse simplex* or *ens*. Since Aquinas wrote after Albert, and since Aquinas had Albert's work before him, the fact that Aquinas composed a *Prooemium* indicates that he was not satisfied with Albert's.[1] The fact that Aquinas explains in his introduction the meanings he attributed to the various doctrines explained much earlier by Albert (among others) would indicate that Aquinas intends his *Prooemium* as a clarification of issues hitherto improperly explained; in Thomas' eyes, the *Prooemium* presents the only acceptable conception of metaphysics.

Yet should not a qualification be made on this last statement? For do we know whether the *Prooemium* presents a sketch of the *totality* of Aquinas' metaphysics, as opposed to a sketch only of those elements which he derived from contact with Aristotle and with Aristotelians? It would seem reasonable to suppose that it presents the outline of the totality; in fact it would seem practically an impossible feat to talk of a conception of metaphysics if by "conception" one meant "half a conception". How would one manage to remain comprehensible, if one spoke of metaphysics by mentioning only some of the essential elements of metaphysics? Thus it does not seem sufficient to see in the *Prooemium* only a few metaphysical truths which Aquinas acquired through his contact with Aristotelianism.

5. THE INTRODUCTORY BOOK OF AVICENNA'S *METAPHYSICS*

Aquinas' *Prooemium* is related not only to the work of Albert the Great, but to that of Avicenna as well. When Avicenna introduced his exposition he too explained the subject matter of metaphysics and the names by which this science is known. The subject matter of metaphysics, *esse* as such, has been treated in sufficient detail in Chapter II, and to that treatment the reader is referred.[2] As far as the names of the science are concerned Avicenna brings forth five: "wisdom", "divine science", "scientia post naturam", "scientia ante naturam", "first philosophy". The term "wisdom", he remarks, denotes that in this science one knows the most noble object comprehensible to man, that is, one knows God. And the term "first philosophy" is given to this

[1] Evidence indicating Aquinas referred constantly to the *Commentary* of Albert, as well as to those of Averroes and Avicenna, will be presented in Chs. III–IV.

[2] Cf. section 1, pp. 23 ff.

science because it studies the first cause of *esse*.[1] But the first cause of *esse* is the most noble object man can know – God; hence "first philosophy" is "wisdom". Still more, this science is called "divine" because it studies God.[2] Thus Avicenna identifies "wisdom" (the knowledge of the most noble object man can know), "divine science" (the knowledge of God), and "first philosophy" (the knowledge of the first cause of *esse*). Then too this science is "post naturam". "Natura" is here understood as the principle of motion and rest had by sensible bodies. Since one studies wisdom after the study of natural science, one can refer to wisdom as "post naturam". Logically speaking however, it could be called "ante naturam" since the objects studied in it are logically prior to natural objects.[3]

That there are resemblances between this Avicennian conception of metaphysics and those of Albert and Thomas is clear. The differences between them, however, are not less evident.

1) Avicenna saw metaphysics as the study of the most knowable object because it studies God. For Aquinas, metaphysics studies the most knowable or intelligible object in so far as it studies God, the first causes, and *ens commune*. Yet a closer analysis of Thomas' *Prooemium* reveals that Thomas would in substance agree with Avicenna; as Thomas explains, if one considers what objects are in themselves, then the most immaterial being (God) is the most intelligible. This is precisely Avicenna's point. Thomas' affirmation that *ens commune* and the first causes are the most intelligible objects must be taken in context; they are the most intelligible objects, not because of what they are, but because, as concerns "common being", it corresponds to the intellect's manner of knowing (universally), and because, as concerns the first

[1] "...et scientia horum [principiorum scientiarum singularium] quaeritur in hoc magisterio, et haec est philosophia prima, quia ipsa est scientia de prima causa esse, et haec est prima causa... et est etiam sapientia quae est nobilior scientia qua apprehenditur nobilius scitum. Nobilior vero scientia, quia est certitudo veritatis et nobilius scitum, quia est Deus..." *Meta.*, T. I, c. 2, fol. 71r, E–F.

[2] Metaphysics is the science which treats: "...de eo quod omnimodo separatum est a natura; et tunc nominabitur haec scientia ab eo, quod est dignus in ea scilicet vocabitur haec scientia, scientia divina." *Meta.*, T. I, c. 4, fol. 71v, C. C. SAUTIER, *Avicennas Bearbeitung der Aristotelischen Metaphysik*, Herder, Freiburg i.B., 1912, p. 47.

[3] "Nomen vero huius scientiae est, quod ipsa est de eo quod est post naturam. Intelligitur autem natura virtus quae est principium motus et quietatis; ... quod vero dicitur post naturam haec posteritas est in respectu quantum ad nos... unde quod meretur vocari haec scientia considerata in se haec est, ut dicatur quod est ante naturam. Ea enim de quibus inquiritur in hac scientia per eandem sunt ante naturam." *Meta.*, T. I, c. 4, fol. 71v, C. Cf. SAUTIER, *Avicennas Bearbeitung...*, p. 50.

causes, they are the source of certitude (cf. footnotes 3–5 on p. 56). – Albert did not like to speak in these terms, as we have seen.

2) Avicenna sees metaphysics as the study of the most certain object in so far as it studies *esse*, for in knowing *esse*, the metaphysician knows the principles of all sciences. For Aquinas, metaphysics merits the title "most certain" because it studies the first causes of being. Albert again did not use this terminology.

3) For Avicenna as well as for Aquinas, metaphysics was "divine science" because it studied God. For Albert, however, it was "divine" because it studies being, the most manifest or divine aspect of things.

4) For Avicenna and, again, for Thomas, the name "first philosophy" is given to metaphysics because it studies the first extrinsic cause of being. For Albert, however, it is "first" because it studies its object universally.

5) Avicenna refers to the science of metaphysics as "ante naturam" in so far as its subject, *esse*, is logically prior to the subjects of the other sciences. Moreover, this sience is called "post naturam" because it is studied after one has mastered the other sciences. Albert's "transphysica" resembles Avicenna's "ante naturam", for the former is the study of principles transcending all physical and mathematical objects in terms of *resolutio*. Finally, Thomas' name "transphysica" ,or "metaphysica", refers to the place this science occupies in the *via resolutionis*; that is to say, first, the metaphysical concepts (*ens* and other *communia*) have a type of static or logical priority in relation to other concepts (similar to Avicenna's "ante naturam"); and second, metaphysics is the last of the sciences to be learned in so far as one resolves concepts and causes back to "being" and to God (partially similar to Avicenna's "post naturam").

6) The subject of metaphysics is called *esse* by Avicenna, and *ens commune* by Aquinas, while Albert uses the expression *simplex esse* or *ens*. Do all three understand the subject matter in the same way? There is not sufficient evidence in their introductions to give the final answer to this question. One can see, however, certain aspects in the three authors which are extremely difficult to reconcile. For example, Avicenna has spoken of *esse* as the intention or meaning had in common by all reality. Nowhere does Avicenna explain that esse expresses each and every aspect of all things. Hence, one wonders whether *esse*, or the aspect common to all beings, is not different than Aquinas' *ens*, the reality of everything. For the latter, *ens* expresses everything about

everything to which it is applied. Would Avicenna say the same about his *esse*?

Nor is it any easier to determine the relationship between Avicenna's *esse* and Albert's *esse simplex*. Is Albert's "form of the whole" the same as Avicennian *esse*? For Avicenna, even number and measure can be considered as *esse*; could Albert consider number and measure as having a "form of the whole"? It is with difficulty that one identifies any two of the three notions given by Albert, Aquinas, and Avicenna.[1]

This brief comparison of these three introductions reveals certain relationships between the three authors. It appears that Aquinas and Albert both wrote in relation to what preceded their own work. For example, one can scarcely fail to notice that 1) Avicenna spoke of metaphysics as the study of the most knowable object, as the divine science, as first philosophy; 2) Albert refuses to speak of metaphysics under the first term, and in relation to the second (divine science) and the third (first philosophy) gives a different view than Avicenna; and 3) that Thomas agrees with Avicenna on all three points. Nor can one fail to note the apparent differences with which the three philosophers explained "being", the fundamental idea of metaphysics. On the other hand, there is an element of continuity among the three writers. They all place a similar relation – at least at this point in our work, it appears similar – between metaphysics and the other sciences: for all three, metaphysics founds all the lower disciplines.

Just as we concluded that for Thomas, Albert's presentation of metaphysics was unacceptable, a similar but more complex judgment must be made concerning the reception accorded Avicenna's doctrines: Albert disapproved totally of parts of them, but adopted others; Thomas disapproved of Albert's disapproval of Avicenna, and hence in large part, returned to Avicenna. – If we turn now to Aristotle, we can trace back even further this metaphysical give-and-take.

6. ARISTOTLE'S INTRODUCTION

Aristotle, too, composed an introduction to his *Metaphysics*, and one can not help but sense a kinship between his introduction and that of Thomas: both are built around a certain idea of intellectual knowledge.

Aristotle begins his *Metaphysics* with a description of the genesis of intellectual knowledge. It is this type of knowledge that most especially

[1] Avicenna uses the word "common" quite regularly. He would appear to be the first to have referred to *unum, multum*, etc. as *communia*. Perhaps, Aquinas copied the term *ens ommune* from Avicenna.

merits the name of "wisdom", for the fundamental characteristic of wisdom is to be a knowledge of the "why" and the cause, and to be such knowledge for the sake of knowledge.[1] Thus wisdom will be the study of the first causes and principles.[2] When Aristotle speaks in this way of "wisdom", he is referring to it in the sense of "philosophy"; he does not, in the first part of his introduction, refer only to metaphysics, that is to the study of being as such, under the term "wisdom".[3] However, Thomas understood Aristotle to speak of metaphysics when he uses the term "wisdom" in this part of the *Metaphysics*.[4] And hence, we too must regard this opening chapter with its emphasis on the genesis of intellectual knowledge, as dealing above all with the genesis of metaphysics.

Chapter 2 of Aristotle's introduction turns to a more precise investigation of the study of science which especially merits the title of "wisdom".[5] In this chapter Aristotle collects the common notions of the wise man;[6] the wise man is he who knows all things, can learn difficult things, is more exact and more capable of teaching the causes, knows what is desirable for itself, and has knowledge through which he rules those less wise than himself.[7] All of these characteristics are had by "him who has in the highest degree universal knowledge".[8] And having the most universal knowledge means knowing all the instances

[1] A 1, 981a24–30; *Vetus*, p. 256, l. 11–18. A 2, 982b11–28: *Vetus*, p. 258, l. 26–p. 259, l. 11. (For the versions of the *Metaphysics* used by Aquinas, see Ch. I, section 1, especially the summary on p. 9.)

[2] A 1, 981b27–982a3: *Vetus*, p. 257, l. 11–20.

[3] V. Decarie, *L'objet de la métaphysique selon Aristote*. Publications de l'Institut d'Études Médiévales, Univ. de Montréal, XVI, Inst. d'Études Méd., Vrin, Paris, 1961, p. 85.

[4] Cf. *In I Meta.*, 1, 1 and 35; *Ibid*, 2, 36.

[5] Decarie, *L'objet de la métaphysique selon Aristote*, p. 87.

[6] The fact that Aristotle simply collects these notions commonly held on the type of knowledge enjoyed by the wise man does not at all imply that Aristotle might not himself have accepted these views as true. As has been explained in a recent article, Aristotle's practice of taking the notion common to his time was an element of his philosophical method. And in the case in question ("wise man") all the elements of the common man's opinion are retained, even more, are elaborated and synthesized by Aristotle in the course of the *Metaphysics*. Cf. G. Verbeke, "Philosophie et conceptions préphilosophiques chez Aristote," *Revue Philosophique de Louvain*, LIX, 1961, pp. 405–30; esp. pp. 416–19. – Aristotle, one may say, was convinced of the value of the opinions universally received. This conviction was based, of course, on doctrinal grounds: "The investigation of truth is in one way hard, in another easy. An indication of this is found in the fact that no one is able to attain the truth adequately, while on the other hand, we do not collectively fail, but every one says something true about the nature of things, and while individually we contribute little or nothing to the truth, by the union of all a considerable amount is amassed.", α1, 993a30–b4.

[7] A 2, 982a8–19: *Vetus*, p. 257, l. 23–p. 258, l. 2. As Thomas very succintly says: "Ex quibus omnibus potest quaedam sapientiae descriptio formari: ut ille sapiens dicatur, qui scit omnia etiam difficilia per certitudinem et causam, ipsum scire propter se quaerens, alios ordinans et persuadens.", *In I Meta.*, 2, 43.

[8] A 2, 982a21–23: *Vetus*, p. 258, l. 4–6.

falling under the universal, knowing what is farthest from sense and hence most difficult, knowing the instances through the fewest number of principles and hence being more exact,[1] knowing the end for which all is done.[2] Thus, Aristotle concludes, if one knows the first principles and causes, one would be truly wise, for to know these principles and causes is to have the most universal knowledge.

This science, it seems to be implied, is the ultimate outcome of the natural wonder in man; beginning from the simplest of problems, man advances bit by bit until he possesses the knowledge which enables him to see the answers to all problems, from the most difficult through to the most simple.[3]

Other than the name "wisdom", this science may also be called "divine", and this for two reasons; first it deals with the divine objects, for God is among the first causes; and secondly, it is a science such as God would have more perfectly than any other intellectual substance.[4]

This, then, was Aristotle's introduction to metaphysics. Its outstanding feature is easily recognizable; rather than build his introduction around the objects studied in metaphysics, Aristotle has placed all the emphasis on the intellect. He who knows the most about everything, he who knows through the most universal causes, this man has wisdom.

Fundamentally, wisdom is characterized as the most universal knowledge open to man. If questioned on the content of this it is knowledge of the "why" and of the cause. But what exactly does this mean? If we judge from Chapter 2 of Aristotle's initial book, we must answer that the wise man knows the intrinsic principles, as well as the end, of all things. If we judge from the remaining chapters of Book A, we would be led to believe that knowledge of the material and formal causes of beings, and that knowledge of the extrinsic principles involves knowing the agent as well as the end of all things. Still more, if we emphasize wisdom to be knowledge of the "why", we can gain more light from

[1] It is interesting to note how Thomas enlarges Aristotle's exposition: A 2, 982a25–28: *Vetus*, p. 258, l. 8–11. *In I Meta.*, 2, 47.

[2] A 2, 982a21–b7: *Vetus*, p. 258, l. 4–25.

[3] "Propter enim id quod est mirari homines et nunc et prius inceperunt philosophari; ex principio quidem pompta deficiencium admirantes, postea secundum parvum sic procedentes, et de majoribus dubitantes, ut deque lune passionibus, et que sunt circa solem et astra, et de omni generatione." A 2, 982b12–17: *Vetus*, p. 258, l. 26–32. "Oportet tamen quodam modo prestruere ordinem ipsius in contrarium nobis que sunt a principio quaestionum. Incipiunt quidem enim, sicut diximus, ab eo quod est admirari omnes si sic se habent, sicut sunt miraculorem per se contingentia nondum considerantibus causam... Oportet enim in contrarium et melius, secundum proverbium, perficere...", A 2, 983a11–19: *Vetus*, p. 259, l. 28–p. 260. l. 2.

[4] A 2, 982b28–983a11: *Vetus*, p. 259, l. 12–28.

Book Z 17; there, "to know the why" appears as knowledge of the essence (or form in matter), as well as knowledge of the end and agent.[1] Yet Aristotle's wise man possesses knowledge of these causes in the most universal way possible: he somehow knows all the individual instances of these causes, for he is said to know all the instances falling under the universal. Thus wisdom appears as the highest of all intellectual knowledge.

If we compare this introductory sketch given by Aristotle with those given by Avicenna, Albert, and Aquinas, we cannot but be struck by a curious fact. Earlier we noted that Aquinas apparently did not agree with Albert, and so returned to the theories of Avicenna on "first philosophy" and "divine science". Now however, it appears that Aquinas follows Aristotle in the use of the name "divine science", as well as in seeing metaphysical knowledge ("being") as the expression of every aspect of reality.

All four spoke of being, but it is difficult to know whether they all understood this in the same sense. The subject of wisdom, Avicenna writes is *esse*; in studying *esse*, the wise man studies the meaning common to all reality. However, Avicenna does not say that *esse* expresses every aspect of all things. When Aristotle speaks of metaphysical knowledge, he appears to make it much richer than Avicenna does. For Aristotle, the metaphysical knowledge of the intrinsic causes of being appears to be the knowledge of the intrinsic causes of all reality.

Albert, on the other hand, spoke of being as the form of the whole, the most basic aspect of anything; being is that which makes a thing real as opposed to merely possible. One wonders whether Albert's idea of "being" is not much poorer than Aristotle's, for the latter's "being" expresses all aspects of realities.

What of Thomas' *Prooemium*? On the surface it is closer to Aristotle than either the introductions of Avicenna or Albert. For Thomas, like Aristotle, builds his introduction around metaphysics as the highest

[1] "The 'why' is always sought in this form – 'why does one thing attach to some other?' Z 17, 1041a10–11. "Thus the inquiry is about the predication of one thing of another... Plainly we are seeking the cause. And this is the essence... which in some cases is the end... and in some cases is the first mover...", Z 17, 1041a25–30. "The object of the inquiry is most easily overlooked where one term is not expressly predicated of another (e.g. when we inquire 'what man is')... clearly the question is why the matter is some definite thing... Therefore what we seek is the cause, i. e. the form, by reason of which the matter is some definite thing...", Z 17, 1041a32–b8. – Thomas does in fact interpret the "why" of A 2 in light of Z 17; this is not stated by Thomas, but a comparison of his expositions reveals the fact; cf. *In I Meta.*, 1, 24–28; *In VII Meta.*, 17, 1649–68. On Aristotle, cf. A. MANSION, "L'objet de la science philosophique suprême d'après Aristote, *Métaphysique E 1*", in *Mélanges de philosophie grècque offerts à Mgr. A. Diès*, Vrin, Paris, 1956, pp. 160–61.

perfection of man's intellect. It is true that Thomas does not expressly speak of the wise man's knowledge of the intrinsic causes of beings. He does, however, explain that knowledge of *ens commune* penetrates all generic and specific notions. Moreover, in our examination into the doctrines presupposed by the *Prooemium*, we discovered that *ens commune* expresses every aspect of every being. Hence, Thomas' wisdom is knowledge of all the intrinsic causes of being, knowledge had in so far as one has the concept of "common being". There is no need to point out the similarity between this notion and Aristotle's "most universal knowledge".

Hence, as far as metaphysics is characterized as an universal knowledge of the intrinsic causes of all beings, Aquinas appears to follow more closely in the steps of Aristotle. But what of knowledge of the extrinsic causes of being? How is that related to the metaphysical knowledge of the intrinsic causes? Of course, Aristotle's introduction does not expressly state that relation. Yet one would assume that the agent and the end are the extrinsic causes which determine somehow the intrinsic causes. It is well known that, as modern commentators on Aristotle explain, God is needed in Aristotle's system only as the first cause of motion.[1] Nevertheless, certain elements of the introductions of Avicenna, Albert, and Aquinas lead us to see that at least two of them understood Aristotle to be introducing God into philosophy as the first cause of being.

First, all three clearly state that God is studied in metaphysics. For Avicenna and Aquinas, this fact explains why this science is called "divine". Although Albert would not accept this explanation, it is clearly that of Aristotle. Secondly, all three of these commentators state that God is the cause of being. And thirdly, Avicenna and Aquinas state explicitly that the fact of studying the first cause of being (God) is the source of the name "first philosophy". Again Albert will have no part of such an explanation. Although Aristotle does not use the name "first philosophy" in his introduction, this is a name used for metaphysics several times in later books. As we shall have cause to note in Chapter V, section 2, A, Aristotle uses the name "first" to indicate that this science studies the first or highest of all beings; moreover, he speaks of this science as the "universal science", the study of being as such. Although Aristotle does not explain why the first science is also the universal science, we shall have cause to note that Thomas and Avicena

[1] E.g.: W. D. Ross, *Aristotle's Metaphysics. A Revised Text with Introduction and Commentary*. Clarendon Press., Oxford, 1958, vol I, p. cliii.

give an explanation, an explanation which they thought faithful to Aristotle's doctrine (sections 1, A, and 2, C). As they explain, the first science must be the universal science, because in studying being *qua* being one discovers God, the first cause of being. Thus, one can have a first science (the study of the first being) only if one studies the universal science, and so discovers the existence of the first being as the cause of being; nor are these two sciences separate, for anything we can know about the first being must be related to its effect, being as such. Thus Avicenna and Aquinas, by their use of the name "first philosophy' refer to their view on the relation of the study of the extrinsic causes to the study of the intrinsic ones, and as well, show their willingness to follow Aristotle's lead. This willingness is shown as well by the use of the name "divine science". Accordingly, Albert's refusal to use either the name "divine science" or "first science" in their Aristotelian senses, reveals either one of two things: first a desire to be independent of Aristotle's system; or two, an understanding of that system which differed from those of Avicenna and Aquinas. Regardless of which of these may be the case, it seems correct to conclude that at least Aquinas and Avicenna believed that Aristotle placed God in the role of the first cause of being. Not only did they believe it, but they accepted his view as correct.

In concluding the previous section (on Avicenna), we noted that Avicenna's ideas on metaphysics apparently were not satisfactory in the eyes of Albert; in like manner, we felt that Albert's theories failed to satisfy Aquinas, and so we believed that Aquinas bypassed Albert and returned to Avicenna. Our brief study of Aristotle however, reveals that we were not quite correct. Avicenna, we now understand, was an Aristotelian, at least as far as the general framework of his metaphysics was concerned. Thus Thomas, in returning to Avicenna, was returning to Aristotle.[1] This very fact of Thomas' return to Aristotle, as revealed by the *Prooemium*, permits us, at least provisonally, to characterize Thomas' metaphysics as "basically Aristotelian". That is to say, Thomas accepted as the framework of his metaphysics the outlines of Aristotle's system, as he, Thomas, understood them.[2]

The very brief comparisons made in this chapter do not permit us to conclude a great deal concerning Thomas' attitudes toward Albert and Avicenna. Obviously, he wasn't satisfied with their commentaries or he

[1] The very fact that Thomas wrote an exposition on Aristotle and not on Avicenna indicates that the former was more truly Thomas' master.

[2] Thus we do not say that Thomas' system is basically identical to the system as actually held by Aristotle. Rather, Thomas accepted what he understood Aristotle to say.

would not have undertaken to write his own. Then too, he did not totally agree with Albert's manner of characterizing metaphysics. More than this we can not conclude at the moment.

Yet as regards Thomas' opinion of Aristotle's metaphysics, an interesting possibility suggests itself. Since Thomas' own metaphysics as presented in the *Prooemium* follows the general lines of Aristotle's, an exposition of Aristotle's *Metaphysics* would have provided an excellent opportunity for Thomas: an opportunity to expose his own thought, basically Aristotelian, under the guise of an exposition of Aristotle's *Metaphysics*.[1]

7. CONCLUSION OF PART ONE

The preliminaries of our study have been completed. The salient points can be summarized as follows.

1) Thomas used five different Latin versions of Aristotle, four of which we do not possess in a complete form. However, in all cases we must attempt to discover the meaning of the text Thomas used. To the extent that this attempt is successful, our conclusions can be definitive.

2) The *Commentary* was composed within a rather short period, and as far as has been determined, was only partially revised. We must naturally be attentive to possible doctrinal differences due to the different chronological periods of composition.

3) Thomas' *Prooemium* presents us a definite, though skeletal, conception of metaphysics; it appears to have been written, moreover, in the light of the writings of Avicenna and Albert. One wonders whether Thomas' entire Commentary and not only its *Prooemium*, will be found to be written in relations to the works of Avicenna and Albert (and Averroes).[2] But whether or not this is found to be the case, it is important to keep well in mind the conception of metaphysics we discovered in the *Prooemium* of Thomas. If the *Commentary* contains Thomas' personal thought, it must be consonant with the metaphysics sketched in the *Prooemium*. That metaphysics, it will be recalled, studies *ens commune*, the *principium* or concept common to all other

[1] "Under the guise of an exposition of Aristotle's *Metaphysics*": to our modern minds this phrase connotes a rather unsavory form of literary activity unworthy of a saint of Aquinas' caliber. But let us beware of attributing to a medieval our modern mores.

[2] We have not considered Averroes' *Prooemium* because this introduction ,which is printed before Book XII of our editions of Averroes, was not translated until several centuries after Thomas' work. Cf. E. BERTOLA, "Le traduzioni delle opere filosofiche arabo-guidaiche nei secoli XII e XIII", in *Scritti in onore de Prof. Mons. Francesco Olgiati*, Vita et Pensiero, Milano, 1961, p. 29. – This *Prooemium* of Averroes is found on fol. 286sqq. of *Aristoteles Opera Omnia*. Vol. VIII. *Metaphysicorum cum Averrois Cordubensis in eosdem commentariis*, Junctas, Venetiis, 1562.

knowledge; in addition however, there are other concepts which must be studied, for these too are assumed by our knowledge: these other concepts are termed "the retinue of concepts following *ens*." Studied as the cause of *ens commune* is God and the other separate substances. Be this as it may, *ens commune* is predicated of God just as it is of other things. This much was said in the *Prooemium* concerning the subject matter of metaphysics; but there was something more as well, a word on method. Thomas mentions the *via resolutionis*. To say that metaphysics is the last science in this *via* means, we discovered, that metaphysical concepts are logically prior to all others; but it means as well, that we arrive at metaphysics by some sort of resolution of concepts and causes to "being" and to God. Such is the conception of metaphysics given in the *Prooemium*. It is by no means complete. Yet it does give us something of a criterion whereby we may judge the legitimacy of the appellations: "Aquinas' metaphysics."

4) Aquinas' metaphysics, as contained in his introduction, can be styled: "basically Aristotelian." It seems well to note that the conception we outlined in point 3) above is not exactly the conception to which reference is made when we note that Thomas' ideas are "basically Aristotelian". The metaphysics in point 3) is somewhat more detailed than the "basically Aristotelian" metaphysics. When we speak of this latter metaphysics, we mean only: (1) that metaphysics knows the intrinsic causes of all things by forming the most universal of all concepts; and (2) that metaphysics studies God, the first cause of being, because knowledge of God flows from knowledge of being as such. Besides these elements, the metaphysics described in point 3) spoke of the *via resolutionis* and the retinue of concepts which follow *ens*. These elements may prove to be "basically Aristotelian." but that has still to be proved. In any case, at this point in our study, we can at least be certain that the general lines of Thomas' metaphysics are identical to those of Aristotles' wisdom. Whether or not this fact will be of importance remains to be discovered.

PART TWO

It is time to turn now to the *Commentary* proper. As was mentioned both in Chapter I and in the Introduction, I shall not presuppose the *Commentary* represents Aquinas' personal thought. In this second part of the study, I want to discover the nature of Aquinas' exposition. To reach this goal, three separate studies must be carried through. First, the doctrine of the *Commentary* concerning the philosophy of being must be discovered. Second, it must be determined whether that doctrine is identical to that of Aristotle's *Metaphysics*; in this respect, it is important to determine whether or not Aquinas thought he was exposing only Aristotle. And finally, Aquinas' work must be related to those of Avicenna, Averroes, and Albert. As will be clear, the data of these three investigations shows what Aquinas was about in the composition of his work.

Because the *Commentary* is a most complex document, I have thought it best to give a three-fold division to this part. In Chapter IV the object of metaphysics will be discussed. From Aquinas' *Commentary* have been chosen two of the most important passages where he speaks of the material and formal object of metaphysics. (Purely for the sake of convenience, I shall refer to the theories of these passages as "Aquinas' doctrines"; this title is intended solely to indicate that they are found in his *Commentary*, not that they express his personal views.) Through the discussion of these texts, the attempt will be made to show how much Aquinas' *Commentary* expects of metaphysics: that is, the view that all things are to be studied in metaphysics, as well as the theory of the completeness, the richness, of the metaphysician's knowledge will be illustrated. In Chapter V much the same goal has dictated the choice of texts to be examined; all of them bear directly on the richness of metaphysical knowledge. However, in addition all of them are implicated in the *Commentary*'s presentation of metaphysics as the queen of the sciences.

In these two chapters, by the investigation into the *Commentary*, we shall realize the importance of the metaphysical knowledge of "being." I shall not, however, attempt to say the last word on the content of the idea of "being" in Chapters IV–V. To grasp the meaning of "being", one must reconstruct the metaphysics operative in the *Commentary*. Yet we can not even begin such a reconstruction before we have some criterion enabling us to decide *which* paragraphs of the *Commentary* were written in the light of *what* metaphysical system. Let me explain.

As noted in the introduction, some historians divide the *Commentary* into paragraphs containing Aquinas' thought, and others containing Aquinas' exposition of Aristotle. But if one compares Aquinas' *Commentary* with the corresponding works of Aristotle, Avicenna, and Albert, one realizes that every paragraph of the former work is to be taken as an expression of the same view of metaphysics. Hence, before attempting to reconstruct the metaphysics operative in the *Commentary* one must show that Aquinas had only one system in mind as he wrote whose so ever system that may be.

To illustrate the unity of the metaphysical system operative in Aquinas' exposition, Chapters IV–V will contain detailed comparisons of the expositions of the five philosophers in question – Aquinas, Albert, Averroes, Avicenna, and Aristotle. Thus, these two chapters will have multiple results. First, we shall see the characteristics attributed to knowledge of "being" by the *Commentary*. Second, we shall understand the difference between this theory of the *Commentary* and the corresponding doctrine of Aristotle. Thirdly, we shall see that it is indeed possible that Aquinas thought the system exposed to be that of Aristotle. Fourthly, we shall recognize the necessity of accepting the *Commentary* as an exposition written in opposition to the *Commentaries* of Avicenna, Averroes, and Albert. Thus, at the close of these two chapters, we shall be authorized to reconstruct the metaphysical system at work in the *Commentary*.

In Chapter VI, then, I shall attempt to evolve the step by step construction of the science of metaphysics. Yet even here, the thought of the *Commentary* will be contrasted to that of Aristotle. Such a comparison will strengthen our impression that Aquinas intended to expose Aristotle's metaphysics.

At the end of Chapter VI we shall be in a position to compare the metaphysics of the *Commentary* with that of the *Prooemium*. This comparison will indicate the possibility that Aquinas not only presents in the *Commentary* what he considers Aristotle's views, but that he accepts those views for his own.

CHAPTER IV

THE OBJECT OF METAPHYSICS

The *Commentary's* doctrine on the object of metaphysics appears to be quite simple: metaphysics is the study of being as such. This doctrine is exposed in several lessons, one notices immediately: *In IV Meta., lectiones* 1–2; *In VI Meta., lectio* 1; *In XI Meta., lectiones* 1 and 3. Yet the more closely one inspects these passages, the more complex the doctrine is seen to become. Hence, in an attempt to do justice to the multiple aspects of the problem, only the few paragraphs from Book IV will be studied, albeit in great detail.

The projected study will be more profitable if one knows in advance Aquinas' theory on the object of metaphysics; such knowledge will enable one to follow with greater comprehension the exposition in the passages to be examined. Roughly speaking, Thomas' theory on the object of this science involves four principal aspects; 1) metaphysics is the study of accidents as well as substance; 2) metaphysics is the study of all substances and of all accidents; 3) the viewpoint proper to metaphysics is such that "being" expresses every aspect of everything other than God; and 4) metaphysics is a study of concepts. Let us consider each of these briefly.

1) Metaphysics is the study of accidents as well as substance. – Here one can speak of a "horizontal conception of being." Metaphysics has a concept of "being" which is predicated of both substances and accidents; in every predication where an accident is called "being," that accident is referred to substance, in so far as substance is also called "being." Because "being" is thus predicated, the metaphysician can study substance in so far as it is said to be "being," and consequently know the most important aspect of accidents in so far as they are called "being."

2) Metaphysics is the study of all substances and of all accidents. – This may be considered a "vertical conception of being." The metaphysical concept of "being," in the sense of "substance," is predicated of all substances; in like manner, "being," in the sense of "accident," is predicated of all accidents. The metaphysician must not only attempt to construct "being" in the sense of the horizontal conception; as well, he must attempt to construct a concept of "being" which can be predicated of all substances and of all accidents.

3) The viewpoint proper to the metaphysician is such that "being" expresses all aspects of everything but God. Nevertheless, Aquinas' words often give the impression that the viewpoint is an univocal one. For example, he will sometimes relate metaphysics to its object in the same way a potency is related to its object; even more, he will be seen to explain that metaphysics resembles a science such as grammar: just as grammar is interested in all words, so metaphysics is interested in all substances. In the context in which these doctrines appear, one is severely tempted to see in metaphysics anything but the traditional analogical viewpoint.

The problem of the formal object of metaphysics, its point of view, and of the concept of "being" which expresses that viewpoint – these can not be totally solved in the early parts of this study. Only in Chapter VI, in the context of a discussion of the predication of concepts, do we actually see the meaning of "being"; only then we can understand why "being" is analogous, not univocal. Until then accordingly, we will not be able to understand fully what Thomas wishes to say when he speaks of "being" as expressing the totality of all things except God.

4) Metaphysics is a study of concepts. – One of the recurrent themes of the *Commentary* is the characterization of metaphysics as the study of *communia*. By *communia* Thomas refers to the concepts and principles which are applicable to every being. Because the "common" principles have as subject and predicate the "common" concepts, metaphysics will be referred to as the "study of concepts," rather than as the "study of concepts and principles." The study of concepts: in the context of the *Commentary* this means that metaphysics must clarify, or fix the meaning of, the concepts common to all things. Of course, these concepts are to be clarified by looking to their application, to their predication of real things.

In so far this aspect of metaphysics is under consideration metaphysics is said to be divided into as many sections as there are transcendental concepts. Accordingly, one should speak of a first section of metaphysics whichs tudies all beings as "being," of a second section studying all things as "one," of a third studying all beings as "thing," and so on. In each of these sections one would treat both substance and accidents as "being," or as "one," or as "thing," etc. In other words, one would treat the horizontal concepts of "being," of "one," of "thing". Yet in like manner however, each of these sections would have to treat all substances and all accidents: the vertical concepts of "being," of "one," and of "thing" must be discussed as well.[1]

Bur does this mean that metaphysics should be called the "study of being as being, as one, and as thing."? One is restrained from answering affirmatively by Aquinas' repeated statements that metaphysics is the study of being as such. Even more, it sometimes seems that the common concepts fall to the perusal of the metaphysician by default: someone must study them and the metaphysician appears to have more right than anyone else to do so; therefore... Yet these concepts such as "one" and "thing" do fall within the domain of the study of "being," a point that will be discussed later. And finally, the method used to study "one" and "thing" is identical to the method used to study "being." – There is much to be said of this problem, but it can best be explained when we view Aquinas' basic insight in Chapter VII. Hence for the moment, let us retain only this: metaphysics studies not only "being," but also "one," and "thing": metaphysics studies the *communia*.

Such then are the four doctrines which are implied in the two lessons we shall examine in this chapter. In our study of *In IV Meta., lectiones* 1–2 we shall have abundant evidence of the first two of these doctrines (the horizontal and vertical conceptions of "being"). The third theory – metaphysics has an analogical viewpoint – is strongly suggested by some of Aquinas' words, but even more strongly, apparently denied by others. The last doctrine, concerning the study of concepts, will be postulated as very probably the explanation of some of Thomas' statements which shall be noted here; in Chapters V and VI these latter two doctrines will again be matter for our examinations. Without more ado, then, let us turn to *In IV Meta., lectiones* 1–2.

[1] Does this mean that the *quarta via* of the *Summa theologiae* (cf. I, 2, 3c) would be the summary and synthesis of the proofs of God involved in each section? Because material beings are "being," is there a first being whose essence is to be? because material beings are "one," is there a first being whose essence is to be one essence? because material beings are "things," is there a first being whose essence is to have a quiddity? This matter will be treated in Chapter VI.

1. THE DOCTRINE OF AQUINAS' *COMMENTARY* ON THE OBJECT OF METAPHYSICS

A. The exposition of In IV Meta,. lectio 1

Aquinas, in this lesson, discusses first the subject of metaphysics, and then turns to the answer given to one of the questions raised in Book III; this procedure, he writes, is actually that of Aristotle.[1]

Thomas' treatment of the first part of this lesson contains three points.

1) Every science studies a subject and the proper or *per se* accidents of its subject; thus, the study of being as such must include a consideration of the proper accidents of being.[2] This study of being as such is not like other sciences, however, for although all sciences study beings, only this science studies beings as being – in so far as they are being.[3] The metaphysical study of being as such does not study the accidents of particular beings, but studies the accidents proper to anything as being. Such a science of being as being is necessary, concludes Thomas. Without the knowledge of things as being, we could not acquire any other knowledge, for all knowledge depends on the knowledge of things *qua* being: the knowledge of this science is related to other knowledge just as more common knowledge is related to less common.[4]

2) A science of being, such as has been outlined, is not the same as any of the particular sciences; for whereas these latter study a definite part (or particular genus) of being, the former science is interested in all being (*universale ens*) and from the viewpoint of being.[5]

3) In the science outlined in Aristotle's *Metaphysics*, it is the first causes and principles of things that are sought. But such causes and principles can only be the causes of the nature of being. Thus the science studied in the *Metaphysics* is the study of all being and *qua* being.[6]

In this lesson Thomas does not identify the first causes which are to be sought; he merely says that they are the first causes of being. But

[1] *In IV Meta.*, 1, 529.

[2] *Ibid.*

[3] *Ibid.*, 530.

[4] "Dicit etiam 'et quae huic insunt per se' et non simpliciter quae huic insunt, ad significandum quod ad scientiam non pertinet considerare de his quae per accidens insunt subiecto suo, sed solum de his quae per se insunt... Sic igitur huiusmodi scientia, cuius est ens subiectum, non oportet quod consideret de omnibus quae insunt enti per accidens, quia sic consideraret accidentia quaesita in omnibus scientiis, cum omnia accidentia insint alicui enti, non tamen secundum quod est ens... Necessitas autem huius scientiae quae speculatur ens et per se accidentia entis, ex hoc apparet, quia huiusmodi non debent ignota remanere, cum ex eis aliorum dependeat cognitio; sicut ex cognitione communium dependet cognitio rerum propriarum.", *Ibid.*, 531.

[5] *Ibid.*, 532.

[6] *Ibid.*, 533.

in an earlier Book he explanied that all four causes are studied. The formal cause alone is studied in so far as metaphysics is interested in what things are; the final cause and the efficient cause are studied in so far as metaphysics studies the first immobile substances: these are the final and moving causes of other beings. And finally, the material cause is studied because material things are included under being.[1] The ability to study the material cause by studying being is explained more clearly by a passage in Book I; there Thomas was discussing metaphysics as the most universal knowledge. Metaphysics, he wrote, will understand everything in understanding its subject, *ens simpliciter*; however, this universal understanding only potentially and not actually, reaches the ultimate determinations of all beings.[2] Thus because all things are understood through the grasp of metaphysical being, even material beings *qua* material are known.[3] (As will appear in Chapter VI, the material cause of material beings will be quite carefully investigated; that investigation takes place in the construction or attainment of the science of metaphysics, whereas in the passages we are here considering, Thomas is speaking about metaphysics as a completed science.)

Thus Thomas' thought on the identity of the causes sought is clear. Clear as well is his statement that metaphysical knowledge plays the role of more common knowledge as compared to the less common knowledge of the particular sciences: "being," the subject of metaphysics,

[1] "Omnis autem substantia vel est ens per seipsam, si sit forma tantum; vel si sit composita ex materia et forma, est ens per suam formam; unde inquantum haec scientia est considerativa entis, considerat maxime causam formalem... Sed quamvis ipsae [primae substantiae] sint immobiles secundum seipsas, sunt tamen causa motus aliorum per modum finis; et ideo ad hanc scientiam, inquantum est considerativa primarum substantiarum, praecipue pertinet considerare causam finalem, et etiam aliqualiter causam moventem. Causam autem materialem secundum seipsam nullo modo, quia materia non convenienter causa est entis, sed alicuius determinati generis... Tales autem causae pertinent ad considerationem particularium scientiarum, nisi forte considerentur ab hac scientia inquantum continentur sub ente. Sic enim ad omnia suam considerationem extendit.", *In III Meta.*, 4, 384. This paragraph gives what Aquinas calls "Aristotle's answer to his question: whether all the causes are studied in this science."

[2] This doctrine is given by Thomas in the context of the discussion of the definition of the wise man. One of the characteristics of the wise man is that his knowledge is certain. Thomas explains this by saying that a science is more certain to the extent it understands less in its considerations. Then to prevent his readers from misunderstanding this, he adds: "Nec illud est contrarium, quia dicitur esse ex paucioribus, cum supra dictum sit, quod sciat omnia. Nam universale quidem comprehendit pauciora in actu, sed plura in potentia. Et tanto aliqua scientia est certior, quanto ad sui subiecti considerationem pauciora actu consideranda requiruntur.", *In I Meta.*, 2, 47. Earlier in this paragraph Thomas mentions the supreme concept, *ens simpliciter*, which is the expression of the subject of metaphysics.

[3] In a later book Thomas brings this out by opposing two aspects of metaphysics; on the one hand there is the study of concepts (*communia*), and on the other, there is the study of the very being of things: "...haec scientia praecipue considerat communia; non tamen quod communia sint principia, sicut Platonici posuerunt. Considerat autem et principia intrinseca rerum, sicut materiam et formam.", *In XI Meta.*, 1, 2167.

says potentially all that one can say of anything; actually, however, it says but little and thus there is still both the possibility and the need of the particular sciences. (Much clarification is needed on these theories, but that must come later.)

Thus far, we have discussed the first part of *lectio* I; in this part Thomas pretended to follow Aristotle through a discussion of the subject of the science sought in the books entitled *Metaphysics*. The second part of this lesson, Thomas says, is the solution of a question earlier raised by Aristotle: "utrum huius scientiae esset consideratio de substantiis et accidentibus simul, et utrum de onmibus substantiis."[1] In the answer to this question, it is important to distinguish again three steps: 1) metaphysics is said to study both substance and accidents; 2) but it studies substance primarily; and 3) it studies all substances.

1) Metaphysics studies everything of which "being" can be predicated; such predication, even though analogical and not univocal, is sufficient to bring all things called "being" under the sway of one science.[2] The things which are analogically called "being" are substance, motion, quality, quantity, properties, etc.[3] To state that "being" is *analogically* predicated of the instances of these categories (of the categories of substance, of quality, etc.) is to say that the instances of accidents *qua* "being" are referred to the instances of substance *qua* "being."[4] Or to put this another way, "being" as predicated always "says" or involves a relation to substance as "being"; the relation said will differ according to the category in which is found the instance related to substance.[5] Thus substance, as "being," is that "which has *esse* through itself," while the

[1] *In IV Meta.*, I, 534.

[2] As Thomas explains, the following is the reason why metaphysics studies both substance and accidents: "Quaecumque communiter unius recipiunt praedicationem, licet non univoce, sed analogice de his praedicetur, pertinent ad unius scientiae considerationem: sed ens hoc modo praedicatur de omnibus entibus: ergo omnia entia pertinent ad considerationem unius scientiae, quae considerat ens inquantum est ens, scilicet tam substantias quam accidentia.", *Ibid.*

[3] *Ibid.*, 539–43.

[4] "Et sicut est de praedictis [exemplis de sanativo et medicativo], ita etiam et ens multipliciter dicitur. Sed tamen omne ens dicitur per respectum ad unum primum. Sed hoc primum... [est] subiectum. Alia enim dicuntur entia vel esse, quia per se habent esse sicut substantiae, quae principaliter et prius entia dicuntur. Alia vero quia sunt passiones sive proprietates substantiae....", *Ibid.*, 539. "Et ad hoc [genus substantiae] sicut ad primum et principale omnia alia referuntur. Nam qualitates et quantitates dicuntur esse, inquantum insunt substantiae;...", *Ibid.*, 543.

[5] "...ens sive quod est, dicitur multipliciter. Sed sciendum quod aliquid praedicatur de diversis multipliciter:... Quandoque vero secundum rationes quae partim sunt diversae et partim non diversae: diversae quidem secundum quod diversae habitudines important, unae autem secundum quod ad unum aliquid et idem istae diversae habitudines referuntur; et illud dicitur "analogice praedicari", idest proportionaliter, prout unumquodque secundum suam habitudinem ad illud unum refertur.", *Ibid.*, 535.

accidents, as "being," will be things such as a "*passio* of that which has *esse* through itself," or a "process or motion toward that which has *esse* through itself," and so on.[1]

To illustrate the difference between analogical and univocal terms, Aquinas notes that things are referred analogically to something which is one *in ratione*, that is, to substance, understood as "that which has *esse* through itself." Yet they are not referred to an *unum in ratione* after the fashion of things receiving univocal predication; rather the *unum in ratione* in analogical predication is also *unum numero, unum sicut una quaedam natura*.[2] That is to say, "John "and "Paul" are both called "man" and thus are both referred to the univocal concept "man," an *unum in ratione*; "John " and "Paul" are in no way referred to one existing thing, to some *unum numero* when they are called "man." Yet accidents, as "being," are referred to some *unum numero*, to some *unum sicut una quaedam natura*; accidents are referred to some existing substance, which too is called "being." This idea is somewhat hazy; the meaning becomes clear, Thomas remarks, if one considers the manner in which all accidents as "being" are referred to substance as "being."[3] If the "whiteness" of a bear is called "being," it is thereby referred to something which is one existing thing: it is referred to the substantial being of which it is an accident – it is referred to the substance of the bear. Thus to be referred to substance is to be referred to *una natura*,

[1] *Ibid.*, 539. Cf. footnote 4, p. 104.

[2] Thomas brings out this doctrine of reference to *unum in ratione*, which is also *unum numero*, in the following paragraph, without mentioning however the identity of that *unum*. From the text of footnote 4, p. 104, as well as from the text of footnote 3 below, it is clear that substance is that *unum*. "Item sciendum quod illud unum ad quod diversae habitudines referuntur in analogicis, est unum numero, et non solum unum ratione,sicut est unum illud quod per nomen univocum designatur. Et ideo dicit quod ens etsi dicatur multipliciter, non tamen dicitur aequivoce, sed per respectum ad unum; non quidem ad unum quod sit solum ratione unum, sed quod est unum sicut una quaedam nature.", *Ibid.*, 536.

[3] Immediately after the text quoted in the previous footnote, Thomas writes: "Et hoc patet in exemplis infra positis. Ponit enim primo unum exemplum, quando multa comparantur ad unum sicut ad finem, sicut patet de hoc nomine sanativum vel salubre.", *Ibid.*, 536–37. The use of the name "healthy" is too famous an example to need repeating. Suffice it to note that all uses of "healthy" refer to the use of that name to designate the proper functioning of an organic body. In paragraph 538 Thomas gives another example, the use of the name "medical". In this case, all predications of the name "medical" depend on the use of that word to indicate the person who has the art of healing. In these two examples of analogical names, all use of the names is referred ultimately to some *unum ratione* which is also an *unum numero* or an *unum sicut una quaedam natura*, In the case of the name "being" the *unum numero* is substance: "Et sicut est de praedictis [exemplis de sanativo et medicativo], ita etiam et ens multipliciter dicitur. Sed tamen omne ens dicitur per respectum ad unum primum. Sed hoc primum non est finis vel efficiens sicut in praemissis exemplis, sed subiectum. Alia enim dicuntur entia vel esse, quia per se habent esse sicut substantiae, quae principaliter et prius entia dicuntur. Alia vero quia sunt passiones sive proprietates substantiae... Quaedam autem dicuntur entia, quia sunt via ad substantiam... Alia autem entia dicuntur, quia sunt corruptiones substantiae.", *Ibid.*, 539.

he says.[1] And because all the accidents are referred to substance when they are called "being," one science can study both substance and accidents.[2] By *unum in ratione* then, Thomas brings out the fact that the name "being" when used refers to "that which has *esse* through itself": any accident as "being" is referred to "that which has *esse* through itself." By *unum sicut una quaedam natura,* he emphasizes that accidents as called "being" are referred to an *existing* substance, also called "being".[3] (It is important to note that, in this discussion, Aquinas has given no hint that "being" as used of *substance* is not an univocal concept.)

It is interesting that nowhere in lesson one does Thomas explicitly state the identity of all the accidents he relates to substance through the analogy of "being." It seems at first rather obvious that by "ac-

[1] Perhaps it is well to note the difference between "*when* we refer something to substance as being," and "*because* we refer something to substance as being." *Whenever* we call an accident "being," we refer it to substance. But *when* we call a substance "being," we do not refer it to anything else. In Chapter VI it will be argued that *because* we call any material substance "being," we must affirm the existence of God. Yet it does not follow that *when* we call a material substance "being," we refer it to God. Just the opposite is true: to call a material substance "being" is to refer it to nothing else at all. This necessity of affirming God because we affirm the being of a material substance involves participation. It does not concern analogy (except in so far as subsequently we must ask how the term "being" fits both God and the material substance).

[2] "...est unius scientiae speculari non solum illa quae dicuntur "secundum unum," idest secundum unam rationem omnino, sed etiam eorum quae dicuntur per respectum ad unam naturam secundum habitudines diversas. Et huius ratio est propter unitatem eius ad quod ista dicuntur; sicut patet quod de omnibus sanatavis considerat una scientia, scilicet medicinalis, et similiter de aliis quae eodem modo dicuntur.", *Ibid.*, 544.

[3] In this section of *lectio* 1, as well as in the greater part of the *Commentary*, Aquinas speaks of but one type of analogy of "being." He deals only with the "horizontal conception of being," the horizontal analogy: "being" as predicated both of substances and accidents. There is of course an occasional reference to another type of analogy of "being," but only in passing. For example *In I Meta.*, 14, 224 speaks of "being" as predicated of God and of creatures. "Being" is said of God "essentially" and of creatures "through participation." (For the meaning of this terminology, cf. L.-B. GEIGER, O.P., *La participation dans la philosophie de S. Thomas d'Aquin*, Bibliothèque Thomiste XXIII, Vrin, Paris, 1953, 2e edit., pp. 458–65). I shall not speak of this latter doctrine on the use of "being" as "analogy of being," and this for two reasons: 1) quite simply, because there is absolutely no need to; and 2) the word "analogy" would tend both to obscure the issue at stake, and confuse the horizontal meaning of analogy. In Ch. VI we shall see that *because* creatures are called "being" there is a Pure *Esse*. This argument is in terms of "participation." To speak of the analogical use of "being" in predications of God and of creatures would be merely a systematic reflection on the proof of God. Thus this analogy of "being" between God and creatures does not found the proof; it is merely a theory expressing the mechanics of the proof. – For a good treatment of this problem: cf. R. MCINERNY, *The Logic of Analogy. An interpretation of St. Thomas*, Nijhoff, The Hague, 1961, pp. 80–90; 144–52; 156–65. For a different approach to the problem, one underlining the intellect's drive toward understanding, cf. D. BURRELL, C. S. C., "A Note on Analogy," *The New Scholasticism*, 1962, XXXVI, pp. 225–32. A study valuable for its collection of texts, cf. G. KLUBERTANZ, S. J., *St. Thomas Aquinas on Analogy. A textual Analysis and Systematic Synthesis*, Loyola Univ. Press, Chicago, 1960. Although the collection of texts is large, it is far from complete, as a glance at the texts from the *Com. on the Meta.* shows.

cidents" he refers only to the nine categories (quantity, quality, etc.).[1] A problem arises, however, when one asks if the *per se* accidents of being referred to in the opening paragraphs of this lesson are the nine categories (cf. *In IV Meta.*, 1, 529). It seems that one can not so identify them, for in a later passage, Thomas refers to a much broader classification of accidents as the "accidentia entis in quantum est ens," in this sense, *unum*, the species of *unum*, these and many others are all accidents of being as such. And in lessons 2–4, Thomas treats the various accidents such as *unum*, and in various ways proves their claims to inclusion in the discussion of the science of metaphysics.[2] One wonders why Thomas placed together in the same lesson first, a proof that metaphysics studies the *per se* accidents, and secondly, a proof that metaphysics studies the nine categories of accidents. Or if he must place these two proofs in the same lesson, why did he not indicate some distinction between the two meanings of "accident of being as such"? But more about this later.

2) In the second step of the second part of lesson one, Thomas makes this point: metaphysics, even though studying both substance and accidents, primarily is a study of substance. The reason given is quite simple: whenever one has many things called by the same name – a name which relates all the things to some first thing — then the science of these many things is principally interested in the first thing. As Thomas explains, it is on the first thing that the other things depend both for their *esse* and their name. Since substance is this first thing, metaphysics will principally study substance.[3]

[1] Cf. *Ibid.*, 539–43; the treatment of the accidents in these paragraphs is not unlike the explanation of the nine categories of accidents in *In V Meta.*, 9, 891–92. In addition, one may compare this passage of Book IV with the explanation of the analogy of being in *In VII Meta.*, 1, 1247, where the accidents are plainly the nine categories of accidents.

[2] "...ad unam scientiam pertinet considerare ens secundum quod est ens, et ea quae per se illi insunt. Et per hoc patet, quod illa scientia non solum est considerativa substantiarum, sed etiam accidentium, cum de utrisque ens praedicetur. Et est considerative eorum, quae dicta sunt, scilicet eiusdem et diversi, similis et dissimilis, aequalis et inaequalis, negationis et privationis, et contrariorum; quae supra diximus esse per se entis accidentia... etiam considerat de priori et posteriori, genere et speci, tot et parte, et aliis huiusmodi, pari ratione, quia haec etiam sunt accidentia entis inquantum est ens." *In IV Meta.*, 4, 587. In lesson two Aquinas relates "simile "and "aequale" to *unum* (cf. *Ibid.*, 2, 561). In lesson three negation and privation are similarly related (cf. *Ibid.*, 3, 565). And finally in lesson four, all these accidents are simultaneously proved to belong to the consideration of *unum* and *ens* (cf. *Ibid.*, 4, 582–85). Aquinas' doctrine on *res* is introduced into the exposition of lesson two; it corresponds to nothing in the Aristotelian text, yet *res* is said to have as much right to be treated as *unum* (cf. *Ibid.*, 2, 553).

[3] "Omnia scientia quae est de pluribus quae dicuntur ad unum primum, est proprie et principaliter illius primi, ex quo alia dependent secundum esse, et propter quod dicuntur secundum nomen; et hoc ubique est verum. Sed substantia est hoc primum inter omnia entia. Ergo philosophus qui considerat omnia entia, primo et principaliter debet habere in sua

3) Thomas' third step is to note that metaphysics studies all substances. If, Thomas writes, all beings are one genus in any way whatsoever, then all the species of that genus fall to the consideration of one general science. This general science does not, of course, study these species according to what is proper to each species; rather, they are studied in so far as they come together in a genus. Thus, all substances in so far as they are beings or substances fall under the investigation of metaphysics.[1] The explanations of this third step are obviously intended by Thomas to introduce a new element. Previously, Thomas had explained Aristotle's doctrine that both substance and accidents are studied. Although in treating this doctrine, he occasionally wrote in the plural "substances and accidents," nevertheless his goal was to prove that accidents are studied in the study of substance, in metaphysics. In this third part, he appears to be broadening the outline of metaphysics by assuring us that all things falling under the genus of substance are to be studied. There is reason to suppose that the word "genus" is used in the ordinary sense of the word; thus the genus of substance, as any other genus, is represented by an univocal concept.[2] In an earlier lesson, Thomas treated the same problem: Does one science study all substances? In that context he explained that Aristotle answers this question in Book IV; there, he said, Aristotle shows that one science studies all substances by studying the common meaning or *ratio* of substance. Is not this statement an open declaration that "substance" is an univocal concept? Does not "common meaning" imply a generic idea? Although the temptation to answer affirmatively may be great, such an answer would be quite incorrect, for as noted "being" expresses the totality of everything. For the moment, however, we are left with the

consideratione principia et causas substantiarum; ergo per consequens eius consideratio primo et principaliter de substantiis est.", *Ibid.*, 1, 546.

[1] "Hic ostendit quod primi philosophi est considerare de omnibus substantiis, tali ratione. Omnium eorum qui sunt unius generis, est unus sensus et una scientia, sicut visus est de omnibus coloribus, et grammatica considerat omnes voces. Si igitur omnia entia sint unius generis aliquo modo, oportet quod omnes species eius pertineant ad considerationem unius scientiae quae est generalis: et species entium diversae pertineant ad species illius scientiae diversas. Hoc autem dicit, quia non oportet quod una scientia consideret de omnibus speciebus unius generis secundum proprias rationes singularum specierum, sed secundum quod conveniunt in genere. Secundum autem proprias rationes pertinent ad scientias speciales... Nam omnes substantiae, inquantum sunt entia vel substantiae, pertinent ad considerationem huius scientiae: inquantum autem sunt talis vel talis substantia, ut leo vel bos, pertinent ad scientiae speciales." *Ibid.*, 547.

[2] Cf. also *In XI Meta.*, 1, 2153 where Aquinas remarks that metaphysics studies all substances by studying all substances in so far as they come together "...in uno genere, quod est ens per se." Much the same doctrine is found in: *In XII Meta.*, 2, 2427.

puzzling statements which appear to be descriptions of the univocal or generical concept of substantial beings.[1]

This then is *In IV Meta., lectio* 1. Thomas began (a) by noting that metaphysics treats being and all its *per se* accidents. His second step (b) was to note that one can study the nine categories of accidents if one studies substance as "being"; and (c) since one has a generic (?) concept of substance, one can study all substances in studying substance as "being."

Before we began the study of this lesson, we noted that Aquinas is guided by four doctrines. The first of these, metaphysics is the study of accidents as well as substance, is explicitly stated in the lesson; thus Thomas was concerned with the horizontal conception of "being" – the concept of "being" which relates accidents to substance or to "that which has *esse* through itself." The second of the doctrines also appeared, for Thomas made it clear that metaphysics studies all substances and all accidents; here we were dealing with the vertical conception of "being" – the concept of "being" (idest, of "substance") which can be predicated of all substances, and the concept of "being" (idest, of "quality, of "quantity", etc.) which can be predicated of any accident. The third doctrine which was offered as Thomas', concerns the viewpoint of metaphysics. This was operative in so far as Thomas spoke of the concept of "substance": metaphysics studies all species of substance, he wrote, and it studies them in so far as they come together in a genus. Yet at the same time Thomas insisted that "being" expresses every aspect of everything. Hence, the viewpoint of the metaphysician does not seem to be an univocal one. The fourth doctrine was only hinted at in this lesson; metaphysics is, it was claimed, a study of concepts. In so far as Thomas, in the opening paragraphs of the lesson (cf. footnotes p. 102), explained that metaphysics studies all the *per se* accidents, he was touching this point. These *per se* accidents, besides including the nine categorical accidents, involve such things as *unum in qualitate, in quantitate*, and so on. Thomas only refers in an off-hand

[1] "Haec autem quaestio determinetur in quarto huius, ubi ostenditur quod ad primam scientiam, ad quam pertinet considerare de ente inquantum est ens, pertinet considerare de substantia inquantum est substantia: et sic considerat omnes substantias secundum communem rationem substantiae;..." *In III Meta.*, 6, 398. – The *ratio communis* of analogical names has been treated recently in two works by the same author: R. McInerny, *The Logic of Analogy...*, pp. 144–52. More especially: "The Ratio Communis of the Analogous Name," *Laval philosophique et Théologique*, XVIII, 1962, pp. 9–34. Mr. McInerny's conclusion is that having a "common meaning" does not make "being" an univocal term; it is not a genus in the strict sense. Rather, there is a common element present in every use of that word. This is the conclusion at which we shall arrive in Ch. VI through the study of the concept "being"; however, the use of "being" shall not be expressed in terms of analogy: cf. footnote 3, p. 106.

manner to metaphysics as the study of concepts in this lesson: lesson two, however, makes the point much more clearly.

B. The exposition of In IV Meta., lectio 2.

According to Aquinas, lesson two contains the partial answer to another question posed earlier by Aristotle: "utrum huius scientiae esset considerare de onmibus istis, quae sunt unum et multa, idem et diversum, oppositum, contrarium et huiusmodi...."[1] The complete answer to this question is exposed in lessons 2–4. All these notions such as *unum* or *multa* are called "communia";[2] moreover, all of them are referred to under the title "per se accidents of being."[3] In the present lesson Thomas explains merely that the science of being must study "unity" (*unum*) and its species; the two lessons which follow are devoted to the discussion of the remaining *communia*. Lesson two, the proof that metaphysics studies *unum*, is divided by Thomas into two parts: first, this science studies unity, and second, it studies the species of unity as well.

The proof that unity is considered in metaphysics is actually a double one. At least Thomas indicates that there are two proofs. Both of them consists in showing that *unum* and *ens* are the same *in re* but different *in ratione*. If he can show this, Thomas implies that it will then be evident that *unum* and *ens* fall under the sway of the same science.[4] This one feels, is a strange supposition: what possible difference can it make if the real object known through concept A is the same object known through concept B? Does the fact that one can consider man through the concept "bundle of energy" and through the concept "living being" give unity to a science which would try to study man through both concepts? Obviously no. What then does Thomas mean by this cryptic statement?

Before explaining the proofs, Thomas notes that unity and being are the same, and are one nature:"...ens et unum sunt idem et una natura." As an example of what is not meant by this expression, Thomas remarks that Socrates and *hoc album* and *hoc musicum* could be

[1] *In IV Meta.*, 1, 534.

[2] "Hic procedit ad ostendendum quod ad considerationem unius scientiae pertinent considerare huiusmodi communia, scilicet unum et multa, idem et diversum...," *Ibid.*, 2, 548.

[3] *Ibid.*, 4, 587.

[4] "Unum autem et ens significant unam secundum diversas rationes... Nihil tamen differt ad propositum, si similiter accipiamus ea dici, sicut illa quae sunt unum et subiecto et ratione. Sed hoc erit "magis prae opere," idest magis utile ad hoc quod intendit. Intendit enim probare quod unum et ens cadunt sub eadem consideratione, et quod habent species sibi correspondentes. Quod manifestius probaretur si unum et ens essent idem re et ratione, quam si sint idem re et non ratione." *Ibid.*, 2, 549.

the same in number without being the same nature; *ens* and *unum* on the contrary signify the same nature.[1] Although *ens* and *unum* are interchangeable names, they are not synonyms such as, for example, "vestments" and "clothes"; rather *ens* and *unum* signify or refer to one and the same real nature, but each according to a different *ratio* or meaning.[2]

Following this brief introduction, Thomas gives the first proof that being and unity are one in reality (*idem re*). Whenever one says "ens homo" and "unum homo," one does not indicate or refer to anything other than one does by saying "homo"; the reality which is signified by each of these terms is the same. Hence *ens* and *unum* are the same in reality.[3] Moreover, writes Thomas, it is evident that unity and being have different definitions: they are different *ratione*. Although terms such as "unity" and "being" and (adds Thomas) "thing" all refer to or signify exactly the same thing in reality, they do so according to different meanings. These three names refer to concepts which express our grasp of any object – a grasp which is a grasp of the totality of the object, but with emphasis on a special aspect. For example, "being" refers to our grasp of an object with emphasis on the object's *esse*; "thing" refers to our grasp of the object but now emphasis is placed on quiddity;and so on.[4]

[1] "...ens et unum sunt idem et una natura. Hoc ideo dicit, quia quaedam sunt idem numero quae non sunt una natura, sed diversae, sicut Socrates, et hoc album, et hoc musicum. Unum autem et ens non diversas naturas, sed unam significant. Hoc autem contingit dupliciter. Quaedam enim sunt quae unum consequuntur se adinvicem convertibiliter sicut principium et causa." *Ibid.*, 548.

[2] "Quaedam vero non solum convertuntur ut sint idem subiecto, sed etiam sunt unum secundum rationem, sicut vestis et indumentum... [Ens et unum] se habent sicut principium et causa, sed non sicut tunica et vestis, quae sunt nomina penitus synonyma." *Ibid.*, 548–49.

[3] "Quaecumque duo addita uni nullam diversitatem afferunt, sunt penitus idem: sed unum et ens addita homini vel cuicumque alii nullam diversitatem afferunt: ergo sunt penitus idem. Minor patet: idem enim est dictum homo, et unus homo. Et similiter est idem dictum, ens homo, vel quod est homo: et non demonstratur aliquid alterum cum secundum dictionem replicamus dicendo, est ens homo, et homo, et unus homo." *Ibid.*, 550. – Thus Socrates, *hoc album*, and *hoc musicum* are all the same existing being (*ens*), yet the concepts of *hoc album* and *hoc musicum* do not refer to the same nature in Socrates; "white" refers to the quality of colour; "musical" to the quality or ability "to make music."

[4] "Nam si [ens et unum] non different ratione, essent penitus synonyma; et sic nugatio esset cum dicitur, ens homo et unus homo. Sciendum est enim quod hoc nomen Homo, imponitur a quidditate, sive a natura hominis; et hoc nomen Res imponitur a quidditate tantum; hoc vero nomen Ens, imponitur ab actu essendi: et hoc nomen Unum, ab ordine vel indivisione. Est enim unum ens indivisum. Idem autem est quod habet essentiam et quidditatem per illam essentiam, et quod est in se indivisum. Unde ista tria, res, ens, unum, significant omnino idem, sed secundum diversas rationes." *Ibid.*, 553. There are several noteworthy aspects of this paragraph. 1) The definitions of "being," "thing," and "one" are said to be different, although all three express the same object, e.g. Socrates. 2) Aquinas says that the name "thing" is imposed from the quiddity, that the name "being" is imposed from the *esse*, and that the name "one" is imposed from indivision. If one were to take these

The second proof that "being" and "unity" are one in reality is explained much more simply by Aquinas. These two (unity and being) are one in reality, he writes, because they are both predicated of the substance of a thing in a *per se* manner: the substance of anything is "unum per se et non secundum accidens."[1] (By "*per se* predication" in this paragraph, Thomas means the predicate is that which the subject is.)[2]

To prove that *unum* and *ens* are predicated in this *per se* manner, Thomas posits the contrary case: if they are not predicated of the substance, then the search to find that which ensures unity and being would involve a regress to infinity. If "unity" and "being" are predicated of the substance of a thing because of some being added to the thing, then they would also be predicated of that added being. And if they are predicated of the added being, they would be predicated because of the addition of something else, which is itself a being. And so on, *ad infinitum*. Thus, one must hold that the substance of a thing is one and being through what it is itself.[3]

statements as implying that "being" means, or is defined as, *esse*; that "thing" is defined as quiddity; and that "one" is defined as indivision; in this case one would be unable to explain how we can conceive *esse*, quiddity, and indivision, and by these three concepts grasp the same object, e.g. Socrates. Quite obviously the concepts which would be defined as *esse*, quiddity, and indivision do not express Socrates; Socrates is not *esse*, nor quiddity, nor indivision, but rather has *esse*, has quiddity, and has indivision. Note: *Ibid.*, 556 where Aquinas says that *ens* does not mean *esse*.

[1] "Quaecumque duo praedicantur de substantia alicuius rei per se et non per accidens, illa sunt idem secundum rem: sed ita se habent unum et ens, quod praedicantur per se et non secundum accidens de substantia cuiuslibet rei. Substantia enim cuiuslibet rei est unum per se et non secundum accidens. Ens ergo et unum significant idem secundum rem." *Ibid.*, 554.

[2] Cf. the discussion of the four modes of *per se* predication in *In V Meta.*, 19, 1054–57. Paragraph 1054 explains the use of a predicate which expresses what the subject is; it is in this sense that Aquinas speaks of *ens* and *unum* in *In IV Meta.*, 2, 554: *ens* and *unum* both express the substance (here: essence) of the subjects of which they are predicated. They express the essence differently, as was explained in paragraph 553. – I shall return to this doctrine of *per se* predication of "being" in Chapter VI, section 1, C where I examine the types of predicates applied in a *per se* manner. – For a treatment of the manner in which *ens* expresses the substance of things, cf. R. McInerny, "Notes on Being and Predication," *Laval Philosophique et Théologique*, XV, 1959, pp. 236–74. For expositions involving the same doctrine, but which are vitiated by the incorrect assumption that *esse* is known in a judgment, see these two articles by J. Owens, C. Ss. R., "The Accidental and Essential Characteristics of Being in the Doctrine of St. Thomas Aquinas," *Mediaeval Studies*, XX, 1958, pp. 1–41; "Unity and Essence in St. Thomas Aquinas," *Mediaeval Studies*, XXIII, 1961, pp. 240–59.

[3] "Quod autem ens et unum praedicentur de substantia cuiuslibet rei per se et non secundum accidens, sic potest probari. Si enim praedicarentur de substantia cuiuslibet rei per aliquod ens ei additum, de illo iterum necesse est praedicari ens, quia unumquodque est unum et ens. Aut ergo iterum de hoc praedicatur per se, aut per aliquod aliud additum. Si per aliquid aliud, iterum esset quaestio de illo addito, et sic erit procedere usque ad infinitum. Hoc autem est impossible: ergo necesse est stare in primo, scilicet quod substantia rei sit una et ens per seipsam, et non per aliquid additum." *Ibid.*, 555.

After giving this proof, Thomas inserts five paragraphs devoted to a critique of Avicenna's ideas on the predication of the concepts "unity" and "being". These paragraphs are important both for their information on 'being" and because they reveal Averroes' influence on Thomas.[1] For Avicenna, Thomas writes, *unum*, and *ens* do not signify or refer to the substance or essence of a thing, but refer instead to something added to the essence. Thus, *ens* refers to the *esse*; it refers to what is other than, and added to, the essence. This is true for Avicenna in all those beings where the *esse* is had or received from some other being.[2]

Then there is Avicenna's doctrine of "unity." The reason why the Moslem philosopher said a thing is "one" through something added to its essence is found in the fact that he identified metaphysical unity and numerical unity. Since the *unum* which is the principle of number follows upon quantity, it signifies something added to the essence or substance. This *unum* was identified with *ens* because both signify an accident inhering in the essence.[3]

Avicenna was wrong in his view on *ens*, notes Thomas, for although a thing's *esse* is different from its essence, nevertheless *esse* is not something added as an accident. Rather, it is as it were constituted through the principles of the essence. Thus, the name *ens*, which is imposed from *esse*, signifies or refers to the same thing as the name *res*, which is imposed from the essence.[4]

[1] As shall be seen in section 2, B of this chapter Averroes was the first of the authors we are studying to criticize Avicenna in this same manner; both Albert and Aquinas give somewhat the same criticism.

[2] "Sciendum est autem quod circa hoc Avicenna aliud sensit. Dixit enim quod unum et ens non significant substantiam rei, sed significant aliquid additium. Et de ente quidem hoc dicebat, quia in qualibet re habet esse ab alio, aliud et esse rei, et substantia sive essentia eius: hoc autem nomen ens, significat ipsum esse. Significat igitur (ut videtur) aliquid additum essentiae." *Ibid.*, 556.

[3] "De uno autem hoc dicebat, quia aestimabat quod illud unum quod convertitur cum ente, sit idem quod illud unum quod est principium numeri. Unum autem quod est principium numeri necesse est significare quandam naturam additam substantiae:... Dicebat autem quod hoc unum convertitur cum ente... quia significat accidens quod inhaerat omni enti, sicut risibile quod convertitur cum homine." *Ibid.*, 557.

[4] "Sed in primo quidem non videtur dixisse recte. Esse enim rei quamvis sit aliud ab eius essentia, non tamen est intelligendum quod sit aliquod superadditum ad modum accidentis, sed quasi constituitur per principia essentiae. Et ideo hoc nomen Ens quod imponitur ab ipso esse, significat idem cum nomine quod imponitur ab ipsa essentia." *Ibid.*, 558. – Siger of Brabent, in his *Quaestiones in Metaphysicam*, refers to paragraph 558 when he wishes to give Aquinas' opinion on the distinction of *esse* and essence; Siger's reference is little more than a rewording: "Ponunt autem quidem, modo medio, quod esse est aliquid additium essentiae rei, non pertinens ad essentiam rei, nec quod sit accidens, sed est aliquid additum quasi per essentiam constitutum sive ex principiis essentiae." *Quaes. in Meta., Intro.*, Q. 7 l. 21–24 (Ed. Graiff, Louvain, 1948, p. 16). Does this citation not permit us to believe that Siger, and hence perhaps, all of Paris, accepted Thomas' *Commentary*, not as an exposition of Aristotle, but as an exposition of Thomas' own personal thought? One can not be certain, of course, but there appears to be at least a possibility that this was so.

It has been noted earlier that *ens* refers to the same thing to which *res* and *unum* refer, (cf. footnote 4, p. 111). In paragraph 553 Aquinas refers to *esse* (in the case of *ens*), to quiddity (in the case of *res*), and to indivision (in the case of *unum*) as to the aspects emphasized respectively by these three names. This doctrine is affirmed more clearly in the text of paragraph 558 just given in footnote 4, p. 113. *Ens* and *unum* (and *res*) are predicated *per se* of the essence of anything (par. 554); this means that they all express the essence: if one could define *ens, unum,* and *res*, one would define the essence of anything. A difficulty apparently arises if we recall that the name *ens* is taken from the *esse* (par. 558); how can *ens* express essence, if *ens* is a name taken from *esse*? No problem at all, Thomas in fact answers. There is a distinction between 1) that which is expressed in a concept (or by a name), and (2) that from which a name is taken (par. 558). Thus, *ens* and *res* are taken respectively from *esse* and essence; they express the same thing, however (par. 558). Thus *ens, res,* and *unum* all express the essence (par. 554), yet each in a different way (par. 553). What can this doctrine mean if not that *ens* expresses essence, but as related to *esse*;[1] if not that *res* expresses essence, but as a quiddity (par. 553);[2] if not that *unum* expresses essence, but as related to indivision?[3]

[1] The doctrine "ens imponitur ab esse" was discussed earlier in footnote 4, p. 111, where it was argued that this doctrine does not imply that *ens* expresses, or means, *esse*. M. Gilson insists on interpretating this sentence in such a way that *ens* means *esse*. Of course, M. Gilson's work is cast against the backdrop of his belief that concepts express only essences and that, hence, one grasps *esse* in a judgment. Cf. E. Gilson, *The Christian Philosophy of St. Thomas Aquinas*, (this is the English equivalent of the last edition of *Le Thomisme*) Random House, N. Y., 1956, Ch. I, esp. pp. 29–45, with footnote 30 of p. 40 deserving special attention. E. Gilson, *Being and Some Philosophers*, 2nd edit. corrected and enlarged, Pont. Inst. of Med. Stud., Toronto, 1952, Ch. VI, p. 190–215, et passim. M. Gilson is followed on this point by most "existential Thomists." For a good, incise criticism of M. Gilson's position on the meaning of "being," cf. R. McInerny, "Some Notes on Being and Predication," *The Thomist*, XXII, July, 1959, pp. 315–335.

[2] "Quiddity" is the name given to the answer to the question: *quid est*? when that question seeks the formal cause of a being; in the case of material beings, the quiddity always expresses or involves matter proportioned to the formal cause. Cf. *In I Meta.*, 12, 183. Thus "quiddity" expresses the essence as a principle of our knowledge of a being.

[3] Fr. Owens has interpretated this section quite differently. He begins by assuming that *ens* (*esse*) can not be an essential predicate except for God; thus *ens* (*esse*) must be classed an accidental predicate. *Ens*, however, in the sense of essence is an essential predicate for all finite beings. With these assumptions Fr. Owens reads *In IV Meta.*, 2, 554–56, where Thomas criticizes Avicenna. This criticism, Fr. Owens remarks, is valid for Aristotelianism, not for Thomism; hence Aquinas is not giving his own thought in these paragraphs. Consequently Fr. Owens refuses to attempt a reconciliation of paragraphs 554–56 (*ens* is predicated of the essence of everything, and not of an accident added to essence) and paragraph 558 (*esse* is not an accident, although added to essence; *ens* and *res* signify the same things). Owens, "The accidental and Essential Characteristics of Being...", pp. 5–8. M. Gilson presents the same type of interpretation; he maintains that in paragraphs 554–56 Aquinas took the position of Aristotle and hence didn't think at all of *esse*. Thus Aquinas said that *ens* is predicated of the essence of anything. Gilson, *Being and Some Philosophers*, pp. 158–59. My approach

So much for Thomas' objections to the Avicennian conception of *ens*. There is still Avicenna's idea of *unum*, however. To this notion Thomas turns next, and rejects it out of hand. Thomas' argument is simple. Avicenna would have been correct when he spoke of *unum* as predicated of essence because of something added to the essence, if he had qualified this *unum* as "mathematical" unity. Avicenna spoke of mathematical unity, Thomas insists. If such unity is to be convertible with "being," that is, if it is to be a *per se* accident of being, it must be caused by the principles of being as such. But it is not. Hence, it is not convertible with *ens*. It follows then, that Avicenna has not proved that *unum* is predicated of essence because of something added to it.[1]

In beginning this lesson I noted that notions such as *unum* were *communia*. Now we see that Thomas calls *unum* a *per se* accident. Moreover, we note that he explains a *per se* accident as:"... ex principiis causetur entis in quantum ens ...".[2] Thus even though *ens* and *unum* and *res* are predicated of the essence of anything, *unum* and *res* are *per se* accidents. In the light of this doctrine, the overall plan of lessons 1–2 is now clear. Lesson one begins with a proof that metaphysics must study the *per se* accidents of being as well as being itself (par. 529–31; cf. footnotes p. 102). Next Thomas begins to treat being; he says first, that being is divided into the ten categories of substance and accidents (par. 534–46; cf. footnotes pp. 104–108); and then, Thomas explains that all substances are studied (par. 547; cf. footnote p. 108). In lesson two, the topic of the *per se* accidents of being comes to the fore; these accidents are *unum* and *res*. – This is the plan of lesson one and of that part of lesson two which has now been discussed. This plan appears all the more important if one rereads lesson 1, par. 529 to

to these paragraphs is this: there is no apparent reason *intrinsic* to the *Commentary* which permits us to separate any paragraph of this lesson from the others. Hence, I have tried to reconcile paragraphs 554–56 with every other in the lesson. It is only after having completed that reconciliation that I ask whether or not the thought of the lesson is that of Aquinas. – One can learn much in this regard from an examination of several texts brought together by Msgr. Grabmann in an article showing that between 1272–1400 philosophers understood Aquinas to be promoting a real distinction between *esse* and essence. In three of the five texts he presents, the medieval writers knew only the treatment of *esse* and essence given in *In IV Meta., lectio* 2; and all three of them understood it as follows: 1) *esse* is not an accident, yet is other than essence; 2) *ens* is predicated of the essence of anything. The problem for them was, not one of distinguishing between Aquinas' exegesis of Aristotle, and Aquinas the Philosopher, but rather one of understanding how one man could propose these two apparently contradictory doctrines. Cf. M. GRABMANN, *Circa historiam distinctionis essentiae et existentiae*, Excerptum ex Acta Pont. Acad. Rom. S. Thomae Aquinatis, nova series, I, 1934, pp. 3–4 (Siger of Brabent; cf. the text in footnote 4, p. 113); pp. 4–7 (Cod. Vat. lat. 2173); pp. 7–9 (Cod. E 1 152 Bibl. Nat. Florent.).

[1] *In IV Meta.*, 2, 559.

[2] *Ibid.*.

lesson 2, par. 560. The doctrines of these paragraphs are subsumed into a unified view of metaphysics once one realizes that the distinction between being and the *per se* acccidents of being is a distinction between the ten categories of being and the common concepts such as *unum* and *res*.

There is one further point made in lesson two. After having defined his views of "being" and "unity" against the background of the discussion with Avicenna, Thomas turns to "unity" and shows that there is a concept of "unity" corresponding to each of the ten categories of being. Every substance is the "same" as itself; every quantity is the "equal" of itself; every quality is "similar" to itself. Thus we can speak of the ten categories of being and the ten categories of that *per se* accident *unum*.[1]

Thomas does not mention *res* again. But on the basis of his earlier statements, we must accept *res* as also divided into ten categories which are *per se* accidents of being, and hence part of the subject matter of metaphysics.

In the closing paragraph of lesson two, Thomas turns from these more detailed considerations of *unum* back to the structure of metaphysics. Thomas explains now that one can distinguish the parts of metaphysics –called "philosophia" in this context[2] – according to the parts of *ens* and *unum*. By "parts" Thomas means the genera of being – sensible and immaterial substances. The argument is as follows: since every predication of *unum* and *ens* involves a reference to substance and since there are various genera of substance which are ordered to one another, there must be an order in metaphysics corresponding to the order among substances. Yet in the order of exposition (*ordo doctrinae*), one must begin from sensible beings, since they are better known to us. In the order of worth and importance, of course the study of the immaterial substances ranks first. There must be, however, some connection or continuation between these two parts of metaphysics – some common ground, let us say. This is so, for both immaterial and sensible

[1] "...quod ex quo unum et ens idem significant, et eiusdem sunt species eaedem, oportet quod tot sint species entis, quot sunt species unius, et sibinvicem respondentes. Sicut enim partes entis sunt substantia, quantitas et qualitas, etc., ita et partes unius sunt idem, aequale et simile. Idem enim unum in substantia est. Aequale, unum in quantitate. Simile, unum in qualitate. Et secundum alias partes entis possent sumi aliae partes unius, si essent nomina posita. Et sicut ad unam scientiam, scilicet ad philosophiam, pertinet consideratio de omnibus partibus entis, ita et de omnibus partibus unius, scilicet eodem et simili et huiusmodi." *Ibid.*, 561.

[2] In the closing paragraph of this lesson, Aquinas refers to "philosophia" as if that were a name proper to the doctrine contained in Aristotle's *Metaphysics*. Cf. the text in the following footnote.

substances fall under the *genera* of *unum* and of *ens*. Hence, in so far as we consided *unum* and *ens*, we unite the various parts of the science.[1] This last part of the lesson should not actually be included within the discussion of metaphysics as the study of the species of unity. Indeed this discussion is totally foreign to the theme of the lesson: metaphysics studies unity as well as being. The corresponding passage of Aristotle's work appears to have been misplaced. Aquinas of course commented on it here as it is found at the end of the treatment of unity. Two aspects of Thomas' exposition of this final passage are extremely significant. First, to maintain some kind of continuity with what has preceded, Thomas has to use the words "partes entis et unius" in two completely different meanings:

> Hic concludit quod philosophi est considerare de partibus unius, sicut de partibus entis. Et primo hoc ostendit. Secundo etiam ostendit, quod secundum diversas partes entis et unius, sunt diversae partes philosophiae, ...[2]

Thomas' words: "Et primo hoc ostendit," refer to a proof that there are ten categories of "unity" just as there are ten categories of "being"; thus, "parts" in this sense means categories. But the words: "Secundo etiam ostendit...," refer to a proof that there are various genera of beings and of "ones," namely the genera of material and immaterial beings; here then, "parts" means genera of substances – a division of what is found in the first of the categories. Thus, to attach this last section of Aristotle to lesson two, Thomas has to resort to employing the same word in totally different senses. A second aspect of Thomas' exposition is just as significant; by attaching the discussion of unity as broken down into genera such as materially "ones" and immaterially "ones" to the discussion of unity as divided into ten categories, Thomas has paralleled his exposition of being in lesson one. In lesson one, it will be remembered, Thomas first explained that metaphysics studies accidents as well as substance (metaphysics studies the ten different species or categories of unity); and then he showed that metaphysics stu-

[1] "...tot sunt partes philosophiae, quot sunt partes substantiae, de qua dicitur principaliter ens et unum et de qua principalis est huius scientiae consideratio et intentio. Et, quia partes substantiae sunt ordinatae adinvicem, nam substantia immaterialis est prior substantia sensibili naturaliter; ideo necesse est inter partes philosophiae esse quamdam primam. Illa tamen, quae est de substantia sensibili, est prima ordine doctrinae, quia a notioribus nobis oportet incipere disciplinam: et de ha [sic.] determinatur in septimo et octavo huius. Illa vero, quae est de substantia immateriali est prior dignitate et intentione huius scientiae, de qua traditur in duodecimo huius. Et tamen quaecumque sunt prima, necesse est quod sint continua aliis partibus, quia omnes partes habent pro genere unum et ens. Unde in consideratione unius et entis diversae partes huius scientiae uniuntur, quamvis sint de diversis partibus substantiae; ut sic sit una scientia inquantum partes praedictae sunt consequentes "hoc," id est [sic.] unum et ens, sicut communia substantiae." *Ibid.*, 563.

[2] *Ibid.*, 561.

dies all the species of being or substance which fall under the genus "being" or "substance" (metaphysics studies all the substances falling under the genera of "being" and "unity").

Before we close our discussion of this lesson, let us note that the fact of studying *unum* explains why metaphysics studies the *communia* such as "negation" and "privation" and all the other contraries. As Thomas writes, all the contraries are reduced to *unum*; hence, metaphysics must study them.[1] As Thomas understands this doctrine, *unum* and *multitudo* are the basic contraries; all others are reduced to them. Moreover, *multitudo* is itself reduced to *unum* since multitude is simply a privation of unity.[2] All of these contraries, and their reduction to "one" is treated in Book X, Thomas concludes.[3]

As should now be clear, the key to lesson two is contained in the opening paragraphs of lesson one. A science must not only study its subject, but the *per se* accidents of its subject as well. It is for this reason that *unum* and *res* must be studied. Anything which is caused by the principles of *ens* as such, and yet which is not expressed by the concept "being" – such a thing must be studied by metaphysics in addition to its study of being. Because it is caused by the principles of being as such, such a thing is a *per se* accident, or an essential attribute, of being as such.[4]

Before we turn our attention away from Thomas' writings, let us note how Thomas has employed in lesson two the four doctrines comprising his theory on the object of metaphysics. In this lesson Thomas speaks of the ten categories of *unum* which correspond to the ten categories of being; in addition, he speaks of the concepts of *unum* and of *ens* which are the genera of material and immaterial substances. In speaking thus, Thomas employs what I have called, respectively, the "horizontal" and the "vertical conceptions of being." Then too,

[1] "Et ad hoc 'principium', scilicet unum, reducuntur omnia contraria 'fere': Et hoc addit quia in quibusdam non est ita manifestum. Et tamen hoc esse necesse est..." *Ibid.*, 561–62. Also cf. *In XI Meta.*, 3, 2198.

[2] "...in omnibus contrariis alterum habeat privationem inclusam, oportet fieri reductionem ad privativa prima, inter quae praecipue est unum. Et iterum multitudo, quae ex uno causatur, causa est diversitatis differentiae et contrarietatis...". *Ibid.*, 562. This is explained in detail in *In IV Meta.*, *lectio* 3.

[3] *In IV Meta.*, 2, 562.

[4] In the introduction to the present chapter, I noted that metaphysics is the study of *unum*, of *res*, as well as of *ens*. We now understand the point of this theory: since the "unity" and "thingness" of an object are caused by the object *qua ens*, and since the "unity" and the "thingness" are not expressed in the concept *ens*, then the metaphysician must devote a section of his science to the study of *unum* and *res*. Supposedly this section will come after the study of *ens*; but will this study evolve proofs for God based on the unity and the thingness of creatures? This is a question for Chapter VI.

Thomas speaks of these genera of *unum* and *ens*, which express what is common to all substance; thus he uses language asking to be understood as implying that metaphysics has an univocal point of view. The fourth doctrine was also present: metaphysics is the study of concepts. These concepts are called *communia*; the concept basic to many of the *communia* appears to be *unum*. Moreover, *unum*, as *ens*, is divided horizontally into ten categories, and vertically included both material and immaterial substances.

This then is Aquinas' teaching on the object of metaphysics. The metaphysician *apparently* takes an univocal point of view, from which he forms his concept of being; yet this concept expresses the totality of everything, it is predicated of the essence of everything! Moreover, this concept is such that it can be predicated of substance; but it must also be such that it is involved in the concepts expressing the nine types of categorical accidents; even more, it is such that it can be predicated of every substance and of every accident falling under one of the categories. Not only does the metaphysician study being, however: he must investigate as well all the essential attributes of being as such. At the root of many of the concepts expressing these attributes, one has the concept of *unum*; yet one must not forget the concept of *res*: one suspects that *res* too will give rise to a number of *communia*.

These concepts, *ens*, *unum*, and *res*, are not defined by Thomas. All of them are predicated of the essence; thus all of them express, in some way, the essence of anything. *Ens* expresses the essence, but in some unexplained fashion, relates essence to *esse*. *Unum* and *res*, in like manner, express the essence, but in relation to indivision and quiddity, respectively.

What exactly is the relation between these three concepts? All of them express the essence of a thing, to be sure. But whereas *ens* expresses the essence *qua* being, the other two concepts express some "*per se*" or "necessary" aspect of the essence. These essential aspects are constituted by the principles of being as such, Thomas says. What are the principles of being as such? First of all, the most obvious principle is essence, to judge from the two lessons examined. Is *esse* also a principle? One wonders at first if *esse* must not be ruled out of court, for Thomas writes: *esse* "... quasi constituitur per principia essentiae" (par. 558). Yet this formula is not the same as: "... per se accidens entis, oportet quod ex principiis causetur entis inquantum ens, sicut quodlibet accidens proprium ex principiis sui subiecti." (par. 559). In addition two types of texts found throughout the *Commentary* distinguish between *esse* and accidents (in the sense of essential attributes). First of all, Thomas notes often that accidents exist in a subject;[1] thus to have an accident, one needs a subject, an *ens*. It is from the principles of the complete being that accidents are caused. On the other hand, Thomas notes that complete being results from the composition of two principles;[2] moreover, an *ens* is composed of essence and of another principle.[3] Hence, it is clear that *esse* and essence are the two principles whose composition results in the possibility of having accidents. Thus because *unum* is caused by the composite of *esse* and essence, and because *unum* expresses essence in relation to indivision, so *unum* expresses a *per se* accident, or an *essential* aspect of being as such. If, for any being, *unum* expresses essence in relation to indivision, it would follow that any essence, *qua* real, has some relation to indivision. But the essence is real in so far as it is joined, or united, or even: related, to the other principle of a composed being – *esse*. Thus, given the composition of essence and *esse*, one would have the *per se* or necessary aspect expressed as *unum*. One would have as well that aspect expressed as *res*. Hence the reality expressed by *ens* founds the necessary or *per se* aspects expressed by *res* and *unum*. Now, we can ask again our question: What is the relation between the concepts *ens*, *res*, and *unum*? Obviously it must parallel the relation between the *realities* expressed by the three concepts. When one has the composition of *esse* and essence, one has the relation of essence to indivision (expressed by *unum*) and the relation of essence to quiddity (expressed by *res*). Thus, one can conclude that *res* and *unum* suppose the concept *ens*. *Res* and *unum* must express *ens*: essence *qua* real; that is, essence *qua* related to *esse*, which is the content of *ens*, must be expressed by *res* and *unum*. Thus, *res* expresses the content of *ens*, but relates that content to quiddity; *unum* expresses the content of *ens*, but relates that content to indivision.

2. ARISTOTLE, AVERROES, ALBERT: THE PREDECESSORS OF AQUINA'S *IN IV META.*, *LECTIO* 1

We have viewed Thomas' exposition in two lessons where the object of metaphysics is discussed. It is necessary to turn now to his predecessors: Is Thomas' thought faithful to the Aristotelian *Metaphysics*?

[1] E.g.: "Et ad hoc [genus substantiae] sicut ad primum et principale omnia alia [entia, seu accidentia] referuntur. Nam qualitates et quantitates dicuntur esse, inquantum insunt substantiae..." *In IV Meta.*, 1, 543.

[2] "Invenitur siquidem et in rebus aliqua compositio; sed talis compositio efficit unam rem, quam intellectus recipit ut unum simplici conceptione." *In VI Meta.*, 4, 1241.

[3] That it has essence is obvious; e.g. *In IV Meta.*, 2, 554; 558. That it has *esse*: "Alia dicuntur entia vel esse, quia per se habent esse sicut substantiae...", *In IV Meta.*, 1, 539.

Does Thomas' dependence upon the work of Avicenna, Averroes, and Albert reveal anything of the nature of the *Commentary*? It is to answering these questions that our attention must now be directed. First, let us compare Thomas and Aristotle, noting the conformity or lack of it between the two philosophers. Then after this comparison, let us turn to the *Commentaries* of Averroes and Albert to discover the extent to which they influenced Thomas. The *Commentaries* of Averroes and Albert, like that of Thomas, are divided into sections which roughly correspond to divisions in Aristotle's text. Hence, Thomas could and did, as we shall see, refer constantly to the explanations of these two forerunners. Avicenna's work, on the other hand, was given a much more personal form; the order of his books and chapters does not closely follow that of Aristotle. Thus, whereas Averroes and Albert served as constant guides to the meaning of Aristotle, Avicenna aided rather by the value of his doctrines. Hence in the present as in the following section, I shall compare the expositions of Averroes and Albert to that of Thomas; and only afterward in a final section, shall I discuss the Avicennian doctrines used or opposed by Thomas in his *Commentary*. (Since we are interested in Avicenna, Averroes, and Albert only in so far as necessary to grasp Aquinas' intention, in Chapter IV–VI I shall not enter into all aspects of their thought.)

A. Aquinas and Aristotle

Thomas gave a three-fold division to his exposition of lesson 1 (Γ1, 1003a 21–2, 1003b 22): 1) metaphysics studies both being and its essential attributes; 2) metaphysics studies accidents as well as substance, for we call them both "being"; yet metaphysics studies primarily substance; and 3) metaphysics studies all substances by gathering them into the genus of "being." When one examines the Aristotelian passage corresponding to Thomas' exposition, one perceives immediately that Aristotle has treated the same three points.[1] In addition, one can not but see that Aristotle explains the second and the third of these points exactly as Thomas does.[2] But what

[1] The *Vetus* is found in *Opera hactenus inedita Rogeri Baconi*, Fasc. XI. The correspondance of Thomas and Aristotle is as follows. 1) Metaphysics studies being and the essential attributes of being as such: *In IV Meta.*, 1, 529–33; Γ 1, 1003a21–22: *Vetus*, p. 303, l. 1–12. 2) Metaphysics studies both accidents and substance: *In IV Meta.*, 1, 534–46; Γ 2, 1003a33–b19: *Vetus*, p. 303, l. 13–p. 304, l. 3. 3) Metaphysics studies all substances as falling under the viewpoint of "being": *In IV Meta.*, 1, 547; Γ 2, 1003b19–22: *Vetus*, p. 304, l. 4–7.

[2] Aristotle explains as follows the third point which implies that "being" will be an univocal concept: "Omnis autem generis et sensus unus unius et sciencia est, ut gramatica, una cum sit, omnes considerat voces. Unde et ipsius quod est secundum quod est quascumque species

of the first point? Does Aristotle distinguish between the "essential attributes" of being and the "categorical accidents"? In addition, does knowledge of being and of these essential attributes represent a more common knowledge in relation to the less common knowledge we have of particular beings and of their essential attributes? These doctrines were present in Thomas' *In IV Meta.*, 529–33; are they found in Aristotle's Γ 1, 1003a 21–32?

There can certainly be no doubt that the "essential attributes" of being are distinguished from the "categorical accidents" in Aristotle. Thomas' exposition bears witness that the *Vetus* speaks of the sicence which studies both being and "ea quae insunt enti per se."[1] Now although Thomas may explain these "ea quae insunt enti per se" by his words "idest entis per se accidentia,"[2] there is nothing in the *Vetus* that would suggest the use of the phrase "per se accidentia" to refer to the essential attributes of being. [3]It appears then, that Thomas chose to translate Aristotle's expression as "per se accidentia"; but in the Aristotle Thomas read there is certainly no linguistic expression which might lead one to confuse these "ea quae insunt enti per se" with the categorical accidents.

On the other hand, there are several lines in Aristotle's Book Γ which would positively indicate these essential attributes are not to be identified with the categories of accidents. For example, Thomas could read in Γ2 that the study of substance must also study all the things such as "unity," "multitude," and so on; this doctrine, Aristotle notes, is the answer to one of the apories.[4] When one turns to that aporie to

speculari unius est sciencie genere, et species specierum." Γ 2, 1003b19–22: *Vetus*, p. 304, l. 4–7. M. Colle notes that here in the last line of this passage one has an introduction to the doctrine Thomas has placed in lesson 2: "unity" also is to be studied. Hence Aristotle is not speaking here either of the categories or of the different substances. The question whether all the substances are to be studied, and whether all accidents as well are to be treated, was solved in speaking of the predication of "being". G. Colle, *La Métaphysique. Livre IV.* Traduction et Commentaire, Vrin, Paris, 1931, pp. 47–48. – In so far as M. Colle decided on grounds of Greek usuage that this final sentence belongs with what follows – the discussion of "unity", I have no quarrel to pick. But I can not agree that the discussion of the predication of "being" solves the problem of whether all substances can be studied. For a view similar to the one presented here, cf. Decarie, *L'objet de la métaphysique selon Aristote*, p. 103, footnote 1; note the quote from Siger of Brabant which substantially agrees with Aquinas' thought.

[1] The *Vetus*, p. 303, l. 1–2 reads: "Est sciencia quedam que speculatur ens secundum quod est ens et que huic insunt per se." Aquinas' words are in *In IV Meta.*, 1, 529.

[2] *Ibid.*, 529.

[3] Nor did Thomas read this phrase "per se accidentia" in the *Arabica*, nor in the *Media* (which gives the same translation as the *Vetus*). For the *Arabica*, cf. *Aristotelis Metaphysicorum ,Libri XIIII...*, fol. 63v, M. For the *Media*, cf. Alberti Magni, *Metaphysica...*, p. 161.

[4] "Manifestum igitur quod vere in *Oppositionibus* dictum est, quoniam unius est de his et substantie est rationem habere (hoc autem erat unum eorum que sunt in *Oppositionibus*),

which Aristotle is referring, one notes that the *Vetus* (as is true of Thomas in his exposition of the aporie) refers to the possibility that metaphysics will consider the essential attributes (the "que huic insunt per se") such ad *idem, alterum, simile,* etc. Such essential attributes, moreover, are referred to in the *Vetus* as "accidentia secundum seipsa substantiis."[1] Thus, not only do we see why Aquinas speaks of the essential attributes as *per se accidentia* in Book IV, but also we see that for Thomas it would be evident that Aristotle uses "que huic insunt per se" to refer to something other than the nine categorical accidents.

But does Aristotle characterize knowledge of being and of its essential attributes as "more common knowledge" in relation to knowledge of a particular being and of its essential attributes? Does Aristotle recognize this "more common knowledge" as necessary for the less common? Of course such a comparison is not in the Aristotelian passage which corresponds to Thomas' words.[2] Thomas' exposition of these points in paragraphs 529–31 is more a synthesis and a summary of Book Γ 1–3 than an exposition of Aristotle's words given in the initial lines of Γ 1. What Thomas has done is to draw freely on what comes later in Aristotle. For example, Thomas refers to geometry to clarify his point, setting up the following proportion: geometry is related to the study of three angles, just as metaphysics is related to the essential attributes of being as such.[3] Much the same comparison is made by Aristotle in a passage occuring a few lines later.[4] Moreover, Aristotle in yet other places implies even more clearly that metaphysics is knowledge of the most common aspects of being, while other sciences study the more particular aspects;[5] thus the implication:

et est philosofi de omnibus posse speculari." Γ 2, 1004a31–34: *Vetus*, p. 305, l. 25–28. The concepts referred to are "unity", "plurality", etc. Cf. Thomas' exposition in *In IV Meta.*, 3, 569.

[1] "...et utrum de substantia speculatio solum sit aut et circa accidencia secundum seipsa substantiis. Ad hec autem, de eodem et altero et simili... et aliis omnibus huiusmodi... cuius sit considerare de omnibus?" B 1, 995b18–27: *Vetus*, p. 286, l. 32–p. 287, l. 4. Cf. *In III Meta.*, 2, 352–54.

[2] "Est sciencia quedam que speculatur ens secundum quod est ens et que huic insunt per se." Γ 1, 1003a21–22: *Vetus*, p. 303, l. 1–2. Cf. *In IV Meta.*, 1, 529–31.

[3] "Dicit etiam 'et quae huic insunt per se' et non simpliciter quae huic insunt, ad significandum quod ad scientiam non pertinet considerare de his quae per accidens insunt subiecto suo, sed solum de his quae per se insunt. Geometra enim non considerat de triangulo utrum sit cupreus vel ligneus, sed solum considerat ipsum absolute secundum quod habet tres angulos aequales etc. Sic igitur huiusmodi scientia, cuius est ens subiectum, non oportet quod considerat de omnibus quae insunt enti per accidens..." *In IV Meta.*, 1, 531.

[4] "Hec autem est neque una que in parte dicuntur ipsa; nequa una enim aliarum considerat universale de eo quod est in quantum est. Sed partem ipsius dividentes circa hanc speculantur accidens ut mathematica scienciarum." Γ 1, 1003a22–26: *Vetus*, p. 303, l. 2–6. Cf. *In IV Meta.*, 1, 532.

[5] Cf. the text of footnote 2, p. 121: Γ 2, 1003b19–22.

the knowledge of the particular sciences presupposes the more common knowledge of metaphysics. Then too, when Aristotle speaks of the propositions or axioms or "truths which clearly hold good for all things *qua* being," he notes that metaphysics studies these axioms in all their generality, and that all other sciences use these axioms in particular applications; that is, metaphysics studies these truths in so far as they express something common to all beings as such, while the particular sciences subsume this common aspect into the context of a particular genus of being.[1] Thomas understood these axioms to be the first principles of demonstration, the first of which is the principle of non-contradiction.[2] (Such an interpretation is not an unjustified one, given the contents of the remaining chapters of Aristotle's Book Γ).[3] In a yet later passage, in Book K, these first principles of demonstration come to the fore again, when Aristotle explains that physics and mathematics are "parts" of metaphysics because the latter science studies the first principles universally, whereas the lower sciences only make a special application of them. An example of a principle or axiom common to all quantity is given: "when equals are taken from equals the remainders are equal." Yet mathematics does not study quantity *qua* beng, and hence does not use this axiom in all its universality.[4] It is important to note that "equal in quantity" is given by Thomas as a species

[1] "Dicendum autem est utrum unius aut alterius sciencie sit de qua in mathematicis vocatis dignitatibus, et de substantia. Manifestum igitur est quoniam uniusque sciencie et que philosofi, et que de his est intencio; omnibus enim inest que sunt, set non generi alicui seorsum proprie ab aliis. Et utuntur quidem omnes, quoniam entis est secundum quod est ens, unumquodque enim genus est ens. In tantum enim utuntur in quantum illis sufficiens est; hoc autem est, quantum subit genus de quo ferunt demonstrationes. Quare quoniam manifestum est quoniam secundum quod sunt encia insunt omnibus (hoc enim ipsis commune est), circa ens secundum quod est ens cognoscentis et de his est speculatio." Γ 3, 1005a21–29; *Vetus*, p. 307, l. 14–26. *In IV Meta.*, 5, 588–91 reveals that Thomas understood these lines as we have explained them, just as one would expect, given the doctrine of the "common" aspect of metaphysical knowledge exposed in *In IV Meta.*, 1, 529–33.

[2] Cf. *In IV Meta.*, 5, 588 where Aquinas notes that the Aristotelian passage quoted in footnote 1 is a proof that metaphysics treats of all these axioms; the rest of Book Γ treats each axiom individually, says Thomas

[3] Ross, *Aristotle's Metaphysics*..., Vol. I, p. 263.

[4] "Since even the mathematician uses the common axioms only in a special application, it must be the business of first philosophy to examine the principles of mathemathics also. That when equals are taken from equals the remainders are equal, is common to all quantities, but mathematics studies a part of its proper matter which it has detached, e.g. lines or angles or number or some other kind of quantitiy – not, however, *qua* bieng but in so far as each of them is continuous in one or two or three dimensions... Physics is in the same position as mathematics; for physics studies the attributes and the principles of the things that are, *qua* moving and not *qua* being (whereas the primary science, we have said, deals with these, only in so far as the underlying subjects are existent, and not in virtue of any other character); and so both physics and mathematics must be classed as parts of Wisdom." K 4, 1061b17–33. For Thomas' understanding of this passage, cf. *In XI Meta.*, *lectio* 4.

of "unity" and this would appear to be Aristotle's mind;[1] hence, the axiom mentioned in Book K involves, as subject and as predicate one of the essential attributes of being. Here then in Book K is an example of how metaphysics, by studying the essential attributes, acquires the most common knowledge; it is an example, too, of how this most common knowledge is subsumed into a particular or less common context. Thus, Aristotle does give, although not in the most unified manner, the doctrine which Thomas attributes to him in the opening paragraphs of *In IV Meta., lectio* 1.

Thomas' exposition in this first lesson must then be viewed as extremely faithful to Aristotle's thought. Let us adopt a terminology which is rapidly becoming accepted: Thomas gives in his *In IV Meta., lectio* 1 the literal menaing of Aristotle, the *verba Aristotelis*. True, the meaning given by Thomas is not completely found in the Aristotelian passage which opens Book Γ; nevertheless, the doctrine given by Thomas is stated elsewhere by Aristotle. Thus Thomas has done little but synthesize Aristotle's though on the object of metaphysics.[2] However, if we recall Thomas' doctrine in lesson two, we must admit that the *ens* of lesson one is to be understood as an essence-*esse* composite. And hence Thomas' exposition in lesson one, while actually being a repetition of the *verba Aristotelis*, was a repetition of a *verba* so thoroughly transformed by the introduction of the doctrine of *esse*, that Thomas' doctrine is more a Thomist meaning *ex sensu proprio* than *verba Aristotelis*. But more about this after we have finished our consideration of lesson two and its predecessors.

B. Aquinas versus Averroes and Albert

When one turns from Thomas to Averroes and Albert, one notices immediately that Thomas depended heavily on the *Commentaries* of

[1] For Aristotle: Γ 2, 1004a17–20. For Thomas, cf. *In IV Meta.*, 3, 567; note earlier reference to "aequale, unum in quantitate": *Ibid.*, 2, 561.

[2] Among contemporary students of Aristotle, several would agree that Aquinas is basically correct in his interpretation of this passage; cf. DECARIE, *L'objet de la métaphysique selon Aristote*, pp. 100–108; A. MANSION, "Philosophie première, philosophie seconde et métaphysique chez Aristote," *Revue Philosophique de Louvain*, LVI, 1958, pp. 179–202; esp. p. 180. Others however maintain that to know being as such one studies the first or separate substance; cf. ROSS, *Aristotle's Metaphysics...*, Vol. I, pp. 251; lxxvii–lxxviii; J. TRICOT, *Aristote. La Métaphysique*, nouvelle édit., Vrin, Paris, 1953, pp. 171–79; OWENS, *The Doctrine of Being...*, pp. 147–57; P. MERLAN, *From Platonism to Neoplatonism*, 2nd revised edit., Nijhoff, The Hague, 1960, pp. 161–70 and 205. Naturally the question of the historical position of Aristotle's thought is entirely outside of the scope of this present study. To us of interest is the question whether Thomas could have legitimately interpretated Aristotle as he did; on this point, one occasionally comes across an author who admits the legality of Thomas' position, e.g. MERLAN, *op. cit.*, p. 169.

his two predecessors. For example, one is forced to admit that Thomas modeled his exposition of the predication of "being" on that of Averroes.[1] Then too, one is faced with a close dependence of Albert on Averroes.[2] Yet all the details of these complex relations of dependence need not detain us. Our interest in Averroes and Albert (as in Avicenna) extends only to those aspects of their works which aid to us to determine the nature of Thomas' *Commentary*. Rather than try to explain what I mean, let us simply turn to a study of the predecessors of Thomas; this point will be much more clearly seen than if I pause here to explain it.

The opening section of Thomas' *In IV Meta., lectio* 1 corresponds to Γ 1 of Aristotle. In these passages both Thomas and Aristotle explain, as we have seen, that metaphysics studies both being as such and the essential attributes of being as such. One can easily recognize that Thomas was but following the footsteps of Averroes and Albert when he separates the discussion of Γ 1 from that of Γ 2; for all three commentators, this initial section deals with the subject of metaphysics. Yet agreement between the three scarcely exceeds this general view of Γ 1.

According to Averroes, metaphysics (or "philosophy") seeks the ultimate principles of being as such. Thus, although it studies sensible beings, it studies them simply as being, and not *qua* mobile.[3] Yet Aristotle's first philosophy, Averroes writes, is the study only of the most noble causes, that is of the formal and final causes of being. This is so because, if metaphysics studies all beings as being, it can study

[1] Compare Thomas' *In IV Meta.*, 1, 534–44 with Averroes' *In IV Meta.*, c. 2, fol. 65r, E sqq; note particularly the examples which Thomas has taken from Averroes. – As far as I can determine, there has been no published study on Averroes' "great Commentary", the one with which we deal here. The most we have is the brief treatments of one or two aspects of the relation between Averroes and Thomas as critics of Avicenna. E.g. A. Forest, *La structure métaphysique du concret selon saint Thomas d'Aquin*, Études de Philosophie Médiévale XIV, Vrin, Paris, 1956, pp. 41sqq; 142 sqq.

[2] This appears more clearly in their expositions corresponding to Thomas' *In IV Meta., lectio* 2. Note for example Averroes' attack on Avicenna in *In IV Meta.*, c. 3, fol. 67v, G – the discussion of the "infinite regress" needed if Avicenna is correct; the same argument is adopted, though in a slightly different context, by Albert: *In IV Meta.*, T. 1, c. 4, p. 166, l. 40–58. – Although there are partial studies of the relation between Averroes 'and Albert's theories of intellect, there exists no examination of the relations between the *Commentaries* on the *Metaphysics* composed by the two philosophers. And it might be added, there is no over-all study of Albert's metaphysics, other than a very brief sketch: cf. De Raeymaeker, "Albert le Grand, Philosophe...," pp. 5–36.

[3] "...cum manifestum est, quod nos quaerimus in hac arte principia ultima, quae sunt principia simpliciter, manifestum est, quod quaerundum est ista principia istorum principiorum sensibilium secundum istum modum, secundum quem quaerimus principia naturae existentis per se. Et intendit, quod principia simpliciter debent quaeri de entibus, quae sunt simpliciter; et, si acciderit quibusdam, ut sint sensibiles non simplices, tunc non quaeruntur ista eis principia, nisi secundum quod sunt entia, v.g. mobilia aut mathematica." *In IV Meta.*, c. 1, fol. 64r, E–F.

only those causes found in all beings. Thus, because all beings are not of the same genus, because they do not all have each of the four causes, all the causes cannot be studied in the science of being as such.[1] This doctrine, Averroes explains, was expressed by Aristotle in Book B (III).[2] Moreover, besides studying being as such, metaphysics studies the essential attributes (the essential accidents) of being *qua* being.[3]

According to Averroes, Book B presents the theory of metaphysics as the study only of the formal and final causes. Apparently, Averroes refers to Aristotle's discussion of the first aporie: "Does one science study all the species of causes?" In his commentary on that passage, Averroes gives the impression that he understood Aristotle to prove that no science can study all four causes.[4] Examining Averroes' exposition more closely, one sees that natural philosophy has the task of studying two of the first causes: the agent or motor cause, and the material cause.[5] Metaphysics, on the other hand, studies the formal cause, and the final cause to which all other causes are ordained.[6] By "formal cause" Averroes understands the substance of a thing. Thus in so far as one knows the characteristics common to any "formal cause," and in so far as one knows the end for which every such cause is had, one knows being as such.[7]

This then is how Averroes understood Γ1 (to which corresponds

[1] "...iam perscrutatus est in tractatu praecedenti de scientia, qua edicitur Philosophia, quae scientia sit, et dixit quod, si scientia omnium causarum est una, illa scientia debet dici Philosophia: deinde declaravit, quod consideratio de omnibus causis omnium entium non est unius scientiae, cum entia sint diversa, quoniam in quibusdam inveniuntur de causis quatuor, quaedam non inveniuntur in quibusdam; hoc enim esset possibile, si entia essent unum genus: deinde induxit ipse sermonem, quod Philosophia forte est illa quae considerat de nobilissima causarum, scilicet de proprio fine, et forma propria..." *Ibid.*, fol. 64r, C–D.

[2] Note the reference to "tractatu praecidenti" in the preceding footnote; this reference governs the entire doctrine expressed in that citation. This becomes clearer when one examines the words which immediately follow the citation: "...incoepit in hoc tractatu declarare hoc secundum demonstrationem..." *Ibid.*, fol. 64r, D.

[3] *Ibid.*.

[4] "Deinde dixit impossibile est enim, ut species causarum, etc., idest et necesse est, ut quaedam scientiae sint proprie ad dandam aliquam causam sine alia, quoniam non quodlibet genus, de quo scientia considerat, habet quatuor causas... Et intendit, quod hoc non invenitur nisi in scientia Naturali tantum... sed si concesserimus, quod omnis scientia dicitur Philosophia, tamen necesse est, ut scientia, quae dicitur Philosophia simpliciter, sit illa, quae considerat in causa finali ultima omnium entium. Omnes autem causae sunt propter istam causam... illud quod dicitur Philosophia, est quae notificat cum causa finali prima causam primam, quae est forma, et substantia... Res enim ut dixit, notificatur multis modis, et magis quod notificatur est per suam substantiam." *In III Meta.*, c. 3, fol. 41v, G–I.

[5] "Scientia enim Naturalis considerat entia de duabus primis causis, scilicet motore, et materia..." *Ibid.*, fol. 41v, I–K.

[6] "...ista autem (scientia seu metaphysica) de duabus causis ultimus, scilicet forma et fine... Finis autem melior aliis causis, quia aliae causae sunt propter illam." *Ibid.*, fol. 41v, K. Also, *Ibid.*, c. 4, fol. 42v, H.

[7] "Res enim, ut dixit notificatur multis modis, et magis quod notificatur est per suam substantiam... Et signum eius, quod scientia substantiae est perfectior omni scientia, est

Thomas' *In IV Meta.*, 1, 529–33). If we turn now to Albert's *In IV Meta.*, T. 1, cc. 1–2, we discover a certain similarity between Albert and Averroes, as well as a difference. Albert explains that metaphysics does not study in an equal way all four causes, since *ens inquantum ens* does not have all four causes. Only mobile beings have all the causes. Being as such is, then, studied in so far as one considers the ultimate end and the first form of being.[1] Of course, the science of being as such, beside investigating its subject, must also study the essential "passiones et praedicta" of being as such.[2]

Albert, as was true also of Averroes, believed that Aristotle's discussion of the first aporie proved that metaphysics studied only two of the first causes.[3] Furthermore one sees a certain influence of Averroes on Albert in the latter's discussion of the first aporie. As if he were depending heaily on Averroes, Albert notes that physics studies the moving and material causes, while metaphysics deals only with the ultimate end and the form.[4] The final cause, Albert says, is that to which all objects and all causes are ordained. Thus the study of the final cause is the highest science.[5] Still in agreement with Averoes, Albert brings the formal cause, or substance, into the consideration of metaphysics. If we consider metaphysics as knowledge of that cause which makes us know a thing to the greatest extent, then metaphysics must be said to study the form or the substance of being as such.[6]

quod cum nos quaesiverimus scire aliquid per demonstrationem, non reputamus ipsum esse scitum, nisi quando nos intellexerimus in nobis, quod scimus illud secundum quid, idest per suam substantiam, et definitionem." *Ibid.*, fol. 41v, I–L.

[1] "...iam dixerimus in praemissis non omnis entis esse omnes causas, quia si omnes causae aeque principaliter essent omnis entis, oporteret, quod omne ens esset eiusdem naturae cum ente transmutabili, in quo conveniunt omnes causae... Diximus autem in praehabitis, quod sapiens, qui dicitur philosophus, est doctissimus causarum; principaliter autem cognoscit finem ultimum et formam primam entis, secundum quod est ens." *In IV Meta.*, T. I, c. 1, p. 161, l. 18–28.

[2] "...scientia quaedam una est et eadem, quae speculari habet ens, inquantum est ens, ut subiectum, et eadem speculari habet eas passiones et praedicata, quae huic enti insunt secundum se sive essentialiter, secundum quod est ens..." *Ibid.*, p. 161, l. 48–52.

[3] Of the doctrine that metaphysics studies the formal and final causes, Albert writes: "Et hoc quidem sine demonstratione dictum est in praehabitis, nunc autem demonstrare intendimus, quod est quaedam scientia quae per extrema et prima principia formae primae et finis ultimi considerat ens, inquantum est ens..." *Ibid.*, p. 161, l. 28–33. For a similar view in Averroes, cf. footnote 1, p. 127. Albert, as Averroes, earlier had expressed this doctrine in commenting on the first aporie: cf. Albert: *In III Meta.*, T. II, c. 1, p. 111–14; Averroes: *In III Meta.*, c. 3, fol. 40v, M–fol. 41v, M.

[4] *In III Meta.*, T. II, c. 1, p. 112, l. 51–53.

[5] "Inquantum enim causa considerata in scientia senior et prior, intantum est principalior et ipsa et scientia, quae eam considerat... Talis autem juste est, quae est ultimi finis et boni eius quod totius universitatis est bonum... Ad hoc enim famulatur efficiens et materia et forma omnia, et ad hoc bonum ordinatur omne quod est..." *Ibid.*, p. 113, l. 7–17.

[6] "Inquantum vera haec eadem sapientia primarum est causarum,... et inquantum diffinita est esse eius quod maxime scibile est secundum suam naturam, maxime etiam faciens

Yet despite the affirmation that metaphysics studies only the formal and final causes in so far as it studies being as such, Albert attributes to metaphysics the study of the efficient and material causes as well, and thus parts company with Averroes. In so far as one studies being as such, one principally studies only the formal and final causes. Yet metaphysics is going to study all beings which fall under the genus of its subject. Thus it studies both caused beings and their cause, the moving or efficient cause; thus too, it must study matter, since one can reduce material being to being as such. In short all those causes which enter into a relation with the final cause of beings are studied.[1]

Apparently Albert is thinking in terms of a distinction between 1) the value of being as such, and 2) the beings which verify that value. To study the value of being as such, one need study the form of substance, for knowledge of this form is the highest knowledge of what is common to all beings; moreover, since the formal cause is related to the ultimate end, the study of the formal cause must take into account this end. On the other hand, Albert says that this science wants to study all beings. Since some of these beings are material, and one of these beings is the first mover, one must take these material and efficient causes into account when one applies the knowledge of being as such to each being.[2]

Thus far I have noted the doctrines given by Averroes and Albert in regard to Aristotle's theory of metaphysics as the study of being as such. As should be evident both Averroes and Albert, in so far as they referred to the causes studied, have given several doctrines not present in Aristotle's Γ1. I have already noted that the essential points of Thomas' exposition of Γ1 correspond to Aristotle's thought. Yet in the discussion of Thomas' work I also noted that he had earlier explained that metaphysics studies all four causes; and I have not yet asked whether this theory represented an Aristotelian doctrine. Instead I have thus far been content to point out that Thomas was giving

scire, sic etiam post considerationem ultimi finis erit etiam considerativa substantiae, quae est forma et quidditas rerum." *Ibid.*, p. 113, l. 21–27.

[1] "Considerat haec scientia etiam efficentem, sed non ita principaliter, sed considerat moventia prima et facientia. Et considerat ultimo loco materiam secundum reductionem in substantiam et ens. Et considerat omnes causas, eo quod est de ente; et id quod consequitur ad ens, inquantum est ens, est causa vel causatum, quia omne ens vel est causa vel causatum. Et sic prima quaestio est determinata, quia causa et causatum sunt generis subiecti huius scientiae, quod est ens. Nec tamen sic uniuntur tantum, quae sunt unius scientiae, sed potius eiusdem scientiae sunt, quae sunt ad finem ultimum unum, qui est finis illius scientiae... ista scientia est de causis omnibus ultimo fini famulantibus." *Ibid.*, p. 113, l. 56–73.

[2] In addition to the preceding footnote see *Ibid.*, p. 113, l. 78–p. 114, l. 13.

the *verba Aristotelis* when he explained in his lesson 1) that metaphysics is the study of being as such, and the study of the essential attributes of being; and 2) that metaphysical knowledge is more common knowledge in relation to the less common knowledge of the particular sciences. However, now that a certain lack of conformity has been disclosed between Averroes, Albert and Aquinas on the subject of the causes studied in metaphysics, it is time to ask if any one of these three commentators truly represents Aristotle's view. Let us review the thought of the three before returning to Aristotle.

Averroes is careful to note that matter and the agent cause are studied by the physician and only by him. The metaphysician, he declares, limits himself to the study of the formal and final causes.

When Albert wrote, he repeated Averroes on some points. If one speaks of those causes which are principally studied, then one must admit that this science works with the formal and final causes only. Yet if one refers to the beings which the metaphysician knows, then one must say that he interests himself in both the material and efficient causes as well.

Both Averroes and Albert referred to these theories in their exposition of Γ1; and both viewed Γ1 as demonstrating what B 2 (the discussion of the first aporie) only stated. Now Thomas says much the same thing on the relation of Γ1 and B: in Γ1 Aristotle begins to demonstrate what in B he had discussed dialectically.[1] In his exposition of Γ1 however, Thomas makes no reference to the identity of the causes studied; he is content to state merely that metaphysics studies the "prima rerum principia et altissimas causas, sicut in primo dictum est."[2] The obvious interpretation to put on these words is that all four causes are studied, for all four were discussed in Book A (1). Yet if one turns to Thomas' discussion of B 2, one finds that Thomas, just as Averroes and Albert, explains fully which causes are to be studied when he comments on the discussion of the first aporie.[3]

Thomas' exposition of Aristotle's first aporie is indeed noteworthy. Aristotle had raised the problem of the identity of the causes metaphysics is to study. Obviously, the answer given to the difficulty will completely determine the science of metaphysics: the object, the method, the relationship to the other sciences – all is affected. As we know, Aristotle gives no answer in B 2; he merely discusses the pros and cons.

[1] *In IV Meta.*, 1, 529.
[2] *Ibid.*, 533.
[3] *In III Meta.*, 4, 384–85.

Averroes, however, saw the discussion of B 2 as a declaration that metaphysics can study only two causes, the formal and final. Albert, perhaps under the influence of Averroes, also saw in B 2 not only a discussion, but the answer of Aristotle. And interestingly, Albert understands Aristotle differently than Averroes did; metaphysics, *qua* the science of being as such, studies the formal and final causes; but *qua* the study of all beings, metaphysics studies the agent and the material cause. Now when Thomas began to compose his exposition of B 2, he distinguished quite clearly between the discussion of the aporie and its solution; the first part of Thomas' exposition explains the discussion, while the last part gives the solution to be pieced together from the doctrines found in other books.[1] Several interesting aspects of the presentation of the solution deserve to be mentioned. 1) The presentation by far exceeds in length the expositions of the solutions given to the other apories. 2) The presentation includes, besides the actual solution, an explicit exposition of how to untie each "knot" of this particular aporie; again such an exposition is given in connection with no other aporie. 3) Aquinas agrees totally with neither Averroes nor Albert. As Albert, so Thomas too says that all four causes are studied. As both Averroes and Albert, Thomas says that to study being as such means above all to study the formal cause; yet unlike his predecessors, Thomas does not hold that one studies especially form because the form is the principle of our knowledge; rather one studies form because beings have their perfection from the formal cause. The efficient cause is studied, says Thomas, and hence disagrees with Averroes; moreover it is studied because the immobile substances are studied and they are the efficient cause of motion: in this Thomas agrees with Albert. As Albert and Averroes, Thomas too attributes to metaphysics the study of the end; unlike his predecessors, Thomas places the reason for his attribution in the fact that metaphysics treats first substances, which move material things by being their end. And finally, Thomas parts company with Averroes, but joins Albert, in nothing that metaphysics studies the material cause due to the fact that some beings are material.[2]

The fact of such disagreement explains both the length of Thomas' exposition of the solution, and the fact of his explicit explanation of

[1] *Ibid.*, 369–83 contains the discussion of the aporie, while *Ibid.*, 384–86 contains the solution synthesized from various Aristotelian doctrines of other books.

[2] Averroes: *In III Meta.*, c. 3, fol. 41v, G–M; cf. footnotes 4–7, p. 127. Albert: *In III Meta.*, T. II, c. 1; cf. footnotes 5, p. 128–1, p. 129. Aquinas: *In III Meta.*, 4, 384–85; cf. footnote 1, p. 103 above and 3, p. 128.

how to unravel the difficulties proposed by Aristotle. Not only does this disagreement explain these aspects, but it goes a long way in illustrating Thomas' dissatisfaction with the expositions of Aristotle written before his own work: it was because he was not satisfied with the work of Albert and Averroes that Thomas wrote his *Commentary*.

These, then, are the theories of Averroes, Albert, and Aquinas; which one, if any, represents Aristotle's thought? Certainly not Averroes, for the efficient cause is needed in Aristotle's system. (I hope it is well understood that, when I search for "Aristotle's thought," I am attempting to discover whether or not Aquinas could have thought Aristotle taught what he, Thomas, puts in his exposition. As I have already mentioned several times, I am not interested in the Aristotle of the modern exegete, but in the Aristotle Aquinas could have *legitimately* taught.[1] In the present case, did Aristotle teach that God was the efficient cause of being, as Aquinas says he did?)[2]

When one turns to Albert and Thomas, it is more difficult to say which one is right. Obviously, the formal cause is studied by metaphysics. But why is it studied? As Albert explains, knowledge of the formal cause gives us the greatest knowledge of a thing. Thomas' answer goes deeper; he explains, implicitly of course, the reason why knowledge of the formal cause is the greatest: because the form of a thing is the source of what the thing is. Thus Thomas says, we study the formal cause of being as such. As Thomas indicates, Book Γ (IV) explains that it is the metaphysician who studies being as such, and not the natural philosopher, since some beings are not natural, that is not material. Now it would follow from this – and this is Thomas' point – that to study what all beings have in common is to study the formal cause.[3] Now it is quite true that Aristotle's Book Γ (IV) contains the doctrine that the natural philosopher can not study being as such because he can not study the first substance.[4] It would indeed appear legitimate

[1] For Averroes doctrine' of the metaphysical knowledge of formal cause, see below the discussion of the final elements of Aquinas' lesson 1.

[2] For the modern view see, e.g. Ross, *Aristotle's Metaphysics*..., Vol. I, p. clii–cliii. For Aquinas' view: *In II Meta.*, 2, 295; *In VI Meta.*, 1, 1164.

[3] "Hanc autem quaestionem Aristoteles in sequentibus expresse solvere non invenitur: potest tamen eius solutio ex his quae ipse inferius in diversis locis determinat, colligi. Determinat enim in quarto, quod ista scientia considerat ens inquantum est ens; unde et eius est considerare primas substantias, non autem scientiae naturalis, quia supra substantiam mobilem sunt aliae substantiae. Omnis autem substantia vel est ens per seipsam, si sit forma tantum; vel si sit somposita ex materia et forma, est ens per suam formam; unde inquantum haec scientia est considerativa entis, considerat maxime causam formalem." *In III Meta.*, 4, 384.

[4] "Unde nullus secundum partem intendencium presumit dicere aliquid de ipsis, si sunt vera aut non, – neque geometra neque arismeticus, set phisicorum quidam hoc merito agentes;

to assume, as Thomas seems to, that only by studying what is common to all substances (namely form) can one study being as such. Thus Thomas' explanation of the first aporie appears to be based on Aristotle's own doctrine.

Albert, on the other hand, can point as well to Aristotelian doctrines in support of his view. Both in his commentary on Γ 1 and on B 2 (the first aporie), Albert refers to a discussion prior to B 2. Apparently, he refers to A 2, for both in Γ 1 and in B 2 he explains that in a prior section the philosopher or wise man is shown to be "doctissimus causarum."[1] That is, the wise man knows the causes much better than other men. On the basis of this doctrine, Albert declares that the philosopher will principally study both the formal and final causes: the final cause, because it is prior to the other causes; the formal, because to know it is to know the most about a thing.[2] Thus Albert, as Thomas, is able to justify his opinion as Aristotelian.

Yet there appears to be a flaw in Albert's reasoning. Or at least, Thomas' words indicate that he wishes to point to a danger inherent in the theory of Albert. We can never know the form of the immaterial substances, Thomas writes. Hence, if one emphasizes that the study of being as such, because it is the highest knowledge, is knowledge of the formal cause, then one runs the risk of not knowing the immaterial beings at all.[3] Hence, as concerns the metaphysical study of the formal cause, both Albert and Thomas can point to Aristotelian doctrines; yet Thomas' views would appear to be better founded in so far as they will not flounder upon contact with the particular difficulties involved in our knowledge of immaterial beings.[4]

solim enim opinati sunt deque tota natura intendere et de ente Quoniam autem est adhuc phisico quedam superiorum (unum enim aliquod genus est entis natura), universalis et circa primam substantiam considerativi et de his utique erit speculatio. Est autem sapiencia quedam et phisica, set non prima." Γ 3, 1005a31–b2: *Vetus*, p. 307, l. 26–34. For Thomas' exposition of this passage: *In IV Meta.*, 5, 593; for Albert's: *In IV Meta.*, T. II, c. 1, p. 173, l. 55 sqq.

[1] *In IV Meta.*, T. I, c. 1, p. 161, l. 24–28; *In III Meta.*, T. II, c. 1, p. 112, l. 72–p. 113, l. 4.

[2] *Ibid.*, p. 113, l. 5–38.

[3] "Primae autem substantiae non cognoscuntur a nobis ut sciamus de eis quod quid est... et sic in earum cognitione non habet locum causa formalis." *In III Meta.*, 4, 384.

[4] It is possible to trace back even further the difference between Albert and Thomas. Ultimately, Albert refuses any proof for the existence of a cause of being; in addition he has a Neoplatonic theory of knowledge, and is able to postulate the possibility of learning about God from studying the light of God in creatures, from studying the *esse simplex*, the *primum creatum*. Cf. G. Sestili, "L'Universale nella dottrine de S. Alberto Magno," *Angelicum*, IX, 1932, pp. 168–86. M. Browne, "Circa intellectum et eius illuminationem apud S. Albertum Magnum," *Angelicum*, IX, 1932, pp. 187–202. – Yet the difference between Albert and Thomas does not lie in the fact that God was the object of metaphysics for Albert, as some have claimed. The difference between the two commentators will be traced further in the two chapters that follow. For the present we must be satisfied that their commentaries on Book Γ 2 are different, and that Aquinas' was apparently directed against Albert's (and Averroes').

But what of Thomas' and Albert's teachings on the other causes? Thomas says metaphysics studies the final cause because the separated substances move material beings as a final cause. Albert says only that to study being as such we need to study the goal of the formal cause of being. Actually there is no difference between the two points of view; both in addition can point to Aristotle's Book Λ 7 for justification, as *de facto* they do.[1] Thus both Thomas and Albert represent Aristotle's view of this issue.

As far as the study of the efficient and the material cause is concerned, it would take us too far afield at the moment to discuss the reasons given by Albert and Thomas. In Chapter V I must treat the efficient cause and its role in bringing first philosophy and universal philosophy into the unity of a single science. In Chapter V also, material causality will be introduced into the discussion of metaphysics as the science of the *communia*. In these contexts it will be more proper to note the Aristotelianism of Thomas and Albert.

Thus far we have examined only the relation between the first part of Thomas' *In IV Meta., lectio* 1, and the corresponding passages of Averroes' and Albert's expositions. Before we turn to the remaining sections of Thomas' *lectio* 1, let me summarize our findings. In the first place, Averroes, Albert, and Thomas were discovered to agree in general on the object of metaphysics: being as such, and the essential attributes. Secondly a large area of disagreement was noted when we asked which causes were involved in the science of being as such: Averroes claimed that only two were studied; Albert and Aquinas, on the other hand, drew all four into the field of metaphysics, although each author had different reasons. Thirdly, when it was asked which of the three authors was faithful to Aristotle, Averroes was quickly eliminated; both Albert and Aquinas were found to base their reasoning on Aristotelian texts, yet on different texts. Fourthly, it was noted that Thomas' exposition appears to criticize Albert's point of view by pointing out difficulties which were sure to arise. Consequently, it appears possible to conclude to the motive behind Thomas' work: a dissatisfaction with prior expositions of the *Metaphysics*. As we proceed in our examination of the relation between Thomas on one hand, and Averroes and Albert on the other, we shall discover still more evidence for this conclusion.[2]

[1] Thomas: *In XII Meta., lectio* 2. Albert: *In XI Meta.*, T. II, c. 6. Since the *Media* contained no Book K (XI), Book Λ (XII) received the title "Book XI"; hence Albert's *In XI Meta.*, is actually the exposition of Book Λ (XII).

[2] Albert too probably wrote because of his dissatisfaction with Averroes' exposition. It

As has been noted, Thomas divides his *lectio* 1 into three parts. In the first part, he explains that metaphysics studies both being as such and its essential attributes. Both the Aristotelian passage corresponding to this first part and the corresponding expositions of Averroes and Albert have been examined. In addition we have noted the doctrines present in the passage of Aristotle corresponding to the last two parts of Thomas' *lectio*. Thus it is time now to turn to those passages in Averroes and Albert which correspond to the remaining two sections of Thomas' work. The Aristotelian passage consits of lines Γ 2, 1003a 33–b 22; these lines should be viewed as giving the answers to two difficulties raised in Book B: such is the opinion of Thomas regarding these lines of Aristotle. First, Thomas believes that Aristotle explains that both substances and accidents are studied (Γ 2, 1003a 33–b 19); then, the Greek shows that all substances are studied (b 19–22). As previously noted, Thomas' exposition is little other than *verba Aristotelis* – if one refers only to *lectio* 1, of course, and does not consider *ens* as explained in *lectio* 2.

When we turn to Averroes' exposition of these lines, we do not find anything like the precision of Thomas' work. In fact, it is somewhat difficult to know exactly what Averroes thought Aristotle intended to say in this passage. Averroes saw fit to include both Γ 2, 1003a 33–b 19 and b 19–22 in the same section; moreover, he makes a clear division between these lines and, on one hand, Γ 1 (metaphysics studies being and its essential attributes), and on the other Γ 2, 1003b 22sqq. (metaphysics studies *unum* and its species). Yet he is not so careful when it comes to dividing Γ 2, 1003a 33–b 19 from lines b 19–22. To be sure, Averroes notes at the end of his exposition of 1003a 33–b 19, that these lines are the answer to Aristotle's problem: Does it belong to one science to consider all substance?[1] Yet the last few lines of this section (b 19–22) appear to have little or no interior unity, and little or no connection with what precedes them, if we are to judge from Averroes.[2]

Thus Averroes sees in the discussion of the predication of "being" of both substance and accidents, the solution to the problem of whether one science can study all substances. Apparently Averroes judged that the discussion of the aporie in Book B also contained Aristotle's solution

has been suggested that Albert himself explicitly explained. Cf. M. GRABMANN, "Die Lehre des heiligen Albertus Magnus von Grunde der Vielheit der Dinge und der lateinische Averroismus," *Mittelalterliches Geistesleben. Abhandlungen zur Geschichte der Scholastick und Mystik*, Band I, Heuber, München, 1936, pp. 259–97.

1 "...et ista est declaratio praedicatae quaestionis, in qua dicitu utrum unius scientiae est consideratio de substantia, aut non." *In IV Meta.*, c. 2, fol. 66r, C–D.

2 Cf. *Ibid.*, fol. 66r, D–E.

although in a less demonstrative fashion.[1] Yet it is a little difficult to discover where Averroes thinks Aristotle gave the solution to the present problem. Certainly Averroes does not seem to answer the problem when he treats Aristotle's discussion of it in Book B.[2] Be this as it may, Averroes in Book IV states that metaphysics studies only the formal and final causes of substance.[3] And he sees in the discussion of the predication of "being" some sort of related doctrine.

Averroes' exposition of the predication of "being" is for the most part quite straightforward. "Being" is predicated of both accidents and substances.[4] In all cases, however, "being" refers primarily to substance.[5] Thus if we study substance, we can have a science of being.[6] Then Averroes explains what he calls "another evident proposition": because the Philosopher is he who knows the first causes of all, and because to study all beings (substances and accidents) is to study substance, the Philosopher must study the principles and causes of substance. And this, he adds, is the answer to the problem concerning the possibility of one science of all substances.[7] Obviously, Averroes has left out a step in his argument. The full argument would seem to be as follows. 1) All substances do not have all four causes; rather they have in common only the formal and final causes. Hence to study all substances is to study these two causes. 2) But "being" when said of accidents refers to substances. 3) Hence, if one studies the formal and final causes of substances, one studies all things *qua* "being." Thus it appears

[1] Cf. the opposition between Book B (III) and Book Γ (IV) mentioned in *In IV Meta.* c. 1, fol. 64r, C–D. Book Γ (IV) demonstrates what was only shown in Book B (III), namely that: 1) all causes are not studied in one science since all beings are not of the same genus, and 2) metaphysics studies only the formal and final causes.

[2] "...et, si scientiae est consideratio de substantia, utrum illa scientia, etc., idest licet dubitatio accidat in hae quaestione sic: quoniam si scientia substantiae fuerit una, contingeret ut omnes substantiae sint unius generis; et sic erunt omnes res generabilis et corruptibiles: quod est inopinabile; existimatur enim quod principia corruptibilium sunt incorruptibilia. Et, si scientiae substantiae sint plures, continget ut naturae substantiarum sint diversae; et si fuerint diversae, aut erunt consimiles, aut nomen substantiae dicetur de eis aequivoce. Si igitur fuerint consimiles, continget ut scientia substantiae dicitur aequivoce pure; et si hoc fuerit, tunc scientiae substantiarum non reducuntur in unam scientiam: quod est inopinabile." *In III Meta.*, c. 4, fol. 43r, B–C.

[3] *In IV Meta.*, c. 1, fol. 64r, C–D.

[4] *Ibid.*, c. 2, fol. 65r, D–v, H.

[5] *Ibid.*, fol. 65v, I–L.

[6] *Ibid.*, fol. 65v, M–fol. 66r, C.

[7] "Deinde dedit aliam propositionem manifestam, et dixit, Si igitur hoc prius sit substantia etc., idest si igitur prius decem entium est substantia, ut post declarabitur: et Philosophus est ille, qui scit primas causas omnium, necesse est ut cognitio principiorum substantiae sit accepta in definitione Philosophi. Dicitur enim, quod Philosophus est ille, qui scit principia et cuasas substantiae; et ista est declaratio praedictae quaestionis, in qua dicitur, utrum unius scientiae est consideratio de substantia, aut non." *Ibid.*, fol. 66r, C–D.

that Averroes views the combination of the doctrines of Books B and Γ as answering this particular problem.

The last few lines of the Aristotelian passage (Γ 2, 1003b 19–22) apparently were not important in Averroes' eyes. This may be partially explained by the poor translation of Aristotle,[1] and partially by Averroes' idea that metaphysics studied but two causes. In any case, when Aristotle says: "Every genus has one science, for example, there is one science of sounds, and it must consider all sounds. And thus we say that the universal consideration of all forms of being belongs to a generically united science,"[2] Averroes remarks: "...it is evident that being is one genus; therefore being has one science."[3] Since what Averroes says here must be compatible with what he says a few lines earlier, we can not take him to be affirming an ontological genus of substance which metaphysics studies. He can only be repeating what he had earlier said: metaphysics studies all substances by studying the formal cause and the final cause of substance.[4] – The last line of the Aristotelian passage (Γ 2, 1003b 22) is very poorly translated in the *Arabica*, and Averroes' thought, a reference to Platonists perhaps, has nothing to do with the issue at hand.[5]

Turning to Albert's exposition of Aristotle's Γ 2, 1003a 33–b 22, we can notice the influence of Averroes. Yet much of what Albert writes is the result of his own reflexion. For example, unlike Averroes, Albert explains the theme of these lines as follows: "Because any one science is about one genus, and because being is not a genus, we must show how it is possible to have one science of being."[6] The crux of the solution lies in the analogy of "being." Because "being" is said of both substance

[1] Cf. the *Arabica* in *Aristotelis Metaphysicorum Libri XIIII...*, fol. 65r, C–D.

[2] Translated from the *Arabica*; cf. *Ibid.*

[3] "Deinde dixit, Sed unumquodque genus, etc. idest et etiam manifestum est, quod quodlibet genus entium habet unam artem, et unam scientiam considerantem in omnibus speciebus illius generis, verbi gratia, quod sonus est unum genus, et habet unam scientiam considerantem in omnibus speciebus sonorum, scilicet Musicum. Deinde dixit Et ideo dicimus, quod consideratio, etc. idest et hoc, quod dixit manifestum est ex praedictis. Posuit enim quod quodlibet unum genus habet unam scientiam; et manifestum est quod ens est unum genus; ergo ens habet unam scientiam." *In IV Meta.*, c. 2, fol. 66r, D.

[4] Or perhaps he means, metaphysics studies the genus *ens* (substance and accidents) and the species, or parts, of metaphysics study the ten species of *ens* (the ten categories).

[5] The Greek has literally: "and the species (are the study) of the species." The Oxford translation: "...and to investigate the several species (of being) is the work of the specific parts of the science (of being)." We read in the *Arabica*: "formae autem sunt formae formarum." (*Aristotelis Metaphysicorum Libri XIIII...*, fol. 65r, D). For Averroes' exposition: *In IV Meta.*, c. 2, fol. 66r, D–E.

[6] *In IV Meta.*, T. I, c. 3, p. 163, l. 37–40. This is Albert's understanding of Aristotle's first aporie; cf. also *In III Meta.*, T. II, c. 1, p. 111, l. 31–48.

and accidents,[1] and because "being" always refers to substance,[2] the study of substance *qua* being is the study of all beings (substance and accidents).[3]

Thus far Albert has scarcely surpassed Averroes' exposition. But in commenting on the last few lines (1003b 19–22), Albert gives us more of the fruit of his own work.[4] All the species of being, that is of substance, are to be studied. Perhaps there are species of being which exist in a separated manner as Plato says; and then perhaps there are no such species. In any case, all the species must be investigated by the science of being.[5]

Let us review briefly the interpretations given of lines Γ 2, 1003a 33–b 22. Aristotle's text seems clearly to distinguish two doctrines. First, there was the discussion of the manner of predicating "being". Because all uses ultimately referred to that use of "being" which refers to substance, one could study all beings by studying substance as "being." A second doctrine appears when Aristotle compares the faculty of sound and the science of sounds to the study of being. Just as all sounds are gathered together in a generically unitary science, so too all beings (that is, all substances, and all qualities, etc.) can be gathered together in the generically unitary science of metaphysics.

When Averroes commented on these lines, he correctly explained how metaphysics studied accidents as well as substance. However, he felt that one already knew that to study substance is to study the formal and final causes of substance. This doctrine he integrated into the present context and announced that to study accidents as well as substance, one studied the formal and final causes of substance. The meaning of the final few lines of Aristotle escaped Averroes.

Albert partially repeated Averroes, yet partially advanced beyond him. The first doctrine of Aristotle was correctly explained: metaphysics, by studying substance, manages to study all beings of which "being" is said. In so far as he followed Averroes on this point, Albert must be said to have been satisfied with the former's exposition. The

[1] *In IV Meta.*, T. I, c. 3, p. 163, l. 40–55.

[2] *Ibid.*, p. 164, l. 18–63.

[3] *Ibid.*, p. 164, l. 64–p. 165, l. 7.

[4] The *Media* version is a good translation, whereas Averroes' was not. For the *Media*, cf. ALBERTI MAGNI, *Metaphysica*..., p. 164.

[5] After discussing the "una scientia", "unus sensus", etc., he writes: "...ideo etiam unius et eiusdem scientiae est speculari entis, inquantum est sens, omne species, quotcumque species habet ens, et species specierum, sive sint separatae, ut Platoni placuit, sive sint in re sive in anima, secundum quod ad ens referuntur. Sic enim unum genus vocamus, quod est unum et primum subiectum...," *In IV Meta.*, T. I, c. 3, p. 165, l. 13–18.

second doctrine of Aristotle – metaphysics studies all substances by studying them from a generical point of view – this was also correctly grasped by Albert. Quite obviously then, Albert regarded the last section of Averroes as totally unsatisfactory.

Finally there was Thomas. More precise than either Averroes or Albert, Thomas interprets both sections of Aristotle correctly, and notes that both sections correspond to a definite Aristotelian aporie. Although Averroes' work would have appeared deficient to Thomas, it is difficult to conclude that Thomas was dissatisfied with Albert's work. Aquinas is clearer, to be sure, but on the interpretation of the latter parts of Aristotle's lines, he agrees substantially with Albert.[1]

As we end this comparative discussion of the expositions of Γ 1, 1003a 21–2, 1003b 22 given by Averroes, Albert, and Aquinas, there are two conclusions that stand out:

1) Albert was not satisfied with Averroes' *Commentary*, and Thomas was satisfied neither with Averroes' nor Albert's; especially is it true that Thomas was unwilling to accept Averroes' writings; yet even Albert could not totally satisfy Thomas, for as will be remembered, Albert and Thomas did not see eye to eye on the subject of why the formal cause was to be studied in metaphysics.

2) All three, Averroes, Albert, and Thomas appear to have desired to explain the true meaning of Aristotles' text.

The outstanding points of this comparison, which permit one to draw these two conclusions, can be schematized as follows.

Aristotle	*Averroes*	*Albert*	*Aquinas*
1) Metaphysics studies being and its essential attributes. (Γ 1, 1003a 21–22)	1) Metaphysics studies being as such as well as its essential attributes. To study the principles and causes of being is to study the formal and final causes of being. (*In IV*	1) Metaphysics studies being as such and its essential attributes. To study being one must study principally the formal and final causes. Because metaphysics is the	1) Metaphysics studies being as such and its essential attributes. Metaphysical knowledge is more common knowledge in relation to the less common knowledge of

[1] From the comparison of Albert's and Aquinas' expositions of Γ 2, this is totally correct. However, if one places the interpretations of this lesson in the context of the total *Commentaries* of Albert and Aquinas, there is a vast difference between their expositions of Γ 2. Aquinas speaks there of a formal object, expressed as "being", from which one must rise to knowledge of the existence of a pure *Esse*. Albert, on the other hand, speaks of a formal object, expressable as "form, the cause of substance", from which one can know something about God without being able to prove Him metaphysically as a cause of being. These theories are treated further in Chapters V–VI.

Aristotle	*Averroes*	*Albert*	*Aquinas*
	Meta., c. 1; *In III Meta.*, c. 3)	highest science, it studies the final cause; because it is the most perfect knowledge it studies the formal cause. Secondarily metaphysics will study the efficient and material causes, because being as such is found both in material, caused beings and in their efficient cause. (*In IV Meta.*, T. I, c. 1)	the particular science. Metaphysics studies all four causes: the formal cause because the form is the source of a thing's perfection; the final cause, because first substances are to be studied and they are the final cause of other beings; the efficient cause, again because one studies first substances; the material cause, because some beings are material. (*In IV Meta.*, 1, 529–33; *In III Meta.*, 4, 384–85)
2) Metaphysics studies accidents as well as substance for both are called "being". Since "being" is predicated primarily of substance, it is substance which is primarily studied. (Γ 2, 1003a 33–b19)	2) Averroes gives the same doctrine as Aristotle, but it is better ordered. This doctrine, he says, is the answer to the aporie: Can one science study all substances? Apparently he means that to study the formal and final causes of substance is to study accidents and all substances. (*In IV Meta.*, c. 2, fol. 65r, D–fol. 66r, D)	2) Albert gives the same doctrine as Aristotle and adds: this is the answer to the aporie: How can we have one science of being, when being is not a genus? (*In IV Meta.*, T. I, c. 3, p. 163, l. 37–p. 165, l. 7)	2) Aquinas gives the same doctrine as Aristotle, adopting much of Averroes' order, and says: This is the solution to the aporie: Can one science study both accidents and substance? (*In IV Meta.*, 1, 534–46)
3) Metaphysics studies all substances falling under the genus of (substantial) being. (Γ 2, 1003b 19–22)	3) Averroes misses the point completely, partially because of a poor translation; perhaps also his view of metaphysics as the study of the formal and final cause influenced	3) Albert appears to understand Aristotle correctly; thus he says that metaphysics is to study all types of substance as falling under a genus. (*In IV Meta.*, T. 1. c. 3,	3) Aquinas says that here we have the answer to the aporie: Does one science study all substance? One science can study all substances if it considers them as coming to-

Aristotle	*Averroes*	*Albert*	*Aquinas*
	his exposition on these lines. (*In IV Meta.*, c. 2, fol. 66r, D–E)	p. 165, l. 8–26)	gether in the genus of substantial being. (*In IV Meta.*, 1, 547)

In the light of this schematic comparison it does not appear unwarranted to conclude that Thomas wished to expose faithfully, correctly, the thought of Aristotle. But since this lesson must be read in the light of lesson two's doctrine of being as composed of *esse* and essence, one is faced with the fact that Thomas is transforming, although one does not know whether consciously or not, Aristotle's thought on the object of metaphysics: the Aristotelian framework remains, but the value, the meaning of that framework is transformed through the introduction of the doctrine of *esse*.

3. ARISTOTLE, AVERROES, ALBERT: THE PREDECESSORS OF AQUINA'S *IN IV META.*, *LECTIO* 2

A. Aquinas and Aristotle

Thomas saw the Aristotelian passage Γ 2, 1003b 22–1004a 9 (to which corresponds *In IV Meta.*, *lectio* 2) as the first part of an answer to another of the questions posed by Aristotle in Book B: Does metaphysics study the *communia – unum, idem*, etc.? This answer extends from Γ 2, 1003b 22 until 1005a 18, Thomas notes. The link between these lines and Aristotle's Γ 1, Thomas appeared to see in the doctrine expressed in the very first verse of Book Γ: "There is a science which investigates being and the attributes which belong to this virtue of its own nature".[1] The initial lines of Γ 2, he says, answer the apories which concern the study of being as being;[2] the remainder of Γ 2 deals with the aporie concerning the study of the attributes of being such as.[3] – Now there can be no doubt that Thomas was correct in seeing Γ 2, 1003b 22–1005a 18 as Aristotle's answer to the fifth aporie: Does one science discuss the essential attributes of substance as well as substance itself?[4]

In the earlier examination of Thomas' thought in lesson two, I noted that this lesson contains two parts: first there are two proofs that metaphysics studies unity as well as being; secondly there is the proof that metaphysics studies all the "parts" of unity. As in the previous

[1] Γ 1, 1003a21–22.
[2] *Ibid.*, a33–b22.
[3] *In IV Meta.*, 2, 548; *Ibid.*, 1, 529 and 534.
[4] The aporie is given in: B 1, 995b18–27; *Vetus*, p. 304, l. 8–33. Cf. Ross, *Aristotle's Metaphysics...*, Vol. I, p. 222.

Section 2 which dealt with lesson one, so now in connection with lesson two one must ask whether Thomas' exposition is a faithful presentation of Aristotle.

As far as the first proof that the science of being must also be the science of unity, Thomas' exposition is certainly more than an elaboration of Aristotle. To be sure,the first five paragraphs of Thomas' exposition of the first proof (paragraphs 548–52) faithfully echo the thought of Aristotle. Aristotle, as well as Thomas, wrote that unity and being, although different in definition, express the same object; hence they are to be studied in the same science.[1] However, Thomas did more than merely repeat this doctrine; paragraph 553 builds on the Aristotelian words in a way unthought of by Aristotle. First of all, Thomas adds the notion *res*, and places it on a par with *ens* and *unum*; *res* too, is to be studied by metaphysics.[2] Secondly, Thomas points out the *id a quo nomen imponitur* – that aspect of things, which as known, supplies the name of the concept through which one knows the thing. The *id a quo* of "unity" is said to be indivision; that of "thing" is quiddity; and that of "being" is *esse*.[3] Thirdly, the mention in paragraph 558 of *esse* must be understood as applicable to the present paragraph.[4] Thus on these counts Thomas' exposition of Aristotle's first proof surpasses the thought of Aristotle. In general, of course, Thomas presents the *verba Aristotelis*: the general framework of Aristotle is accepted, but *broadened* by the introduction of *res*, and then transformed in so far as the object known as "one," "thing," and "being" becomes a composite of *esse*

[1] "Si igitur ens et unum idem et una natura est in consequendo adinvicem sicut principium et causa set non sicut una ratione ostensum...; idem enim est 'unus homo' et 'homo', et cum sit homo et homo et non alterum aliquod demonstratur secundum dictionem replicatum quod 'est homo' et 'homo' et 'unus homo'...", Γ 2, 1003b22–32; *Vetus*, p. 304, l. 8–14. There is no mention in these lines that the study of being is the study of unity, as Thomas indicates when the comments: *In IV Meta.*, 2, 549. However, the general context concerns the objects to be studied.

[2] Apropos of the transcendentals, it has been remarked that Aristotle certainly teaches that "being" and "unity" can be predicated of anything. Even "good" was spoken of as applicable to something within each of the categories, through not to everything within each. Such hints were developed by scholastics as the theory of the transcendentals. Yet such a doctrine "though based on hints in Aristotle, has no Aristotelian authority". Cf. W. Ross, *Aristotle*, Methuen, London, 1949, 5th ed. revised, p. 156.

[3] *In IV Meta.*, 2, 553: cf. footnote 4, p. 111. Aristotle would have answered in the affirmative if he had been questioned on the correctness of Thomas' words concerning the *id a quo* of "unity"; Δ 6, 1016b1–9 reads: "In general those things the thought of whose essence is indivisible, and cannot separate them either in time or in place or in definition, are most of all one, and of these especially those which are substances. For in general those things that do not admit of division are called one in so far as they do not admit of it; ...the things that are primarily called one are those whose substance is one...". Thomas exposition: *In V Meta.*, 7, 865–8, 869.

[4] The text of paragraph 558 is found in footnote 4, p. 113.

and essence. Thus these paragraphs 548–53 are to be considered as something other than purely "Aristotelian."

Thomas' methodology in treating the second proof is similar to that he used in exposing the first one: Thomas is basically true to Aristotle's theories. Aristotle wrote: "And if, further, the substance of each thing is one in no merely accidental way, and similarly is from its very nature something that is."[1] According to Thomas, this means that, since *unum* and *ens* are predicated *per se* of the substance (that is, essence) of anything, the reality which is *unum* is also *ens*.[2] Since Aristotle's short and concise phrase comes after the discussions of the predication of "being" and of "unity," Thomas is certainly justified in introducing "predication" into his explanation. (As shall become evident in Chapter VI, for Thomas the "study of predication" is central to metaphysics.)

Although Thomas' exposition in so far as its framework is concerned remains true to Aristotle's Γ 2, 1003b 32–33, the doctrines explained by Thomas in the context of that framework are totally foreign to Aristotle's writing. Aristotle mentioned only that the substance of anything is both *unum* and *ens*; this is the entirety of his thought. Thomas, however, introduces two elements: first a proof that *unum* and *ens* are predicated of the substance of anything (paragraph 555); and secondly a lengthy discussion of Avicenna (paragraphs 556–60).[3] I have already discussed at length these two additions in examining Thomas' exposition in lesson two, and they need not be reconsidered here. Suffice it to point out that the larger part of these additions involves the discussion of *esse* as distinct from essence. Thus, since the doctrine of *esse* is introduced into the framework of Aristotle's doctrine, the doctrine at issue (the predication of *unum* and *ens* of the substance of anything) takes on a new meaning.

Even though Thomas thus transforms Aristotle, one must admit that the metaphysics resulting from this transformation is still in the line of Aristotle's system: the overall framework remains that of Aristotle.

So much for the two proofs that metaphysics studies unity as well as being. There remains still the discussion which Thomas called "the conclusion that metaphysics studies the 'parts' of unity, just as it does the 'parts' of being": Γ 2, 1003b 33–1004a 9. As was mentioned earlier in our study of Thomas' explanations, the word "part" is used here in two totally different meanings. First, Thomas spoke of the "parts" or

[1] Γ 2, 1003b32–33. The *Vetus*, p. 304, l. 19–18 reads: "...amplius autem, uniuscujusque substantia unum est et non secundum accidens...".

[2] *In IV Meta.*, 2, 554.

[3] Cf. footnotes 3, p. 112 and 2–3, p. 113.

categories of unity which correspond to the ten categories of being (paragraph 561).[1] Then, he turns to the "parts" or sections of metaphysics, each of which studies a "part" or species of substantial unity and of substantial being (paragraph 563).[2] By using the word "part" in these ways, Thomas did not disrupt the unity he attributed to Γ 2, 1003b 22–1005a 18: even the lines in question (1003b 33–1004a 9) further the discussion of metaphysics as the study of unity. In addition, such an exposition parallels that of lesson 1; whereas lesson 1 spoke of an horizontal plane of being (the ten categories of being) and of a vertical plane of being (the genus of substance), so lesson 2 speaks of an horizontal plane of unity (the ten categories of unity) and of a vertical plane of unity (the genus of substance – the genus of that which especially has unity).

Does Thomas' thought reflect that of Aristotle? Basically it does.[3] For first, Aristotle notes that there are as many species of unity as there are of being.[4] This can rightly be interpretated as indicating the ten categories of unity. Then, speaking "vertically," Aristotle says that there will be as many parts of philosophy as there are types of substance; one can distinguish several different sciences corresponding to these types of substance.[5] Thomas' reading of this second notion is obviously historically incorrect. Aristotle, by "parts of philosophy"

[1] Cf. footnote 1, p. 116.

[2] Cf. footnote 1, p. 117.

[3] On the point of reducing all contraries to "unity" there is room for discussion. The *Vetus*, p. 304, l. 21–24 reads: "De quibus quod 'quid est' ejusdem sciencie est genere considerare – dico autem sic, de eodem et simili et aliis hujusmodi; fere autem omnia reducuntur contraria in principium hoc...". Γ 2, 1003b34–1004a1. Yet Thomas explains that *all* contraries are reduced to *unum* and not "fere omnia": "...'fere'. Et hoc addit, quia in quibusdam non est ita manifestum. Et tamen hoc esse necesse est...," *In IV Meta.*, 2, 561–62. In the *Arabica*, and in Averroes, Thomas could find this same universal reduction mentioned: "Universaliter omnia contraria attribuuntur huic primae scientiae"; cf. *Aristotelis Metaphysicorum Libri XIIII...*, fol. 66v, K; Averroes: *In IV Meta.*, c. 3, fol. 67v, I. However in the *Media* Thomas again could read the "fere omnia"; cf. ALBERTI MAGNI, *Metaphysica...*, p. 167. What Thomas appears to have done is to read the present passage from Book Γ in connection with the aporie in B 2, 996a20–21: "How could it belong to one science to recognize the principles if these are not contrary?" as well as with the restatement of the aporie in K 1, 1059a20–23: "but one might ask the question whether Wisdom is to be conceived as one science or as several. If as one, it may be objected that one science always deals with contraries, but the first principles are not contrary." In Thomas' expositions of these two passages, one can see a connection between his thought and the theory of the reduction of all contraries to "unity"; cf. *In III Meta.*, 4, 370–71; *In XI Meta.*, 1, 2147–49. Thomas saw no contradiction between the lines from Γ 2 and those of K 1 as some moderns do: cf. L. ELDERS, "Aristote et l'objet de la métaphysique," *Revue Philosophique de Louvain*, LX, 1962, p. 174.

[4] "...quare quecunque vere unius species sunt, tot sunt et entis." Γ 2, 1003b33–34: *Vetus*, p. 304, l. 20–21.

[5] "Et tot partes philosophie sunt quot vere substantie quare necesse est quandam primam et continuam ipsis. Sunt enim mox genera habencia ens et unum; unde et sciencie sunt consequentes hec." Γ 2, 1004a1–6: *Vetus*, p. 304, l. 27–30.

means metaphysics (first philosophy and universal philosophy) and physics; as modern commentators maintain, this reference to the several parts of philosophy is a misplaced passage in our texts of Aristotle. Thomas, however, had not thought of such a possibility, and hence attempted to interpret these "parts of philosophy" in terms of "parts of metaphysics." He succeeded thus in integrating this passage into its context.[1]

And so we come to the end of this study of the Aristotelian passage corresponding to Thomas' lesson 2. The conclusion which should be drawn has already been twice mentioned: Thomas has retained the Aristotelian framework but transformed it from within principally through the introduction of *esse* as a principle of being. At present, however, we do not know whether this transformation is a conscious one.

B. Aquinas versus Averroes and Albert

When we examine the expositions of Averroes and Albert given in relation to Aristotle's Γ 2, 1003b 22–1004a 9, we come to much the same conclusion reached in the earlier examination of their expositions which correspond to Thomas' lesson 1: Averroes and Albert were both used and opposed by Aquinas in his own work.

Averroes' exposition begins with a treatment of unity; he shows why *unum* is studied in the science of being. This explanation is not of any great merit; its doctrine is the same as Aristotle's Γ 2, 1003b 22–33, and Thomas' *In IV Meta.*, 2, 548–54.[2] It is very difficult to see any evidence that Thomas was guided by Averroes on this point. Yet immediately following this discussion, Averroes lauches into a direct attack on Avicenna, and here one must acknowledge an influence on Thomas' lesson 2.

Averroes begins by remarking:

Avicenna autem paeccvit multum in hoc, quia existimavit, quod unum et ens significant dispositiones additas essentiae rei. Et mirum est de isto homine,

[1] That the passage is incorrectly placed in our texts, cf. Ross, *Aristotle's Metaphysics...*, Vol. I, pp. 256–57; Decarie, *L'objet de la métaphysique selon Aristote*, pp. 104–106. Prof. Merlan sees this "parts of philosophy" as an anticipation of the doctrines of E 1, the divisions of the sciences; cf. Merlan, *From Platonism to Neoplatonism*, p. 165. As has already been mentioned, Prof. Merlan views the Aristotelian metaphysics of Γ and E 1 as the study of separate substance; yet he admits the legitimacy of viewing this passage as if Aristotle were teaching that metaphysics should study all things from the point of view of "being" (p. 169). As is revealed from a comparison of Thomas' exposition of these "parts of philosophy" (*In IV Meta.*, 2, 563) with that of E 1 (*In VI Meta.*, 1, 1164–65), that is exactly how Thomas understansd Aristotle: metaphysics studies all substances as "being".

[2] For Averroes' exposition cf. *In IV Meta.*, c. 3, fol. 66v, K–fol. 67v, B.

quomodo erravit tali errore; et iste audivit Loquentes in nostra lege, cum quorum sermonibus admiscuit ipse suam scientiam Divinam.[1]

"Peccavit multum" and "mirum est de isto homine, quomodo erravit tali errore": these are rather strong words. Their strength appears the greater when one compares them to Thomas' sentence: "Sciendum est autem quod circa hoc Avicenna aliud sensit."[2] As the following considerations reveal, this difference between Averroes' and Thomas' words are indicative.

After his introductory attack on Avicenna, Averroes begins a resume of Avicenna's views. In short, as Averroes understood him, Avicenna said that if "being" and "unity" refer to the same thing, then it would be a waste of words to say: "Being is one." Such a sentence is a pure tautology. Hence Avicenna said that "being" and "one" refer to the essence disposed or perfected by something other than the essence itself.[3] The fault with this view, Averroes says, is that it overlooks a simple fact: since "being" and "one" do not have the same meaning, the fact that they refer to the same thing does not make them synonymous.[4]

Following this general attack, Averroes turns to the Avicennian idea of "unity." The fault of Avicenna is twofold: 1) he thought that the name "one" signified something added to the essence;[5] 2) he mistook the principle of number for mathematical unity.[6] Thus Avicenna said that "unity" signifies an accident.[7]

After finishing his exposition of Avicenna, Averroes refutes the former's views on "one" by giving this argument: "if a thing were one through something added to its nature..., then nothing would be one through itself... but through something added to its essence. And... [if] we seek that through which is had the unity of a thing, [namely that] through which the [first] thing is one – if that [second thing] is

[1] *Ibid.*, fol. 67r, B. On Averroes' criticism of Avicenna, cf. Roland-Gosselin, *Le "De ente et essentia"...*, pp. 157–59; Gilson, *Being and Some Philosophers*, pp. 51–56.

[2] *In IV Meta.*, 1, 556.

[3] "...et dixit, quod unum et ens reducuntur ad essentiam dispositiam. Et iste homo ratiocinatur ad suam opinionem, dicendo quod, si unum et ens significant idem, tunc dicere ens est unum esse nugatio, quasi dicere unum est unum, aut ens est ens... (Ens et unum referuntur ad) dispositiones diversas essentiae additas." *In IV Meta.*, c. 3, fol. 67r, B–C.

[4] "Et hoc non sequeretur nisi dicerimus, quod dicere de aliquo quod est ens et unum, quod significant eandem intentionem et eodem modo. Nos autem diximus, quod significant eandem essentiam, sed modis diversis, non dispositiones diversas essentiae additas...," *Ibid.*, fol. 67r, C. Cf. Forest, *La structure métaphysique du concret...*, pp. 144–45.

[5] "...existimavit, quod hoc nomen unum significat intentionem in re carente divisibilitate, et quod illa intentio est alia ab intentione, quae est natura illius rei." *Ibid.*, fol. 67r, D.

[6] *Ibid.*, fol. 67r, D–E.

[7] "Unde opinatus fuit iste, quod hoc nomen unum significat accidens in entibus...," *Ibid.*, fol. 67r, E.

one through some [third] thing added to its, the question must be asked again [concerning the third thing], and so on ad infinitum."[1]

Such is Averroes' presentation and refutation of Avicenna.[2] We have already noted the contrast between Averroes' and Thomas' references to Avicenna. It is interesting to note that Thomas adopts the argument (based on the regress to infinity) used against Avicenna; Thomas, however, gives it as the proof that *ens* and *unum* are predicated of the substance (essence) of anything.[3] Then too Thomas agrees with Averroes' view of the Avicennian "one": there is a confusion between mathematical and metaphysical unity.[4] Yet despite this dependence of Thomas on Averroes, it is a dependence in detail, and not in spirit: although Thomas attacks Avicenna and even uses some of Averroes' thoughts, Thomas' attack is strongly mitigated in language; in addition, Thomas adopts from Avicenna the idea of *esse* as other than essence. – One sees then that Thomas knew his own mind, and thus one must admit that the *Commentary* of Averroes was decidedly unsatisfactory in Thomas' eyes, at least in the passage in question.

When we turn to Albert, we discover that much the same relation existed between Albert and Averroes as between Thomas and Averroes; of more importance, however, is the presence in Thomas' work of somewhat the same attitude towards Albert as towards Averroes.

In beginning our study of Averroes' exposition of Γ 2, 1003b 22–33, we noted that there was nothing to distinguish his exposition of metaphysics as the study of unity from the exposition of Aristotle. The same can be said of Albert's work, although in three ways Albert unifies and makes more precise the Aristotelian exposition. What is important about these additions is that Thomas took a position in regard to each of them.

[1] "Quoniam, si res esset unum per aliquam rem additam suae naturae, sicut credit Avicenna, tunc nihil esset unum per se, et per suam substantiam, sed per rem additam suae substantiae. Et illa res, quae est una, si dicitur quod est una per intentionem additam suae essentiae, quaeretur etiam de illa re, per quam fit una, et per quid fit una: si igitur fit una per intentionem additam illi, revertitur quaestio, et procedetur in infinitum." *Ibid.*, fol. 67v, G. Cf. Forest, *La structure métaphysique du concret...*, p. 144.

[2] Averroes does not signal out the Avicennian notion of "being" for a special attack at this time. Perhaps he felt that simple explanation of Aristotles' words was sufficient; this may have been the case for he had in the *Arabica* a doctrine contrary to Avicenna: "...dicimus, quod substantia cuiuslibet unius communis est esse eius." cf. *Aristotelis Metaphysicorum Libri XIIII...*, fol. 66v, I. When Averroes explained this he wrote: "...substantia cuiuslibet unius, per quam est unum, est suum esse, per quod est ens." Cf. *In IV Meta.*, c. 3, fol. 67v, H. – In a later passage (*In X Meta.*, c. 8, fol. 257r, E–v, K) Averroes is very careful to point out the root of Avicenna's mistaken opinion on *esse*; the passage will be discussed in section 4, B below.

[3] *In IV Meta.*, 2, 555; cf. footnote 3, p. 112.

[4] *Ibid.*, 557; 559–60. Cf. footnote 3, p. 113 for par. 557.

Book Γ 2, 1003b 22–33 explains that the study of being is the study of unity because "being" and "unity" signify the same nature. Albert agrees with this doctrine of course.[1] Yet he notes that "unity" is the first among all the essential attributes of being as such; and since it was proved in the opening section of Book Γ that metaphysics studies being and its essential attributes, it follows that it will study "unity".[2]

This doctrine of metaphysics as the study of being and its essential attributes is of course correctly attributed to Aristotle.[3] It is, too, the doctrine Thomas gives in opening his exposition of Book Γ (IV).[4] Yet Thomas, unlike Albert, makes no reference to it when he begins his exposition of Γ 2, 1003b 22–33; Thomas prefers to see these lines as part of Aristotle's answer to the fifth aporie.[5] Yet Thomas fundamentally agrees with Albert, for Thomas gives as a conclusion to the treatment of the fifth aporie the doctrine of Albert noted above.[6] It seems possible, then, to see here a slight influence of Albert on Thomas.

A more evident influence concerns the distinction between the *id a quo* and the *id ad quod* of names. In making this distinction in the context of Aristotle's discussion of the predication of *ens* end *unum*, Albert makes more precise the Aristotelian doctrine. But as well, he gives a hint to Thomas.

Every name, Albert writes, is imposed *on* something, and is imposed *from* some quality or mode of quality.[7] By "from some quality or mode of quality," Albert means that every name is imposed in relation to some affirmation or negation of a quality as in a subject.[8] This is true both of *ens* and of *unum*. Both *unum* and *ens* are imposed on the same object: they both signify or refer to the same thing – to the "nature of the form." *Ens* is a name signifying the substantial form as giving *esse*, or as being the act of matter. *Unum* signifies the same substantial

[1] *In IV Meta.*, T. I, c. 4, p. 165, l. 37–p. 166, l. 58.

[2] "Nunc igitur tempus est, ut secundam partem a principio huius libri inductae propositionis declaremus, quod videlicet haec eadem scientia est de his quae insunt enti, secundum quod est ens, sicut superius diximus. Primum autem inter haec quae sunt entis, inquantum est ens, unum est, quod convertitur cum ipso." *Ibid.*, p. 165, l. 31–37. – The earlier sections to which he refers are: *Ibid.*, c. 1, p. 161, l. 46–53; and p. 161, l. 55–p. 162, l. 41.

[3] Γ 1, 1003a21–22.

[4] *In IV Meta.*, 1, 529–31; cf. footnote 4, p. 102 for par. 531.

[5] *Ibid.*, 548; cf. footnote 2, p. 110.

[6] *Ibid.*, 4, 587; cf. footnote 2, p. 107.

[7] "...ens et unum non sunt nominabilia sive ostensibilia nomine uno, ideo, quia omne nomen, sicut alicui imponitur, ita ab aliqua imponitur qualitate vel modo qualitatis." *In IV Meta.*, T. 1, c. 4, p. 165, l. 63–66.

[8] "...qualitatem tamen large accipit pro qualitate et modo qualitatis sive per affirmationem sive per negationem dictae...," *Ibid.*, p. 165, l. 68–70.

form, but as "terminating" the thing – that is, as making the thing undivided in itself and divided from another.[1]

This doctrine is given by Albert as a type of preamble to the proofs that "unity" and "being" express the same object, but differently.[2] In Thomas' exposition, the distinction between *nomen imponitur ab...* and *nomen imponitur ad...* is every bit as clear, although it comes at the end of each proof rather than as a preamble. There is no direct exposition of this theory in Thomas lesson 2, to be sure; even more this distinction had appeared in Thomas' work long before the *Commentary*.[3]

There is still a third point upon which Albert adds to Aristotle's Γ 2, 1003b 22–33. Averroes had criticized Avicenna's views on "being" and "one." As an argument against the Avicennian theory of "one," Averroes notes that Avicenna would be forced into an infinite search for the source of a thing's unity.[4] Thomas, as we noted, adopts this argument; he takes it out of the anti-Avicenna context, however, and uses it to bolster Aristotle's theory.[5] In doing this, Thomas was but copying Albert, for the latter too uses it only to bolster the Aristotelian doctrine that "one" and "being" are predicated of the substance of anything.[6]

There are these three points, then, in which Thomas was influenced by Albert: 1) the discussion of unity as studied by metaphysics rests on the prior doctrine that metaphysics studies the essential attributes of being; 2) the distinction of the *id a quo* and *id ad quod* of names is used to clarify Aristotle's teaching on the predication of "unity" and "being"; and 3) the separation of Averroes' infinite-regress argument

[1] "Et similiter est de modis entis et unius. Ens enim est a forma; forma autem duo facit per suam essentiam et non per accidens: unum quidem, quia dat esse per hoc quod est actus, alterum autum est, quod terminat per hoc quod est terminus entis; terminat autem per hoc quod facit indivisum in se et ab aliis divisum. Indivisum autem in se est nondivisum et divisum ab aliis est per hoc quod est non-alia, et sic terminatio formae in eo quod est. Nomen ergo entis est naturae formae per hoc quod dat esse, et nomen unius est eiusdem naturae per hoc quod est terminus, nec addit super ens naturam, sed modum, qui consistit in negatione consequenti hanc affirmationem, qua dicitur hoc ens esse ens." *Ibid.*, p. 166, l. 1–15.

[2] For the proofs, cf. *Ibid.*, p. 166, l. 16–58.

[3] For example, note the texts prior to the *Commentary* discussed in: McInerny, *The Logic of Analogy*, pp. 54–57.

[4] *In IV Meta.*, c. 3, fol. 67v, G; cf. footnote 1, p. 147.

[5] *In IV Meta.*, 2, 555; cf. footnote 3, p. 112.

[6] "Amplius autem uniusquiusque rei substantia unum est quiddam non secundum accidens aliquod, sed per seipsam. Si enim secundum accidens aliquod aliud a se esset una, illud accidens aut esset unum aut non. Si non esset unum, tunc quod non esset unum, faceret substantiam unam, quod non potest esse. Si autem esset unum, tunc aut illud per seipsum esset unum aut per aliud. Et non potest dici, quod per seipsum, quia datum est, quod unum addit naturam aliam, qua id est unum, quod unum dicitur. Si autem per aliud est unum, tunc iterum quaeritur de illo alio, et ibit in infinitum. Igitur dare oportet, quod uniuscuiusque rei per se unum est non secundum aliud accidens aliquod." *In IV Meta.*, T. 1, c. 4, p. 166, l. 40–53.

from its Avicennian context and the use of the argument to strengthen Aristotle's thought. Thus from what has been seen of Albert's exposition of Γ 2, 1003b 22–33, one must concede that it was rather satisfactory in Thomas'eyes: "rather satisfactory," but not totally. As was noted, Thomas prefers to see these lines of Aristotle as the answer to the fifth aporie.

When one compares Albert's exposition of Avicenna's errors with Thomas', one must note a further reason which prevented Thomas from agreeing whole-heartedly with Albert. Albert devotes an entire chapter to the exposition and refutation of Avicenna. Unlike Averroes' very harsh judgment, that of Albert is mild: "Et facile est per haec quae dicta sunt, excusare dicta Avicennae, quia pro certo, si quis subtiliter dicta sua respiciat, dicere intendit hoc quod hic dictum est."[1] In his mildness Albert was the forerunner of Thomas.[2]

Yet even though Thomas felt much more "at home" in Albert's leniency toward Avicenna, he could no more accept Albert's exposition than he could that of Averroes: for Albert, as Averroes, rejects the distinction of *esse* and essence.[3] Hence despite a great deal of agreement between the works of Albert and Thomas, Thomas' exposition yet rises above that of his predecessor in so far as the doctrine of the *esse*-essence composition is affirmed.

So much for the proofs that metaphysics studies unity. Thomas adds a second section to his lesson 2, in which section he makes two points: 1) there are ten categories of unity, all to be studied by metaphysics; and 2) metaphysics studies all the species of substances as falling under the genus of *ens* and *unum*. There is no need to examine the exposition of the corresponding lines of Aristotle given by Averroes and Albert. Fundamentally, Albert and Thomas are in agreement.[4] Averroes has a different theory because of a poor translation of Aristotle.[5] There is nothing of value to learn from a comparison of these expositions. Hence

[1] *Ibid.*, c. 5, p. 167, l. 65–68. Compare this with Averroes' words: "Avicenna autem peccavit multum in hoc, quia existimavit, quod unum et ens significant dispositiones additas essentiae rei. Et mirum est de isto homine, quomodo erravit tali errore; et iste audivit Loquentes in nostra lege, cum quorum sermonibus admiscuit ipse suam scientiam Divinam." *In IV Meta.*, c. 3, fol. 67r, B.

[2] Thomas wrote: "Sciendum est autem quod circa hoc Avicenna aliud sensit." *In IV Meta.*, 2, 556.

[3] For Averroes see: *In IV Meta.*, c. 3, fol. 66v, Ksqq. For Albert: *In IV Meta.*, T. I, c. 5, p. 166sqq. For Thomas: *In IV Meta.*, 2, 558.

[4] For Albert: *In IV Meta.*, T. I, c. 4, p. 166, l. 59–66; *Ibid.*, c. 6, p. 167, l. 75–p. 168, l. 15. For Aquinas: *In IV Meta.*, 2, 561–63.

[5] *In IV Meta.*, c. 3, fol. 67v, H–I; *Ibid.*, c. 4, fol. 68v, I–M. For the errors in *Arabica*, cf. *Aristotelis Metaphysicorum Libri XIIII...*, fol. 66v, I–K.

we can conclude our comparative study of Averroes', Albert's, and Thomas' expositions of Γ 2, 1003b 22–1004a 9.

The conclusions which may be drawn on the basis of these comparisons are approximately the same as those permissable at the end of the previous section.

1) Albert was not at all satisfied with Averroes' treatment of Aristotle; Thomas was not totally satisfied with either Averroes' or Albert's expositions.

2) Thomas certainly wished to adhere to the general pattern and limits of Aristotelian metaphysics.

3) However, Thomas changed the innermost meaning of that metaphysics by inserting the doctrine of *esse* as distinct from essence. – The principle elements of the comparative study may be schematized again as follows.

Aristotle	*Averroes*	*Albert*	*Aquinas*
Being and unity are the same in *re* but they do not have the same definition. There are two proofs of this: a) "one" and "being", if said of man, do not add anything not expressed by "man"; b) the substance of anything is one and is from its very nature. The context implies that he is proving that one is to be studied by metaphysics. (Γ 2, 1003b 22–23)	He gives the same doctrine as Aristotle, and then attacks Avicenna's theory on *ens* and on *unum*; he gives this as an argument against Avicenna's *unum*: if Avicenna is right, we are involved in an endless search for the source of unity. (*In IV Meta.*, c. 3, fol. 66v, K–fol. 67v, H)	He gives Aristotle's doctrine, but adds precision to it: a) he notes that unity is an essential attribute of being and hence should be studied; b) he distinguishes between the *id a quo* and the *id ad quod* of names; c) he removes from its Avicennian context the argument involving an infinite regress, and uses it to prove Aristotle's point. Albert then lists and refutes the theories of Avicenna; but he is very mild in his judgment of him. (*In IV Meta.*, T. I, c. 4, p. 165, l. 31–p. 166, l. 58; c. 5, p. 166 sqq.)	This section of Aristotle is an answer to the aporie concerning the study of unity; this section is divided into three parts: a) introduction: *ens* and *unum* are the same in *re* but not in *ratione* and hence are to be studied in the same science; b) the first proof that they are the same is the first one of Aristotle; into this context the distinction of *id a quo* and the *id ad quod* is introduced; *res* is spoken of as on the same level as *unum* and *ens*; c) the second proof is given; this is the second one of Aristotle; Thomas places

Aristotle	*Averroes*	*Albert*	*Aquinas*
			the Averroist argument against an infinite regress into this context; then Avicenna is explained and refuted with no word of condemnation; Thomas notes that Avicenna's *esse* is a part of being, but *ens* does not mean *esse* as Avicenna said; again the *id a quo* and the *id ad quod* appear. (*In IV Meta.*, 2, 548–60)

On the basis of such a comparison we are permitted to conclude that Aquinas' *In IV Meta.*, *lectio* 2 does not agree with Aristotle's passage. Moreover, *lectio* 1, since it must be read in the light of the *esse*-essence composition of beings presents also a doctrine not found in Aristotle. But is Aquinas conscious of the change he has made? We have seen how carefully Aquinas reads Aristotle, how he correlates widely scattered passages of Aristotle. In the light of such evidence one is perhaps inclined to feel that Aquinas is unaware of the transformation he has affected. But at the present, it is much too early to be certain.

4. AVICENNA'S AND AQUINAS' EXPOSITIONS OF THE OBJECT OF METAPHYSICS

Thus far we have studied the relation of Thomas' work to three of his predecessors: Aristotle, Averroes, and Albert. There was however a fourth philosopher present in Thomas' thought as he wrote: Avicenna. The latter, unlike Averroes and Albert, did not mold his metaphysical writings according to the plan of Aristotle's *Metaphysics*. Despite this fact, the *Metaphysics* of Avicenna was used by Thomas as a guide when the latter set out to write his *Commentary*.[1] Hence, before closing this study of Aquinas' views on the object of metaphysics, we must ask how those views are related to those of Avicenna.[2]

[1] E.g. note Thomas' appeal to Avicenna's ideas in: *In V Meta.*, 2, 766–69; *In VI Meta.*, 1, 1165; *Ibid.*, 3, 1192–93; and so on. In ten separate passages Thomas refers by name to Avicenna.

[2] I shall refer primarily to Avicenna's *Metaphysica*. Thomas knew and could have referred

A. Avicenna's metaphysics

In Chapters II–III various Avicennian doctrines on the science of metaphysics were discussed. Metaphysics, it was noted, is a study of God, of being as such, and of the causes; yet only being is studied as the subject. It is this notion that must now be examined more closely, but before we begin, it is necessary to know exactly what Avicenna means by such terms as *ens* and *esse*. One must have a good grasp of the meanings attributed to these words; otherwise one will most certainly misunderstand the most general of Avicennian statements concerning metaphysics, not to mention Thomas' criticism of Avicenna.

The best method of discovering the Avicennian meaning of *ens* and *esse* is to follow his exposition in *Meta.*, T. I, c. 6–7. These chapters open with his famous theory of *ens* as the first notion had by the intellect. *Ens, res, necesse*: the ideas corresponding to these words are the first man has, Avicenna writes.[1] They are so well known to man, so basic an element of his knowledge, that they can not be defined or explained through any other notions. As Avicenna remarks, any attempt to explain these notions through other ones would involve us in a vicious circle, since all other notions are to be explained through *ens, res,* and *necesse.*[2] An example of how some falsely try to explain *ens*: "certitudo entis est vel quod est agens vel patiens."[3] In other words, some would say, the reality of *ens* is either that of an agent or that of a patient. A thing, as realized, is either patient or agent.[4] Such an explanation, insists Avicenna, is actually no "explanation" at all, for *ens* is better known than anything else. Everybody knows what it means to be real, to be *ens*; but it is not so easy to discover whether a given *ens* is an agent or a patient. In fact, one must "argue" to the realization that any given object is an agent, or a patient, as the case may be.[5]

to the *De Anima* or to parts of the *Logyca*. Yet the *Metaphysica* presents Avicenna's systematic metaphysics. Thus when Thomas turned to Avicenna the metaphysician it was primarily to the *Metaphysica* that he turned.

[1] "...ens et res et necesse talia sunt quod statim imprimuntur in anima prima impressione quae non acquiritur ex aliis notioribus se...," *Meta.*, T. I, c. 6, fol. 72r, A. Cf. GOICHON, *La distinction de l'essence et de l'existence...*, pp. 2–5; SAUTIER, *Avicennas Bearbeitung der Aristotelischen Metaphysik*, p. 52.

[2] *Ibid.*, cf. the lines immediately following the quotation of footnote 1. See GOICHON, *La distinction de l'essence et de l'existence...*, pp. 7–8.

[3] *Meta.*, T. I, c. 6, fol. 72r, A.

[4] As noted earlier, Mlle. Goichon explains that *certitudo* can have three meanings: 1) the ontological truth of a thing: a thing realized conformably to its essence; 2) the truth in our intellects: the exact conception of a thing; and 3) quiddity. Cf. GOICHON, *La distinction de l'essence et de l'existence...*, p. 48. In the present case, the *certitudo entis* would appear to involve the first meaning, the realization of essence.

[5] "Sed tamen ens notius est quam agens vel patiens. Omnes enim homines imaginantur certitudinem entis, sed ignorant an debeat esse agens vel patiens, quod et mihi quodque

After the opening discussion of the futility of all attempts to explain the meaning of *ens* through anything better known, Avicenna turns to *res*, noting that it too can not be explained through anything else. In the course of the exposition he makes it clear that the following words are synonyms: *res, quid, aliquid, illud, id, ens.*[1] On the other hand, Avicenna notes that we have an idea of *ens* which is different from our idea of *res*: "Dico ergo quod intentio entis, et intentio rei imaginantur in animabus due intentiones."[2] To what does the "ergo" refer? This sentence follows the parallel explanation of why one can not explain the meanings of *ens* and *res*; in the discussion of *res*, there is the identification of *res* with many other notions, among which is *ens*. Yet *res* has been said to be different from *ens*! The key to Avicenna's thought lies in the distinction of *esse* and essence: *res* is the same as *ens*, when *ens* means essence, but different from *ens* or *esse*. Hence, the "ergo" must be an introduction to a new point: the "ergo" refers to the distinction between *esse* and essence which Avicenna immediately begins to explain.

Everything has "certitude" or quiddity by which it is what it is; for example, a triangle has a quiddity by which it is a triangle. This fact of having a quiddity is what is called *esse proprium*. By *esse proprium*, Avicenna explains that he means the "intention" or understanding of an "affirmed esse." That is to say, when one consciously affirms (or judges) the reality of a quiddity, one affirms the *esse* proper to the quiddity. The word *ens*, Avicenna is quick to note, refers not only to the quiddities of things, but also to their proper *esse*'s. Thus a thing's quiddity is not its *esse*.[3]

usque nunc non patuit, nisi argumentatione tantum; qualis est ergo iste qui id quod est manifestum laborat facere notum per proprietatem quae adhuc opus est probari, ut constet esse illius." *Meta.*, T. I, c. 6, fol. 72r, B–fol. 72v. – One may explain the affair a little differently: we can't explain *ens* through other concepts such as "agent", because *ens*, as the poorest of all concepts, is included in "agent"; cf. Goichon, *La distinction de l'essence et de l'existence...*, pp. 2–5; 13–15.

[1] "...res vel aliquid vel quid vel illud, et haec omnia multivoca sunt nomina rei...... cum dicis quod res est id de quo vere potest aliquid enunciari idem est quasi diceres quod res est res, de qua vere potest aliquid enunciari, nam id et illud, et res eiusdem sensus sunt. ...ens vero et aliquid sunt nomina multivoca unius intentionis." *Meta.*, T. I, c. 6, fol. 72v, B–C.

[2] *Ibid.*, fol. 72v, C. Cf. Sautier, *Avicennas Bearbeitung der Aristotelischen Metaphysik*, pp. 52–53; Paulus, *Henri de Gand...*, pp. 24–25.

[3] "...unaquaeque enim res habet certitudinem qua est id quod est; sicut triangulus habet certitudinem, qua est triangulus, et albedo habet certitudinem qua est albedo, et hoc est quod fortasse appellamus esse proprium; nec intendimus per illud nisi intentionem esse affirmativi, quia verbum ens signat etiam multas intentiones, ex quibus est certitudo qua est unaquaeque res, et est sicut esse proprium rei. Redeamus igitur et dicamus quod de his quae manifesti sunt est hoc quod unaquaeque res habet certitudinem propriam quae est eius quidditas, et notum est quod certitudo cuiusque rei quae est propria ei est praeter esse quod multivocum est cum aliquid." *Ibid.*, Let us note how this passage is being understood. 1) A thing's *certitudo*

Avicenna continues: When one speaks of the "certitude" or quiddity, then one can choose two ways of speaking: 1) one can say that the quiddity is found or exists in the mind, or in an individual object; and one can say as well that the quiddity is considered as common both to the quiddity in the mind and to the quiddity in an individual object; 2) on the other hand, one can say that a quiddity is a quiddity; for example: "Horseness is horseness." Now if we speak of quiddity in the first way, what we may say makes sense, and will be understood; but if we speak of quiddity in the second way, we are wasting our time by stating the obvious.[1] In like manner, Avicenna continues, if one takes the useless statement: "Horseness is horseness," and substitutes the general terms "certitude" and "thing", and thus says: "Certitude is thing," one is still wasting one's time.[2] Avicenna is thus saying that, if one understands a quiddity, and then tells oneself that this quiddity is a quiddity, one is only telling oneself what one already knows. However, one is not wasting one's time if one speaks of

is its quiddity as is explicitly stated. 2) *Esse* is not quiddity as Avicenna says. 3) "hoc est quod fortasse appellamus esse proprium; nec intendimus per illud nisi intentionem esse affirmativi"; to what does "hoc" refer? It can not refer to *certitudo*, as this is distinct from *esse*; hence it must refer to "habet certitudinem qua est id quod est". I have translated this as "the fact of having a quiddity". 4) "intentionem esse affirmativi": "an affirmed *esse*", this is translated. The remainder of Avicenna's chapter bolsters this reading, since he begins immediately afterwards to distinguish between a) the knowledge of quiddity and b) the knowledge of *esse*, or the knowledge that a quiddity does or can exist. 5) "quia verbum ens signat etiam multas intentiones ex quibus est certitudo qua est unaquaeque res": I feel that Avicenna desires to refer to his doctrine that the ten categories of substance and accidents are divisions of *ens* taken as a quiddity, whereas *ens* (here: *esse*) can also be used to refer to the *fact* of having a quiddity. 6) "certitudo cuiusque rei quae est propria ei est praeter esse quod multivocum est cum aliquid": what does the "esse quod multivocum est cum aliquid" mean? Supposing this translation to be correct, note that Avicenna does not identify the meaning of *esse* and of *aliquid*. He had earlier identified *ens*, *aliquid*, and *res*; cf. footnote 1, p. 154. All Avicenna appears to affirm here is that *esse* has as many uses or meanings as does *aliquid*; or perhaps he wishes to say that for every *aliquid*, there is an *esse*. – Although this text appears central to Avicennian metaphysics, no treatment of Avicenna treats it. This lack is especially noteworthy as regards SAUTIER, *Avicennas Bearbeitung der Aristotelischen Metaphysik*; although the author follows Avicenna's work, at times, chapter by chapter, there is practically nothing about the distinction of *esse* and essence, and nothing at all about the text just discussed. For a similar lacuna, cf. O. CHAHINE, *Ontologie et théologie chez Avicenne*, Adrien Maisonneuve, Paris, 1962.

[1] "Quoniam cum dixeris quod certitudo talis rei est in singularibus vel in anima vel absolute, ita ut communicet utrisque, erit tunc haec intentio apprehensa vel intellecta. Sed cum dixeris quod certitudo huius vel certitudo illius est certitudo, erit superflua enunciatio et inutilis." *Ibid.* Avicenna is thinking here of his famous distinction between 1) *natura absoluta*, 2) *natura* existing in the mind, 3) *natura* existing in an individual. Cf. GOICHON, *La distinction de l'essence et de l'existence...*, pp. 31–44; PAULUS, *Henri de Gand...*, pp. 69–74. Avicenna's distinction also appears in *Logyca*, T. III, fol. 12r, A–B; this was used by the early Thomas in *De ente*, c. 3, p. 24, l. 1–p. 25, l. 4 (édition Roland-Gosselin).

[2] "Si autem diceres quod certitudo huius est res, erit etiam haec enunciatio inutilis ad id quod ignorabamus." *Meta.*, T. I, c. 6, fol. 72v, C.

quiddity in the first way mentioned; that is, one expresses something hitherto unexpressed by stating: "Quiddity is."[1]

What Avicenna is trying to point out is quite obvious to us today: it is one thing to understand a quiddity; it is roughly the same thing to know we understand a quiddity; it is quite another thing to know that quiddity exists in reality or in the mind, or that it can exist in either of these. Avicenna is thus distinguishing between two acts of knowledge. It would be wrong to believe, however, that Avicenna does not intend to present this distinction between *esse* and essence as a real one. He intends the knowledge of quiddity and the knowledge of existence to refer to different aspects of a being; that this is his meaning will be clear when we note his ideas of the possible and the necessary.[2]

Avicenna's exposition then takes up the case of statements about non-being. We needn't follow the entire argument; a brief review of the conclusion is sufficient. There he gives the example of the proposition: "There will be a future resurrection" (*Resurrectio erit*). To make such a statement four intellectual acts must be made: 1) the understanding of "resurrection"; 2) the understanding of "there will be" (*erit*); 3) the predication of "resurrection" of "there will be"; and 4) the understanding that in the future someone can say of "resurrection": "It is." What Avicenna means can be restated thus: one must understand the idea of a future resurrection; and then one must affirm an actual existence of the quiddity "future resurrection" in the sense that it is true that in the future someone can predicate the word "is" of "resurrection."[3] By this explanation Avicenna is again able to distinguish between existence and essence. That is to say, because the idea of "future resurrection" has within it no note of actual existence, actual existence must be something which is other than real quiddities, something which is needed to have a being.[4]

[1] "Quod igitur utilius est dicere hoc est, scilicet ut dices quod certitudo est res, sed haec res intelligitur ens; sicut si diceres quod certitudo huius est certitudo *quae* est." *Ibid.*. Italics added. Note that Avicenna says "certitudo *quae* est" and not "certitudo *qua* est".

[2] Cf. GOICHON, *La distinction de l'essence et de l'existence...*, pp. 134–48; GILSON, *Being and Some Philosophers*, pp. 76–81.

[3] "Si dixeris quod resurrectio erit, intellexisti resurrectionem et intellexisti erit, et praedicasti erit, quod est in anima de resurrectione. Sed haec intentio non potest esse vera, nisi de alia intentione intellecta etiam quae intelligatur, ut in hora futura de ea dicatur intentio tertia intellecta quae est intentio, scilicet est, et secundum hanc considerationem similiter est in praeterito." *Ibid.*.

[4] The context makes this evident, as do the final words of the section: "Iam igitur intellexisti nunc qualiter different, et id quod intelligitur de esse, et quod intelligitur de aliquid, quamvis haec duo sint comitantia." *Ibid.*. Also the scathing denunciation which is next given: "Significatum est tamen mihi esse homines qui dicunt quod aliquid est aliquid, quamvis non habeat esse... Isti autem non sunt de universalitate eorum qui cognoscunt." *Ibid.*.

Esse (or the fact of actually having a quiddity here and now) and quiddity: these then are the constituents of things for Avicenna. As for *ens*, it has been noted already that it is a term which includes both *esse affirmativum* (that is, *esse proprium*) and quiddity.[1] Avicenna brings out more clearly the quidditative meaning of *ens* when he notes that *ens* is not a genus, nor is it predicated in the same way of everything of which it is predicated. Rather, first of all it is predicated of the quiddity which is in substance, and then of the other categories. Finally it is predicated of the essential accidents of being.[2]

If we turn now to examine our knowledge of things as "in esse," we note immediately that it divides into two classes: the possible and the necessary. The possible is that whose *esse* is neither necessary nor impossible; if it were impossible, then of course our knowledge would not contain the idea of possible. (The possible is, then, not what a Thomist would identify by "contingent," but rather by "intrinsically or essentially possible.")[3] On the other hand is the necessary: that whose *esse* is necessary, that whose *esse* is seen as necessary.[4]

Avicenna's distinction of existence and essence is not only mental but extra-mental as well. Although Avicenna does not expose his argument in the following way, such an argument would express his theory. 1) One must think of beings as composed of existence and essence, two distinct aspects. 2) Moreover, if existence is not distinct from essence in extra-mental beings, then existence would be part of the essence; if

[1] But it would be wrong to say it is predicated of quiddity, if *ens* is taken in the sense of an extra-mental here and now thing.

[2] "...quamvis ens, sicut scisti non sit genus, nec praedicatum equaliter de his quae sub eo sunt, tamen est intentio in qua conveniunt secundum prius et posterius; primum autem est quidditati, quae est in substantia, deinde ei quae est post ipsum; postquam autem una intentio est ens secundum hoc, quod assignavimus; sequitur illud accidentalia, quae ei sunt propria sicut supra docuimus..., *Ibid.*. Note that *ens* is predicated of *quidditas* which *is* in substance. This statement follows on Avicenna's theory that the first principle of thought is based on *esse*, on existence, and not on quiddity. Cf. *Meta.*, T. I, c. 9, fol. 74r, B.

[3] Cf. *Ibid.*, c. 7, fol. 73r, Col. A.

[4] "Dicimus igitur quod ea quae cadunt sub esse possunt in intellectu dividi in duo. Quorum unum est quod cum consideratum fuerit per se eius esse non est necessarium, et palam est etiam quod eius esse non est impossibile, alioquin non cadet sub esse...... alterum est, quod cum consideratum fuerit per se eius esse erit necesse." *Ibid.*, fol. 73r, A. Cf. Sautier, *Avicennas Bearbeitung der Aristotelischen Metaphysik*, pp. 53–54. – Avicenna's "being" is an univocal concept, at least by implication, it has been said; cf. E. Gilson, "Avicenne et le point de départ de Duns Sciot," *Archives d' histoire doctrinale et littéraire du Moyen Age*, II, 1927, pp. 100–117; Paulus, *Henri de Gand...*, p. 55. But M. Gardet does not agree; the real distinction he says prevents Avicenna from conceiving "being" as univocal. That is, the difference between the possible being, which may have *esse*, and the necessary being, which must have *esse*, precludes a concept containing them both; cf. L. Gardet, *La pensée religieuse d'Avicenne (Ibn Sīnā)*, Études de phil. médiévale XLI, Vrin, Paris, 1951, pp. 55–57. Mlle Goichon is completely in agreement with M. Gardet; cf. Goichon, *La philosophie d'Avicenne...*, p. 24, footnote 1.

existence is part of the essence of every being, then every being is intrinsically necessary – every being is totally perfect, necessary, uncaused, and moreover, there is only one being.[5] Since these conclusions obviously do not hold for the world around us, Avicenna could say, all beings of our world have essences distinct from their existences.

As has been pointed out, the *esse proprium* of an object is known when we affirm the reality of a quiddity. The idea of *esse* is, then, the idea of what is known when we judge: "There will be a resurrection." *Esse* is the actual, the here-and-now, possession of quiddity. Obviously the knowledge expressed in the Avicennian judgment of existence is, then, knowledge of the quiddity as found outside the mind here and now: it is knowledge of a *de facto* realization of a quiddity. Thus we affirm the "here-and-now" of John's humanity, which needn't exist "here-and-now" but could be only in the past – such a here-and-now quiddity is a possible. Avicenna's doctrine is thus intelligible: in the judgment of existence we express the realization of a quiddity in the "here-and-now."[1]

It must follow that for Avicenna one can make such judgments of existence without referring to the materiality of a being. If one refers, for example, to "white," not as a particular material accident of color, but as known in a judgment of "here-and-now" presence, then one does not refer to materiality. Thus, to say: "It is white" is to judge of the *de facto* realization of whiteness. As far as the judgment is concerned, one contacts the *de facto* realization, abstraction made from the material accident of color itself. If one wished to consider material objects such as "white," "bodies," etc. in so far as one judges their here-and-now reality, then one would have an immaterial consideration (formal object) of material beings (material object). It is precisely this type of "gathering together" that Avicenna assigns to metaphysics.

Considering the sciences such as mathematics and physics, Avicenna remarks that neither of these particular sciences has studied substance under the formality of *ens*. Nor has any of them studied measure or

[1] "Dicemus igitur quod necesse esse per se non habet causam, et quod possibile esse per se habet causam; et quod necesse per se est necesse omnibus suis modis; et quod impossible est ut esse eius quod est necesse esse sit coequale ad esse alterius; ...sequitur quod necesse esse nonest ... multiplex[;]... constat quod si necesse esse per se haberet causam perfecto non esse necesse esse per se; manifestum est igitur quod necesse esse non habet causam." *Meta.*, T. I, c. 7, fol. 73r, A–B.

[2] The judgment of existence is a judgment of the here-and-now of an essence: most of the existential Thomists avoid this mistake. E. g. GILSON, *Being and Some Philosophers*, p. 167. However, some fail to distinguish between the "here and now" and Aquinas' *actualitas omnium formarum* which is *esse*; e.g. J. MARITAIN, *A Preface to Metaphysics. Seven Lectures on Being*, Sheed & Ward, New York, 1948, pp. 17–22

number in so far as they have *esse*. Still more, none of them has studied immaterial substances in so far as they have *esse*. The reason such topics have not been taken up in the particular sciences is easy to discover: such a study must be made by a science interested in things separated from matter both in their existence and in their definition. Such a science is not to be found within the number of particular sciences.[1] Yet the study which takes the viewpoint of *esse* can carry through such investigations: the study of things in their here-and-now realizations will be in no way involved with matter.

Thus it is that the key conception of Avicennian metaphysics is the expression: "Nam estimatio est expoliatio a sensibilibus." In other words, metaphysics is to consider that aspect of things which is basic to anything; metaphysics is to consider that aspect which underlies even the materiality of material beings: metaphysics studies the *esse* or the here-and-now realization which is common to all things.[2]

An interesting aspect of this theory is the use of the word "common": *esse* or being is "common" to material and immaterial substances, to body, to number, to measure. In another example in this same context the word "common" is used thus: measure is a common name because by it one understands either the dimension which constutites a natural body, or else continuous quantity which is said of things such as a line or surface; if one studies the "common" aspect of measure expressed by the common name "measure," then one grasps the being or the *esse* of measure. One grasps what it means to be here-and-now a measure: one grasps what one affirms by: "It is a measure."[3] Thus for Avicenna,

[1] "Deinde consideratio de substantia inquantum est ens vel est substantia, vel de corpore inquantum est substantia, et de mensura et numero inquantum habent esse, et quomodo habent esse, et de rebus formalibus, quae non sunt in materia vel si sunt in materia, non tamen corporea, et quomodo sunt illae, et quis modus est magis proprius illis separatim per se debet haberi; non enim potest esse ut sit alicuius scientiarum de sensibilibus, nec alicuius scientiarum de eo quod habet esse in sensibilibus." *Meta.*, T. I, c. 2, fol. 70v, A. "...divinae scientiae non inquirunt nisi res separatas a materia secundum existentiam et diffinitionem." *Ibid.*, c. 1, fol. 70r, B. "Manifestum est enim ex dispositione huius scientiae quod ipsa inquirit res separatas omnino a materia." *Ibid.*, fol. 70r, C.

[2] After the list of items studied in metaphysics (cf. the initial quote of the preceding footnote) Avicenna continues: "Nam estimatio est expoliatio a sensibilibus... ...manifestum est enim quod esse substantiae in quantum est substantiae tamen, non pendet ex materia; alioquin non esset substantia nisi sensibilibus. Numquam etiam accidit esse in sensibilibus et in non sensibilibus. Unde numerus inquantum numerus est, non pendet ex sensibilibus, nec ex insensibilibus... Subiectum autem logicae secundum se manifestum est esse praeter sensibilia; manifestum est igitur quod haec omnia cadunt in scientia, quae profitetur id cuius constitutio non pendet ex sensibilibus. Sed non potest ponit eis subiectum commune, ut illorum omnium sint dispositiones et accidentalia communia nisi esse." *Ibid.*, c. 2, fol. 70v, B.

[3] "...mensura etiam commune nomen est, quia mensura vel intelligitur dimensio, quae constituit corpus naturale vel intelligitur quantitas continua, quae dicitur de linea et superficie et corpore terminato... Nulla autem earum [idest, dimensionum corporis naturalis] est

there appears to be a link between what is "common" to things and that aspect of things which is their *esse*: *esse*, the here-and-now, is precisely that which is common to all things.

The word "common" recurs again in a passage where Avicenna begins to be more meticulous in listing the items to be studied. These things such as substances, numbers, measure, and so forth, are to be divided into predicaments; thus we have substance, quantity, etc. What is *common* to all these predicaments? Only *esse*.[1] Besides the predicaments, which all share in the "common" intention of *esse*, metaphysics must study *unum, multum, conveniens*; these realities are "common" to all sciences, they are used by science – yet are studied by none of the particular sciences. These are the common accidents of *esse* as such.[2] Thus metaphysics for Avicenna is the study of common *esse* and of the common accidents of *esse* as such.

Metaphysics, as we can now realize, contains under its material object such diverse items as material and immaterial substances, number, measures, the notions used in all science: *unum, multum*, etc.. It is because metaphysics can take the point of view of *esse* that it can simultaneously study all these different things; it is by studying the here-and -now realization of their quiddities that metaphysics can in a single breath speak of God, man, number, measure, and the essential attributes of things.[3]

separata a materia, mensura vero secundum primam acceptionem quamvis non sit separata a materia est tamen principium essendi corpora naturalia. Nec tamen ob hoc potest esse, ut constitutio eorum [corporum] pendeat ex ea [dimensione seu mensura], quasi ipsa [dimensio] det eis [corporibus] constitutionem ipsam; tunc enim procederet in esse ipsa sensibilia, sed non est ita... ...extremitates etiam sunt sub mensura, inquantum materia perficitur per illas, et postea concomittantur. Postquam autem ita est, tunc figura non habet esse nisi in materia, nec est prima causa in esse veniendi ad effectum; mensura vero secundum acceptionem secundam [idest, quantitas continua quae dicitur de linea, etc.] consideratur secundum esse suum, et secundum sua accidentalia. Sed consideratio de ea secundum esse suum, de quomodo essendi, et de qua divisione essendi sit, non est etiam ut consideratio de eo, quod pendet ex materia." *Ibid.*.

[1] "Quaedam enim eorum [idest: numerus, mensura, substantia materialis, etc.] sunt substantiae, et quaedam quantitates et quaedam alia praedicamenta, quae non possunt habere communem intentionem qua certificent nisi intentionem essendi." *Ibid.*.

[2] "Similiter etiam sunt res, quae debent diffiniri et verificari in anima, quae sunt communes in scientiis; nulla tamen earum [scientiarum] tractat de eis; sicut est unum inquantum est unum, et multum inquantum est multum, conveniens et inconveniens... De his enim mentionem tamen faciunt [particulares scientiae] et inducunt diffinitiones eorum, nec tamen loquuntur de modo essendi eorum, quia haec non sunt accidentalia propria alicui subiectorum aliarum scientiarum particularium, nec sunt de rebus quae habet esse nisi proprietatis esse essentialiter. Nec sunt etiam [particulares scientiae] de proprietatibus, quae sunt communes omni rei sic, ut unumquodque eorum [accidentium] sit commune omni rei; nec possunt esse propria [accidentia] alicuius praedicamenti, nec possunt esse accidentalia alicui nisi ei, quod est esse inquantum est esse. Igitur ostensum est tibi ex his omnibus quod ens inquantum est ens commune est omnibus his...," *Ibid*, fol. 70v, B–C.

[3] It is difficult to understand why no study of Avicenna mentions that the formal object

This then is Avicenna's metaphysics. Four important points beg for comparison with the writings of Aristotle, Averroes, Albert and Thomas 1) Man knows the common aspects of all things, 2) when he knows the *esse*, or here-and-now of quiddities; 3) the quiddities of things can be divided into the categories or predicaments; and 4) the ideas common to all sciences can be studied by metaphysics because these too can be seen under the light of *esse*. As is evident, these four points revolve around the doctrine of *esse*.

B. The fundamental criticisms of Avicenna given by Averroes, by Albert, and by Aquinas

When Averroes read Avicenna's *Metaphysics*, he would have had no difficulty in recognizing some elements of Avicenna's work as authentically Aristotelian: *ens* is not a genus, but is predicated of substances and accidents according to priority and posteriority; *unum, multum* and the other essential accidents of being are studied in metaphysics, while all the other sciences use such terms; and finally, metaphysics studies what is separate from matter both in existence and definition. These doctrines Averroes could discover in Avicenna and thus recognize the Aristotelian character of the latter's work.[1] Averroes would, of course, have noted the manner in which Avicenna employs the last-named doctrine of Aristotle: metaphysics is the study of what is separated from matter. Avicenna, as we have seen, saw metaphysics as the study of what falls under *esse*; metaphysics studies whatever can be affirmed as here and now. Thus, metaphysics completely prescinds from materiality.[2]

To attack such a theory, Averroes had but to point to some fallacy in the idea or theory of *esse*; once the distinction of *esse* and essence is discredited, Avicenna's system falls to pieces. Hence it is that we discover Averroes attacking Avicenna's views on the subject of *ens*. As

of his metaphysics is *esse*. Mlle. Goichon, to be sure, has called the distinction of *esse* and essence the central doctrine of the Avicennian system, but she does not point out how *esse* is the point of view. Cf. GOICHON, *La distinction de l'essence et de l'existence...*, pp. 2–5.

[1] The first two of these doctrines are found in Aristotle: Γ 1, 1003a21–2, 1003b19 (cf. section 2, A of this chapter) and Γ 3, 1005a21–29. The third doctrine, metaphysics studies what is separate from matter, is expressed by Aristotle thus: "...the first science deals with things which both exist separately and are immovable." E 1, 1026a15–16.

[2] Avicenna, one supposes, would have read E 1, 1026a15–16 in its context: "Now, we must not fail to notice the mode of being of the essence and of its definition...," E 1, 1025b28–29. Aristotle compares the being and definition of the subject of physics, mathematics, and metaphysics. But Aristotle had already explained that material substances, *unum, multum*, etc. are to be studied (Γ 1–2). Hence Avicenna would have attempted to explain how all these objects could be studied by the "science of the immaterial"; his answer is given in terms of the affirmation of *esse*: the formal object of metaphysics.

has already been pointed out, Averroes in his *In IV Meta.*, c. 3 launches a rather vehement attack on Avicenna's theory. Avicenna says that *ens* and *unum* do not signify the substance (or essence) of a thing, but rather some added nature, wrote Averroes.[1] Now although Averroes' manner of expressing himself leads one to believe that he is primarily thinking of Avicennian *esse* as if it were a nature, that is not the most important aspect of his attack on Avicenna. The essence of Averroes' opposition to the theory of *esse* is revealed in his exposition of Book I (X). Aristotle explains in the corresponding passage that "unity" denotes the same thing as "being"; the doctrine is obviously the same as that given in Book Γ 2.[2] In this context Averroes returns to the attack on Avicenna begun earlier in *In IV Meta.*, c. 3. The same charge is brought against Avicenna: Avicenna said that *ens* and *unum* signify a disposition, an accident added to a thing.[3] The source of Avicenna's error, writes Averroes, was a fundamental misunderstanding of the difference between *ens* as referring to the categories, and *ens* as referring to the truth of our knowledge. When we speak of that *ens* which refers to truth, we are speaking of an act; we do not thus refer to the categories of substance and accidents.[4] This failure to understand the difference between *ens* as true and *ens* as category explains why Avicenna could not understand that *ens homo* and *unus homo* are not synonymous, Averroes concludes. He the explains: Avicenna thought that if to refer to a real man, one need only say: *ens*, or: *res*, or: *unus homo*, then one would only be saying: "Man is man." Let us explain. If "one man" signifies "man" (the quiddity), then to say: "This man is one" is to set up a proposition where the quiddity "man" can be substituted both for the subject and for the predicate: "This man (or man) is one (or man)." Hence: "Man is man." It is because Avicenna thought this would be the consequence of identifying *ens*, *res*, and *unum* with the quiddity of anything that he introduced the distinction of *esse* and essence.[5]

[1] Cf. footnotes 1, 3, 5, 7, p. 146; 1, p. 147. PAULUS, *Henri de Cand...*, pp. 239; 271–75.

[2] Cf. I 2, 1053b25–1054a19.

[3] "Et Avicenna dixit quod ens et unum significant intentionem additam essentiae rei. Non enim opinatur, quod res est ens per se, sed per dispositionem additam ei: ut dicimus aliquid esse album; unum igitur et ens apud ipsum significat accidens in re." *In X Meta.*, c. 8, fol. 257r, E.

[4] "Sed deceperunt eum [Avicennam] duo, scilicet quia opinabatur quod unum... [etc.] Et etiam fuit deceptus, quia ignoravit differentiam inter hoc nomen ens, quod significat genus, et quod significat verum. Quod enim significat verum, est actus, et quod significat genus, significat unumquodque decem praedicamentorum multipliciter." *Ibid.*, fol. 257r, F–v, G.

[5] "Et non intendit [Aristoteles], quod ista nomina unum et ens sunt idem omnibus modis, sed intendit, quod significant idem secundum subiectum et diversa secundum modum; et ideo deceptus fuit Avicenna cum dixit si haec nomina unum et ens essent synonyma cum

Hence it would appear that the basis of Averroes' opposition to Avicenna is the difference between *ens* as true and *ens* as category. When we speak of that act by which we express the truth of a proposition, we are dealing with the *ens* which is expressed by the *est* of a proposition. Thus when we say: "Socrates is musical," where "is" means that our knowledge of Socrates is true, then "is" or *ens* deals only with truth.[1] Thus in our judgment that Socrates is white, we reach only the truth of our knowledge; in such a judgment we are not dealing directly with the being of Socrates which can be placed in one of the categories; We are not, therefore, dealing with the metaphysical being of objects; we are dealing only with our acts of knowledge – we are stating that our knowledge is true.[2]

In brief, what Averroes reproaches to Avicenna, is to have introduced into metaphysics that *ens* which signifies the truth of propositions: to study all things in so far as they are affirmed to be is to study all things in so far as we can express the truth of our knowledge of them. That is hardly Aristotelian metaphysics, Averroes would say, and of course he would be correct.[3]

Albert, too, was not content with Avicenna's theories of *unum* and *ens*. However, Albert's *In IV Meta.*, T. I, c. 5, the discussion of Avicenna, uses none of the Averroistic attacks on Avicenna. Rather Albert

hoc nomine res, non esset verum dicere res esse unam, verbi gratia, et si unus homo significaret hominem, tunc dicere istum hominem demonstratum esse unum, non esset enuntiabile compositum ex praedicato et subiecto: sed dicere hominem esse unum, esset sicut dicere hominem esse hominem." *Ibid.*, fol. 257v, I–K. Although Averroes does not mention here that Avicenna's *esse* is implicated in the latter's theory of the predication of *unum* and *ens*, this would follow from the text given in the preceding footnote; that earlier text sets the tone of the later discussion represented by the text cited in this footnote.

[1] "hoc nomen ens etiam significat illud, quod significat dicere aliquid esse verum. Cum enim dicimus aliquid esse, demonstramus ipsum esse verum, et quando dicimus non esse, demonstramus ipsum non esse, scilicet ipsum esse falsum... In propositione autem composita, sicut dicimus, quod Socrates est musicus, aut Socrates non est musicus: in quaestione autem simplici, sicut dicimus, utrum Socrates sit aut non sit, et ulterius hoc nomen ens non significat nisi verum... Sed debes scire ulterius, quod hoc nomen ens, quod significat essentiam rei, est aliud ab ente, quod significat verum...," *In V Meta.*, c. 14, fol. 117r, D–F.

[2] "Qui enim intelligit de ente illud quod est commune decem praedicamentis, dixit quod collocatur in quaestionibus generis: et qui intellexit de ente illud, quod intelligitur de vero, dixit quod collocatur in quaestionibus accidentis." *Ibid.*, fol. 117v, G. GILSON, *Being and Some Philosophers*, p. 56. FOREST, *La structure métaphysique du concret...*, p. 143.

[3] Averroes refutation' of Avicenna is much more explicit in his *The Incoherence of the Incoherence* (usually known as *Destructio Destructionis*); e.g. "But the whole of his discussion is built on the mistake that the existence of a thing is one of its attributes. For the existence which in our knowledge is prior to the quiddity of a thing is that which signifies the true." p. 392, l. 8–11. Cf. *Averroes' Tahafut Al-tahafut.* (*The Incoherence of the Incoherence*). Translated from the Arabic with introduction and notes by S. Van den Bergh. Vol. I, p. 302, l. 14–p. 305, l. 5; p. 391, l. 8–p. 393, l. 4. – This work was not known to Thomas since it was not translated into Latin until 1328; cf. BERTOLA, "Le traduzioni delle opere filosofiche arabo-guidaiche...," p. 31.

goes directly to Avicenna's *Metaphysica* and uncovers seven arguments proving that *unum* and *ens* do not signify the substance of a thing. This is followed by a refutation, a refutation owing nothing to Averroes. Albert's attack, unlike that of Averroes, does not touch the source of Avicenna's theory on the predication of *ens* and *unum*.[1] In his exposition in Book I (X) Albert makes no reference to Avicenna's theory but is content to expose Aristotle's thought.[2] This is most certainly a sign that he was either not at all interested, or perhaps even failed to grasp the significance of Averroes' attack. We need not pause over Albert's criticism of Avicenna, as it had no influence on Thomas' work.[3]

Thomas' attack on Avicenna parallels those attacks found in Averroes' *Commentary*: *In IV Meta.*, 2, 556–60; *In X Meta.*, 3, 1981–82. The earlier of the two we have already studied; Thomas adopts Averroes' argument involving the reduction to an infinite regress, but takes it out of its anti-Avicennian context; in addition, he agrees with the doctrine that *ens* expresses the essence of a thing; yet Thomas retains the distinction of *esse* and essence.[4] As we have noted, Thomas' procedure ultimately points to a fundamental disagreement between Aquinas and Aristotle; yet this disagreement, resulting in the introduction of *esse*, enables Thomas to enlarge and transform, without destroying, the Aristotelian system.

But without examining Thomas' second attack on Avicenna (*In X Meta.*, 3, 1981–82), we can not correctly understand what Thomas has done either to Aristotelian metaphysics, nor to the Avicennian theory of the real distinction. Even the briefest reading of paragraphs 1981–82 reveals how Thomas has followed Averroes' criticism. Avicenna was deceived by the multiple uses of the word *ens*, writes Thomas, practically copying Averroes' expression. There is a difference, unnoticed by Avicenna, between the *ens* which signifies the truth of a proposition, and the *ens* which is divided into the ten predicaments. The *ens* which signifies the composition of a proposition is an accidental predicate. That is, when we say (judge): "Socrates is white," the "is" is our way of denoting that we have correctly joined "Socrates" and "white"; such a joining is made by the intellect today, instead of yesterday or tomorrow. What we do in a judgment is express that our present know-

[1] Cf. p. 166–67 of Albert's *Metaphysics*.

[2] Cf. *In X Meta.*, T. I, c. 7.

[3] That is, Albert's attack had no influence on Thomas' treatment of Avicenna. However, Albert's explanation of the genesis of our knowledge of "one" appears to have inspired Thomas' discussion of this problem. Compare Albert's *In IV Meta.*, T. I, c. 5, p. 167, l. 18–24 with Thomas' *In IV Meta.*, 3, 566.

[4] Cf. *In IV Meta.*, 2, 555–58; cf. footnotes 3, p. 112; 2–4, p. 113.

ledge is true. It is completely accidental to Socrates, however, whether we know him today or tomorrow. "Is white" may be an actual accident of Socrates; yet the fact that we know it is (or the fact that we say our knowledge is true), this is totally accidental to the reality of whiteness in Socrates. Hence, when Avicenna tried to study all things "here-and-now realized," he was actually studying them as "here-and-now known by us as realized." Metaphysical *ens* is not that type of *ens*, Thomas writes; metaphysical *ens* is that being which is divided into the ten predicaments, and which expresses thus the natures of the ten genera of beings.[1]

Thus it is that Thomas categorically rejects the Avicennian metaphysics which studied all things as realized here and now. Yet Thomas accepts the distinction of *esse* and essence. How he manages to distinguish *esse* from essence, and what exactly he means by *esse*, these are topics to which we shall return in Chapter VI.

Now there can be no doubt that Averroes and Thomas, in attacking Avicenna for having confused the two senses of "being" were but following pure Aristotelian doctrine.[2] Moreover, in pointing out that metaphysics is not interested in the being signified by the "is" of a proposition, but rather in the being of the categories, they were but recalling a theory expressly taught by Aristotle.[3] Hence, it is obvious that Thomas desired to preserve Aristotle's doctrines on this point: the being studied in metaphysics is not that being known in a judgment. Yet we are faced with the fact that Thomas retains the distinction of

[1] "Similiter etiam deceptus est ex aequivocatione entis. Nam ems quod significat compositionem propositionis est praedicatum accidentale, quia compositio fit per intellectum secundum determinatum tempus. Esse autem in hoc tempore vel in illo, est accidentale praedicatum. Sed ens quod divitur per decem praedicata, significat ipsas naturas decem generum secundum quod sunt actu vel potentia." *In X Meta.*, 3, 1982. This text should be read in connection with *In V Meta.*, 9, 895–96 where Thomas distinguishes between *ens* which posits something in reality, and the *ens* which states only the truth of a proposition; the "is" of this latter case is an accidental predicate. Then too, note the very clear expression in the following text: "Excludit ens verum et ens per accidens a principali consideratione huius doctrinae; dicens quod compositio et divisio, in quibus est verum et falsum, est in mente, et non in rebus. Invenitur siquidem et in rebus *aliqua compositio*; sed *talis compositio efficit unam rem, quam intellectus recipit ut unum simplici conceptione*. Sed illa compositio vel divisio, qua intellectus coniungit vel dividit sua concepta, est tantum in intellectu, non in rebus." *In VI Meta.*, 4, 1241. Italics added.

[2] Δ 7, 1017a22–35.

[3] E 4, 1027b17–1028a6. For example: "But since the combination and the separation are in thought and not in the things, and that which is in this sense is a different sort of "being" from the things that are in the full sense (for the thought attaches or removes either the subject's 'what' or its having a certain quality...), that which is accidentally and that which is in the sense of being true must be dismissed... Therefore let these be dismissed, and let us consider the causes and the principles of being itself, *qua* being." 1027b29–1028a6. For Averroes' exposition of this passage: *In VI Meta.*, c. 8, fol. 152r, Esqq. For Thomas': *In VI Meta.*, 4, 1241–44; cf. the last text in footnote 1 above.

esse and essence in spite of his rejection of Avicennian *esse*. And as we know, *esse* is not to be found in Aristotle's writings; there is, to be sure, a distinction between "what a thing is" and "that it is" in Aristotle's logical writings.[1] But such a distinction is exactly what Averroes and Thomas have been attacking as irrelevant to metaphysics. Hence, the *Thomist* doctrine of *esse* as distinct from essence is apparently an un-Aristotelian doctrine.[2]

Thus for the moment, but one conclusion can be made. The facts are these.

1) Avicenna saw metaphysics as the study of all things from the point of view of their *esse*, or their here-and-now, *de facto* realization.

2) Averroes rejected this theory on the basis that the so-called "knowledge" of this *esse* is actually knowledge of an act of man; the *est* of *Socrates est albus* signifies not an aspect of Socrates but rather signifies the truth of our knowledge.

3) Albert, too, rejects Avicenna but not on the same basis as Averroes; Albert rather merely refutes the claim that *ens* does not express the essence of a thing.

[1] *Post. Anal.*, II, 1, 89b24sqq; see also II 7, 92b3sqq.

[2] *In IV Meta.*, 2, 553; cf. footnote 4, p. 111. M. Gilson's view of Aristotle and of Thomas suffers from a slight misappraisal of what it means to understand an object. Aristotle's metaphysics, M. Gilson explains, was interested only in the essences of things. Aquinas', on the contrary surpassed Aristotelianism by placing emphasis on judgment. To understand is for M. Gilson to direct attention to the essence of an object; to use the famous distinction, it is not only to *mean* an essence, it is also to *refer to* an essence. But beings have an existence as well as an essence, and it is to Aquinas' credit that his metaphysics underlines knowledge of existence had in the judgment. Cf. GILSON, *Being and Some Philosophers*, pp. 202; 2–5; 44. Concerning such a view, two points can be made. 1) The mature Aquinas rejected knowledge of *esse* in a judgment (cf. Chapter VII below). 2) Aristotle and Aquinas would have been satisfied with the view that equates the activity of understanding (the act of the passive intellect) with possession of the essense of a thing; yet they would not have agreed that to understand (= to be referring to an object) is to *refer* only to the *essence* existing in the object. Rather, they would both maintain that one understands (or refers to) the entire, particular object from the viewpoint of the universal essence. E.g. "man" *means* essence, but when we know John as "man", we *refer to* the individual object which is John. Thus even though Aristotle may have considered an individual thing to be only a concrete essence, it remains that his attention was directed to the totality of things. This is the spirit behind Aristotle's affirmation that "being", expressing the reality of a thing, expresses the same things which "man" would express; there is no difference between "He is a man" and "He is an existent man", in so far as the object known is concerned. Γ 2, 1003b22–24. Cf. COLLE, *La Métaphysique. Livre VI...*, pp. 40–50. The real difference between Aquinas and Aristotle is primarily in their conception of the principles of really existing objects. (There is another important difference to be discussed in Chapter VI, a difference regarding the act of the possible intellect; Aquinas distinguishes two elements in regard to meaning – essence, or the "material" content of meaning, and *esse*, or the "formal" aspect of the content of meaning. That is, an essence is conceived as "actualized"; this "actualization" is our knowledge of the effect of *esse*.) For a criticism of M. Gilson's theory of knowledge see: G. VAN RIET, "Philosophie et existence. A propos de "L'être et l'essence" de M. Etienne Gilson," in *Problèmes d'Epistémologie*, Nauwelaerts, Louvain, 1960, pp. 144–69.

4) Aquinas finally is guided by Averroes in his refutation of Avicenna's metaphysics of *esse*; such an *esse* is an accident of man's intellect, not of the objects of our knowledge, Thomas declares. Hence, to know such an *esse* is to know ourselves, not objects. Yet Thomas keeps the distinction of *esse* and essence, and thus is un-Aristotelian to that extent.

C. Aquinas' use of several less important aspects of Avicenna's metaphysics

The most important aspect of our study of Thomas' attitude toward Avicenna has been completed. However, there were several individual elements of Avicenna's theories which, one might have noted, appear also in Thomas' writing. For both Avicenna and Aquinas equate *res* and *ens*; and both refer to the essential accidents as *communia*; moreover, both note that all the particular sciences use them, although only metaphysics treats them.

As we have noted, Aquinas attributed to metaphysics not only the study of *ens* and *unum*, but the study of *res* as well; *res*, he wrote, is predicated of everything of which *ens* is predicated. Such a notion as *res* is not found in Aristotle, as was remarked, nor in Averroes or Albert. Hence it must be Avicenna who gave Thomas the impetus to develop this doctrine. For Avicenna, *res* and *ens* are the first ideas of the intellect;[1] moreover, they are said to be synonymous.[2] Yet this last aspect (they are synonymous names) was not accepted by Thomas; *ens* and *res*, although said of all things, although the same in things, nevertheless have different definitions.[3] Now it should have been evident to Thomas that this doctrine of *res* is not in Aristotle; nor was it in the *Commentaries* of Averroes and Albert, as Thomas could have seen. Why then did Thomas take a hint from Avicenna, and on the basis of this, develop and introduce a doctrine of *res* into the context of Aristotle's Metaphysics? Are we to maintain that this was Thomas' method of exposing Aristotle's thought? At least it is evident that Thomas was interested in preserving the main outlines of Aristotelian metaphysics; else, he should not have insisted that *res* be introduced upon the same basis as *unum*: *res* is not synonymous with *ens*, but rather like *unum*, is the same as *ens in re*, while different *in ratione*.

Thomas and Avicenna have a second point in common: the use of

[1] *Meta.*, T. I, c. 6, fol. 72v, B–C.
[2] *Ibid.*, cf. footnote 1, p. 154.
[3] *In IV Meta.*, 2, 553; cf. footnote 4, p. 111.

communia to refer to the concepts used by all the particular sciences. Moreover, both note that none of these particular sciences treat these *communia*. Hence, metaphysics does this by relating them to its formal object. Here Aquinas parts company with Avicenna, however; whereas the latter sees all the *communia* such as *unum* and *multum* as accidental attributes of *esse*, Thomas relates the *communia* to *ens*, to the composite of *esse* and essence.[1] Thus the basic disagreement between Aquinas and Avicenna on the formal object of metaphysics has its repercussions in this doctrine. The idea of metaphysics as the study of *communia* is in Aristotle, even if the word "communia" is not.

5. CONCLUSION

At the end of the investigation into the object of metaphysics according to Aquinas, Albert, Averroes, Avicenna, and Aristotle, what can we say of the nature of Thomas Aquinas' *Commentary on the Metaphysics*? What kind of a book is it? What is its goal?

First, from our comparison of Aristotle and Thomas, we concluded that the *Commentary* presents a metaphysics which is basically Aristotelian, that is, its framework or its general characteristics are those set by Aristotle. Thus Thomas explains that metaphysics is the study both of being as such and of the essential attributes of being as such. Moreover, metaphysics as the study of being comprises both the study of accidents and the study of substances, although primarily it is interested in the latter. As the study of the essential attributes of being, metaphysics must examine such things as "unity," "multiplicity," and so oni the concept of "unity" is at the base of all these attributes. The concept of "unity," finally has the same extension as "being." In so far as Thomas' *Commentary* speaks of the object of metaphysics within this framework, then it presents obviously an Aristotelian metaphysics. The *Commentary* gives, then, the *verba Aristotelis*. (That is, of course, it gives the *verba* as they would be understood by someone who considered the *Metaphysics* of Aristotle as a compact unit, built around one and only one conception of metaphysics.)

Be this as it may, it is still evident that this Aristotelian metaphysics has been changed from within. Thus, although Aquinas speaks of the study of being as such, "being" is not understood as the "essence" of Aristotle's philosophy; rather, "being" is the result of an *esse*-essence

[1] Avicenna: *Meta.*, T. I, c. 2, fol. 70v, B–C; cf. footnote 2, p. 160. Thomas: *In IV Meta.*, 1, 529–31; *Ibid.*, 4, 587.

composition. Hence, and this is our second conclusion, the *Commentary* contains a transformed Aristotelian metaphysics.

The comparison we made between, on the one hand, Aristotle and Thomas and, on the other, Averroes and Albert, gave further credence to these two conclusions. Thomas adopted many of Averroes' expositions, as we have noted. Yet basically, Averroes' *Commentary* was unsatisfying for Thomas; Averroes' *Commentary* not only did not explain the Aristotelian framework of metaphysics so dear to Thomas, but as well Averroes failed to transform Aristotle from within through the addition of the doctrine of *esse*. Aquinas' work must be viewed as intended to fill the lacunae resulting from Averroes' failure.

Albert, finally, is much closer to the Aristotelian view of metaphysics which Thomas gives in his work. Yet even Albert failed to grasp the reasons for which metaphysics studies formal causality; for Albert, the formal cause is the source of our knowledge. For Thomas and Aristotle, however, we can never know the formal cause of immaterial beings. Hence, wrote Thomas in opposition to Albert, we study formal causality because it is the source of a thing's perfection. Albert, as Averroes, needed correction and completion, too, for Albert, as Averroes, did not accept the Avicennian distinction of *esse* and essence. The correction and completion of Albert's work was to be accomplished by Thomas' *Commentary*.

When we compared Avicenna's work to the Thomist *Commentary*, we discovered yet further evidence that Thomas wished to salvage the Aristotelian metaphysical framework. In Avicenna's eyes there was a fundamental rupture between the entity and the unity of things. Thoms explicitly refutes such a theory, thus retaining Aristotle's viewpoint. Yet this was not the only point upon which Thomas opposed Avicenna. No one could doubt that Avicenna transformed Aristotle's metaphysics: if there were ever an "existential Aristotelianism," it is to be found in the *Metaphysica* of Avicenna. Yet this transformation and this existentialism are basically incorrect, Thomas would have said. Avicenna might be correct in distinguishing *esse* and essence in beings; yet he was certainly far from the truth when he mistook the being of a judgment for metaphysical *esse*. Such was Thomas' opinion on this particular theory of his Moslem predecessor. Thus Thomas set out to write his own *Commentary*, correcting in the process the damage done to the Aristotelian framework of metaphysics by Avicenna.

Yet despite the transformation of Aristotelian metaphysics accomplished by the use of *esse*, it would be rash to affirm that Aquinas was conscious that this introduction involved a transformation. Have we not seen his great care in defending several Aristotelian doctrines against Averroes, Albert, and Avicenna? Perhaps, all the discussion of *esse* was only Aquinas' way of making certain that the allure of the distinction introduced by Avicenna between essence and *esse* did not lead others away from Aristotle. True, *res* was introduced by Aquinas, as well as *esse*. But as we shall see in the following two chapters, there is reason to believe that these introductions, in Aquinas' mind, were in accordance with Aristotle's wishes.

Before we turn to further investigation of the *Commentary*, let us recall briefly Aquinas' theory on the object of metaphysics.

1) Metaphysics is the study of accidents as well as substance. – Here we can speak of a "horizontal conception of being." Metaphysics has a concept of "being" which is predicated of both substances and accidents; in every predication where an accident is called "being," that accident is referred to substance, in so far as substance is also called "being." Because "being" is thus predicated, the metaphysician can study substance in so far as it is said to be "being," and thus he will know the most important aspect of accidents in so far as they are called "being."

2) Metaphysics is the study of all substances and of all accidents. – Here we have a "vertical conception of being." The metaphysical concept of "being," in the sense of "substance," is predicated of all substances; in like manner, "being," in the sense of "accident," is predicated of all accidents. The metaphysician must not only attempt to reconstruct "being" in the sense of the horizontal conception; as well, he must attempt to reconstruct a concept of "being" which can be predicated of all substances and of all accidents.

3) The viewpoint proper to the metaphysician is such that the expression of his formal object, "being," represents the totality of everything. Thus, the metaphysician sets out primarily to study all substances as falling under his concept "being." Within any substance one must distinguish the two aspects of *esse* and essence, aspects expressed somehow as "being." – Yet on the other hand, Thomas has compared the metaphysical viewpoint to an univocal one by speaking of "substance" as a genus much like the genus "word."

4) Metaphysics is a study of the *communia*. The first of these is *unum*, but *res* must be considered as well. Both of these *communia* have ten species or genera corresponding to the ten categories of being. Just as

metaphysics regards all substances through the univocal concept of substantial being, so too it must regard all substances as *unum* and as *res*. (In the introduction to the present chapter, the study of *communia* was equated with "the study of concepts." That this equation is justified will appear more clearly in the two chapters that follow.)

These then are the principal aspects of the *Commentary's* doctrine on the object of metaphysics. As will be evident to the reader, the foregoing statement of these aspects leaves unsolved many questions posed both by Thomas' language and by his theories; for example: Does metaphysics prove the distinction of *esse* and essence, or is it just going to assume it? If "being" ("substance") is univocal, how does it express all substances? What will it mean to say :"God is a being"? What will be the relation of metaphysics to other sciences if metaphysics is to study the common concepts used by the particular sciences? Thomas' opinion on such riddles can only be exposed after a great deal of further study of the *Commentary* has been completed. To that study let us now turn, looking first of all for the *Commentary*'s doctrine on the relation of metaphysics to the other, less common, sciences.

CHAPTER V

THE RELATION OF METAPHYSICS TO THE OTHER SCIENCES

The last chapter revealed the manner of composition employed by Aquinas in his *Commentary*; the investigations in the present chapter will reveal much the same thing. This chapter in a very true sense is a continuation of the last; although Chapter IV dealt with the object of metaphysics, and the present chapter with the relation obtaining between metaphysics and the other sciences, the two chapters complement one another, for the nature of the object determines the relations of metaphysics to the other sciences.

In our investigations into the object of metaphysics, we discovered metaphysics to be the study of both substance and accidents, although primarily of substance; moreover, it was seen to be the study of all substances. In reading *In IV Meta., lectio* 1, we noted that metaphysics, when presented as the "study of the *per se* accidents of being," is not simply an investigation into the nine categories of accidents; rather as was mentioned, it studies the *communia*, the concepts common to all things. An examination of those passages of the *Commentary* which treat the relation holding between metaphysics and the other sciences will both verify and clarify these conclusions.

1. THE THOUGHT OF AQUINAS

To discover the doctrine of the *Commentary* concerning the relations of the lower sciences to metaphysics, one can not study one or two lessons after the manner adopted in the preceding chapter. Rather here we must take doctrines from many widely scattered paragraphs. Hence in the exposition to follow a synthetic method has been chosen: the doctrines will be exposed under the following three headings: 1) the "universal science" and the "first science"; 2) the study of *communia*; and 3) metaphysics as the "lord" of the sciences.

But before beginning the exposition, let us note the identity of the studies referred to as the "other sciences" to which metaphysics bears certain relations. In the *Commentary*, whenever Thomas speaks of of these relationships of metaphysics to "other sciences," he is speaking of metaphysics as the queen of all the sciences. Sometimes he will refer to all sciences other than metaphysics as "particular sciences," that is, as studies of a particular type or genus of being.[1] As opposed to these particular sciences, metaphysics is the "universal" or "common" science, the study of all being as such.[2] It is extremely clear that by the opposition of "universal science" and "particular science," Thomas intended to distinguish metaphysics from all other intellectual disciplines. It is against the backdrop of this doctrine, that we must search for the meaning of the "universality" of metaphysics.

A. The "universal science" and the "first science"

In reading the *Commentary* one comes across this distinction often enough to realize that it was a fully conscious one, chosen for a definite purpose. For example, Thomas notes that physics studies "natural things," that is, things which have in themselves a principle of motion. Yet such things are only one type of being, only one genus of "universal being"; since all beings are not "natural" in this sense, the physical study of these natural beings is not the universal science, or the study of all beings *qua* being. Rather, the first science, that is, the study of the highest being, will be the study of being as such.[3]

"First science": the study of the highest being – this is a common doctrine of the *Commentary*. Thomas affirms more than once that the study of the highest being deserves the name "first philosophy" or "first science."[4] In like manner, he explains that the first philosophy, because it is first, that is because it studies the highest being, must also study all being as such; it must be the study of *ens commune* or of *ens inquantum ens*: it must be the universal sicence.[5]

But what precisely is the reason why one science must be both "first"

[1] *In VI Meta.*, 1, 1147.

[2] *In IV Meta.*, 1, 532; *In VI Meta.*, 1, 1170; *In XI Meta.*, 7, 2266–67.

[3] "Antiqui enim non opinabantur aliquam substantiam esse praeter substantiam corpoream mobilem, de qua physicus tractat. Et ideo creditum est, quod soli determinent de tota natura, et per consequens de ente... Hoc autem falsum est; quia adhuc est quaedam scientia superior naturali: ipsa enim natura, idest res naturalis habens in se principium motus, in se ipsa est unum aliquod genus entis universalis. Non enim omne ens est huiusmodi: cum probatum est in octavo Physicorum, esse aliquod ens immobile... Et quia ad illam scientiam pertinet consideratio entis communis, ad quam pertinet consideratio entis primi, ideo ad aliam scientiam quam ad naturalem pertinet consideratio entis communis... Physica enim est quaedam pars philosophiae: sed non prima, quae considerat ens commune...". *In IV Meta.*, 5, 593.

[4] *In VI Meta.*, 1, 1162–63; 1170. *In IV Meta.*, 2, 563. *In XI Meta.*, 7, 2266–67.

[5] *In VI Meta.*, 1, 1170.

and "universal"? Since the reason why a science is first is not the reason why a science is universal, why must the first science be the universal science? Aquinas does not explicitly place great emphasis on the doctrine which answers this question, although he clearly puts forth his view of the matter. (As we shall see however when we study the writings of Averroes and Albert on this point, Thomas was actually engaged in a polemic against their views of the relation of first and universal science. The very words of Thomas – and these words are our sole interest at the moment – although revealing the existence of this polemic, do not show the depth of it.)

The first science must be at once the study of the first being and the study of all beings as being, Thomas writes, "because the first beings are the principles of other beings."[1] Now metaphysics, in studying immaterial beings (the highest beings), is going to be *de facto* a study of the causes of sensible beings as being. These immaterial beings are *maxime*, or "especially," or "to the highest degree," beings; as the highest of all beings, they are the causes of lower beings. Now not only are they the causes of motion, however; as the highest example of being, they are the causes of the being of other beings. Moreover, one of these highest beings will be the highest of all – although the proof of this does not interest us as yet, and hence this being will be the cause of the being of all other beings: it will be the first cause of being as such.[2] As is obvious from the text given in footnote 2, Thomas has very definite commentators in mind when he explains this theory; their identity will be discussed below. Suffice it for the present to note that Thomas' thought is clear: the study of the first being is the study of being as such.

The argument has been presented here much as Thomas exposed it. As is evident, it does totally prove what one might wish; actually it proves only this: if one studies the highest being in so far as it is the cause of the being of all things, then naturally one studies at the same

[1] "Si autem est alia natura et substantia praeter substantias naturales, quae sit separabilis et immobilis, necesse est alteram scientiam ipsius esse, quae sit prior naturali. Et ex eo quod est prima, oportet quod sit universalis. Eadem enim est scientia qua est de primis entibus, et quae est universalis. Nam prima entia sunt principia aliorum". *In XI Meta.*, 7, 2267.

[2] "Necesse vero est communes causas esse sempiternas. Primas enim causas entium generativorum oportet esse ingenitas, ne generatio in infinitum procedat; et maxime has, quae sunt omnino immobiles et immateriales. Hae namque causae immateriales et immobiles sunt causae sensibilibus manifestis nobis, quia sunt maxime entia, et per consequens causae aliorum, ut in secundo libro ostensum est. Et per hoc patet, quod scientia quae huiusmodi entia pertractat, prima est inter omnes, et considerat communes causas omnium entium. Unde sunt causae entium secundum quod sunt entia, quae inquiruntur in prima philosophia, ut in primo proposuit. Ex hoc autem apparet manifeste falsitas opinionis illorum, qui posuerunt Aristotelem sensisse, quo Deus non sit causa substantiae caeli, sed solum motus eius.", *In VI Meta.*, 1, 1164.

time all other things as being. But need one study the first or the highest being as the cause of being? Thomas writes that we must seek a first substance, a substance in relation to which all other substances are substances.[1] Hence Thomas' thought is clear: if one studies the immaterial beings (the first science), one thereby studies all beings as being (the universal science); but even more important, Thomas is saying that the highest being is proved to be the cause of being as such and not just of motion.

The somewhat sketchy notion of Aquinas can be completed if one recalls the order he places in metaphysics in the final paragraph of *In IV Meta., lectio* 2.[2] As was noted in Chapter IV, this paragraph teaches that metaphysics begins with a study of sensible substances and only afterward turns to the immaterial beings. The connection between the two parts lies in the formal object of metaphysics: by considering sensible substances as "being," metaphysics rises to a consideration of their causes as "being."[3] By being first a study of being as such (universal science), metaphysics becomes a study of the first being (first science).

B. The study of communia

Even if we admit that metaphysics, since it is first, must be universal, what consequences does this involve? Fundamentally, there are two: first, the metaphysical concepts will give the content of all concepts employed by the lower sciences; secondly, metaphysics will become the ruling science and as such will establish the limits and the possibility of other sciences. That this is indeed Thomas' thought will become clear as we study Thomas' writings on the study of *communia*, and on the role of metaphysics as the "lord" of the sciences. Just as in the preceding sub-section it was noted that the depth of Thomas' polemic on the identity of the first and the universal science can not be

[1] Cf. *In X Meta.*, 3, 1967–73. In Chapter VI we shall return to this necessity of seeking the first substance.

[2] *In IV Meta.*, 2, 563. Cf. Chapter IV, section 1, B; footnote 1, p. 117 quotes paragraph 563 at length.

[3] "Unde in consideratione unius et entis diversae partes huius scientiae uniuntur, quamvis sint de diversis partibus substantiae; ut sic sit una scientia inquantam partes praedictae sunt consequentes "hoc", id est unum et ens, sicut communia substantiae." *In IV Meta.*, 2, 563. Cf. *In XII Meta.*, 2, 2427. "In cognitione enim harum substantiarum [immobilium] non pervenimus nisi ex substantiis sensibilibus, quarum substantiae simplices sunt quodammodo causae. Et ideo utimur substantiis sensibilibus ut notis, et per eas quaerimus substantiae simplices... Et etiam patet, quod illae substantiae comparantur ad istas in via doctrinae, sicut formae et aliae causae ad materiam. Sicut enim quaerimus in substantiis materialibus formam, finem et agentem ut causas materiae; ita quaerimus substantias simplices ut causas substantiarum materialium." *In VII Meta.*, 17, 1671. Cf. *Ibid.*, 1661.

seen except in consequence of the study of Averroes and Albert, so too much the same is true for Thomas thought on the *communia;* it is only when we see his thought in relation to Averoes, Avicenna, and Albert that we can realize what he intends to say.

First of all, let us note that the particular sciences treat only a particular type of being, and that from a particular point of view; for example, arithmetic treats only number, and this from the point of view of number.[1] Opposed to these particular sciences would be the universal science with its universal point of view; thus Thomas speaks of metaphysics as a study of all substances as substance, and of particular sciences as studies of particular types of substance or of accidental being.[2] Accordingly one can compare metaphysics to the other sciences by comparing its universal formal and material objects to particular formal and material objects.

In the preceding sub-section, it was pointed out that the study of the first being must be the study of all being as such and vice versa. Yet, it is quite another matter to ask for the value of universal science in relation to the particular sciences: apart from its value as an aid in knowing the highest being, what is the value of the study of being as such? Thomas' answer is straightforward. Every science, he writes, has something in common with every other: they all use a number of common principles and concepts. All of them, for example, use the principle of non-contradiction, some of the concepts of causes, the concepts of essence, the concept of *per se* or proper accident, the concept of *unum*, and so on. Each science, however, uses these concepts and principles in different ways. For example, although in all sciences the principle of non-contradiction is expressed as: "the same thing cannot be and not be at the same time and under the same respect," the concept or word "thing" will be understood differently in each science. For the natural philosopher, the word "thing" will mean "material thing," or "spatial thing," etc. And when this principle is used in mathematics, "thing" will be understood as "quantified thing." Finally, in metaphysics the "thing" of the principle of non-contradiction means "being." Thus, each science takes this principle and uses it in the context proper to its discipline. Since every science uses this principle – as well as other principles and concepts – it is evident that somewhere these must be studied. As Thomas explains, no particular science has any more right to study them than any other particular

[1] *In VI Meta.*, 1, 1147. *In IV Meta.*, 1, 530.
[2] *In IV Meta.*, 1, 547.

science. Hence, the universal science must study these common concepts and principles – the *communia.*[1]

Thus for Thomas the study of being as such – the study of the *communia* – is a necessary link in the intellectual life of mankind; there must be a study which clearly formulates the common principles and concepts which are to be used by all the particular sciences. Yet as Thomas points out, and as we must not forget, the universal science is not only interested in these *communia*, but it studies the intrinsic principles of being as well, for example, matter and form, essence and *esse.*[2] As we have already noted, the study of being as such (the universal science) is the study of immaterial beings (the first science); indeed if metaphysics were interested only in the common concepts and principles of the sciences, it would never rise to knowledge of the highest being.[3]

But what precisely is the relationship between the study of *communia* and the study of the intrinsic principles of things (the study of real beings as being)? Thomas neither poses nor explicitly answers such a question; from his various expositions in the *Commentary*, however, it appears that the concepts numbered among the *communia* are merely those concepts by which the metaphysician expresses his knowledge of intrinsic principles of things – matter and form, essence

[1] One must synthesize the following texts to arrive at this doctrine. "Necessitas autem huius scientiae quae speculatur ens et per se accidentia entis, ex hoc apparet, quia huiusmodi non debent ignota remanere, cum ex cognitione communium dependet cognitio rerum propriarum." *In IV Meta.*, 1, 531. "Utuntur autem principiis praedictis [idest: primis] scientiae particularis non secundum suam communitatem, prout se extendunt ad omnia entia, sed quantum sufficit eis: et hoc secundum continentiam generis, quod in scientia subiicitur, de quo ipsa scientia demonstrationes affert. Sicut ipsa philosophia naturalis utitur eis secundum quod se extendunt ad entia mobilia, et non ulterius.", *Ibid.*, 5, 591. "Est autem veritas, quod una scientia principaliter considerat ista principia, ad quam consideratio pertinet communium, qui sunt termini illorum principiorum, sicut ens et non ens, totum et pars, et alia huiusmodi; et ab ea aliae scientiae huiusmodi principia accipiunt.", *In XI Meta.*, 1, 2151. "Quia particulares scientiae quaedam eorum quae perscrutatione indigent praetermittunt, necesse fuit quamdam scientiam esse universalem et primam, quae perscrutetur ea, de quibus particulares scientiae non considerant. Huiusmodi autem videntur esse tam communia quae sequuntur ens commune (de quibus nulla scientia particularis considerat, cum non magis ad unam pertineant quam ad aliam, sed ad omnes communiter), quam etiam substantiae separatae quae excedunt considerationem omnium particularium scientiarium." *Ibid.*, 2146. "Illa, quibus utuntur omnes scientiae, sunt entis inquantum huiusmodi: sed prima principia sunt huiusmodi: ergo pertinent ad ens inquantum est ens. Rationem autem, quare omnes scientiae eis utuntur, sic assignat; quia unumquodque genus subiectum alicuius scientiae recipit praedicationem entis.", *In IV Meta.*, 5, 590–91. Cf. also: *In III Meta.*, 6, 398. *In IV Meta.*, 1, 530.

[2] "Et veritas est, quod haec scientia praecipue considerat communia; non tamen quod communia sint principia, sicut Platonici posuerunt. Considerat autem et prinipia intrinseca rerum, sicut materiam et formam.", *In XI Meta.*, 1, 2167.

[3] "In cognitione enim harum substantiarum [immobilum] non pervenimus nisi ex substantiis sensibilibus, quarum substantiae simplices sunt quodammodo causae. Et ideo utimur substantiis sensibilibus ut notis, et per eas quaerimus substantias simplices." *In VII Meta.*, 17, 1671. Cf. also *Ibid.*, 1661.

and *esse*. Numbered among the *communia* would appear to be those principles whose subject and predicate are common concepts. Thomas', thought appears as a synthesis of several texts: 1) the *principia communia* studied by the metaphysician are known to all – because their subjects and predicates are the *communia* – common concepts – known by all men, e.g. *ens, non-ens, idem, diversum,* etc;[1] 2) the *communia* such as *ens, unum, res,* etc. are all studied by the metaphysician;[2] 3) the *commune ens* is the expression of the formal object of metaphysics;[3] 4) the *commune unum* has a right to be studied in metaphysics because it expresses the same object as *ens*; and the *communia idem, multum,* etc. are studied because of their dependence on unum;[4] and 5) the *commune ens* is identical with the first concept had by man, and hence appears to be the concept *per se notum* and so used in the first of the common principles.[5] Thus it appears correct to maintain that the study of being as such is the science which clearly formulates the *communia*, for the *communia* are the concepts expressing man's knowledge of being as such.[6]

The study of being as such was seen to be the study of the highest being, the cause of being. And the study of being as such is seen to be the study of the *communia*. Moreover, the *communia* are concepts and principles used in particular ways by the particular sciences. The duties

[1] "Quod autem huiusmodi principia communia pertineant ad considerationem primae philosophiae, huius ratio est quia cum omnes primae propositiones per se sint, quorum praedicata sunt de ratione subiectorum; ad hoc quod sint per se notae quantum ad omnes, oportet quod subiecta et praedicata sint nota omnibus. Huiusmodi autem sunt communia, quae in omnium conceptione cadunt; ut ens et non ens, et totum et pars, aequale et inaequale, idem et diversum, et similia quae sunt de consideratione philosophi primi. Unde oportet, quod propositiones communes, quae ex huiusmodi terminis constituuntur, sint principaliter de consideratione philosophi primi.", *In XI Meta.*, 4, 2210. *Ibid.*, 1, 2151.

[2] Cf. *In IV Meta., lectiones* 2–4.

[3] *In IV Meta., lectio* 1 is devoted to showing that metaphysics studies both substance and the nine types of accidents, as well as all substances and all accidents falling under those ten categories; this lesson revolves around the doctrine that *ens* is predicated of the ten categories. *In IV Meta., lectiones* 2–4 are devoted to showing that the *communia* such as *unum, multa, idem,* etc. are studied; the proof lies in the fact that the last two of these common concepts mentioned depend on *unum*, and that *unum* is convertible with *ens*. This concept of *ens* in lesson 1 is spoken of as the expression of *universale ens qua tale* and hence is opposed to the concept *ens quantum*, the expression of quantified beings as quantified (par. 532). Moreover, the *commune* known as *ens* which is in every concept and in our first principle according to *In XI Meta.*, 4, 2210, is the same *ens* which is said to be our first concept in *In X Meta.*, 4, 1998, *In IV Meta.*, 6, 605, and *In I Meta.*, 2, 46; this same *ens* is the expression of the metaphysical point of view in *In I Meta.*, 2, 47 where it is called *ens simpliciter*. – These and many other texts make this the most evident doctrine of the *Commentary*: *ens commune* = *ens quod primo cadit in intellectu* = the concept expressing the formal object of metaphysics.

[4] Cf. preceding footnote.

[5] Cf. footnote 3. An author who agrees: L. De Raeymaeker, "L'idée inspiratrice de la métaphysique thomiste", *Aquinas*, III, 1960, pp. 64–65.

[6] Aquinas says of *homo communis*: it is the concept of man considered as a concept. Cf. *In VII Meta.*, 11, 1536.

of metaphysics are beginning to be evident: metaphysics, by expressing its knowledge through the *communia*, gives the general concepts and principles of the lower or particular sciences; through the *communia* it expresses the intrinsic principles of things. For confirmation and completion of this doctrine, let us turn to what Aquinas has to say of metaphysics as the "lord" of all the sciences.[1]

C. Metaphysics as the "lord" of sciences

The main doctrine on this point is found in the opening lesson of the discussion of Book E (VI). The particular sciences, Aquinas notes, do not study particular types of being *qua* being, but only according to the particularity of the type under consideration.Thus the study of number is a study of number as number and not as being.[2] This means, of course, that none of the particular sciences will begin by demonstrating that beings of the type A have an essence, and that this essence is defined as "X". The ability to determine that every being has an essence, and that a particular type such as A has an essence, belongs exclusively to metaphysics. The particular sciences can only assume that their subjects exist, that their subjects have essence; on the basis of these assumptions, the particular sciences set out to show the accidents of their subjects.[3]

Let us note clearly this doctrine. A particular science does not prove its subject has an essence. Rather, assuming that its subject exists, and

[1] One may be indisposed to accept the assertion of the identity of *ens commune, ens quod primo cadit in intellectu*, and the concept expressing the metaphysical point of view. It will be argued that the *ens* represented by the intellect's first concept can not be the highly sophisticated concept of the metaphysician. In a certain sense, such an objection is correct: in our first concept of *ens* we certainly do not have the same understanding as in the metaphysical concept. But it remains to clarify what we mean by "the same understanding". It is my contention that the pseudo-definition of the first concept of *ens* is identical to the pseudo-definition of the sophisticated metaphysical concept; and that definition may be expressed as "that which is". It remains quite true, however, that we do not yet realize what it is in substantial beings which most perfectly corresponds to our idea of "that which is". The entire initial section of metaphysics is nothing other than the attempt to discover the content of our idea of substantial being, and consequently those aspects of substantial beings which correspond to our concept. Cf. Chapter VI, section 1, C. See *In I Meta.*, 2, 45–46.

[2] *In VI Meta.*, 1, 1147.

[3] "De quolibet enim ente inquantum est ens, proprium est metaphysici considerare. Et, quia eiusdem est considerare de ente inquantum est ens, "et de eo quod quid est", idest de quidditate rei, quia unumquodque habet esse per suam quidditatem, ideo etiam aliae scientiae particulares "nullam mentionem", idest determinationem faciunt de eo "quod quid est", idest de quidditate rei, et de definitione, quae ipsam significat. Sed "ex hoc", idest ex ipso quod quid est ad alia procedunt, utentes eo quasi demonstrato principio ad alia probanda... Et sicut nulla scientia particularis determinat quod quid est determinat quod quid est, ita nulla earum dicit de genere subiecto, circa quod versatur, est aut non est. Et hoc rationabiliter accidit; quia eiusdem scientiae est determinare quaestionem an est, et manifestare quid est.", *Ibid.*, 1147–48.

assuming that its subject has an essence, the science begins to "make manifest through sense knowledge" the essence of its subject.[1] As Aquinas implies, to know the existence of the subject of a particular science is the task of the metaphysician who asks and answers the question: *an est*? Moreover, only the metaphysician can determine what is the essence of a particular subject studied by a particular science: only the metaphysician shows the *quid est,* the quiddity. Aquinas' reason is quite simply given: to ask *an est*?, one must know *quid est*; since the metaphysician knows the *quid est,* he would properly ask: *an est*?[2]

What exactly does Thomas mean? Why does the metaphysician know the *quid est* and the *quia est* (the answer to *an est*?) about all things?[3] In what acts of knowledge does he know the *quid est*? and the *quia est*? Thomas very evidently would consider these questions as touching on the nature of metaphysics: three times he appears to link the study of *ens inquantum ens* and the knowledge of *quid est* and *quia est*; e.g.:

> Oportet enim quod quid est accipere ut medium ad ostendendum an est. Et utraque est consideratio philosophi, qui considerat ens inquantum ens. Et ideo quaelibet scientia particularis supponit de subiecto suo, quia est, et quid est, ut dicitur in primo Posteriorum; et hoc est signum, quod nulla scientia particularis determinat de ente simpliciter, nec de aliquo ente inquantum est ens.[4]

As Thomas says here, the very fact that a science supposes *quid est* and *quia est* of its subject – the quiddity and the existence of its subject – this fact is a sign that that science does not study *ens simpliciter* nor any being as being. In other words, the fact of supposing the quiddity and existence of one's subject means that one is not engaged in metaphysics. It follows, does it not, that the metaphysician discovers both the quiddity and the existence of any being?

How would the metaphysician make these discoveries? – Although this question may appear to be outside the interest of the present chapter, it is not actually so. – Since the context of Aquinas' exposition

[1] I have simplified Thomas' thought somewhat on this score: the exact statement of his theory would only cause unnecessary complication, however, "Ipsum autem quod quid est sui subiecti aliae scientiae faciunt esse manifestum per sensum; sicut scientia, quae est de animalibus, accipit quid est animal per id quod "apparet sensui", idest per sensum et motum, quibus animal a non animali discernitur. Aliae vero scientiae accipiunt quod quid est sui subiecti, per suppositionem ab aliqua scientia, sicut geometria accipit quid est magnitudo a philosopho primo. Et sic ex ipso quod quid est noto per sensum vel per suppositionem, demonstrant scientiae proprias passiones, quae secundum se insunt generi subiecto, circa quod sunt.", *Ibid.*, 1149; cf. *Ibid.*, 1150.

[2] "Et sicut nulla scientia particularis determinat quod quid est, ita etiam nulla earum dicit de genere subiecto, circa quod versatur, est, aut non est. Et hoc rationabiliter accidit; quia eiusdem scientiae est determinare quaestionem an est, et manifestare quid est.", *Ibid.*, 1151.

[3] That *quia est* answers *an est*?, see: *Ibid.*,

[4] *Ibid.*.

on this point concerns metaphysics as the ruling science – metaphysics as a completed discipline, ruling others – the metaphysician must discover the *quid est* and the *quia est* of the subjects of other science by acting as a metaphysician: the metaphysician, because he possesses his metaphysical science, because he views things from the metaphysical point of view, discovers the *quid est* and the *an est* of everything. The metaphysician, Aquinas implies, views an object as "being," and thus discovers the *quid est* and the *an est*.

Our problem then is the following: What does it mean to view an object as "being" and thus to discover the *quid est*? What type of activity is this "viewing"? What type of concept is "being"? Thomas'words indicate the manner of attacking this problem: "...eiusdem scientiae est determinare quaestionem an est, et manifestare quid est. Oportet enim quod quid est accipere ut medium ad ostendem an est."[1] Thomas has made an interesting choice of words: "determinare" expresses man's activity with regard to the question *an est*?; "manifestare" expresses the activity concerning *quid est* or the *quod quid est*; again, one takes the *quid est* as the middle term (*medium*) to show *an est*. As we shall see this choice of words was deliberate, for *quid est* and *quia est* involve two different faculties of man.[2]

In another part of the *Commentary* Aquinas speaks at length on the meaning of the questions: *quid est*? and: *an est*? Actually, one should distinguish four questions, Thomas writes, *quid est*?, *an est*?, *propter quid est*?, and *quia est*? The questions *quid est*? and *propter quid est*? can be identified, however.[3] It is the same, for example, to ask: "What is man?" and: "Why is Socrates a man?" When one asks: "What is man?" (*quid est*?), one seeks the quiddity because of which "man" is predicated of objects falling under the class "man," and because of which "man" belongs to material objects such as Socrates. And when one asks: "Why is Socrates a man?", one wishes to know that because of which "man" belongs to the flesh and bones of Socrates.[4] Thus it is evident

[1] *Ibid.*.

[2] The same doctrine is found, but in shorter version, in: *In XI Meta.*, 7, 2249–51.

[3] "Cum enim sint quatuor quaestiones, ut in secundo *Posteriorum* habetur, scilicet an est, quid est, quia est, et propter quid: duae istarum, scilicet quid, et propter quid, in idem coincidunt, ut ibi probatur.", *In VII Meta.*, 17, 1651.

[4] "Sed dicendum, quod quaestio quid et propter quid in idem quodammodo incidunt, ut est dictum. Et ideo quaestio quid est, potest transformari in quaestionem propter quid. Quaestio enim quid est, quaerit de quidditate propter quam id, de quo quid est quaeritur, praedicatur de quolibet suorum subiectorum, et convenit suis partibus. Propter hoc enim Socrates est homo, quia convenit ei illud, quod respondetur ad quaestionem quid est homo. Propter hoc etiam carnes et ossa sunt homo, quia quod quid est homo est in carnibus et in ossibus. Idem ergo est quaerrere quid est homo, et quaerere propter quid hoc, scilicet Socrates est homo? vel propter quid hoc, scilicet carnes et ossa sunt homo?", *Ibid.*, 1663.

that the two questions– *quid est*? and *propter quid est*? – are the same.

When one asks: *quid est,*? Thomas writes, one must already know the answer to: *an est*?[1] This doctrine, seemingly strange, is mentioned both by Aristotle and Aquinas.[2] Its strangeness – only apparent – lies in the fact that the question: *quid est*? seems to presuppose absolutely no other knowledge on the part of the questioner.[3] Nevertheless, it does presuppose something; for example, before one can ask: *quid est homo*? one must know that there is a man.[4] Suppose this man is Socrates; then the question: *quid est homo*? is no different from: *propter quid Socrates est homo*?[5] Thus it is that: *quid est homo*? is actually a question asking for the "causa, quae est forma in materia"; in asking "what is man?" one seeks to discover the formal cause which makes a definite glob of matter into Socrates, a man.[6] And so we can say that to ask the question: *quid est*? or the question: *propter quid est,*? we must fiist know something;[7] we must know at least that something exists and that it is called by a certain name. Thomas uses the example of "e-clipse." We can ask: "What is an eclipse?" or "Why is this event an eclipse?" But to ask this, we must know that there is a certain event such as the blackout of the sun's light which is called "eclipse." Thus when we ask: "What?" or "Why?" and are seeking the formal cause, we need to know the answer to: *an est*?[8]

We have thus far noted the identity of *quid est*? and *propter quid est*? when we seek the formal cause. But we can also seek to discover the "causa ipsius formae in materia"; that is, we can seek the final or the efficient causes.[9] In cases where we seek the final and efficient causes,

[1] Speaking of the case where *quid est* and *propter quid est* both ask for the form, Thomas writes:"Secundum enim quod propter quid est idem ei quod est quid, oportet esse manifestum an est.", *Ibid.*, 1651.

[2] As Thomas notes in: *Ibid.*.

[3] Cf. *Ibid.*, 1662–63.

[4] "Quoniam vero in hac quaestione, qua quaeritur quid est homo, oportet habere notum existere verum hoc ipsum quod est esse hominem (aliter nihil quaereretur): ...", *Ibid.*, 1666.

[5] Immediately after the passage in footnote 4, Thomas continues: "...sicut cum quaeritur propter quid sit eclipsis, oportet esse notum, quia est eclipsis: palam est, quod ille qui quaerit quid est homo quaerit propter quid est.", *Ibid.*.

[6] "In quaerendo autem propter quid de aliquo, aliquando quaeritur causa, quae est forma in materia. Unde cum quaeritur, propter quid tonat ? Respondetur, quia sonitus fit in nubibus: hic enim constat quod aliud de alio est quod quaeritur. Est enim sonitus in nubibus, vel tonitruum in aëre.", *Ibid.*, 1656.

[7] Cf. *Ibid.*, 1650–51; 1655; 1664–66.

[8] "Nam esse est praesuppositum ad hoc quod quaeritur quid est, quia est praesuppositum ad propter quid; sicut cum quaerimus quid est domus? idem est ac si quaereremus propter quid haec, scilicet lapides et ligna, sunt domus? propter haec scilicet "quia partes domus existunt id quod erat domus esse", idest propter hoc quod quidditas domus inest partibus domus.", *Ibid.*, 1666; also 1651.

[9] "Aliquando autem quaeritur causa ipsius formae in materia quae est efficiens vel finis; ut cum quaerimus propter quid haec, scilicet lapides et lateres, sunt domus?", *Ibid.*, 1657.

we already know the formal cause; we already know the answer to: *quid est*? (and to: *propter quid est*? when this latter question is the same as the former). Then too, we know already the answer to: *an est*?, for this knowledge was presupposed for the question: *quid est*? Thus, for example, to ask for the final or efficient causes of man (*propter quid Socrates est homo*?), we have to know that there is something called "man" (*an est homo*?), as well that "man" means "rational animal" or "body-soul composite" (*quid est homo*?). But this is not sufficient: we must know too the answer to: *quia est homo*? – The difference between: *quia est*? and: *an est*? appears to be the following. When one is seeking the formal cause of man (*quid est homo*?), one must know that there exists something called "man" (*an est homo*?). But when one seeks the final or efficient cause of man (*propter quid est homo*?) one must know that a certain object is called "man," and one must see, or grasp, "man" in this object. Thus to know only: *an est homo*? is still to be ignorant of what "man" means. To know both: *an est homo*? and: *quid est homo*? is to know both that object A is called "man," and that "man" means "body-soul composite" or something like this. But one does not know the answer to: *quia est homo*? until one grasps or understands object A through the concept "man." Thus it appears these three questions are distinct. And after having the answers to all three, one can ask: *propter quid homo*? in the sense of asking for the final or efficient causes of man A.[1]

It is quite true, as Thomas says, that these four questions can be reduced to three: *propter quid est*?, *an est*?, *quia est*?. But of course, one must remember that *propter quid est*? can be a question seeking the formal cause, and thus presupposing: *an est*?; or that *propter quid est*? can seek the final and efficient causes, and thus it presupposes *quid est*?, *an est*?, and *quia est*?.

This theory would easily involve Thomas in great difficulties were he to state it as absolutely universal; that is, were Thomas to characterize as impossible the grasp of *any* answer to *quid est*? prior to the knowledge of *an est*?, then he would have to say that man's very first concept is preceded by intellectual knowledge of the *an est*. Thomas recognized this difficulty, and hence in the context of this discussion of these four

[1] "Sicut autem quaestio quid est, se habet ad quaestionem an est, ita quaestio propter quid, ad quaestionem quia. Cum igitur quaeritur propter quid oportet illa duo esse manifesta. Secundum enim quod propter quid est idem ei quod est quid, oportet esse manifestum an est. Secundum autem quod propter quid distinguiter a quid est, oportet esse manifestum quia. Et ideo dicit, quod cum quaeritur propter quid, oportet existere manifesta entia ista duo: scilicet ipsum quia et ipsum esse, quod pertinet ad quaestionem an est.", *Ibid.*, 1651.

questions, he mentions the exceptions to the rule: the *communia*, both concepts and principles, are outside this rule.[1] Actually it is not even correct to consider the *communia* as "outside" the rule; rather they found it. It is because all men have the initial concepts and principles known as *communia*, that they can seek new or further knowledge. Because we know that an object is itself, we can ask: *quid est res ipsa*?. Thus, as soon as men know the identity of a material object, they ask for its formal cause: *quid est*?[2]

This doctrine needs further study to insure that we correctly locate both the first activity of man and its content. As Thomas notes, the first principles are known to all because their subjects and predicates, the *communia* are known to all.[3] These *communia* are our first concepts; and *ens* is the first of all.[4] All men "naturally" form the concept "being" as their first concept, and immediately follow this conception by their first judgment: "impossibile est esse et non esse simul."[5] Now as has already been remarked, this concept of *ens* is the metaphysical concept. (Again, the problem which arises from the difficulty men experience in arriving at an understanding of that concept need not concern us here.) Thus it appears that men, in their first intellectual activity

[1] After the discussion of the relative positions of knowledge of the answers to the four questions, Aquinas explains that we must always ask: "Why is A 'B'?", that we can never ask: "Why is A 'A'?" in the sense of *Quid est A*?. That is, once we know A is 'A', we can not ask for the quiddity of A: "Est enim una ratio et una causa in omnibus, quam impossible est ignorari; *sicut nec alia communia, quae dicuntur communes animi conceptiones, ignori possibile est.* Huius autem ratio est, quia unumquodque est unum sibiipsi. Unde unumquodque de se praedicatur. Nisi forte aliquis velit assignare aliam, dicens, quod ideo homo est homo, et musicum est musicum... quia unumquodque est individsibile ad seipsum. Et ita non potest de seipso negari, ut dicatur homo non est homo. Unde oportet ut de se affirmetur. Sed haec ratio non differt a primo, quam diximus; scilicet quod unumquodque unum est sibiipsi... *Unde non potest quaeri quasi ignoratum, sicut nec alia principia communia.*", *Ibid.*, 1652–54. Italics added.

[2] "Unde relinquitur, quod semper quaeritur propter quid hoc sit aliud hoc. Et hoc subsequenter manifestat; dicens, quod si aliquis quaereret propter quid tale animal est homo? hoc quidem igitur palam quod non quaeritur propter quid homo est homo. Et sic patet quod aliud de aliquo quaeritur propter quid existit, non idem de seipso. Sed cum quaeritur aliquid de aliquo propter quid existit, oportet manifestum esse, quia existit. Nam si non sit ita, ut scilicet si non sit manifestum quod existit, nihil quaerit.", *Ibid.*, 1655. In paragraphs 1652–55 Thomas is quite evidently speaking of *propter quid* in so far as that question can sometimes be the same as *quid est*; he is speaking of *propter quid* as a question seeking the formal cause not the final nor the efficient causes.

[3] *In XI Meta.*, 4, 2210.

[4] *In I Meta.*, 2, 47; *In IV Meta.*, 6, 605.

[5] "Ad huius autem evidentiam sciendum est, quod, cum duplex sit operatio intellectus: una, qua cognoscit quod quid est, quae vocatur indivisibilium intelligentia: alia, qua componit et dividit: in utroque est aliquod primum: in prima quidem operatione est aliquod primum, quod cadit in conceptione intellectus, scilicet hoc quod dico ens; nec aliquid hac operatione potest mente concipi, nisi intelligatur ens. Et quia hoc principium, impossibile est esse et non esse simul, dependet ex intellectu entis... ideo hoc etiam principium est naturaliter primum in secunda operatione intellectus...", *In IV Meta.*, 6, 605. That the formation of *ens* and the first principle are "natural", see: *Ibid.*, 599 and 604.

conceive an object as "being." Following this conception, they judge their conception to be true; this judgment is expressed as the first principle. It is consequent to these two acts, these two activities of knowing *communia*, that man can ask: *quid est?*[1]

Let us note carefully what this doctrine implies. First of all, the concept of "being" is formed prior to any knowledge of: *an est?* Secondly, our first judgment tells us that there is, or exists, a being; thus we know the answer to: *an est hoc ens?* And so thirdly we can ask for the formal cause of this material being by our question: *quid est hoc ens?*

At the end of this long discussion of *quid est?*, *an est?*, and so on, let us remind ourselves where we are going, as well as where we have been. This sub-section opened by noting that metaphysics as the ruling science must inform the particular sciences that their subjects exist, and that they have an essence; on the basis of this information, the particular sciences can use sense knowledge to illustrate, or "fill out" the essence of their subjects. As a justification of this theory, Aquinas notes that metaphysics studies the quiddity or the *quid est* of being as such, as well as the *quid est* of any being *qua* being; moreover, as he writes, the science studying the *quid est* would use this *quid est* to show *an est.* Thus Aquinas is seen to maintain that the metaphysician looks at any being as having a quiddity, and then shows the *an est* or the existence of that being; whereupon a particular science may begin to make a universal study of that being. This emphasis laid on *quid est* and on *an est* led us then to investigate these doctrines. We saw that actually there are four questions involved in science; besides these two, there are: *propter quid est?* and *quia est?* As was noted, *quid est A?* was a way of seeking the formal cause of a thing known as "A"; but of course we have to know that an object called "A" exists (*an est*) before we can seek the formal cause of A.

Although no mention was made at the moment of our investigation, this last doctrine apparently contradicts the one governing the relation of metaphysics and other sciences. For now we have learned that we must answer: *an est?* before we can ask: *quid est?*; earlier, in our study of *In VI Meta., lectio* I we noted that metaphysics uses knowledge of *quid est* to determine *an est.* Surely a contradiction, even though it was not noted at that time. – However, nothing could be further from contradiction than the theory of Aquinas on this point. In the context of Aquinas' discussion of the relative priority of our knowledge of *quid est*

[1] This doctrine will be more intelligible after the discussion of the meaning of *ens* in Ch. VI. section I, C.

and *an est,* we noted a certain restriction placed on the priority of our ledge of *an est,* a restriction which founded as well as illustrated the extension of Aquinas' theory. The *commune ens,* Thomas noted by implication, precedes all knowledge of *an est.* The second intellectual act of man, the judgment that a thing was properly conceived as *ens,* serves as the knowledge of *an est.* After these two acts, man can ask: *quid est hoc ens?* and so begin his life-long search for further understanding.

Thus Thomas has not contradicted himself. The metaphysician must first, conceive an object as "being," then judge that he has properly or truly conceived this object. Following these two acts, a particular science can begin to investigate this being by asking: *Quid est?* and by referring to sense knowledge in its quest for the formal cause of this type of being.

(As we had reason to note in the preceding chapter, the judgment that an object is here and now realized – or that it is a being – is not knowledge of metaphysical *esse.* As we have just seen in the present chapter, man's first judgment amounts to answering the question: *an est?* And prior to the question: *an est?* man understands the object as "being." Yet have we not seen Thomas say that to understand "being" is to know quiddity only (*quid est*)? Where then is man's knowledge of metaphysical *esse,* if it is found neither in the act of knowing *an est* nor in that of knowing *quid est?* – Actually, we have not seen Thomas state that in the act by which we know *quid est,* we know only an essence. As I shall attempt to illustrate both later in the present chapter as well as in the following one, in the apprehension of the *quid est* man knows both essence and *esse.*)[1]

[1] From what has been said, it is evident that we shall not in the course of our study place any great emphasis on "separation". Many authors apparently concede to the negative judgment, or separation, the honor of being the instrument used to *form* the concept of *ens* (as opposed to discovering completely the meaning of that concept). E.g. GEIGER, *La participation dans la philosophie de S. Thomas d'Aquin,* pp. 315–21. To abstract, "c'est, par un acte de l'intelligence, en sa simple appréhension, sans affirmer ni nier, considérer dans la réalité un aspect sans tenir compte de l'autre ou des autres." (p. 317). Such an act can not attain, or produce, the concept of being, we are told; so we must "...recourir à l'acte où S. Thomas voit l'opération characteristique de la métaphysique. Il l'appelle *separatio* ou jugement négatif." (p. 318). The present study differs considerably from P. Geiger's view. First, the texts already quoted concerning *ens quod primo cadit in intellectu,* as well as those treating the relations between *quid est* and *an est,* do not coincide with P. Geiger's view. Secondly, he bases his thought on *In Boeth. De Trin.,* q. 5, a. 3c, a document giving much importance to another doctrine closely related to the "separation": the judgmental knowledge of *esse.* As shall be illustrated in Ch. VII, even the Aquinas of the first writings could not have meant what is often found in the unfortunate "secunda operato respicit esse rei". Thirdly, I do not understand how P. Geiger's view on separation is compatible with the Thomist theories on *resolutio* also found in the *De Trinitate*; cf. Ch. III, section 3 above. Finally, I am far from

Let us pause briefly to coordinate what we have discovered in our investigations into the role of metaphysics as the ruling science. In our studies we discovered that the *communia* are the concepts and principles expressing the metaphysician's knowledge of all things at being. We remarked that these *communia* are used by the particular sciences, but in particular or limited ways; yet metaphysics alone is to study these *communia*. We also noted that metaphysics begins by studying being as such (the study of the meaning to be given to *communia*) and rises from this study to God. Just what exactly are the steps by which one goes to God from the metaphysical knowledge of being as such (expressed in the *communia*) has not yet been discovered.

On the other hand, our detailed study of *quid est* and *an est* and their use in metaphysics informed us that in the concept *ens* the metaphysician, in actually knowing an object studied by a particular science, knows the essence proper to that object. In addition, he answers the question: *an est*? in relation to the essence of that object. For example, in knowing Socrates as "being," the metaphysician expresses the essence of Socrates, and consequently judges that the essence is as he has known it. On the other hand, the particular science, physics, accepts the essence of material beings like Socrates from the metaphysician; consequently, physics uses sense knowledge to illustrate this essence as capable of motion.

This doctrine is not as clear as we may wish. Nor can it be so before we grasp the meaning given to the concept *ens* by Aquinas. We must wait until Chapter VI before we can investigate in a complete manner that meaning, however. At the moment we must turn to the thought of Aquinas' predecessors on the relation of metaphysics to the other sciences.

2. "UNIVERSAL SCIENCE" AND "FIRST SCIENCE" IN THE PREDECESSORS OF AQUINAS

After this exposition of Thomas' thought on the relation of metaphysics to the other sciences, we need now investigate the writings of

convinced that the passages in the *De Trinitate* which treat the separation, intend to present that operation as the metaphysical instrument of forming the concept of "being"; after all, that passage attributes *separatio* to *scientiae divinae sive metaphysicae*, which is not the same as if Thomas had written *scientiae universalis sive metaphysicae*: divine science is the last part of universal science – divine science begins after one has acquired the idea of "being", and from it, proved the existence of God; note too, that the *De Trinitate* attributes abstraction to metaphysics as well as to the other sciences. For authors who agree with P. Geiger, see the references in: R. SCHMIDT, "L'emploi de la séparation en métaphysique", *Revue Philosophique de Louvain*, LVIII, 1960, pp. 373–93.

his predecessors. Our procedure will be to take each division used in the exposition of Thomas and to discuss in that context the works of Aristotle, Avicenna, Averroes, and Albert, as well as the influence of these works on Aquinas.

A. The "universal science" and the "first science" in Aristotle's work

In the *Commentary* Aquinas clearly distinguishes metaphysics as the "universal science" from metaphysics as the "first science": it is universal because it studies all being from the viewpoint of being, whereas it is first because it studies the highest of all types of beings, the immaterial beings. Yet it is also Thomas' doctrine that the first science *must* be the universal science. The link between the first and the universal science is had in the fact that man knows immaterial beings (the highest beings) only as causes of material beings *qua* being; thus for man, the study of the highest beings must begin with a study of beings as such: the science of beings as being (the universal science) must discover and discuss the first cause of the being of things (the first science).

In making much of this distinction between "universal" and "first" science or philosophy, Aquinas is true to the thought of Aristotle.[1] For Aristotle, the first science is the study of the immaterial or highest being;[2] the two passages of the *Metaphysics* where this doctrine appears – E 1, 1026a10–32; and Γ 2, 1004a2–6 – are two of the passages upon which Thomas comments by giving his synthesis of first and universal science. On the other hand, as Aristotle notes, there is a study of being as such; this science is undoubtedly higher than physics or than any other science, for these latter sciences study a particular genus of being.[3]

[1] The objection put forth by some modern commentators on Aristotle need not detain us on this point. As already noted, even one of the authors who insists on making the "study of being as such" into a study of the principles of being and non-being, of one and multitude, as principles found in the divine beings, would nevertheless admit the legitimacy of Aquinas' view, given the latter's suppositions as to the unity of Aristotle's thought. Thus, although maintaining that being "as such" is to be taken as elements or principles existing in things and not as a point of view, Prof. Merlan would concede the possibility of a reconciliation such as Aquinas has made. Cf. MERLAN, *From Platonism to Neoplatonism*, pp. 169–70.

[2] E 1, 1026a10-32. E.g. "...the first science deal with things which both exist separately– and are immobile." (l. 15–16) "...if there is an immovable substance, the science of this must be prior and must be first philosophy..." (l. 29–30) Cf. also Γ 2, 1004a2–6. For a view of these texts similar to that of Aquinas, see A. MANSION, "L'objet de la science philosophique suprême d'après Aristote, *Métaphysique E 1*", in *Mélanges de philosophie grècque offerts à Mgr. A. Diès*, Vrin, Paris, 1956, pp. 160–66.

[3] "Quoniam autem est adhuc phisico quedam superiorum (unum enim aliquod genus est entis natura), universalis et circa primam substantiam consideratur et de his utique erit speculatio. Est autem sapiencia quedam et phisica, set non prima." Γ 3, 1005a33–b2: *Vetus*, p. 307, l. 30–34. Msgr. Mansion points out that "first philosophy" in this context is not the science of the highest being, as in Γ 2, 1004a2–6 or E 1, 1026a10–32; rather "first philosophy" denotes here that this is a science of all beings as such and thus is prior to the studies of the

Thus in Aristotle, as in Thomas, first science is distinguished from the universal study of being as such.

We must thus consider that Thomas' doctrine presents the *verba Aristotelis*, the actual meaning of Aristotle, in so far as the universal study of being is distinguished from the first science, or the study of the highest being. But Thomas links these two studies in one science, in metaphysics. What of Aristotle? Obviously, he too links the two studies. In the passage where he mentions that the ancient naturalists were wrong in thinking their treatment of natural being was a study of all being, he explains that there is a higher science which deals with the primary being as well as with being as such.[1] Thus, again Thomas must be credited with fidelity to the doctrine of Aristotle.

When we turn to the reason why the first science must be the universal science, we are confronted with a more difficult problem. Aristotle does not seem to give a reason; yet Thomas does. An investigation of Thomas' words will reveal a great deal. As I have had occasion to remark already, Thomas would have presumed Aristotle's *Metaphysics* to be a unit, with the various books arranged in more or less their proper order, and so on. The treatment of the doctrine in question – the link of the first and the universal sciences – is an example of Thomas' presupposition at work.

On the first occasion that he comes upon Aristotle's doctrine on first philosophy (Γ 2, 1004a2–7), Thomas appears to have had a correct text of Aristotle.[2] Moreover, on this occasion Thomas presents a faithful exposition of the words before him (*In IV Meta.*, 2, 563). Aristotle speaks of various distinct parts of philosophy in this text, our contemporary students of Aristotle maintain; he speaks implicitly of physics as well as of metaphysics.[3] Thomas, however, sees this section as an integral part of Book Γ (IV), which, he notes, explains the subjects to be treated in metaphysics; thus Γ 2, 1004a2–7 was, for Thomas, a continuation of the explanation that all substances were studied in

particular genera of beings. Cf. MANSION, "Philosophie première, philosophie seconde et métaphysique chez Aristote", pp. 171–74. *In IV Meta.*, 5, 593 reveals that Thomas understood "prima" to refer to a study of all beings as such, to refer to the prior science; yet into this context, Aquinas introduced the identity of the study of being as such and the study of the highest beings.

[1] Cf. Γ 3, 1005a33–b2, quoted in the preceding note.

[2] "Et tot partes philosophiae sunt quot vere substantie, quare necesse est quandam primam et continuam ipsis. Sunt enim mox genera habencia ens et unum; unde et sciencie sunt consequentes hec. Est enim philosophus sicut mathematicus dictus; et namque hec habet partes, et prima quedam et secunda est scientia et alie consequenter in disciplinis.", Γ 2, 1004a2–7: *Vetus*, p. 304, l. 27–33.

[3] E.g. Ross, *Aristotle*, pp. 256–57.

metaphysics. Thus whereas Aristotle is distinguishing first philosophy from secondary philosophies, Thomas understands this as a distinction between the first and secondary parts of *metaphysics*. So Thomas explains: 1) that metaphysics studies all substances, but particularly the highest substances, and 2) that there is an unity to this study because metaphysics gathers all these substances under the common concepts *ens* and *unum*.[1]

This attitude, this procedure, is typical of Thomas. A few lines later, he comes across Aristotle's doctrine of the universal science. In this passage (3, 1005a33–b2) Aristotle explains that the existence of an immaterial being requires that the study of all being *qua* being belong to a science higher than physics, to the science which studies the immaterial being. Here Aristotle is emphasizing metaphysics as universal; by implication, of course, he identifies the universal science with the first science. Thomas in his *Commentary* (*In IV Meta.*, 5, 593) does not depart from Aristotle's meaning, although he does give a more complete doctrine by stating categorically that the science which studies the highest being is superior to physics, and that to this superior science belongs the consideration of *ens commune* or being as such.[2] Thus Thomas changes the emphasis slightly; Aristotle had spoken primarily of the science of all being, and said almost as an aside, that this science studies the highest being. Since the context of the discussion for Thomas concerns the study of first principles, the principles common to all being, Thomas emphasizes that this study belong to the science of the highest being.[3] Yet Thomas must be seen here as exposing the *verba Aristotelis*; his exposition is more ordered than Aristotle's, but this fact takes nothing away from the Aristotelian character of Thomas' words.

[1] "Hic ostendit partes philosophiae distingui secundum partes entis et unius... Et, quia partes substantiae sunt ordinatae adinvicem, nam substantia immaterialis est prior substantia naturaliter; ideo necesse est inter partes philosophiae esse quamdam primam. Illa tamen, quae est de substantia sensibili, est prima ordine doctrinae, quia a notoribus nobis oportet incipere disciplinam: et de ha [sic.] determinatur in septimo et octavo huius. Illa vero, quae est de substantia immateriali est prior dignitate et intentione huius scientiae, de qua traditur in duodecimo huius. Et tamen quaecumque sunt prima, necesse est quod sint continua aliis partibus, quia omnes partes habent pro genere unum et ens. Unde in consideratione unius et entis diversae partes huius scientiae uniuntur, quamvis sint de diversis partibus substantiae; ut sic sit una scientia inquantum partes praedictae sunt consequentes "hoc", id est unum et ens sicut communia substantiae.", *In IV Meta.*, 2, 563.

[2] Cf. Γ 3, 1005a33–b2, quoted in footnote 3, p. 188. Thomas' exposition reads: "...est quaedam scientia superior naturali: ipsa enim natura, idest res naturalis habens in se principium motus, in se ipsa est unum aliquod genus entis universalis... Et quia ad illam scientiam pertinet consideratio entis communis, ad quam pertinet consideratio entis primi, ideo ad aliam scientiam quam ad naturalem pertinet consideratio entis communis...", *In IV Meta.*, 5, 593.

[3] *Ibid.*, 588–93.

When we investigate the third occasion upon which one of the notions is mentioned, it should not surprise us to witness Thomas' manner of exposition. By this point in his reading of the *Metaphysics*, Thomas has discovered that the first philosophy studies the highest or immaterial being; he has seen as well that the universal science is the study of all being as such; and finally, he knows that Aristotle identified the first and the universal sciences. When Thomas, with these doctrines behind him, begins to comment on E I, what must have been the opinion already formed of Aristotle's linking first and universal science? Thomas, we can easily speculate, would have thought in a pattern somewhat like this: for Aristotle, all knowledge comes through the senses, and hence if we know the immaterial being we know it from its sensible effects; if Aristotle studies the immaterial beings, he studies them first as the causes of sensible beings; if Aristotle studies *in one science* the immaterial beings and all being as being, he must be considering the immaterial beings as the causes of the being of things.

It seems safe to consider that Thomas could have reasoned in the manner stated above. That he did so reason appears from his treatment of E I. Aristotle, in explaining the difference between physics, mathematics, and metaphysics, wrote that the study of an eternal, immaterial being could not belong to physics or to metaphysics; then, in a sentence which must have appeared as a digression to Thomas, Aristotle wrote: "Now all causes must be eternal, but especially these; for they are the causes that operate on so much of the divine as appears to us."[1] Thomas read this passage in the *Moerbecana*; if we judge from the words written by Thomas, the translation was basically accurate.[2] In connection with this sentence of Aristotle, Thomas gives the latter's explanation of why the first science is the universal science – that is, what Thomas understood as the latter's explanation. Thomas' reasoning is as follows. The common causes of all things must be eternal. Moreover, since we can not have an infinite regress in our search for the first causes of generation, the first causes must be ungenerated. These causes, immobile and immaterial, are the causes of what we see in the sensible world. Since these are the highest beings, they are the causes of the being of sensible beings. (As Aristotle had noted in his Book α, Thomas says, sensible things have a cause of their being.) Thus it is evident that the study of these immaterial beings is the first science,

[1] E I, 1026a16–18. Cf. MANSION, "L'objet de la science philosophique suprême d'après Aristote, *Métaphysique E I*", pp. 161–66.

[2] Thomas in *In VI Meta.*, I, 1164 refers to these causes as "eternal", as "common causes", as the "causes of the sensible world which is manifest to us."

is higher than physics or mathematics; thus it is also evident that this first science considers the common causes of all beings *qua* being.[1]

There appears to be no doubt that Thomas thinks he is giving Aristotle's thought. The closing words of his exposition of these lines evidences this belief: "Ex hoc autem apparet manifeste falsitas opinionis illorum, qui posuerunt Aristotelem sensisse, quod Deus non sit causa substantiae caeli, sed solum motus eius."[2] Whoever these other philosophers may be, Thomas clearly indicates that he believes they are wrong. Thus it follows that Thomas is conscious of giving, if not the *verba Aristotelis*, at least the *intentio* of the Greek philosopher: Thomas may not be able to point to an explicit phrase in Book E to prove his point, but by the reference to Book α Thomas gives the logical conclusion of Aristotle's premisses. Thus, Thomas maintains that Aristotelian metaphysics include a proof of God as the cause of being. By his exposition, Thomas has protected Aristotle against misinterpretation.

Although K 7 is not to be considered as an integral part of the *Metaphysics* if one follows contemporary rules of exegesis, Thomas did not adopt such a sophisticated view.[3] Thus when he arrived at K 7, 1064b6–

[1] "Necesse vero est communes causas esse sempiternas. Primas enim causas entium generativorum oportet esse ingenitas, ne generatio in infinitum procedat; et maxime has, quae sunt omnino immobiles et immateriales. Hae namque causae immateriales et immobiles sunt causae sensibilibus manifestis nobis, quia sunt maxime entia, et per consequens causae aliorum, ut in secundo libro ostensum est. Et per hoc patet, quod scientia quae huiusmodi entia pertractat, prima est inter omnes, et considerat communes causas omnium entium. Unde sunt causae entium secundum quod sunt entia, quae inquiruntur in prima philosophia, ut in primo proposuit." *Ibid.* In the end of this same lesson (par. 1170) Thomas mentions the identity of first and universal science in commenting on E 1, 1026a27–32, and notes that Aristotle in Γ 1 affirms their identity.

[2] *Ibid.*, Judging solely on the words of the *Commentary*, one must admit that Aquinas believed there was a proof of the cause of being in Aristotle's *Metaphysics*. Perhaps an exhaustive study of Aquinas' writings will demand that this view be qualified. For example, in *Sum. theol. I*, 44, 1–2 and *De pot.*, 3, 5, Aquinas remarks that both Plato and Aristotle affirmed God to be the cause of all being. However, these passages are to be related to *De subs. sep.*, c. 7; in the context of an abridged "history of metaphysics", Aquinas remarks that Plato and Aristotle should *have* posited God as the cause of all; they should have done this, as it followed from some of their theories, Aquinas notes. Cf. A. Pegis, "A Note on St. Thomas, *Summa Theologica I*, 44, 1–2." *Mediaeval studies*, VIII, 1946, pp. 159–68. In a work much closer to the *Commentary on the Metaphysics*, *In VIII Phys.*, 2, 4; *Ibid.*, 21, 14, Thomas affirms that Book α of the *Metaphysics* contains a proof of the cause of being. Even more, the latter work reconciles Aristotle's intention in stating the eternity of motion in his proof of the cause of being: *Ibid.*, 2, 3–17. Cf. R. Jolivet, *Essai sur les rapports entre la pensée grecque et la pensée chrétienne. Aristote et Saint Thomas ou l'idée de création. Plotin et Saint Augustin ou le problème du mal. Hellénisme et christianisme*, Vrin, Paris, 1955, novelle édit., pp. 3–84; esp. 3–17. – In the light of such texts from Aquinas' works, one must reject M. Gilson's view that "causa substantiae" (par. 1164) and "causa esse" (In *II Meta.*, 2, 295) do not mean a "cause of being" in the sence of a creator. Cf. Gilson, *Being and Some Philosophers*, pp. 70–71. – Perhaps an exhaustive study of Aquinas' works will reveal that Thomas felt only that Aristotle *should* have posited the cause of being, but in any case, one can not doubt that "being" in "cause of being" means just what it pretends to mean.

[3] Owens, *The Doctrine of Being in the Aristotelian Metaphysics*, pp. 35–38.

14 Thomas considered this passage as an another discussion of the first-universal science: since this science studies the immaterial being, is it thus the universal science? Thomas wrote his exposition on the basis of the *Moerbecana*; if we judge from Thomas' words, this version was exact.[1] To be noted is that Thomas repeats Aristotle's thought: the study of the highest being is the universal science.[2] But then Thomas adds on his own: "Nam prima etia sunt principia aliorum."[3] This sentence appears to be given as the reason why the first science is the universal science: if we study the immaterial beings, we can only do so in so far as we see them as the cause of being or of motion; if we study being as being, we must study first beings as the causes of being as such. Thomas is not of course giving the *verba Aristotelis* when he presents the reason behind Aristote's thought. Yet one can reconstruct the method of Thomas as follows: Book XI (K) was written by Thomas after the passage from Book VI (E) mentioned immediately above;[4] since the *intentio Aristotelis* had been explained in the earlier book, here only the vaguest reference need be made.

I have made quite detailed what appeared to be only a minor point. This detail will stand us in good stead, however, when we pass in review the opinions of Albert and Averroes; as will appear, these two commentators must be numbered among those who have misunderstood Aristotle by refusing to accredit his metaphysics with a proof of God as the cause of being. Thus Aquinas is writing of first science and of universal science in the context of a polemic against these two of his predecessors.

As we have discovered in our examination of this problem, Thomas must be viewed as proposing the *verba Aristotelis* when he explained first science as the study of the first being; when he wrote that the universal science is the study of being as such, and that the universal science and the first science are the same in practice, he must again be said to expose the *verba*. On the other hand, it was the *intentio Aristotelis* given when he presented the reason why wisdom must be both first and universal: one can study the first being only as the cause of being as such.[5]

[1] Cf. *In XI Meta.*, 7, 2266–67.

[2] "Si autem est alia natura et substantia praeter substantias naturales, quae sit separabilis et immobilis, necesse est alteram scientiam ipsius esse, quae sit prior naturali. Et ex eo quod est prima, oportet quod sit universalis.", *Ibid.*, 2267.

[3] *Ibid.*.

[4] Cf. Ch. I, section 2, the chronology of the composition.

[5] As we noted in Ch. IV, "being" means an *esse*–essence composite for Aquinas; hence already we can suspect that Thomas' representation of metaphysics as the study of the first

So much for the conformity of Thomas and Aristotle. We must now examine the writings of Avicenna, Averroes, and Albert to discover their interpretations of the relations between first and universal science. Since Avicenna does not mention all aspects of this problem, it is best to study his theory last of all.

B. Averroes and Albert on the identity of "first science" and "universal science"

Aristotle in Γ 2, 1004a2–6 had written of the parts of "philosophy." As has been argued, Aquinas legitimately understood "philosophy" in the sense of "this philosophy now being studied," that is "metaphysics." And so he explained that there are various parts of metaphysics in so far as we divide the metaphysical study of sensible beings from that of immaterial beings. Yet this science has a unity, for all these beings are regarded as "being" and as "one."[1]

This explanation of Thomas is not without historical roots: it is found totally in Averroes; even Thomas' examples have been copied.[2] Interestingly enough, Albert presents much the same doctrine. Yet there is a difference in the latters' thought, and it is an important one, for it reveals how Thomas follows Averroes' legitimate interpretation and how he opposes it to Albert's incorrect one.

When Albert treats these few lines of Γ 2, he uses "first philosophy" exactly as in his introductory chapters (which we studied in Chapter III, section 4). The most noble, the most manifest aspects of things, are studied in metaphysics, Albert had said in the early chapters of his work. Such things are *ens*, the *partes entis* (the accidents), and their principles.[3] It is because metaphysics studies these most noble aspects in their totality that it is the "first science," that it merits being called "first philosophy."[4] It is in the light of this theory that Albert comments on these verses of Γ 2. To be sure, Albert follows Averroes in all respects other than in this: the meaning given to the title "first philosophy."[5]

causes of being as such is not going to be the same as Aristotle's. As shall be explained in Ch. VI, the proof of the existence of the first cause of being rests on the composite character of being: because Thomas realizes that the correct way of conceiving "being" is "actualized essence", he is forced to posite a pure actuality. Thus in the conclusion of Ch. VI, we shall be in a position to note that Thomas' presentation of the *intentio Aristotelis* in his exposition of the identity of first and universal science, is actually a presentation of a transforming meaning when placed in the over-all context of the *Commentary*.

[1] For Aristotle, cf. footnote 2, p. 189. Thomas: *In IV Meta.*, 2, 563; cf. footnote 1, p. 190.

[2] Cf. *In IV Meta.*, c. 4, fol. 68v, I–M.

[3] Cf. *In I Meta.*, T. I, c. 2, p. 5, l. 51–54; footnote 2, p. 79.

[4] *Ibid.*, p. 5, l. 55–58; footnote 4, p. 79.

[5] "Quia tamen speculatur philosophus de ente et primum ens, ex quo alia pendent et entia nominantur, est substantia, quot sunt substantiae, tot sunt partes philosophiae primae."

Thus we note a very large area of agreement between Thomas on one hand, and Aristotle, Averroes, and Albert on the other. Yet we note, too, the presence of an un-Aristotelian doctrine in Albert's *Commentary*: the meaning given to the title "first philosophy." Thomas' acceptance of Aristotle and Averroes rather than of Albert can have but one explanation: Thomas wished to adhere to Aristotelian metaphysics as he understood it.

Only a few lines later than the text just considered from Γ 2, Averroes, Albert, and Thomas found another, yet related doctrine in Aristote's writing. In Γ 3, 1005a33–b2, Aristotle speaks once more of the study of being as such and the first of the sciences; only the study of the highest type of being – of that kind higher than physical being – can be an universal investigation of the principles common to being as such.[1]

Thomas for his part does little more than emphasize the identity in one science of the study of the principles of being, the study of the highest being, and the study of being as such.[2] When one reads Averroes' exposition, one is conscious that he too has repeated Aristotle; there are re-arrangements of Aristotle's thoughts, to be sure, but that thought is faithfully, clearly repeated by Averroes. One does not, however, have the impression that Thomas has molded or modeled his exposition on that of Averroes; there is nothing common to the two expositions in the line of literary form.[3] And finally, Albert too repeats

In IV Meta., T. I, c. 6, p. 168, l. 1–5. "Ex his igitur patet, quod necesse est aliquam esse primam philosophiam habentem istas substantias et similiter habentem cognoscere unum et ens. Omnia enim ista recta ratione existunt habentia pro generibus primis unum et ens. Dico autem unum et ens esse genera, eo quod sunt prima subiecta, ad quae omnia alia se habent per informationem et additionem... Ex quo prima philosophia est de his, necesse est, quod etiam scientiae librorum, in quos ista philosophia dividitur, sequantur hoc, sicut diximus.", *Ibid.*, l. 16–26.

[1] Cf. footnote 1, p. 188; "first science" in this context does not mean the divine science as was explained in the same note, despite the fact that Thomas added here information about divine science.

[2] *In IV Meta.*, 5, 593; cf. footnote 2, p. 190.

[3] Averroes writes: "Ista enim principia sunt communia omnibus generibus entium,... et quod est tale, debet considerari ab eo, qui considerat de ente simpliciter, scilicet Philosopho." *In IV Meta.*, c. 7, fol. 72v, H–I. "...quia scientia istorum [primorum principiorum] est universalis, nullus loquentium in scientia particularibus intromisit se ad loquendum de eis, sed tamen quidam Naturales locuti sunt de istis. Existimant enim, quod sua scientia est universalis, quae debet considerare de ente secundum quod est ens... Sed, cum aliqua scientia fuerit magis alta... et magis communis scientia Naturali: et illa debet considerare de ente abstracto, et de entibus abstractis, quoniam scientia Naturalis non considerat nisi in quibusdam generibus entium, scilicet mobilibus: necesse est ut ista scientia perscrutetur de istis principiis... Et intendit per primam substantiam primum principium substantiarum, et est Deus.", *Ibid.*, fol. 72v, M–fol. 73r, A.

the same Aristotelian doctrine. He too did not copy Averroes, however, now does he appear to have served as a model for Thomas.[1]

Thus all three commentators read Aristotle in the same fashion. Although we have not noted Thomas oppose his thoughts to those of his predecessors, we at least have again discovered that the lack of opposition occures in relations to a passage correctly explained by his predecessors.

In E 1, 1026a10–32 we find Aristotle dealing again with the "first science," with the study of the immaterial or highest being. Thomas' commentary on this passage revealed, as I have argued, that he felt the connection between the first science and the universal science was to be found in two doctrines: "causality" and "knowledge." The first being is the cause of being as such, he felt; moreover, to know the first being is to know the cause of being in so far as this cause is a cause.[2] As we remarked as well, Thomas in exposing his commentary on this passage was opposing what he felt an incorrect view: the view, held by some, that Aristotle's God is not the cause of being as such.[3]

Such a doctrine is not totally foreign to this passage of Aristotle's E 1, as we have earlier remarked: Aristotle here connects 1) the study of the immaterial substance, 2) the first philosophy, and 3) the study of being as being. Moreover, the immaterial substances are referred to as "eternal," and as "the causes that operate on so much of the divine as appears to us."[4]

When we turn to Averroes' *Commentary*, we recognize again a basic fidelity to Aristotle's doctrine. Yet the fidelity is not total. Metaphysics

[1] "Igitur speculatio de his principiis [primis] est illius philosophi qui cognoscit ens, inquantum est ens... Sed quidam physicorum ignorantes proprium subiectum de his volebant determinare... Sed quoniam... aliquis philosophus superior est physico, eo quod natura est determinatum quoddam genus..., ideo erit de his principiis perscrutari illius qui theoricat ea quae sunt entis universalis, et qui theoricat sive speculatur ea quae sunt circa primam substantiam, quae est ommum entium principium,et hoc est deus ipse.", *In IV Meta.*, T.II c. 1, p. 173, l. 47–p. 174, l. 2.

[2] *In VI Meta.*, 1, 1164; cf. footnote 1, p. 192. Also see: *In XI Meta.*, 7, 2267, the explanation of K 7, 1064b6–14. Averroes and Albert did not know Book K (XI).

[3] "Ex hoc autem apparet manifeste falsitas opinionis illorum, qui posuerunt Aristotelem sensisse, quod Deus non sit causa substantiae caeli, sed solum motus eius.", *In VI Meta.*, 1, 1164.

[4] "But if there is something which is eternal and immovable and separable, clearly the knowledge of it belongs to a theoretical science, – not, however, to physics... nor to mathematics, but to a science prior to both... the first science deals with things which both exist separately and are immovable. Now all causes must be eternal, but especially these; for they are the causes that operate on so much of the divine as appears to us... but if there is an immovable substance, the science of this must be prior and must be first philosophy, and universal in this way, because it is first. And it will belong to this to consider being *qua* being – both what it is and the attributes which belong to it *qua* being." E 1, 1026a10–32.

says Averroes, is the study of the "abstract" or immobile beings. These immobile beings are not the heavenly bodies, but are their causes.[1] Moreover, precisely because this science studies the first being, it will be "universal science" – the knowledge of what is common to all beings.[2] But does Averroes understand "knowledge common to all being" as "knowledge of being as such"? It appears he does not, and thus we reach the Aristotelian doctrine rejected by Averroes.

Metaphysics studies all four causes, Aristotle says (as was argued in the preceding chapter); thus to study being as such is to know all four causes. Averroes did not accept this doctrine but regarded metaphysics as the study of the first form and first final cause.[3] The consequences of Averroes' doctrine on that point begin to be felt in his interpretation of Book E (VI). – Aristotle's E 1 ended with the statement: "And it will belong to this (first philosophy) to consider being *qua* being – both what it is and the attributes which belong to it qua being."[4] Now if Averroes had connected this sentence with the discussion of the first science as the study of the first or immobile beings, he would have been forced to see Aristotle's metaphysics in a different light. E 1, after discussing the manner of definition used in the three philosophical disciplines, asks if the highest science is universal, or whether it deals with only one genus of being.[5] It is as a conclusion to this problem that Aristotle gives the statement, quoted above, which identifies "first philosophy" and the "study of being as such." Aristotle thus opts for the difficult theory: this first philosophy studies one genus (immobile being), but at the same time is universal, studying being as such. Averroes' exposition of this passage is difficult, but it appears that he means that first philosophy is universal because it studies the first cause: it studies being as such only by studying the first causes of being as such, the first form and the first final cause. Thus metaphysics is actually a study only of immobile substances.[6]

[1] "Scientia enim Naturales considerat de rebus mobilibus, Mathematica autem de rebus abstractis secundum definitionem, non secundum esse. Sed consideratio de abstractis naturis est scientiae altioris istis.", *In VI Meta.*, c. 2, fol. 146r, E–F. "Deinde dixit, et quae non moventur, etc., et intendit, quod res immobiles necesse est ut sint aeternae magis quam aeternae mobiles divinae, scilicet corpora coelestia. Sunt enim causae istorum, scilicet quod substantia abstracta est causa corporum coelestium.", *Ibid.*, fol. 146v, G.

[2] "...sed, si scientia ista fuerit universalis, dignius ut fit communis omnibus entis... et manifestum est quod, si aliqua substantia immobilis est, quod ista substantia est prima, et quod scientia istius est scientia universalis, et Philosophia prima.", *Ibid.*, c. 3, fol. 146v, M–fol. 147r, A.

[3] Cf. Ch. IV, section 2, B.

[4] E 1, 1026a31–32.

[5] E 1, 1026a23–32.

[6] "Idest, Et rectum est quaerere, postquam declaravit talem scientiam esse, quae dicitur

The introduction given by Averroes to his Book VI appears to bear out our reading of the Arab's thought. Metaphysics, Averroes remarks, is knowledge through causes. Thus the study of all beings *qua* being amounts to an investigation into their causes. But since all beings come together in *ens simplex*, or substance, that is in pure form, the causes of beings as being must be simple or immaterial being; they are the first form and the first end. Metaphysics thus is able to know all being as such by studying the immaterial causes of being.[1] Hence, because the first form and the first end is God, and because metaphysics knows all being by studying the first form and the first end, metaphysics can be said to study whatever is defined in terms of "God." Natural philosophy defines objects by using the notion of "nature"; and in like manner, metaphysics defines by using some common notion. But the notion used in metaphysics is not "nature," but "God," or what is the same "first form and first end."[2]

Such then is Averroes' explanation of Aristotle's E 1, 1026a10–32. Metaphysics for the commentator is a science directly of immaterial substance, and secondarily of being as such; that is, in so far as the

prima Philosophia, utrum ista scientia fit universalis pluribus generibus, aut eiusdem generis, aut eiusdem naturae. Deinde dicit, Non est enim una species, idest apparet enim quod Mathematica non est unius generis, sed diversorum generum. Geometria enim est naturae diversae a natura, cuius est Astrologia et utraque est Mathematica. Deinde dicit, et scientia universalis communis est omnibus idest sed, si scientia ista fuerit universalis, dignius ut sit communis omnibus entibus. Deinde dicit Si igitur non fuerit, etc. idest et manifestum est, quod si non esset hic alia scientia a sensibili, quod non esset scientia prior Naturali. Deinde dicit Et si aliqua substantia mobilis, etc. idest et manifestum est quod, si aliqua substantia immobilis est, quod ista substantia est prima, et quod scientia istius est scientia universalis, et Philosophia prima." *In VI Meta.*, c. 3, fol. 146v, L–fol. 147r, A. One could presumedly interpret this text in a different manner; but if it is placed in conjunction with the text given in note 2 below, there is no doubt that for Averroes, God is the object of metaphysics. Cf. PAULUS, *Henri de Gand...*, p. 48; SALIBA, *Étude sur la métaphysique d'Avicenne*, p. 64. For a different view, one making being as such the object of Averroes' metaphysics, see: E. GILSON, *History of Christian Philosophy in the Middle Ages*, Random House, N.Y., 1955, p. 220.

[1] "...declaratum est... quod ista scientia habet considerare de omnibus speciebus entium, secundum quod sunt entia, et dat supra hoc testimonia, et dixit ...idest manifestum est, quod iam declaratur scientiam esse perscrutantem de ente secundum quod est ens... [idest] secundum illud, in quo conveniunt, scilicet inquantum sunt entia. Et, cum posuit, quod necesse est ut ens simpliciter habeat causas simplices, incoepit declarare hoc, et dicit ...idest per hoc, quod credimus, quod scire aliquid constat ex scientia causarum, apparet quod omnis scientia, et omnis ars habeat causas, de quibus perscrutetur. Sed haec est scientia perscrutans de ente simplici: ergo habet perscrutari de causis ei simplicibus." *In VI Meta.*, c. 1, fol. 144r, D–F. This last sentence is the climax of Averroes' introduction to Book VI and thus he passes from it to a discussion of metaphysics as the study of divine objects; cf. footnote 2.

[2] "...modi Philosophiae speculativae sunt tres, scilicet scientia rerum mathematicarum, et naturalium, et divinarum, scilicet substantiarum, in quarum definitione accipitur Deus. Et intendebat quod, quemadmodum naturalia sunt illa, in quorum definitione accipitur natura, ita divina sunt illa, in quorum definitione accipitur Deus: et voluntaria sunt illa, in quorum definitione accipitur voluntas.", *Ibid.*, c. 2, fol. 146v, G–H.

immaterial substance, is the cause (final and formal) of beings, one knows being as such by studying its causes.

Turning now to Albert, we are faced with a profound dependence on Averroes. Yet this does not mean that Albert has merely copied the Arabian's works. Just the contrary, Albert's exposition reveals the result of a great deal of thought regarding Averroes' notion of "defining immaterial being through the notion of God."

Metaphysics, Albert explains, is the study of the immobile being, for it is the study of the causes of *esse* in so far as *esse* is real.[1] These causes of *esse* are of course God and other immaterial substances. Metaphysics studies God and other divine beings, it is true. Nevertheless, it knows these divine objects in so far as they are the principles or causes of being as such.[2] It is because metaphysics thus studies the effect of the divine beings, namely because it studies *esse* in which God is reflected, that metaphysics knows being as such. The metaphysician knows what reflects God, what is reduceable to God, and hence he studies the "divine": *esse*. Thus, Albert says that being (*esse*) is defined in terms of the divine, in terms of God.[3]

This theory of Albert recalls, as it should, his doctrine of metaphysics as primarily the study of the formal and final causes of being. Yet secundarily metaphysics knows the efficient and material causes in so far as all beings fall under the genus of being.[4]

As is evident, Albert is close to Averroes. Both fix the metaphysician's

[1] "Prima vero philosophia, quae ab utrisque diversa est, et circa immobilia simpliciter est et circa simpliciter separabilia. Immobilia vero entia sunt simpliciter causae omnes quae sunt sempiternae... Tales autem maxime sunt causae esse, secundum quod est.", *In VI Meta.*, T. I, c. 2, p. 305, l. 16–20.

[2] "Ex hoc autem quod diximus primam philosophiam esse circa divina immobilia et separata et simplicia, dubitabit fortasse aliquis, utrum ipsa sit scientia universalis..., an forte sit circa aliquod genus subjectum unum et circa naturam unam divinam, et sic sit scientia particularis... Sed ad hoc solvitur, quod non est idem modus in mathematicis et physica, quae sunt circa unum ens determinatum, et in divina scientia, quia geometria et astrologia circa aliquam naturam sunt, cujus principia non sunt omnium principia neque sunt principia entis, secundum quod est ens... Illa vero quam divinam vocamus, est communis omnium, quae licet sit de deo et divinis, est tamen de his, secundum quod illa sunt principia universi esse per hoc quod sunt principia entis vere, secundum quod est ens...", *Ibid.*, c. 3, p. 305, l. 63–p. 306, l. 3.

[3] "...in prima philosophia omnia dicuntur divina, eo quod in diffinitione eorum cadit deus. Quia... omnia exeunt a primis principiis divinis et in ipsis sunt sicut artificiata in mente artificis. Et sicut artificiata resolvuntur ad lumen intellectus primi activi et per ipsum diffiniuntur, ita omnia resolvuntur ad lumen separatarum substantiarum, et ipsae separatae substantiae resolvuntur ad lumen intellectus dei, per quod subsistunt, et per ipsum sicut per primum principium diffiniuntur." *Ibid.*, p. 305, l. 38–48. "Ista autem sapientia desiderabilior est inter theoricas, eo quod, sicut diximus, ipsa stat in lumine intellectus divini, quod scito nihil invenitur ultra quaerendum...", *Ibid.*, p. 305, l. 57–60. "Est igitur ipsius de ente in, quantum est ens, speculari... Omnia enim haec reducit ista in principium universi esse cujus intellectus causa est entis... in quantum est ens simpliciter.", *Ibid.*, p. 306, l. 16–21.

[4] Cf. Ch. IV, section 2, B; footnote 3, p. 128–2, p. 129.

attention on God, in so far as being is defined in terms of Him. For Averroes, because metaphysics attends directly to God and the immaterial beings as the causes of being as such, it deserves to be called the science of being as such; in knowing immaterial substance, the metaphysician knows the first form and final goal of all beings. For Albert, however, the metaphysician attends to God in so far as he finds God in *esse simplex*. For Albert, *esse* is able to reflect God, as was explained in Chapter II. It was because of this ability on the part of *esse* to reflect something of the perfection of God that Albert says that the study of being as such is primarily a study of the formal, but also final cause.[1]

Averroes can, then, be considered as a God-centered metaphysician. Does such an approach represent Aristotle's views in E 1? When one reads only Aristotle's remarks in this particular chapter, one could possibly be tempted to agree that it does. Aristotle divides the three speculative sciences according to the division of their definitions: physics defines material being through the use of matter; mathematics defines material aspects of beings without using matter; and first philosophy defines immaterial beings without recourse to materiality. Moreover, says Aristotle, these immaterial beings are the causes of whatever we see as caused by divine beings. And finally, Aristotle affirms that the study of such beings is the study of being as such. It does not seem incorrect to see in Averroes a legitimate synthesis of these doctrines.

Yet on the other hand, Thomas views Aristotle's metaphysics as being-centered; God, in Thomas' eyes, comes into metaphysics in so far as one discovers He is needed as the cause of being as such. And as I have argued above in point A of the present section, Thomas can legitimately be considered as giving an Aristotelian doctrine.

[1] "Esse enim, quod haec scientia considerat, non accipitur contractum ad hoc vel illud, sed potius prout est prima effluxio dei et creatum primum, ante quod non est creatum aliud." *In I Meta.*, T. I, c. 1,p. 3, l. 1–4. Cf. Ch. III, section 4. Thus in Albert's view, the object of metaphysics is not God as has been maintained by: M. SCHOOYANS, "La distinction entre philosophie et théologie d'après les commentaires aristotéliciens de saint Albert le Grand", *Revista da Universidade Catholica de São Paulo*, XVIII, 1959, pp. 273–76. As is obvious from the texts in note 3, p, 199. Albert has introduced a bit of Neoplatonism into his *Commentary*. M. Schooyans gives many texts to illustrate these aspects of Albert's work, cf. *Ibid.*, pp. 274–76. Unfortunately M. Schooyans neglects to quote the text given in this note (perhaps the key text for Albert's *Commentary*), and hence concludes that the object is God. It is not God, however, but is *esse*, which is a radiation of God's light. Since *esse* is so conceived, if one takes into account only the passages where Albert speaks of the study of *esse*, the study which "stat in lumine intellectus divini", naturally one will think that Albert means one is to study God directly and see creatures in His light. But actually the situation is just the opposite: through a study of the divine light in things, one can learn to know something of God. Thus if one studies the first form of things ,the cause of substance, one learns something about God.

Obviously Aristotelian metaphysics is not both God-centered and being-centered. That is, metaphysics can not both know being by knowing God, and know God by knowing being. Actually the problem is dependent on the solution of a prior problem: does the philosopher prove God in physics, and then describe Him in first philosophy? or does he prove Him in physics, and through the study of being as such prove that the prime mover is the cause of being? Averroes and Albert have chosen the first notion of philosophy; Thomas, the second.[1] – The present chapter is not the place to discuss the Aristotelian character of these choices; such a discussion properly belongs to the following chapter which will discuss the steps taken in constructing a metaphysics. At the moment, what is important is that Thomas has interpreted E 1 in a manner radically different from that chosen by Averroes and by Albert.

When one reads Aquinas' exposition of E 1, it is with difficulty that one can be certain that Aquinas directs any statement against Averroes and Albert. To be sure, Aquinas points out that Aristotle's E 1 must be connected with the doctrines of Books A and α: E 1 states that metaphysics studied those immaterial beings which are the causes of sensible beings; and α said that these beings are the causes of the being of things and not only of their motion; and finally A said that metaphysics seeks the causes of being as such. Hence, says Aquinas, it is not correct to say that Aristotle taught that God was only the cause of motion and not also of being: "Ex hoc autem apparet manifeste falsitas opinionis illorum, qui posuerunt Aristotelem sensisse, quod Deus non sit causa substantiae caeli, sed solum motus eius."[2] These

[1] Averroes: *In XII Meta.*, c. 5, fol. 292v, L–fol. 293v, L; *Ibid.*, c. 6, fol. 294r, C–D. Albert: *In XI Meta.*, T. I, c. 3. Aquinas: *In XII Meta.*, 2, 2427; cf. Ch. VI, section 1, A.

[2] *In VI Meta.*. 1, 1164. "The world of Aristotle and of Averroes ...is, and there is nothing more to be said. Obviously, it would be a foolish thing to speak of creation on the occasion of such a world, and, to the best of my knowledge, Thomas Aquinas has never spoken of the Aristotelian cosmos as of a created world; on the other hand, Averroes and his disciples have always maintained that, in the doctrine of Aristotle, God is not merely the Prime Mover of the world, but that he also is its Prime Maker. Nothing could have been better calculated than this subtle distinction between Mover, Maker, and Creater, to help us in ascertaining the true nature of Aristotelian being. If the Gcd of Aristotle were nothing more than the Prime Mover of the world, He would, in no sense of the word "being", be the cause of its being. A merely physical cause, such as God, would not be a metaphysical cause. If, as Averroes, Thomas Aquinas and many Averroists have said, the God of Aristotle is the Maker of the world, the reason for it is that He actually is, for all beings, the cause of their very being. They owe Him, not only to move if they move, to live if they live, but to be... Still this is not yet a created universe... As a World-Maker, the God of Aristotle can insure the permanence of substances, but nothing else..." GILSON, *Being and Some Philosophers*, pp. 70–71. To prove that Aquinas has called Aristotle's God the "Maker of the World", M. Gilson cites *In VI Meta.*, 1, 1164. It appears to me that the "causa substantiae caeli" of par. 1164, the "causa esse" of *In II Meta.*, 2, 295, the "agens divinum quod influit esse sine motus, est

statements of Thomas echo his earlier remark that it is not true that Aristotle taught that God is only the cause of motion: "...etsi sint incorruptibilia, tamen habent causam non solum quantum ad suum moveri, ut quidam opinati sunt, sed etiam quantum ad suum esse, ut hic Philosophus expresse dicit."[1] Moreover, it foreshadows, and gives substance to, a later statement that one can seek an agent cause of *esse* as well as of motion.[2] But who are these other commentators who have taught that Aristotle's God is a God of motion and not of being? Against whom is Thomas directing these statements?

Perhaps Thomas refers to Albert. Albert as a Christian theologian, would have believed in the God of creation. As a philosopher did Albert prove God was the cause of being? If he did, he has certainly left no trace of his proof in his exposition of Book Λ. There we discover that Albert intends to give only "opiniones Peripateticorum de istis insensibilibus et immobilibus substantiis"; whether or not these theories are correct is a problem which, Albert notes, is to be decided by others.[3] Interestingly enough, Albert's list of opinions contains no proof of a cause of being. – In like manner, if one investigates Albert's exposition of Book α corresponding to Thomas' affirmation that Aristotle proves a God of being, one finds that there, too, Albert has steered clear of the problem. Albert, instead of noting that those beings which are the causes of truth are also the causes of being, has taken refuge in a doctrine of knowledge: just as metaphysics knows the principles of truth, so too does it know the principles of being.[4] As is evident, Albert can say one knows the principles of being without stating whether one reaches this knowledge from being as such. Thus Albert skirts the problem; perhaps the God studied in metaphysics is the cause of being, but the question is: how do we learn to know that God? – Elsewhere however, one feels that Albert became more courageous and explained

causa non solum in fieri, sed etiam in esse" of *In VII Meta.*, 17, 1661, and the "si non fuerit mundus aeternus, necesse est quod fuerit productus in esse ab aliquo praeexistente" of *In XII Meta.*, 5, 2499, are none of them referring to what Gilson terms "a Maker of the world". And of these four statements of the need for a cause of being, not a single one can be shown to be a statement which Aquinas knew was not implied in the Aristotelian *Metaphysics*.

[1] *In II Meta.*, 2, 295.

[2] *In VII Meta.*, 17, 1661.

[3] "...relinquentes aliis judicium, quid verum vel falsum sit de his quae dicunt.", *In XI Meta.*, T. II, c. 1, p. 482, l. 28–29. Albert's Book XI is an exposition of Book Λ (XII). – This "let others judge" should be connected with Albert's statements to the effect that his expositions of Aristotle are intended to present solely the view proper to Aristotle; there is nothing of Albert in them, or at least so Albert thought. Cf. M. GRABMANN, "Die Lehre des heiligen Albertus Magnus...", pp. 295–97. M. GRABMANN, *Methoden und Hilfsmittel des Aristotelestudiums in Mittelalter*, Sitzungsbericht der Bayer. Akad. d. Wiss., philos.-histor. Abteilung, 5, München, 1939, pp. 39–42.

[4] Cf. *In II Meta.*, c. 4, p. 95, l. 28–41.

the physical and metaphysical methods of studying God. Metaphysics, he says, considers God as the cause moving things by His form and power; this motion, caused by God, is the instrument by which all being flows from God. Physics, on the other hand, studies God only as the cause of motion.[1] Yet such a statement still falls short of an *affirmation* that Aristotle proves only a God who causes motion.

Thomas affirms the existence of an Aristotelian proof of God as the cause of being in his exposition of E 1 and α 1. In both these places Albert has made no reference to such an Aristotelian theory; rather he has referred to metaphysics only under the terms of "study of God." If we look now at Thomas' exposition which corresponds to the passage in which Albert gives the doctrine of the different ways of treating God in physics and metaphysics, we find once again a statement which could be directly opposing Albert's views. Albert notes that physics studies God as first mover, while metaphysics views all beings as flowing from God through the instrumentality of motion. Thomas, for his part, does not mention such a theory; yet in the corresponding point of his *Commentary* he digresses for a moment to explain the relation of physics and metaphysics. Physics studies earth-bound beings as well as the heavenly bodies from the point of view of matter, Thomas writes; metaphysics, however, studies both mobile substances and God in so far as they have something in common: in so far as they come together in "being" and in "substance."[2] The fact that this theory, different from that of Albert, occurs at the identical point in which Albert's is found, must indicate, I believe, that Thomas directs this exposition against Albert.

Thus it would seem reasonable to view Thomas' attack on "quidam" and "opinio illorum" as, at least, an attack directed toward Albert. The point at issue in Aristotle's E 1 is metaphysics as "first philosophy." It deserves that name because it studies the first being, Aristotle says.

[1] "Sed physicus considerat eas, [idest, omnes quattuor causas] prout sunt principia mobilis, primus autem philosophus reducit efficientem in forman primam et finem ultimum, et sic ipse est causa universi esse et forma et finis. Et si accipiatur per motum prima causa, hoc non est, ut sciatur, inquantum est movens talem motum, sed potius in quantum ipse ambit virtute et forma sua mobile et motum, quod est instrumentum fluxus totius entis ab ipso...", *In XI Meta.*, T, I, c. 3, p. 463, l. 1–9.

[2] "...substantiae sensibiles, sive sint corruptibiles sive perpetuae, pertinent ad considerationem naturalis philosophiae, quae determinat de ente mobili. Huiusmodi enim substantiae sensibiles sunt in motu. Substantia autem separabilis et immobilis pertinet ad considerationem alterius scientiae, et non ad eamdem, si tamen nullum principium sit commune substantiis: quia si in aliquo conveniant, pertinebit utrarumque substantiarum consideratio ad illam scientiam, quae illud commune considerat. Et ideo naturalis scientia considerat solum de substantiis sensibilibus, inquantum sunt in actu et in motu. Et ideo tam de his etiam quam de substantiis immobilibus considerat haec scientia, inquantum communicant in hoc quod sunt entia et substantiae.", *In XII Meta.*, 2, 2427.

"True," writes Thomas, "but we must not forget that metaphysics as the science of being rises to the level of 'first philosophy' in so far as the first being is discovered as the cause of being. This is Aristotle's opinion, and hence those philosophers are wrong who refuse to see in the *Metaphysics* a God who causes being." – Although we can not discover in Albert the explicit statement that Aristotle's God is not proved as the cause of being, Albert does systematically disagree with Thomas' views on the very nature of metaphysics. Thus by implication, Albert would affirm that Aristotle's proof of God is not a proof of the cause of being. And hence Thomas could have had Albert in mind when he commented on E 1 and on α 1.[1]

Much the same thing can be said of Thomas' relation to Averroes. First, by implication, Averroes makes the proof of God a proof of the cause of motion in his exposition of Book Λ. Here Averroes gives a clear indication of his view on the role of God in Aristotelian metaphysics: God can be considered only as the eternal, immobile principle of the eternal, mobile substances. For Averroes, the study of physics proves that corruptible and generable substances have a principle. Metaphysics begins at this point; but it is necessary for the metaphysician to speak of temporal substances – to repeat much of physics. Only after the metaphysician has mentioned that these temporal, generable substances are ruled by eternal, mobile substances, can he show that these eternal, mobile substances have as their principle an immobile, eternal substance.[2] In the proof and in the discussion of the first, immaterial motor, there is not the slightest hint that he may be a creator of being,

[1] For a discussion of Albert's evolution on the problem of creation, see: HANSEN, "Zur Frage der anfangslosen und zeitlichen Schöpfung bei Albert dem Grossen", pp. 167–88. As the author explains, physics, using a *demonstratio quia*, proves the existence of God as cause of motion; metaphysics can only show what God is from the beings in the material world. In short, Albert was unwilling to admit an absolutely certain proof of the cause of being in his *Commentaries* on the *Physics* and the *Metaphysics* (p. 177). Cf. also: P. A. ROHNER, O. P., *Das Schopfungsproblem bei Moses Maimonides, Albertus Magnus und Thomas von Aquin.* Beiträge zur Geschichte der Philosophie des Mittelalters, B. XI, H. 5, Aschendorf, Munster i. W., 1913, pp. 47–48; 55–59; 69–72. SCHOOYANS, "La distinction entre philosophie et théologie d'après les commentaires aristotéliciens de saint Albert le Grand", pp. 22–23.

[2] "Et illud, super quod debemus substentari in imaginatione est, quod substantia est duobus modis, modo aeterno, et non aeterno ut declaratum est in scientia Naturali. Et quod sua intentio in hoc tractatu est perscrutari de substantia, quae est principium substantiae aeternae: cum declarata sint principia substantiae generabilis et corruptibilis in naturalibus. Sed, quia sermo eius in hac scientia est de principiis substantiae simpliciter, necesse fuit exponere considerationem bipartitam, scilicet in principiis substantiae non aeternae, et principiis substantiae aeternae. Et ideo divisit hunc tractatum primo in duo. Quorum primum est de principiis substantiae non aeternae...: et posuit hoc de naturalibus. Secundum autem est de principiis aeternae propriis. Et, quia principia substantiae sunt substantiae, necesse fuit loqui de substantia, quae est principium aeternae substantiae mobilis, et declaravit eam esse immobilem et aeternam.", *In XII Meta.*, c. 29, fol. 313v, G–I.

as well as the prime mover.[1] Thus, Averroes sees Aristotle's God as proved to be a cause only of motion. – In other places, however, Averroes' words come close to attributing to God a causality in relation to being. For example in his exposition of α 1, he writes: Physics shows that the first among beings is the cause of all others; thus this first being must be first in both being and truth, and all other beings are "being" and "true" through the being and truth of the first being.[2] However, in another place, Averroes notes that the eternal first mover of physics is not only the moving cause, but also principle by being the first form and end, that is, the first substance attracts other beings as the goal of their motion.[3] Such a notion of the first immobile substance as the principle of being is far from a notion of a cause of being.[4]

Thus despite the interpretation one is inclined to put on his explanation of Book α 1, Averroes in other places would imply that Aristotle deals only with the causes of motion (the efficient and final cause). Besides the passages already noted, the exposition of *In XII Meta.*, c. 5 is extremely clear regarding this doctrine.[5] This passage must be noted for it corresponds to the one in which Albert put forth his theory on the difference between physics and metaphysics, just as it corresponds to that passage which contains Thomas' views on the same subject. Thomas and Albert make explicit statements: for Albert, physics studies God as the first mover, while metaphysics views all beings as flowing from God through the instrumentality of motion;[6] for Thomas, who makes no reference to Albert's views, physics studies mobile being as mobile, while metaphysics studies whatever falls under the common notion of "substantial being."[7] An examination of Averroes' exposition reveals that he too paused to explain the difference between physics and metaphysics. Physics, Averroes writes, studies the principles of all

[1] Cf. *In XII Meta.*, cc. 29–35, fol. 313r, E–fol. 318r, C.

[2] "...et cum declaratum est quod causa in quolibet genere entium est magis digna in esse, et in veritate, quam illa, quorum est causa in genere illa, manifestum est quod si est haec prima causa omnium entium, *ut declaratum est in scientia naturalium*, quod illa causa est magis digna, et in esse, et in veritate, quam omnia entia. Omnia enim entia non acquirunt esse et veritatem nisi ab ista causa. Est igitur ens per se, et verum per se: et omnis alia sunt entia, et vera per esse et veritate eius.", *In II Meta.*, c. 4, fol. 30r, C. Italics added. Note the similar doctrine: "...cum fuerint plura, quae habent nomen idem propter aliud; illud aliud est dignius habere illud nomen, et erit causa illorum.", *In VIII Meta.*, c. 7, fol. 215v, L.

[3] "Et intendit quod, cum huic fuerit iunctum quod declaratum est in Physicis, scilicet hoc esse primum motorem aeternum, et absolutum ab omni materia, et declaravit post, quod hoc non solummodo est principium tamquam motor, sed tamquam forma et finis..." *In X Meta.*, c. 7, fol. 257r, A–B. Cf. *In XII Meta.*, c. 5, fol. 292v, I–fol. 293v, M.

[4] *In XII Meta.*, c. 36, fol. 318r, E–fol. 319r, D.

[5] Cf. fol. 292v, I–fol. 293v, M.

[6] *In XI Meta.*, T. I, c. 3, p. 463, l. 1–9. Cf. footnote 1, p. 203.

[7] *In XII Meta.*, 2, 2427; cf. footnote 2, p. 203.

mobile substance.[1] Metaphysics, on the other hand, must accept the results of physics; that is, metaphysics must begin with the fact of matter and form as the composing principles of natural beings, as well as with the fact that there exists an eternal, immobile and immaterial first mover.[2] Consequent to the acceptance of these discoveries of physics, the metaphysician begins his own proper work: he considers the principles of substance without pausing at their temporal or eternal character; that is, he studies the principles *qua* principles of all substances. His first step is to consider the principles *qua* temporal substances; then, he will turn to the principles of eternal substance. Thus far he has been dealing with information received from physics. At this point, however, the method proper to metaphysics comes to the fore: the metaphysician considers these principles of eternal substances in so far as the principle of principles is itself a substance, a first form, a first final cause. After this discussion, he can ask the further question of the unicity of this principle of principles.[3] In short, Averroes remarks, physics tells us that there exist principles of mobile substance, and metaphysics considers these principles in so far as they are principles of substance as such.[4] Yet despite the fact that metaphysics studies these principles as principles of substance, it does not prove these principles are the causes of being, Averroes warns. In fact, no proof of God is valid save the physical proof based on motion.[5]

[1] "...quomodo est rectum dicere, quod Naturalis non declarat nisi principia substantiae generabilis et corruptibilis, cum non solummodo consideret de substantia generabili et corruptibili, sed etiam de non generabili et non corruptibili? Considerat enim de ente mobili sive generabili sive non.", *In XII Meta.*, c. 5, fol. 293r, A.

[2] "Et dicemus nos quidem, quod Philosophus... accepit pro constanti, hoc quod declaratum est in naturalibus de principiis substantiae generabilis et corruptibilis, scilicet quod declaratum est in primo Physici scilicet quod est compositum ex materia et forma: ... et quod declaratum est in Octavo scilicet quod movens aeternam substantiam est abstractum a materia.", *Ibid.*, fol. 293r, E–F.

[3] "Sed, quia modus considerandi in hac scientia est de principiis substantiae, in eo quod est substantia, sive aeterna sive non, incoepit declarare in hoc tractatu de principiis substantiae non aeternae, et fecit rememorationem de hoc, quod declaratum est ex eis in naturalibus... Deinde post hos incoepit declarare principia aeternae substantiae, et posuit etiam hoc, quod declaratum est in naturalibus, et consideravit de eis consideratione propria huic scientiae. Verbi gratia quod est substantia et prima forma, et quod est principium et finis. Deinde considerat de hac substantia immobili, utrum sit una, aut plures...", *Ibid.*, fol. 293v, G.

[4] "Sic igitur intelligenda est communicatio istarum duarum scientiarum in consideratione de principiis substantiae, scilicet quod Naturalis declarat ea esse secundum quod sunt principia substantiae mobilis. Philosophus autum considerat de eis secundum quod sunt principia substantiae in eo quod est substantia, non substantia mobilis. Et etiam Philosophus declarat abstractum esse, quod est principium substantiae sensibilis: sed postquam accepit pro concesso a Naturali huiusmodi esse substantiam.", *Ibid.*, fol. 293v, H–I.

[5] "...principia subiecti Naturalis [idest motor primus, materia, forma] non demonstrantur nisi per res posteriores in scientia Naturali. Et ideo impossible est declarare aliquid abstractum esse, nisi ex motu. Et omnes viae quae reputantur esse ducentes ad primum motorem esse praeter viam motus aequaeliter, sunt insufficientes.", *Ibid.*, fol. 293r, C. This text is found in

This then was Averroes' position. Precisely at the same point in his *Commentary* Albert explains, yet much more simply, a somewhat similar theory: physics studies God as the prime mover, while metaphysics views all beings as flowing from God through the instrumentality of motion. Thomas, at the parallel point in his *Commentary*, explains that physics studies mobile being as mobile, while metaphysics studies all substances from the point of view of "substantial being." So it is that, whereas the systems of Albert and of Averroes allow no place for a proof that being as such has a cause, Thomas' system demands such a proof as an integral, an essential part: the universal science, studying being as such, becomes the first science when the proof of the immaterial cause of being is constructed. Thus it seems extremely possible that Thomas, in his exposition of E 1, is referring to Averroes and to Albert (and possibly to others) when he writes: "From this appears manifestly the falsity of the opinion of those men who say that Aristotle held that God is not a cause of the heavenly substances, but only of their motion."[1]

A comparison of the expositions of a passage from Aristotle's Book Z, as contained in the works of Averroes, Albert, and Aquinas, confirms the view explained above. In Z 17 Aristotle mentions that one can seek the efficient cause in the case of substantial change.[2] Both Averroes and Albert repeat this doctrine with no additions; the over-all context of their explanations leads one to wonder if they do not wish to affirm that one can not seek an efficient cause of being as such.[3] Thomas, however, interrupts his exposition of this passage to make it clearly understood that Aristotle is speaking here only of an efficient cause which causes substantial changes through motion; however, there is, remarks Thomas, an efficient cause of being as such, which is a divine agent,

the context of an attack on Alexander and Avicenna. Against Alexander, Averroes notes that the only possible way to prove the existence of the prime mover, or of an immaterial being, is from motion (cf. the text just quoted). If there were any other argument, it would be a part of first philosophy; yet this is impossible, because first principles cannot be proved, Averroes concludes. Then turning to Avicenna, Averroes notes that the former felt that physics could not prove its own principles (that is, the prime mover, the distinction of matter and form), but that metaphysics must do this. This too is incorrect, Averroes remarks; physics says once and for all that there is a prime mover – an immaterial mover. Metaphysics must explain how this immobile, eternal substance is the principle of the substance of other things. Cf. *Ibid.*, fol. 293r, D–E. See Ch. II, section 2.

[1] *In VI Meta.*, 1, 1164. Cf. *In II Meta.*, 2, 295.

[2] Z 17, 1041a31–32.

[3] Averroes: "...in quibusdam rebus potest quaeri ...primus motor. Sed ista causa, scilicet agens, quaeritur maxime in rebus generabilibus et corruptibilibus.", *In VII Meta.*, c. 59, fol. 207v, G–H. Albert: "In quibusdam vero hac eadem quaestione quaeritur: 'quid movit primum', quod est causa movens;... Sed causa movens in fieri quaeritur et corrumpi factorum et corruptorum...", *In VII Meta.*, T. V, c. 8, p. 385, l. 39–43.

and which imparts being independent of any motion.[1] Why should Thomas have paused to interject this explanation? Certainly, he wishes to preserve Aristotelian metaphysics from any possible misunderstanding. But was this the extend of his gloss of this passage? Did he not wish, as well, to correct the false impression already, *de facto* promulgated by Averroes and Albert?

Our itinerary in the past several pages has taken us through many texts; it is best to note clearly what we have accomplished.

1) We began by examining Averroes' exposition of E 1, 1026a10–32. Aristotle speaks in this passage of the first philosophy as the study of the first or immaterial being; yet this first philosophy must be also the study of being as such. Thomas, as we noted, places the key to this theory in a proof of the first being as the cause of being as such; that is, metaphysics begins to study being as such, and later becomes first philosophy when it proves the existence of the first being which is the cause of the being of other things. Averroes' exposition of this passage reveals a different theory; metaphysics studies the first, immaterial being, and by knowing that being, realizes the nature of being as such. Averroes' first philosophy is God-centered. And finally Albert's interpretation of this passage of E 1 reveals a somewhat similar God-centered metaphysics. Yet for Albert, metaphysics is not studying God as its object, but rather studies his first creature, *esse simplex*, and from this learns something about God.

2) Our second step was to note that Aquinas' exposition is actually a refutation of those who say that Aristotle had no proof of God as the cause of being; we noted too that in exposing Book α 1 Thomas refers to this same erroneous theory. We wondered if Aquinas was referring to Averroes and to Albert as the exponents of this false view of Aristotle. Examination of Albert was our first step to discover if he held the view opposed by Aquinas; we noted that nowhere does he prove God to be the cause of being. Moreover, we discovered that in Book α Albert fails to see in Aristotle a proof of God as the cause of being. And finally, in his exposition of Book Λ, we found that Albert gives a theory of metaphysical method which appears to make it impossible to prove God to be the cause of being. When we turned to Aquinas' exposition of Book Λ, we found an exposition parallel to Albert's; Thomas words, however,

[1] "Hic autem loquitur Philosophus in substantiis sensibilibus. Unde intelligendum est quod hic dicitur, solum de agente naturali, quod agit per motum. Nam agens divinum quod influit esse sine motu, est causa non solum in fieri, sed etiam in esse.", *In VII Meta.*, 17, 1661.

leave room for the proof of the cause of being in the Aristotelian search for metaphysical knowledge.

3) Thirdly, we turned to Averroes' *Commentary*. In his explanation of Aristotle's proof and discussion of the prime mover in Book Λ, Averroes leaves no room for a proof of the cause of being. Yet paradoxically enough, in discussing Book α, Averroes notes that God must be the cause of being. Finally when we turned again to his commentary on Book Λ, we noted that he too has a digression on metaphysical method at the same point at which Albert and Aquinas presented their theories. What is important is that Averroes denies the value of a proof of God as the cause of being.

Accordingly, I believe that one must see in Thomas' reference to those who misunderstand Aristotle, at least a reference to Averroes and to Albert. Thomas, moreover, indicates that his theory is none other than that of Aristotle; yet the introduction of *esse* into metaphysics – into the exposition of Aristotelian metaphysics – leaves no room for doubting the un-Aristotelian character of that exposition; this point becomes yet clearer in the following chapter in the context of a discussion of the connection between the proof for the cause of being and the *esse*-essence composition of being as such.

What is our conclusion then from this study of Albert's and Averroes' doctrines on first science and universal science? What is the relation of their doctrines to those of Thomas?

In our study we centered our investigation on three passages of Aristotle.[1]

1) Γ 2, 1004a2–6. Averroes and Thomas were totally in agreement on the interpretation of this passage: there are various parts of metaphysics in so far as one divides the study of sensible beings from that of immaterial beings, yet there is a unity to the science, for all beings are regarded as "one" and as "being". Such an interpretation is true to Aristotle it was noted. Albert's explanation tallies with that of the other two commentators, if one excepts the meaning he gives to the term "first philosophy": metaphysics is first because it studies *esse simplex*, or the most manifest properties of things in their totality.

2) Γ 3, 1005a33–b2. All three commentators read Aristotle in an identical manner: the first science, that is the study of the first being, is the universal investigation of being as such.

[1] Aristotle speaks of "first" and "universal science" in a fourth place: K 7, 1064b6–14, but Averroes and Albert did not know this book.

3) E 1, 1026a10–32. The first philosophy is the study of the immaterial being, Aristotle wrote; yet this science is at the same time universal, studying all being *qua* being. Thomas repeats this doctrine, clarifying and synthesizing it as well: metaphysics discovers the immaterial being as the cause of being as such. Averroes has reversed the being-centered metaphysics seen in Aristotle by Thomas. Metaphysics, he writes, is universal because it is first: because it studies the immaterial being, it knows being as such. And Albert explains, at least by implication, that one can not prove God to be anything but a cause of motion. Thomas in his work takes note of these doctrines, I feel, and in an indirect manner attacks them: Aristotle proves God as the cause of being, he writes, thus touching a point which can not be integrated into the conceptions of metaphysics enunciated by Averroes and Albert.

The conclusions to be drawn from this investigation are not difficult to formulate. 1) Aquinas wished to retain in his *Commentary* what he understood to be the framework of Aristotelian metaphysics; 2) thus he uses, or opposes, the works of Averroes and Albert when they aid, or obstruct, this desire to express an Aristotelian metaphysics. 3) De facto, the Aristotelian metaphysics of the *Commentary* has been transformed by the introduction of *esse*, as noted in Chapter IV. Our investigations into the doctrines concerning "first and universal science" hint at the importance of the introduction of *esse* as a principle of being as such, for the universal science rises from a study of being as such to the first cause of being. But "being" expresses a substance somehow in terms of an essence related to its *esse*, as noted in the preceding chapter. Hence, it would seem that the connection between "universal" and "first" science is dependent, to some extent, on the *esse*-essence composition as expressed by "being." As shall be seen in Chapter VI, this is indeed the case; it is from beings as composed of essence and an actuality to which they have no right, that Aquinas argues to the first cause of being. Thus, *esse* is the core of the *Commentary*, the center of its metaphysics. 4) Does it follow that Aquinas wrote his exposition to give anything other than Aristotle's metaphysics? As mentioned, Aquinas is very insistent that Aristotle proved God as the cause of being; from a study of being as such, Aquinas claims, Aristotle rose to God, the first cause of being. Precisely because Aquinas was so insistent on this point, one would expect that the proof of God present in the *Commentary* would be that of Aristotle. For this reason, it appears that one must accept as possible the fact that Aquinas consciously felt he was giving Aristotelian metaphysics when he exposes the *Commentary* around *esse*. Of course, it does not thereby follow that Aquinas did not accept the metaphysics of the *Commentary* as his own; but up to the present point, we have no positive evidence of that.

C. Avicenna's theory on "universal science" and "first philosophy"

In our discussion of Avicenna, it is well to confine ourselves to his general theory of "universal" and "first science." As is to be expected much of Avicenna's doctrine on this point was involved in the theories we examined in connection with the object of metaphysics in Chapter IV, as well as with the Avicennian introduction to metaphysics discussed in Chapter III. In Chapter II, we noted that for Avicenna, metaphysics is called "first philosophy" because it studies the first cause of *esse*: it studies God, the most noble object of man's knowledge. Yet God is not the subject of metaphysics, Avicenna maintains; one assumes the existence of the subject in any science, and in metaphysics we cannot assume that there is a God, a first cause.[1]

One will look in vain for the term "universal science" in Avicenna's *Metaphysics*, I believe. Whether or not it occurs elsewhere, it is in any case not at all a common title for metaphysics. Yet the name "particular science" is used several times to designate all those sciences other than metaphysics.[2] Although Avicenna does not explicitly state what it means to be a particular science, from his discussions it appears that a science is particular in so far as it limits its studies to some particular aspects of reality.[3] Hence, it would follow that metaphysics is going to study all objects. But from what point of view? As was explained in Chapter IV, the metaphysician places himself at the point of view of *esse*: he studies all things as here and now realized.[4]

Thus for Avicenna the study of all things *qua esse* is the (universal) science opposed to the particular sciences. Moreover, this science is called "first science" not because of its subject matter (*esse* as such) but because the cause of its subject is the first being.[5]

[1] E.g., the texts in footnotes 3–4, p. 25.

[2] E.g. "...nec accidentalia propria alicui subiectorum harum scientiarum particularium ...", *Meta.*, T. I, c. 2, fol. 70v, B. "...comparatio huius scientiae ad alias scientias particulares est sicut...", *Ibid.*, c. 3, fol. 71r, B.

[3] See e.g. *Ibid.*, c. 1, fol. 70r, A: the discussion of physics and mathematics.

[4] Cf. Ch. IV, section 4, A.

[5] Avicenna has spoken of "universal" and "first science" in several of his works but apparently not in the *Metaphysics*. Although there was an evolution in his thought, at the time of his *Metaphysics*, his view was as explained above. Cf. A.-M. Goichon "Introduction. L'évolution philosophique d'Ibn Sina", in *Ibn Sina* (*Avicenne*). *Livre des directives et remarques*. Traduction avec introduction et notes par A.-M. Goichon, Vrin, Paris, 1951, pp. 20–23.

Now it is obvious that such a conception has strayed slightly from the paths of the doctrine characterized earlier as Aristotelian. True, Avicenna, as Aristotle, calls the science of the first being the "first science." Moreover, both place this first being within the bounds of metaphysics in so far as the first being is the cause of that aspect studied by metaphysics. Yet in naming the aspect primarily studied by metaphysics, Avicenna differs from Aristotle. For whereas Aristotle rejects any metaphysical consideration of being as logical truth, Avicenna mistakes being as logically true for metaphysical being.[1] Surely we needn't refer again to Thomas' rejection of Avicennian metaphysics of existence. This rejection would explain why Avicenna's general theory of the "first philosophy" and the study of being as such could not be accepted by Thomas. As far as I have been able to discover, however, Thomas nowhere refers specifically to Avicenna's general theory of metaphysics.[2]

By way of concluding this section on the theories of "universal science" and "first science" in the predecessors of Aquinas, I shall limit myself to a few general statements. First, a study of Aristotle revealed that Thomas' *Commentary* presents a faithful picture of Aristotle's thought; thus when Thomas explained "first science" as the study of the highest being and "universal science" as the study of being as such, he was giving the *verba Aristotelis* (idest, a medieval's view of the *verba*); and when Thomas linked first and universal science on the basis of the doctrine of God as the cause of being, he was presenting what to him appeared as the *intentio Aristotelis*. Thus our first conclusion: Thomas wished to preserve the metaphysical framework he thought was Aristotle's. Secondly, a study of Averroes and Albert illustrated how Thomas opposed and/or used his predecessors' works in so far as this opposition and/or use aided the preservation of what he understood as the framework of Aristotelian metaphysics. And finally, our study of

– None of the books which reveal Avicenna's evolution on this subject were translated into Latin, however; hence, Thomas knew but the one stage in his thought, that of the *Metaphysics*. For the Latin translations of Avicenna, see M.-T. d'Alverny, "Avicenna Latinus", *Archives d'histoire doctinale et littéraire du Moyen Age*, XXVIII, 1961, pp. 281–316.

[1] Cf. Ch. IV, section 4, A–B.

[2] There is another aspect of Avicenna's thought which could possibly be mentioned: the relation of physics to metaphysics. Averroes attacked this theory in *In XII Meta.*, c. 5, fol. 293v, D–E. It was after attacking Avicenna, that Averroes gives his own view; this view of Averroes (and the corresponding one of Albert) were the theories rejected by Thomas in *In XII Meta.*, 2, 2427. Thomas makes no mention of Avicenna's theory that physics can not possibly prove the existence of God, yet he implicitly attacks it, just as much as he opposes those of Albert and Averroes, when he gives his own views on metaphysics and physics in par. 2427.

Avicenna revealed only that Thomas, by attacking the former's theory of *esse*, had insured there would be no danger to Aristotelian metaphysics from Avicenna's quarter.

3. ARISTOTLE, AVERROES, AVICENNA, AND ALBERT ON METAPHYSICS AS THE STUDY OF *COMMUNIA*

As will be remembered, Thomas links the universal study of being with the study of the *communia*, those concepts and principles used by all sciences. The *communia*, as formulated by the metaphysician, appear to be the concept's expressing man's knowledge of being as such. The other sciences take these *communia*, but not according to the universal extension of the metaphysical concepts. Rather, the *communia*, as used by the particular sciences, express the concepts and principles proper to each particular science. What relation is there between Thomas' thought on these points and the theories of his predecessors?

A. Aristotle's doctrine

In the exposition of Thomas' thought, three points were brought forth. 1) The particular sciences treat only of particular types of beings and from particular points of view, whereas the universal science treats all beings as such. The texts which were noted in this context (*In VI Meta.*, 1, 1147; *In IV Meta.*, 1, 530; 547) must be considered as faithful expositions of Aristotle (E 1, 1025b7–18; Γ 1, 1003a21–22; Γ 2, 1003b19–22). Thomas' exposition is faithful, although more detailed and precise; Thomas, for example, explicitly notes the particular points of view of the particular sciences, whereas Aristotle can more correctly be said to "hint" at these limited viewpoints.

2) It was also brought out that, for Thomas, the universal science uses the *communia* universally, while other sciences use them particularly. Is Thomas exposing the *verba Aristotelis* when he gives this doctrine? This is a difficult question to answer. Let me divide the answer into several parts.

a) Aristotle does explain that each science will use the first principles as far as its subject matter will allow, yet metaphysics studies them in themselves, without restrictions.[1] Thus in so far as Thomas explains

[1] "Manifestum igitur est quoniam uniusque sciencie, et que philosofi, et que de his est intencio; omnibus enim inest que sunt, set non generi alicui seorsum proprie ab aliis. Et utuntur quidem omnes, quoniam entis est secundum quod est ens, unumquodque enim genus est ens. In tantum enim utuntur in quantum illis sufficiens est; hoc autem est, quantum subit genus de quo ferunt demonstrationes. Quare quoniam manifestum est quoniam secundum quod sunt encia insunt omnibus (hoc enim ipsis commune est), circa ens secundum quod est ens cognoscentis et de his est speculatio.", Γ 3, 1005a21–29: *Vetus*, p. 307, l. 16–26.

that the first principles are used or studied without any restriction by metaphysics, and used in a limited way by the particular sciences, he is giving Aristotle's explicit doctrine.[1]

b) By what right does metaphysics study these first principles? In the text from Γ 3 just cited in footnote 1, p. 213, Aristotle explains that these principles must be studied by the science of being for these are principles of being as such. Thomas in his *Commentary* gives the same doctrine.[2] Yet elsewhere Thomas gives a second reason: since the common concepts – the concepts found among the *communia* – are the subjects and the predicates of the first principles, these principles must be studied by that science which studies the common concepts.[3] Is this the doctrine of Aristotle? There are two questions here which I shall treat as points c) and d): c) does Aristotle teach that metaphysics treats concepts called *communia*, which are used in a particular way by the particular sciences just as the first principles are used? and d) do these *communia* appear in the first principles as subjects and predicates?

c) Does Aristotle explain that the *communia* are treated universally by metaphysics, and used particularly by other sciences? Thomas first presents a more or less well-defined doctrine in *In IV Meta.*, 1, 531. Here he explains that the *per se* accidents of being as such are studied. The particular sciences treat particular accidents of particular beings. Most important of all, Thomas explains that the science of being and of its *per se* accidents is necessary because all the knowledge of the particular sciences depends on this metaphysical knowledge. The implication is clear: the universal knowledge of metaphysics is clearly presupposed by more particular knowledge.[4] As Thomas indicates, this doctrine is his explanation of these words of Aristotle: "et quae huic [that is, *enti*] insunt per se."[5] By these words Aristotle does, of course, admit that the *per se* accidents of being are treated. As I argued in the preceding chapter, these essential accidents or attributes are not only the nine categorical accidents, but as the general context shows, attributes such

[1] Cf. *In IV Meta.*, 5, 591; cf. footnote 1, p. 177.

[2] Cf. *In IV Meta.*, 5, 590–91.

[3] Cf. *In XI Meta.*, 1, 2151; cf. footnote 1, p. 177. *Ibid.*, 4, 2210; cf. footnote 1, p. 178.

[4] "Dicit etiam 'et quae huic insunt per se' et non simpliciter quae huic insunt, ad significandum quod ad scientiam non pertinet considerare de his quae per accidens insunt subiecto suo, sed solum de his quae per se insunt. Geometra enim non considerat de triangulo utrum sit cupreus vel ligneus, sed solum considerat ipsum absolute secundum quod habet tres angulos aequales etc... Necessitas autem huius scientiae quae speculatur ens et per se accidentia entis, ex hoc apparet, quia huiusmodi non debent ignota remanere, cum ex eis aliorum dependeat cognitio; sicut ex cognitione communium dependet cognitio rerum propriarum.", *In IV Meta.*, 1, 531.

[5] Γ 1, 1003a21–22: *Vetus*, p. 303, l. 1–2.

as "unity," "sameness," "prior," etc. Aristotle does indeed hold that metaphysics is to treat these common concepts and that other sciences are to use them.[1] Hence it does not seem unjustified to view as Aristotelian Thomas' exposition of metaphysics as the study of the *communia*.

d) For Aristotle, do these *communia* serve as subject and predicate of the first principles? Thomas expresses this idea several times: *In XI Meta.*, 1, 2151; *Ibid.*, 4, 2210; *In IV Meta.*, 5, 595. The Aristotelian passages corresponding to these three expositions do not contain the doctrine given by Thomas: K 1, 1059a23sqq; K 4, 1061b17sqq; Γ 3, 1005a19–b8. Yet in *In IV Meta.*, 5, 595 Thomas more or less shows us how he evolved this doctrine from Aristotle's words. Aristotle holds that the first principles of syllogisms are to be studied by metaphysics, writes Thomas – and this is evidently the doctrine found in Γ 3. To understand the ultimate reason behind Aristotle's doctrine: "Ad huius autem evidentiam sciendum..."[2] – one should note the nature of *per se notae* propositions. These are the propositions which are known the minute one knows the terms of the propositions. Now this is the doctrine of the first book of Aristotle's *Posterior Analytics*, notes Thomas. However, Boethius would add (cf. *De hebdomadibus*) that we should distinguish between propositions *per se* known to all, and those *per se* known only to the well educated. The propositions which are the first principles of syllogism are of the first type – they are known to all. The reason is that their terms (subject and predicate) are known to all men. These terms are contained in every other concept; they are the most common ideas of all. And finally, they are had by all men. This is indeed consonant with Aristotle's thought, for in the first book of the *Physics* he explains that all men know the most common things first, and from these most common concepts rise to more precise knowledge. Thus since Aristotle maintains all men immediately have these most common concepts, he would maintain that all men know the first principles which use those concepts as subject and predicate. Moreover, since these most common notions are considered by metaphysics, the first

[1] "Et propter hoc non est geometre speculari quid contrarium aut perfectum aut unum aut ens aut idem aut alterum sit, set aut ex conditione. – Quod quidem igitur unius sciencie sit ens secundum quod est ens speculari, et que insunt ipsi secundum quod est ens, manifestum est, et quod non solum substantiarum, set illorum que insunt, ipsa speculative sit, et eorum que dicta sunt, et de 'priori' et 'posteriori', et 'genere' et 'specie', et 'toto' et 'parte', et aliis hujusmodi.", Γ 2, 1005a11–18: *Vetus*, p. 307, l. 5–13. – Another exposition of this same doctrine is given by Thomas in *In III Meta.*, 6, 398; there he refers forward to Book Γ where he says we have this doctrine expressed by Aristotle.

[2] *In IV Meta.*, 5, 595.

principles must also be treated. Thus it is that Thomas understands Aristotles' metaphysics.[1]

Thomas has quite correctly interpreted Aristotle's thought as given in the *Posterior Analytics*[2] and in the *Physics*.[3] Moreover, as was noted above in point c), Aristotle does indicate that the *communia* such as "unity," "sameness," etc. are studied by metaphysics. Hence it would appear that these doctrines expressed by Thomas are indeed *verba Aristotelis*. However, the synthesis of them exposed by Thomas would appear to be more the *intentio* of Aristotle than his exact *verba*.[4]

3) The third element in Thomas' exposition of metaphysics as the study of *communia* is the relationship between the study of *communia* and the study of the intrinsic principles of real beings. I noted that

[1] "...philosophi erit considerare de omni substantia inquantum huiusmodi, et de primis syloogismorum principis. Ad huius evidentiam sciendum, quod propositiones per se notae sunt, quae statim notis terminis cognoscuntur, ut dicitur primo *Posteriorum*... Sed contingit aliquam propositionem quantum in se est esse per se notam, non tamen esse per se notam omnibus, qui ignorant definitionem praedicati et subiecti. Unde Boëtius dicit in libro de *Hebdomadibus*, quod quaedam sunt per se nota sapientibus quae non sunt per se nota omnibus. Illa autem sunt per se nota omnibus, quorum termini in conceptionem omnium cadunt. Huiusmodi autem sunt communia, eo quod nostra cognitio a communibus ad propria pervenit, ut dicitur in primo *Physicorum*. Et ideo istae propositiones sunt prima demonstrationum principia, quae componuntur ex terminis communibus... Et quia huiusmodi communes termini pertinent ad considerationem philosophi, ideo haec principia de consideratione philosophi sunt." *Ibid.*

[2] It is difficult to know exactly what passages of the *Posterior Analytics* Aquinas refers to. His *In I Post. Anal.*, 5, 50 reads: "...quaelibet propositio, cuius praedicatum est in ratione subiecti, est immediata et per se nota, quantum est in se. Sed quarundam propositionum termini sunt tales, quod sunt in notitia omnium, sicut ens, et unum, et alia quae sunt entis, in quantum ens: nam ens est prima conceptio intellectus. Unde oportet quod tales propositiones non solum in se, sed etiam quoad omnes, quasi per se notae habeantur. Sicut quod, non contingit idem esse et non esse..." Yet the text Aquinas is explaining reads thus: "...one basic truth which the pupil must know if he is to learn anything whatever is an axiom.", I, 2, 72a16–17. Thomas obviously reads this text in the light of the *Metaphysics* as he informs us in par. 49 where he refers to "IV Metaphysicae" as containing the proof that the principle of non-contradiction is necessarily had by all men. – What Thomas has done seems to be as follows: 1) Aristotle's Meta., Γ 3, 1005a19–b5 explains that the principle of non-contradiction uses terms such as "being"; 2) Aristotle's *Post. Anal.*, I, 2, 72a16–17 explains that such a principle must be known by all men; and 3) the entire context of Aristotle's *Metaphysics* assumes that "being", the metaphysical concept, is the first concept of man: cf. section 4, A of this chapter. Hence Thomas has concluded that for Aristotle, the terms referred to as "essential attributes" and the terms used by all sciences as far as each science's subject matter permits, these terms express the metaphysician's concepts. (The English translation of Aristotle's *Posterior Analytics*: *The Works of Aristotle Translated into English*, Vol. I, Clarendon Press, Oxford, 1955.)

[3] *Phys.*, A 1, 184a16–b5. Here there is no problem as Aristotle clearly notes that we must begin from more common, universal knowledge, and proceed to less common, more particular knowledge of each thing. See W. D. Ross, *Aristotle's Physics. A Revised Text with Introduction and Commentary*, Clarendon Press, Oxford, 1955, pp. 456–57. See Thomas' exposition: *In I Phys.*, 1, 6–8.

[4] Cf. the complementary doctrine of *In I Meta.*, 2, 45–46. There Thomas attempts to reconcile the Aristotelian doctrines that "being" is the first idea, and that metaphysics knows the most difficult things to know.

Thomas does not explicitly pose, nor does he answer this question. Rather he seems to take it for granted that his reader will see the relationship, for the *communia* (the concepts) would appear to be the concepts expressing the metaphysician's knowledge of the intrinsic principles of things. I feel this is Thomas' thought, and I pointed to the following doctrines as confirmation of this belief. 1) The *principia communia* studied by the metaphysician are known to all men because their subjects and predicates are the *communia* (concepts) known to all, e.g. *ens*, *idem*, etc.[1] – As was just pointed out, this doctrine can be classed as *intentio Aristotelis*. 2) The *communia* such as *ens*, *idem*, are studied by the metaphysician.[2] – A *verba Aristotelis* in Γ 2, as noted in point c) above. 3) The *commune ens* is the expression of the formal object of metaphysics.[3] – This would seem to be Aristotelian, perhaps an *intentio*, for the first principles are principles of being as such, as noted in point b) above. 4) The *commune unum* has a right to be studied because it expresses the same object as *ens*, and the *communia idem*, *multum*, etc. because of their dependence on *unum* are brought into metaphysics.[4] – This is certainly Aristotle's thought in Γ 1–2. 5) The *commune ens* is identical with the first concept had by man, and hence appears to be *the* concept *per se notum* and so used in the first of the common principles.[5] – This would be an *intentio Aristotelis* according to Thomas' exposition of *In IV Meta.*, 5, 595 examined above. – Accordingly, this third aspect of Aquinas' explanation of *communia* can be classed as an *intentio Aristotelis*.

Thus we see that the theory of metaphysics as the study of those concepts and principles used particularly by other sciences, must be considered as essentially a theory latent in Aristotle: although it remained unexpressed, it was operative in the *Metaphysics*. It represents Aristotle's intention. However, if we reflect on the fact that the most important of the concepts formulated by metaphysics, and used by the other sciences, is "being," then we see that Thomas has transformed Aristotle's intention by transforming "being." Aristotle's study of the *communia* has been given a new meaning, thanks to the new meaning attributed to "being."

[1] *In XI Meta.*, 4, 2210. Cf. footnote 1, p. 178.
[2] *In IV Meta., lectiones* 2–4.
[3] Cf. the exposition given in footnote 3, p. 178.
[4] *Ibid.*
[5] *Ibid.*

B. Avicenna and the study of communia

The theory of Aristotle on the study of *communia* can be summarized as follows: 1) the particular sciences take a particular point of view from which to study particular beings; metaphysics is the universal science, the study of being as such; 2) the universal science uses *communia* universally, while the particular sciences use them in a restricted or particular way; 3) the *communia* express the intrinsic principles of real beings (that is, the essence, or as Aquinas explains, the essence related to *esse*). Our question now concerns Avicenna's views on these subjects.

As was mentioned in discussing Avicenna's theories of "first science" and "universal science," a particular science is said to be the study of a particular being from a particular point of view; the universal science, it seems, is the study of all things from the viewpoint of *esse*.[1] Such a theory is close to the first point of Aristotle mentioned above, if one make allowance for the difference between Avicennian *esse* and Aristotelian being.

The second point of Aristotle's theory appears also in Avicenna's work. There are certain "common notions" found in all sciences, writes the latter; for example, *unum, multum*. The particular sciences speak of these *communia*, certainly; yet they do not discuss the mode of being (*de modo essendi*) had by the reality expressed by the *communia*. Rather they are content to arrive at the definitions of the *communia* by induction.[2] The reason why the particular sciences limit their treatment of the *communia* is quite clear: a science can speak of its subject and of the proper or essential accidents of its subject; yet *unum, multum* – expressed by the *communia*, these are the essential accidents only of *esse*. Hence only the universal science, or the science which studies *esse*, can treat these common notions universally; only metaphysics studies *unum qua unum*, and so on.[3] – Again we have a rough correspondance between Aristotle and Avicenna; the same doctrine divides them, however: Aristotelian "being" versus Avicennian *esse*.

The third point of Aristotle's theory on the *communia* was that these common concepts express the intrinsic principles of beings. For Aris-

[1] Cf. Ch. IV, section 4, A.

[2] "Similiter et sunt res, quae debent diffiniri et verificari in anima, quae sunt communes in scientiis; nulla tamen earum tractat de eis: sicut est unum inquantum est unum: et multum inquantum est multum, ...De his enim mentionem tamen faciunt et inducunt diffinitiones eorum: nec tamen loquuntur de modo essendi eorum...", *Meta.*, T. I, c. 2, fol. 70v, B.

[3] After the passage in the preceding footnote, Avicenna continues: "...quia hae non sunt accidentalia propria alicui subjectorum aliarum scientiarum particularium; nec sunt de rebus quae habent esse nisi proprietates esse essentialiter.", *Ibid.*.

totle, the common notion of *ens* was the expression of the formal object of metaphysics. Moreover, *ens* is the concept which serves as the subject and predicate of the first judgment, the first principle of thought. Thus, for Aristotle every real aspect or thing is *ens*; the common notion of *ens* expresses the reality of everything.

Now such is not the case for Avicenna. *Ens* is a versatile predicate, to be sure: it can be predicated of anything.[1] Yet for Avicenna there is a dichotomy in the use of the word *ens*. *Ens* can express both quiddity and existence.[2] It is not *ens* as quiddity that governs the principle of excluded middle: it is not because of a thing's quiddity that the first principle is valid. Rather, the impossibility of doing other than affirming and denying follows on *esse*, on existence. Such an impossibility is, Avicenna remarks, a property of existence.[3] Thus we are forced to admit a deviation of Avicenna's thought from the doctrine set down by Aristotle.

Thomas, one can easily discover, refers to Avicenna in the context of a discussion of the *communia*. After explaining the Aristotelian doctrine in E 1 – the study of separated beings belongs to a science other than physics or metaphysics – Thomas quickly notes that metaphysics must not only study separate being, but also sensible beings *qua* being.[4] Such a doctrine implies the theory mentioned earlier that the study of being as such will develop into a study of the cause of being. Hence, by the present reference to metaphysics as the study of all things as being, Aquinas is merely making certain that the doctrine of E 1 does not induce a misunderstanding in the minds of his readers. However, the very statement which prevents this misunderstanding itself raises a further problem: E 1 teaches not only that the study of the separated beings is higher than physics and mathematics; it implies as well that the entire subject matter of metaphysics is separated from matter.[5] How does one

[1] "Ens enim talis naturae est, quod potest predicari de omni, sive illud sit substantia sive aliud.", *Meta.*, T. I, c. 9, fol. 74v, B.

[2] Cf. the discussion of Ch. IV, section 4, A.

[3] "Ex dictionibus autem veris illa est digniora dici vera, cuius certitudo est semper. Sed quae dignior est ad hoc est illa, cuius certitudo est prima; . . .inter affirmationem et negationem non est medium. Et haec proprietas non est de accidentalibus alicuius rei nisi de accidentalibus esse inquantum habet esse communiter in omni quod est.", *Meta.*, T, I. c. 9, fol. 74r, A. Also note that *unum* is not said of quiddity: *Meta.*, T. III, c. 2, fol. 78r, A.

[4] "Advertendum est autem, quod licet ad considerationem primae philosophiae pertineant ea quae sunt separata secundum esse et rationem a materia et motu, non tamen solum ea; sed etiam de sensibilibus, inquantum sunt entia, Philosophus perscrutatur.", *In VI Meta.*, 1 1165.

[5] Cf. the context of E 1, a comparison of the different ways of defining used by the various speculative sciences: "Now, we must not fail to notice the mode of being of the essence and

reconcile this Aristotelian doctrine with the framework of metaphysics as Thomas saw it? Thomas has little difficulty; he writes: we can say with Avicenna that the *communia* are separated beings since they do not require matter for existence, although they may sometimes exist in matter.[1]

Does not this reference to Avicenna amount to attributing to metaphysics a consideration of *communia* according to *esse*? That is, is not Thomas accepting Avicenna's theory of metaphysics as a study of the *communia* according to their here-and-now existence? It would seem that it is only by emphasizing the "here-and-now reality" that Thomas is able to avoid considering the materiality often involved in the *communia*. Yet such is not the case; Thomas adopts a position different from that of Avicenna, and the last words of his sentence emphasize this fact: "sicut mathematica."[2] Thus metaphysics is said to study only separated beings; yet so far as the *communia* are concerned, they are said to be "separated beings" in the same way mathematical entities are.[3]

One needn't look far afield for Thomas' meaning, for only a few paragraphs earlier he had discussed the object of mathematics. This latter science, he noted, is like physics in that it studies material and mobile beings; yet it is unlike physics for it studies these beings from a point of view which abstracts from matter and motion.[4] Thus mathematics, like physics, is said to study the *non separata*; the objects mathematics studies do have their *esse* in matter. Yet Thomas had appealed to mathe-

of its definition, for without this, inquiry is idle.", E 1, 1025b28–30. It has been pointed out that Aristotle emphasizes both the manner of definition (the formal object) and the fact of immersion or separation from matter on the part of the object studied; thus his division of the sciences is not accomplished solely on the part of the formal object. MANSION, *Introduction à la Physique Aristotélicienne*, pp. 133–37. Aquinas, however, places all the weight of the divisions of the sciences on the formal object, on the manner of defining; cf. *In VI Meta.*, 1, 1156. The reason why Aquinas could place all the importance on this one aspect is found ultimately in "being", the concept which can express even the materiality of a material object.

[1] "Nisi forte dicamus, ut Avicenna dicit, quod huiusmodi communia de quibus haec scientia perscrutatur, dicuntur separata secundum esse, non quia semper sint sine materia; sed quia non de necessitate habent esse in materia, sicut mathematica.", *Ibid.*, 1, 1165.

[2] Thomas' rejection of Avicennian *esse*, discussed in Ch. IV, section 4, B, would indicate that he can not here accept the Avicennian meaning of *esse* without contradicting himself.

[3] As an attentive reading of par. 1165, footnote 1 above, will show, the expression "sicut mathematica" must be read "as mathematical entities", and not "as the science of mathematics".

[4] "...scientia mathematica speculatur quaedam inquantum sunt immobilia et inquantum sunt separata a materia sensibili, licet secundum esse non sint immobilia vel separabilia. Ratio enim eorum est sine materia sensibili, sicut ratio concavi vel curvi. In hoc ergo differt mathematica a physica, quia physica considerat ea quorum definitiones sunt cum materia sensibili. Et ideo considerat non separata, inquantum sunt non separata. Mathematica vero considerat ea, quorum definitiones sunt sine materia sensibili. Et ideo, etsi sunt non separata ea quae considerat, tamen considerat ea inquantum sunt separata.", *Ibid.*, 1161.

matical entities as an example of objects which need not necessarily have *esse* in matter! He can have had but one idea in mind: a mathematical entity, for example, a circle, can be considered as existing in matter (the circular table top); or it can be considered as an intelligible circle. Thus, a circle as any other mathematical object need not exist in matter, for the intelligible entity does not so exist. But when we say that a given mathematical entity (a circle) is separated from matter in *esse* because it need not exist in matter, we are certainly not talking about a circular table top; for such a circle must exist in matter. Nor are we talking about the intelligible circle, for that circle can never exist in matter. Rather we are talking about what can exist both in matter and separated from matter: we are talking about the intelligible content of circle, abstraction made from all real existence. We are speaking of an object *absolute considerata*. We are thus speaking of an idea, of course, but not *qua* idea, which is a real existing thing. Rather, we are talking about the content of an idea, a content which although necessarily found either in matter or in the intellect, need not be found exclusively in one or the other.[1]

If this then is how Thomas compares *communia* to *mathematica*, he is openly saying that metaphysics studies the content of those concepts used by all science.[2] Because metaphysics studies the content of these common concepts, it can be said to study what is separated from matter. – This is a somewhat flimsy way of explaining away the difficulty involved in the acceptance of the Aristotelian theory of metaphysics as the study of the immaterial *qua* immaterial, but nevertheless the difficulty is explained right out of court![3]

But how does Thomas' position compare with Avicenna's? In the first place, Avicenna would have metaphysics study the *communia* as here and now realized. Thomas has already rejected this Avicennian "here-and-now reality" as being of no philosophical (or metaphysical) interest. Moreover, Thomas, although repeating an Avicennian expression, gives it new meaning. The *communia* studied do not ne-

[1] This distinction between an essence considered "in itself", or as existing either "in the mind" or "in an individual", is of course of Avicennian origin. Cf. AVICENNA, *Logyca*, T. III, fol. 12r, A–B. The same distinction is at work in Avicenna's proof of *esse* as distinct from essence; cf. Ch, IV, section 4, A. Thomas used this distinction in his earlier years in *De ente*, c. 3, p. 24, l. 1–p. 25, l. 4 (Roland-Gosselin). For the Aristotelian character of the *mathematica* as existing neither in matter, nor independent of matter, cf. MANSION, *Introduction à la Physique Aristotélicienne*, pp. 142–54; 166–74; esp. pp. 150–51; 153–54; 171–72.

[2] Metaphysics is the study of concepts: this is not an unusual doctrine for Thomas' *Commentary*. Cf. *In IV Meta., lectiones* 1–4; *In XI Meta.*, 1, 2146; and *passim*.

[3] How the study of the *communia* can be the study of sensible beings will appear in Ch. VI, section 1.

cessarily exist in matter, said Avicenna; but this means that we should consider them in their *esse* – in their here-and-now. But Thomas says in effect, Avicenna was right in saying these *communia qua esse* are separate from matter; but Thomas implies, when I repeat that phrase, I mean that the *communia*, in their content, are separate from the existence they have either in the mind or in matter – in this the *communia* resemble the *mathematica*. And so, Thomas in his own inimitable way simultaneously repeats and changes Avicenna's words. Evidently, Aquinas wishes to preserve what he understands as the Aristotelian framework or outline of metaphysics.

Let us pause for a moment and note what new information we have just discovered bearing on the doctrine of *esse*. Although this may appear to be a digression, nothing could be more central to the discussion of Thomas' doctrine on the *communia*.

Metaphysics studies *communia*, Thomas insists, but not as existing in sensible beings, nor as existing in immaterial beings. One should realize that in Thomas' expressions "existing in sensible beings" and "existing in immaterial beings," or even "existing in the mind" – in such expressions the emphasis can not be placed on "in sensible beings" or on "in the mind": one can not thus separate the *locus* of existence from existence itself. One can not pretend that Thomas is rejecting on behalf of metaphysics the *locus* of existence and not existence itself. Thomas says, in effect: metaphysics studies the content of the *communia*, and pays no attention to any of the possible here-and-now existences in which the *communia* may be found. Thus, one can not say that metaphysics studies the content of the *communia qua* existing, although it pays no attention to the type of existence had. One can not thus interpret Thomas: his "sicut mathematica" rejects such an interpretation.

Implicit in Thomas' exposition of the manner of studying sensible being as being, or as expressed in *communia*, is another repetition of this rejection of Avicennian *esse* (existence): the existence of things does not interest the metaphysics I am exposing, Thomas is saying.

If this is so, to what does Thomas refer when he speaks of *esse* as other than essence in book IV? I am not interested in Avicenna's *esse*, we have seen Thomas state. Yet Thomas clearly accepts the distinction of *esse* and essence; and what is more, it appears correct to see in Thomas' words an affirmation that a thing is constituted as such through the composition of *esse* and essence.[1] If a being is constituted as such through

[1] The rejection of Avicenna's *esse*: *In IV Meta.*, 2, 556 and 558; *In X Meta.*, 3, 1982. For

this composition, must not metaphysics study both principles of the composition? – Still more, the intellect grasps this *esse* together with essence whenever it grasps a thing as *ens* (that is, as substantial *ens*).[1] Again, this same *ens* is the chief among the *communia* which are to be studied.[2] All the other *communia* are to be related to *ens*.[3] In the light of these doctrines, what are we to make of Thomas' rejection of Avicennian *esse* as metaphysically uninteresting?

The answer is quite simple, and forces itself upon us: we must admit a distinction between Thomist and Avicennian *esse*. *Esse* in Avicenna's eyes was the here-and-now realization of an essence. *Esse* for Aquinas is not the here-and-now of an essence; rather *esse* is one of the principles that constitutes a being as a being. *Esse* for Aquinas is something expressed in the first among the *communia* – *ens commune*.

To return to our point of departure – metaphysics as the study of *communia* – it appears that we must view metaphysics as the study of the content of ideas and, in the first place, as the study of the content of the idea *ens commune*. Metaphysics, as the universal science, is to study the *esse*-essence composition of beings in so far as one has that composition expressed in the concept of *ens commune*, and in so far as the other common concepts are reduced to the concept of *ens*.

C. *Averroes and Albert on the study of communia*

On one aspect of their theories on this point, Averroes and Albert appear very close to Aristotle and to Aquinas: universal philosophy studies *per se* accidents (that is, what is expressed in *communia*) as such, while the particular sciences merely use these accidents in a particular way.[4] Yet there are differences between Averroes and Albert on the one hand, and Thomas and Aristotle on the other. The latter pair explain that since these essential attributes (expressed in *communia*) are to be

a statement that a thing's essence is not its *esse*: *In IV Meta.*, 2, 558. For a statement that a being is "that which has *esse*": *In IV Meta.*, 1, 539. That a being is made a thing through the composition of the two principles, which principles can only be essence and *esse*, see: *In VI Meta.*, 4, 1241.

[1] "Invenitur siquidem et in rebus aliqua compositio; sed talis compositio efficit unam rem, quam intellectus recipit ut unum simplici conceptione.", *In VI Meta.*, 4, 1241. In this passage Aquinas opposes the composition of subject and predicate in a judgment to that composition that makes a thing to be a thing; this composition can not be that of matter and form, but is rather that of *esse* and essence. – Obviously the intellect forms a concept of *ens*, else the discussion of the predication of *ens* would be nonsensical. Yet, as Thomas explains, one calls a thing *ens* because it is "that which has *esse* on its own". Cf. *In IV Meta.*, 1, 539. It would appear thus that the concept *ens* expresses both the *esse* and essence of an object.

[2] Cf. the texts of footnote 1 and 3, p. 178.

[3] Cf. the conclusion of Ch. IV, section 1.

[4] Averroes: *In IV Meta.*, c. 8, fol. 73v, K–M. Albert: *In IV Meta.*, T. II, c. 1, p. 173, l. 12–48.

studied by metaphysics, then the first principles also are studied; the key to this reasoning is found in the doctrine of the *communia* as subject and predicate of the first principles. This doctrine is missing in Averroes and Albert.[1] For his part, Averroes connects first philosophy and the study of the first principles in two ways. 1) Whoever wishes to study a genus of being must first study the principles of knowledge of that genus; the first philosopher wants to know being as such and hence must study the principles of knowledge as such.[2] 2) Since first philosophy must consider the first substance which is the highest of substances, it must consider the most true, namely the first principles of syllogism.[3] Albert's explanation is different from that of Averroes: since these first principles are the principles of being as such, they must be studied in the science of being as such.[4]

This is as far as Averroes and Albert carried the discussion of the *per se* accidents. Aristotle and Aquinas, very understandably, had more to say: because the formal object of metaphysics is expressed by *ens commune* and the other *communia*, metaphysics will study the propositions employing the *communia* as subject and predicate. As is evident, Aristotle and Aquinas show by this theory the being-centeredness of their metaphysics. Since Averroes and Albert accepted a metaphysics defining its objects in terms of God it is not surprising that they do not express the theory given by Aristotle and Aquinas.

In the context of the discussion of *communia*, Thomas referred to Avicenna as we have noted (*In VI Meta.*, 1, 1165). When we look to the corresponding passages of Albert and Averroes, we find the reason why Thomas, through his reference to Avicenna and to *esse*, paused to make clear his interpretation of first philosophy as the "study of the separated in existence." When we look to the corresponding passages of Albert and Averroes, we find a doctrine dependent on their views of metaphy-

[1] Aristotle: Cf. section 3, A above. The main texts of Thomas are: *In XI Meta.*, 1, 2151; *Ibid.*, 4, 2210.

[2] "...quemadmodum oportet habentem cognitionem alicuius generis, ut habeat posse ad sciendum principia cognitionis illo genere, et ordines eorum in cognitione, sic oportet considerantem in ente secundum quod est ens habere posse ad sciendum principia cognitionis in eo quod est cognitio, non principia alicuius cognitionis.", *In IV Meta.*, c. 8, fol. 73v, L–M.

[3] "...manifestum est, quod, quemadmodum Philosophus debet considerare de prima substantia, quae est altior omnibus substantiis, sic habet etiam considerare de rebus, quae sunt magis verae, quam altae, scilicet de principiis syllogismi.", *Ibid.*, fol. 73v, K–L.

[4] "...omnium demonstrationum principia insunt omnibus, inquantum communicant in ente, et hoc est, inquantum entia sunt. Igitur speculatio de his principiis est illius philosophi qui cognoscit ens, inquantum est ens.", *In IV Meta.*, T, II, c. l, p. 173, l. 44–48. Aquinas adopts this argument in his *In IV Meta.*, 5, 590–91, but of course it is just a rewording of Aristotle's Γ 3, 1005a19–29.

sics as defining all things in terms of God.[1] At the same point in his exposition of E 1, Aquinas explained how metaphysics could study sensible being *qua* being and still be said to be studying the separated in *esse*. Again then, we are faced with an example of Aquinas' opposition to Averroes and to Albert.

And thus we come to the close of our investigation of the doctrines held by Thomas' predecessors concerning the study of *communia*. All four (Aristotle, Avicenna, Averroes, and Albert) held that the *per se* accidents of being as such (expressed in *communia*) were to be studied. Only Aristotle, however, had given the same doctrine we found in Aquinas' work: 1) metaphysics studies and uses the knowledge of *per se* accidents in an universal manner, while the particular sciences use it to the extent demanded by their subject matters; 2) the concepts of these *per se* accidents express the intrinsic principles of real beings. (The second doctrine is only the *intentio Aristotelis*, while the first is his *verba*).

Our study of Avicenna revealed that his theory had little in common with that of Aquinas. They both styled the *per se* accidents "communia"; they both maintained that the *communia* could be said to be "separata secundum esse et secundum rationem." Yet Avicenna, we noted, understood *secundum esse* as the "here-and-now actuality" or existence of the *per se* accidents. Thomas, we discovered in the preceding chapter, rejected such a formal object as being devoid of metaphysical interest. When Thomas called the *communia* "separata secundum esse sicut mathematica," he referred to the content of the *communia*; metaphysics, he was implying, studies the content of these *communia* by abstracting from any of the here-and-now existences open to them. – As a type of side-effect of this discussion, we realized that the metaphysical *esse* Aquinas speaks of on occasion, is not here-and-now existence; rather it is a constituting principle of a being, and is part of the content of *ens commune*.

And finally, our brief examination of Averroes and Albert revealed only that they did not explain, with Aristotle and Thomas, that the *communia* express the formal object of metaphysics. This was understandable in the light of the object given to metaphysics by Averroes and Albert: the immaterial beings (Averroes) and *esse simplex*, the radius of divine light (Albert). Thus both of these two commentators explained that metaphysics defines all objects in terms of God; this

[1] Averroes: *In VI Meta.*, c. 2, fol. 146v, G–H; cf. footnote 2, p. 198. Albert: *In VI Meta.*, T. I, c. 3, p. 305, l. 38–48; l. 57–60; p. 306, l. 16–21. cf. footnote 3, p. 199.

they explain in the passages corresponding to Aquinas' explanation of metaphysics as the study of the *separata secundum esse.*

It appears that there is only one legitimate conclusion to draw from these facts: Thomas' *Commentary* presents the Aristotle understood by Thomas; the *Commentary* is, moreover, written in a polemical spirit against (at least) the earlier expositions of Averroes, Avicenna, and Albert.

4. AQUINAS' PREDECESSORS ON METAPHYSICS AS THE "LORD" OF ALL SCIENCE

A. Aristotle's view

Metaphysics, Thomas holds, has an important role to fulfill in relation to the other sciences. If we consider the hierarchy of sciences in their logical dependence on one another, then metaphysics is in first place: metaphysics is the ruling science. It is metaphysics that has the obligation of stating the existence of a certain type of being; to metaphysics too belongs the right to declare that a particular type of being has an essence. The lower sciences begin from these metaphysical statements and show, through sense knowledge, the essence of the particular types of being they are interested in. The metaphysician, we argued, understands an object as "being," judges his understanding to be correct, and then permits a particular science to study the essence of the object in question. Thus the particular science begins from the answer to the question: *an est?*, and asks: *quid est?*.[1]

The part of Thomas' exposition contained in *In VI Meta., lectio* 1 is basically quite true to the thought of Aristotle. As we discovered in the preceding section in the context of our examination into metaphysics as the study of the *communia,* Aristotle regards the particular sciences as the studies of particular types of being *qua* particular, and not *qua* being.[2] Aristotle, as Thomas, points out that these particular sciences must assume that the genus of their subject exists; it is for metaphysics to show that a thing is.[3] In like manner, these particular sciences do not "offer any discussion of the essence of the things of which they treat";[4] it is for metaphysics to show the "what it is" as well as "that it is."[5]

[1] Cf. *VI Meta.,* 1, 1147–51; see footnotes 3, p. 179–2, p. 180.

[2] E 1, 1025b3–10.

[3] The particular sciences "...omit the question whether the genus with which they deal exists or does not exist, because it belongs to the same kind of thinking to show what it is and that it is.", E 1, 1025b16–18.

[4] E 1, 1025b10.

[5] Cf. footnote 2.

Accordingly, Aristotle regards metaphysics as the science which determines both the existence and the essence of any type of being.[1] Thomas in his exposition explains why it belongs to the metaphysician to determine both essence and existence: the metaphysician, as the student of being as such, knows the essences of things; only he who knows the essence of a thing can know its existence, for the *quid est* is used as a medium to show *an est*.[2] Since we do not have the version of Aristotle Thomas was using at this point (the *Moerbecana*), we can not judge with absolute accuracy how much of this doctrine Thomas found in the text of Aristotle; in the Greek text at any rate, there is no mention of using the *quid est* as a medium to show *an est*.[3] From the references Thomas gives, however, we do know that he introduced here into the explanation of the *Metaphysics* several doctrines of the *Posterior Analytics*. Thomas explains that no particular science can proceed without supposing the *quid est* and the *quia est* (or *an est*) of its subject. He notes that we read this doctrine in the first book of the *Posterior Analytics*.[4] Perhaps Thomas is referring to *Posterior Analytics* I 1, 71a1–11 where Aristotle remarks that all scientific reasoning begins from pre-existing knowledge, a fact we can see from looking at the various sciences.[5] Perhaps too, Thomas is thinking of I 1, 71a11–17 where Aristotle notes that we must both know *what* things are and *that* they are before we begin a particular science.[6] Thus it appears proper to recognize Thomas' exposition as faithful to Aristotle, as far as concerns the knowledge of *quid est* and *an est* supposed by the particular sciences. Aristotle, how-

[1] Fr. Owens would object to my terminology, particularly to "existence". Cf. OWENS, *The Doctrine of Being in the Aristotelian Metaphysics*, pp. 170 and 172. That the word "existence" is the correct word to use, cf. Ross, *Aristotle's Metaphysics...*, Vol. I, pp. 351–52.

[2] *In VI Meta.*, 1, 1151.

[3] Cf. footnote 3, p. 226.

[4] After mentioning that the *quid est* is used as a *medium* to show the *an est*, Aquinas writes: "Et ideo quaelibet scientia particularis supponit de subiecto suo, quia est, et quid est, ut dicitur in primo Posteriorum...", *In VI Meta.*, 1, 1151.

[5] "All instruction given or received by way of argument proceeds from pre-existent knowledge. This becomes evident upon a survey of all the species of such instruction. The mathematical sciences and all other speculative disciplines are acquired in this way, and so are the two forms of dialectical reasoning, syllogistic and inductive; for each of these latter makes use of old knowledge to impart new, the syllogism assuming an audience that accepts its premisses...".

[6] Aquinas' exposition of these lines makes no mention of metaphysics: Cf. *In I Post. Analy.*, 2, 1–7. Interestingly enough, Aquinas points out (as does Aristotle) that the particular sciences must know the meaning of a term before they can conclude something. Of course, the question arises: how does one know that the term is properly applied to any object? If one does not know this, then of course, there would be no value in concluding something from the meaning of the term. Thomas does not raise this problem however, which indicates that some science prior to a particular science should raise it, some science not prior in time, but in the logical order of dependence.

ever, does not explain in these passages from the *Posterior Analytics* that the metaphysician must use *quid est* to show *an est*. But this doctrine would be implied by his thought in Z 17 of the *Metaphysics*.

After we considered Aquinas' thought on metaphysics as expressed in Book VI, we turned to *In VII Meta., lectio* 17 for his discussion of the four questions involved in knowledge. There Thomas stated we must know *an est* (or *quia est*) before asking *quid est*? (or *propter quid est*?); yet I argued, he exempts metaphysics from the rule: the metaphysician grasps an object as "being" and then judges *an est*, and so opens a field of inquiry to a particular science. Now this doctrine must be accepted as basically that of Aristotle in Book Z 17; and to the extent that Aristotle exempts metaphysics from the rule of the priority of *an est* (or *quia est*), he would of course by implication attribute to metaphysics the same role as Thomas gives it in *In VI Meta., lectio* 1.

In Z 17 Aristotle speaks of the question "why" (*propter quid est*) as identical to "what" (*quid est*.)[1] In addition, the fact or the existence of an object is said to be known before one asks "why" or "what".[2] Thomas' explanation of these questions is much more ample than that of Aristotle, for, as Thomas declares, he is introducing information from the second book of the *Posterior Analytics*.[3]

On the exemption of metaphysics from the rule of the priority of *an est* over *quid est*, Aristotle has only the vaguest of references. In speaking of the possible meanings of a question such as: "Why is a man a man?," Aristotle notes that one could answer: "Because each thing is inseparable from itself, and its being one just meant this." Such an answer is correct, of course, but as Aristotle says, this answer, this truth does not advance our knowledge one iota for this truth is common to all things. In short, it does not tell us anything at all.[4]

[1] "The object of the inquiry is most easily overlooked where one term is not expressly predicated of another (e.g. when we inquire 'what is man'), because we do not distinguish and do not say definitely that certain elements make up a certain whole. But we must articulate our meaning before we begin to inquire; if not, the inquiry is on the border-line between a search for something and a search for nothing. Since we must have the existence of the thing as something given, clearly the question is *why* the matter is some definite thing; e.g. why are these materials a house? Because that which was the essence of a house is present... Therefore what we seek is the cause, i.e. the form, by reason of which this matter is some definite thing; and this is the substance of the thing.", Z 17, 1041a32–b9.

[2] "Now 'why a thing is itself' is a meaningless inquiry (for <to give meaning to the question 'why'> the fact or the existence of the thing must already be evident...)", Z 17, 1041a14–16.

[3] Cf. *In VII Meta.*, 17, 1651.

[4] "(...but the fact that a thing is itself is the single reason and the single cause to be given to all such questions as 'why the man is man, or the musician musical', unless one were to answer 'because each thing is inseparable from itself, and its being one just meant this'; this, however is common to all things and is a short and easy way with the question.)", Z 17, 1041a16–20.

Aristotle admits that man has knowledge of a thing's self-identity before he asks *quid est?* (or *propter quid est?*). This knowledge is expressed in a judgment of course; it amounts to knowledge of *an est.* Moreover it concerns the unity of a thing, and unity is known after knowledge of "being," and after the first judgment. Knowledge of unity is acquired through a manipulation, through a use, of our knowledge of "being."[1] As has been suggested, Aristotle implies our concept of "being" is our first one.[2] Hence, if knowledge of unity is exempt from the rule of the priority of *an est* over *quid est,* Thomas does not seem unjustified in implying that knowledge of all the *communia* are exempted from this rule. Hence for Aristotle, as for Aquinas, metaphysics would be prior to laws governing *an est* and *quid est.*[3]

Thus Thomas does appear to be giving an Aristotelian conception when he explains the role of metaphysics: 1) Thomas maintains the *verba Aristotelis* in so far as he gives to metaphysics the duty of stating the existence and the essence of the particular sciences; 2) Thomas would appear to give the *intentio Aristotelis* in so far as he assigns this role to metaphysics "because man must use *quid est* to show *an est*"; 3) in exempting metaphysics from the law of the priority of *an est* over *quid est,* Thomas gives the *intentiio Aristotelis.*[4]

[1] Aquinas could have developed this doctrine from texts such as these: "For in general those things that do not admit of division are called one in so far as they do not admit of it; e.g. if two things are indistinguishable *qua* man, they are one kind of man; if *qua* animal, one kind of animal; if *qua* magnitude, one kind of magnitude.", Δ 6, 1016b3–6; the discussion of "unity" as said of the essence of things in Γ 2, 1003b22 sqq.: *Vetus*, p. 304, l. 4 sqq. Aquinas would have been aided in his interpretation of this passage from Γ 2 by Albert's explanations in *In IV Meta.*, T.I, c. 4, p. 165, l. 62–p. 166, l. 15; in this text, Albert explains that one denies division in entity to discover unity.

[2] "But we *can* inquire why man is an animal of such and such a nature. This, then, is plain, that we are not inquiring why he who is a man is a man. We are inquiring, then, why something (that it is predicable must be clear; for if not, the inquiry is an inquiry into nothing). Z 17, 1041a20–24. "But we must articulate our meaning before we begin to inquire; if not, the inquiry is on the border-line between a search for something and a search for nothing. Since we must have the existence of the thing as something given, clearly the question is *why* the matter is some definite thing...", *Ibid.*, 1041b2–5.

[3] Mlle. Mansion writes of Aristotle's thought: "D'une part il est impossible de connaître l'essence d'une chose quand on ne sait même pas si elle existe, mais d'autre part,... une saisie de l'existence d'une chose qui ne nous livrerait rien de ce qu'elle est, serait parfaitement illusoire." Aristotle remarks, concerning our knowledge of a thing as real, that "...la manière dont nous connaissons l'existence détermine notre aptitude à connaître l'essence (*Anal. Post.*, II, 93a28–29). Qu'est ce à dire sinon que la saisie de l'existence d'une chose est déjà une appréhension confuse de son essence? Comment saurions-nous que c'est cette chose-là qui existe, si nous ne savons pas du tout ce qu'elle est? Ainsi entendue, la connaissance de l'existence est la base qu'il faut pour entreprendre des recherches approfondies sur la nature de l'objet, recherches qui aboutiront à la définition explicite de son essence...", S. MANSION, *Le jugement d'existence chez Aristote*, Desclée, Louvain-Paris, 1946, pp. 53–54.

[4] According to Mlle. MANSION, *Le jugement d'existence chez Aristote*, pp. 214–72, any attempt to understand "what "an object is must be preceded by a grasp of the reality of the object, and thus Aristotle's view of knowledge would be as follows. First, man understands

B. Avicenna on metaphysics as the ruling science

Turning to Avicenna, we find it more difficult to discover his position vis-a-vis the role of metaphysics in the schema of sciences. However, if one emphasize the consequences of the doctrines of Aristotle and Aquinas, rather than the terms in which the doctrines are expressed, then one can discover an echo of this problem in Avicenna's work.

The terms in which Aristotle and Aquinas treated this matter included such things as: *quid est?*, *an est?*, *medium* to prove *an est*, and so on. Such items are almost totally absent from Avicenna's *Metaphysica*. However, the same is not true of the consequences of the theories of Aristotle and Aquinas: metaphysics knows the essence of any being, and thus instructs the lower sciences what to look for in their work. Such a theory does have a place in Avicenna, although there its meaning is somewhat distorted.

Divine science seeks out the first causes of both natural and mathematical beings; as well, this science studies whatever can be said to depend on such first causes; and finally, the first of all causes and the principle of all principles, God, is also treated.[1] However, Avicenna warns, the subject of this science is not God.[2] Rather, we are to investi-

confusedly the essence and the existence of an object. Through further intellectual effort, man can understand even more clearly the essence of the object in question. Finally, one must judge whether or not the grasp of the essence was a correct one. Aristotle's sciences are a discussion of the essential world, to be sure; but none the less they presume an intellectual contact with existing objects; one must have an initial contact with real individuals else one would be lost in a world of nominal definitions. – Aquinas' *Commentary* has adopted some, but perhaps not all of Aristotle's theory. From the *Commentary*, one sees that Aquinas would call the expression of the initial intellectual grasp of an object "being". Aristotle would do the same. But as Mlle. Mansion agrues, the "being" of Aristotle would be knowledge of an existing essence. But Aquinas' "being" is not a grasp of an *esse*-essence composite as existent. We believe that Aquinas' "being" is only a grasp of "actualized essence" in a real individual, but not *qua* real, not *qua in* the individual. In this first operation of the intellect, one would not be conscious of the validity of this knowledge. Only in the reflexion following apprehension and preceding judgment would Aquinas consider it possible to grasp as existent the object apprehended as "being". Only in this reflexion would Aquinas attribute to man a grasp of the here-and-now existence. Thus Aristotle and Aquinas do not totally agree. True, they both appear to attribute these steps to the knowing process: 1) the grasp of "being"; 2) the attempt to understand more clearly the essence; 3) the judgment expressing the truth of the act of knowing the essence (the validity of step n. 2). Aquinas' initial idea of "being" certainly has a different content than Aristotle's, however; Aristotle would grasp essence; Thomas, *esse*-essence composite. Aristotle, moreover, would say he knows the essence is found here and now in *that* object; Aquinas would not admit this point, I believe. In Ch. VI, section 1, C, the knowledge of the validity of the initial grasp of "being" is involved in the distinction of essence and *esse*. There that knowledge of the validity of "being" is attributed to an operation which occurs after the grasp of "being", and not to the very grasp of "being".

[1] "Iam etiam audisti quod scientia divina est in qua quaeritur de primis causis naturalis esse, et doctrinalis esse, et de eo quod pendet ex eis, et de causa causarum et de principio principiorum quod est deus excelsus.", *Meta.*, T. I, c. 1, fol. 70r, B.

[2] *Ibid.*, cf. Ch. II, section 1.

gate whatever is separated from matter in esse.[1] Then too, the subject of metaphysics is not one of the causes, nor any combination of them.[2] Rather, the subject is *esse*, for in metaphysics we consider all the causes as "having *esse*."[3]

In explaining these doctrines, Avicenna hints that he conceives metaphysics as knowledge of what is essential to everything studied. What is only a hint becomes clear as Avicenna goes deeper into his science. As soon as one realizes that the formal object of metaphysics is *esse*, one sees that metaphysical knowledge does in some sense express the essential or necessary aspects of everything falling under the science's sway.

Because he examines things from the point of view of "here and now realized," Avicenna's metaphysician can draw into the unity of one science such diverse items as causes, substance, number, matter, body, form, and so on. As was argued earlier, Avicenna thought he grasped all these objects and aspects of objects from the point of view of "judged as existent" – he feels he grasps them all as being "here and now realized". Thus, it does not matter whether a given substance is material or not; Avicenna, by emphasizing his viewpoint of *esse*, can study any given object as separated from matter.[4] Hence, it would seem to follow that Avicenna studies the barest outlines of the objects investigated in metaphysics. Take a tree, for example. If one knows it in a concept, one knows its "treeness," its quiddity, Avicenna explains.[5] In such a concept one would know both the formal cause and the material cause of a tree. Much the same would be true if we looked at the tree through the concept of quidditative *ens*.[6] However, if we emphasize the knowledge acquired by man when he knows the existence of the tree, or the existence of quiddity in the tree, then we have the point of view of Avicennian metaphysics. As Avicenna says, what man knows in the judg-

[1] Metaphysics is the science which treats "...de eo quod omnimodo separatum est a natura; et tunc nominabitur haec scientia ab eo, quod est dignius in ea, scilicet vocabitur haec scientia, scientia divina.", *Meta.*, T. I, c. 4, fol. 71v, C. "Non enim potest esse ut metaphysica sit alicuius scientiarum de sensibilibus, nec alicuius scientiarum de eo quod habet esse in sensibilibus...", *Ibid.*, c. 2, fol. 70v, A–B.

[2] *Ibid.*, c. 1, fol. 70r, B.

[3] "Sed non potest poni eis subiectum commune ut illorum omnium sint dispositiones et accidentalia communia nisi esse. Quaedam enim eorum sunt substantiae et quaedam quantitates et quaedam alia praedicamenta quae non possunt habere communem intentionem quae certificentur nisi intentionem essendi.", *Ibid.*, c. 2, fol. 70v, B.

[4] *Ibid.*, c. 2, fol. 70v, A–D; see also footnote 3.

[5] *Ibid.*, c. 6, fol. 72v, C. Cf. Ch. IV, section 4, A.

[6] *Ens* is the poorest of concepts from the point of view of meaning; hence, if one knows a tree through the concept of quidditative *ens* (*ens* as a concept), one knows only that the tree's essence involves matter and form. That *ens* is the poorest of concepts for Avicenna, see: GOICHON, *La distinction de l'essence et de l'existence...*, pp. 2–5; PAULUS, *Henri de Gand...*, p. 25.

ment is the "here-and-now" of objects, or of aspects of objects. To draw substances, causes, bodies, form, matter, number, etc. into the unity of a science, Avicenna emphasizes the viewpoint of the "here-and-now" of something – a viewpoint expressing what would seem to be the least a-mount of knowledge possible about the something known as "here-and-now." The metaphysical viewpoint is expressable then as "here-and-now of X." Whatever replaces "X" will not be known in its totality. For example, to know a body as "here-and-now" is to know only: "there must exist a material and formal cause which constitute the essence of the body."

Hence, whereas Aquinas and Aristotle would point out how "being" expresses implicitly the total reality of everything known as "being," Avicenna has a much less rich point of view. When Aristotle and Aquinas know an individual tree as "being," they know everything about its essence. But when Avicenna knows the same tree as *esse*, he knows only the existence of a material and a formal cause, abstraction made from the kind of matter and form in question; because he studies only what is "separated from matter," he is considering the tree as known to exist. Because he studies the "separated," he must forget about the individual matter and the individual form of the tree, emphasizing only the "here-and-now" of a form and a material cause.[1]

Thus the doctrine of Aristotle is admitted, if in diluted form, into the Avicennian system. Certainly there is no need to underline the connection between the change wrought by Avicenna in Aristotle's doctrine of "being" and the doctrine of *esse*. Quite evidently, it is that *esse* which is the heart of the Avicennian metaphysics.

C. *Averroes and Albert on metaphysics as the first of the sciences*

When one questions Averroes and Albert on this issue, one notices that parts of their expositions agree with those of Aristotle and Aquinas. For example, the discussion of the relative order, and of the relationship, between the four questions (Aristotle's Z 17) – here one discovers little difference between the various expositions, other than the fact that Thomas' is much clearer.[2] Averroes and Albert, though less clear, point out that the common notions are known as belonging to all things

[1] Another example is Avicenna's definition of "body": "...substantia in qua potest ponit dimensio quocumque modo volueris incipere et illa a qua primum inceperis erit lognitudo.", *Meta.*, T. II, c. 2, fol. 75r, A.

[2] Averroes: *In VII Meta.*, c. 59, fol. 207r, A sqq; *Ibid.*, c. 60, fol. 208r, E sqq. Albert: *In VII Meta.*, T. V, cc. 8–10. Aquinas: *In VII Meta., lectio* 17. – I do not mean to imply that

prior to asking a question such as *quid est*? about a particular thing.[1] When one studies their interpretations of the very important opening chapter of Book E, one notes again what appears to be a fundamental agreement. For Thomas and Aristotle, metaphysics shows "what" a thing is (essence) and "that" it is (existence).[2] The other sciences, says Aristotle, study the essences of certain types of beings and show the essential attributes of these types.[3] Or as Thomas has it, the other sciences will show only the essences of particular beings by referring them to the sense knowledge we have of such essences.[4] Averroes has read this Aristotelian doctrine in much the same way, although he introduces a few notions of his own: there will be a study of *ens simplex*, as well as a science which accepts by sense what exists, and another which assumes its subject exists.[5] By this Averroes most certainly does not mean that metaphysics demonstrates that a substance exists. One can demonstrate only an accident.[6] Yet metaphysics can, by a judgment, point out the existence of substances.[7] Thus Averroes, as Aristotle and Thomas, would appear to agree fundamentally on the doctrine that the metaphysician knows a thing as "being," and then judges this knowledge to be correct. When one studies Albert's *Commentary*, one notes no important difference between it and Averroes', or between it and Thomas'.[8]

Yet despite these areas of agreement, or rather of apparent agreement, Averroes and Albert are definitely opposed to Aristotle and Aquinas on the more profound aspects of metaphysics' role. For example, Albert openly declares his hostility to any theory making physics a science subalterned to metaphysics.[9] As Albert explains, if one

these three expositions have the same doctrine on every point: e,g. Averroes (*In VII Meta.*, c. 59, fol. 207v, G–H) affirms that one seeks an efficient cause only in the case of substantial change; Albert (*In VII Meta.*, T. V, c. 8) implies the same doctrine; Thomas, however (*In VII Meta.*, 17, 1661) explicitly points out that one can seek an efficient cause of being as well. Yet these differences do not affect their similar explanations of the four questions of scientific procedure.

[1] Averroes: *In VII Meta.*, c. 59, fol. 207r, D–E. Albert: *In VII Meta.*, T. V, c. 8. Aquinas: *In VII Meta.*, 17, 1652–54.

[2] Aristotle: E 1, 1025b16–18. Thomas: *In VI Meta.*, 1, 1147–51.

[3] E 1, 1025b3–13.

[4] *In VI Meta.*, 1, 1148–49; cf. footnotes 3, p. 179, 1, p. 180.

[5] "...manifestum est aliquam scientiam esse perscrutantem de ente simplici. Deinde dicit sed quaedam, etc. idest sed quaedam scientiarum perscrutantur de ente proprio, ponendo illud esse sub sensu: et quaedam perscrutantur de ente, concedendo illud esse, ut Arithmaticus, qui ponit unum esse.", *In VI Meta.*, c. 1, fol. 144v, H.

[6] *Ibid.*, fol. 144v, I.

[7] *Ibid.*, fol. 144v, L.

[8] Cf. *In VI Meta.*, T. I, c. 1.

[9] "...si sic physicus acciperet a metaphysico, oportet quod physica subalternaretur

science is subalterned to another, then the concept of the subject of the subalterned science must be contained within the concept of the subject of the higher science. And such a situation does not hold for the relation of metaphysics and the other sciences.[1]

In Averroes, one discovers much the same view. In commenting on Aristotle's doctrine of particular sciences as studies of "parts of being," Averroes explains that these sciences will study a part of being, dividing that part, as it were, from being itself.[2] An example of a particular science would be mathematics, which studies number, magnitudes, which are distinct from being.[3] These statements, however, are not clear enough to warrant an identification of the views of Averroes and Albert. However, Averroes' Book XII supplies the doctrine needed to see clearly the similarity of the two commentators.

Arguing against the Avicennian theory that metaphysics proves the principles of mobile being, Averroes remarks that this is quite incorrect; just the contrary, physics tells metaphysics that mobile beings are dependent on a prime mover, on a pure form.[4] It is up to the metaphysician, however, to explain that this mover is the formal and final principle of sensible substances.[5] In other words, physics studies the principles of mobile being, whereas metaphysics expresses the first form and ultimate goal of substances by speaking of God.[6] Thus *ens mobile*, or the physician's primary concept, would express the intrinsic principles of mobile being; *ens simpliciter* of the metaphysician, however, ex-

primae philosophiae, quod a principio huius sapientiae falsum esse ostendimus.", *In XI Meta.*, T.I, c. 3, p. 462, l. 78–80. *In IV Meta.*, T. I, c. 1, p. 162, l. 38–41.

[1] "Nec etiam propriae et determinatae scientiae cuiusdam entis isti subalternantur, quia ad subalternationem non requiritur, quod subiectum sit sub subiecto tantum, sed quod medium, quod est causa propter quid, sub alterius scientiae medio concludatur vel contineatur.", *In I Meta.*, T. I, c. 3, p. 5, l. 25–30.

[2] "...est scientiarum particularium consideratio de accidentibus, quae accidunt alicui parti partium entium, accipiendo illam partem quasi distinctam ab ente...", *In IV Meta.*, c. 1, fol. 64r, E. This explanation parallels Albert's *In IV Meta.*, T. I, c. 1, p. 162, l. 38–41, which is a denial that particular sciences are subalternated.

[3] *In IV Meta.*, c. 1, fol. 64r, E.

[4] *In XII Meta.*, c. 5, fol. 293r, D; *Ibid.*, c. 6, fol. 295r, D.

[5] "Naturalis enim habet dare causas mobilis substantiae materiales et moventes; formales autem et finales non potest. Sed Philosophus dat eas demonstrando, quod primum movens, quod iam declaratum est esse in scientia Naturali, est principium substantiae sensibilis secundum formam et finem.", *Ibid.*, fol. 294v, K–L.

[6] "Secundum igitur hunc modum quaerit Philosophus elementa substantiae sensibilis, idest elementa entis, in eo quod est ens. Declarat igitur in hac scientia, quod ens non materiale, quod iam declaratum est esse movens substantiam sensibilem, est substantia antecedens substantiam sensibilem: et quod est principium eius secundum formam et finem. Secundum hoc igitur intelligendum est, quod Philosophus quaerit de principiis substantiae naturalis, scilicet de prima forma et fine. De causa autem movente et materiali est Naturalis. Et in hac scientia ponit has duas causas esse principium perscrutationis de duabus aliis causis.", *Ibid.*, fol. 294v, L–M.

presses the first form and ultimate goal of all beings. *Ens simpliciter*, in Averroes' view, obviously does not contain *ens mobile* and *ens quantum* in its meaning.

Hence, Averroes and Albert are both at odds with Aristotle and Aquinas on this point. Whereas for the latter pair metaphysical concepts contain the concepts proper to the lower sciences, Averroes and Albert reject such a theory. Actually, Averroes and Albert do not hold exactly the same doctrine on the meaning of *ens*. True, they both speak of all the objects studied in divine science as defined in terms of God; they speak too of "being" as expressing the first form and first goal of things. Nevertheless, they both do not accept the same theory of metaphysics. For Averroes, to define things in terms of God, implies that one studies God and hence knows being as such. For Albert, in the context of a Neoplatonistic theory, if one studies the divine light shining forth in *esse simplex* – in beings – then one learns to know something about God. Yet both of these philosophers ultimately say the same thing: *ens* does not express totally the object to which it is applied.[1]

Our investigations into the role of metaphysics in the schema of the sciences have once again underlined the differences between the five philosophers we are studying. Aristotle and Aquinas agree in giving to metaphysics a double prerogative: metaphysics knows the essence of any being, and as well judges the existence of that being. Thus the metaphysical concept of "being" expresses the essence of anything when known as "being."

Avicenna has somewhat the same view of metaphysics, although he transformed this theory, as well as every other Aristotelian doctrine he adopted, by placing it in the context of *esse*. Thus, Avicenna's metaphysician would never know the entirety or the totality of each thing when studied as "here-and-now." Rather, he would know only what is required for existence. Avicenna's metaphysician would know something with not much more intelligible content than a Kantian category.

Averroes and Albert returned somewhat to Aristotle's position by refusing Avicennian *esse*. However, their adherence to theories of "being' as defined in terms of God prevented their Aristotelianism from being perfect. "Being," they explained, expresses the first form and the first end of anything known as "being." "Being" does not thus express the essence of any material being known by the metaphysician.

Once again then, Aquinas' *Commentary* marks a return to Aristotle's

[1] Cf. above, section 2, B.

position. The same conclusion is thus forced upon us: Aquinas' work, its conception of metaphysics, is directed against the mis-interpretations of Aristotle which were current in his time; Thomas' *Commentary* is intended to present Aristotle's doctrine on metaphysics as the first of the sciences.

5. CONCLUSION

Our study of the doctrines implying a relation between metaphysics and the other, lower sciences has brought us one step further in our quest for the nature of Aquinas' *Commentary*.

The investigation of the relation between the "universal science" and "first science" revealed that neither Avicenna, nor Averroes, nor Albert had correctly presented the theories of Aristotle. Averroes and Albert, of course, were further from their Greek master than Avicenna. For the former pair, the "first science" is the study of what is defined in terms of God. For Averroes, this means that knowledge of God, the first form and the ultimate end, is automatically the knowledge of being as such, or "universal science." This conception is obviously a reversal of Aristotle's theory in which being, not God, occupies the center-ground. Albert, on the other hand, places the emphasis on the "universal science": if we know created *esse simplex*, then we know God, then we have "first science." Avicenna, finally, retained Aristotle's being-centered perspective even more than Albert, albeit Avicenna too has introduced some radically un-Aristotelian theories, notably that of *esse*. It is in direct opposition to these three types of interpretations that Thomas placed his *Commentary*. Avicenna he rejected by explaining that the *esse* in question is logical truth, thus not a fit candidate for the formal object of metaphysics. Averroes and Albert were much more directly opposed. Do they shy away from a proof of the first cause of being? Very well, Thomas calls their opinions false, and heaping philosophical coals on their heads, points out how Aristotle's work implies a proof for such a cause.

The question of the legality of a metaphysical study of *communia* illustrated well to what extent Aquinas wished to preserve Aristotle's system, as he understood it. The *communia* are studied because they are the expression of the metaphysician's understanding of his subject. Thus wrote Thomas, echoing what was operative in Aristotle's writing, even though not explicitly proposed. Avicenna too had accepted this theory; however, in his work this doctrine becomes transformed by his understanding of his metaphysical viewpoint, the "here-and-now realization." This Avicennian transformation of Aristotle, this existential Aristotelianism, also felt the weight of Thomas' displeasure; and by his opposition, Aquinas revealed how he himself had transformed Aristotelianism: beings are *esse*-essence composites. Finally, Averroes and Albert, those advocates of a study of God-defined objects, also explained that *communia* are to be studied; however, these authors do not maintain that these common notions express the grasp of even material beings *qua* being – an understandable omission.

Thirdly, we sought to understand in what sense metaphysics can be said to be prior to all other sciences. In the minds of Aristotle and Aquinas, this concept of "being" is the expression of the entirety of an object. Hence, to the metaphysician falls the task of deciding what things are "being," and so, what things are to be studied in their particularity by other, lower disciplines. Avicenna did not speak directly of this directive role of metaphysics; however, by implication he presents a semi-Aristotelian doctrine. Metaphysics, in Avicenna's mind, will give the formal aspects of all things. Averroes and Albert had something to add to the discussion, again an un-Aristotelian theory. In their works, *ens* expresses the first form of being; but Averroes means "God" when he speaks of such a form; and Albert, though meaning *esse simplex*, nevertheless views this *esse* as an expression of God's light, an expression which can enable us to know God. Accordingly, concepts of *ens* such as that expressed by Averroes, and such as that expressed by Albert, are not concepts capable of expressing the totality of each and every being. In short they are not Aristotelian "being."

Hence, our comparative study of these five authors reveals that the four commentators were all engaged in a systematic exposition of Aristotle: Avicenna weaves his theory of *esse* throughout his work; Averroes and Albert, though each in his own way, view all from the side of God-defined objects; and Aquinas works with the *esse*-essence distinction in mind. True, the Thomist theory of *esse* appeared explicitly only in the discussion of the *communia*. However, by implication it taints the explanation of "universal and first science" as well as that of "metaphysics as the ruling science." It taints "universal and first science" because the connection between the two aspects, or parts, of metaphysics is found in "being": Thomas argues from an *esse*-essence composite, expressed as "being," to the cause of being. It taints as well "metaphysics as the ruling science," because it is "being," the expression of an *esse*-essence composition which is capable of expressing the totality of anything.

There is but one conclusion: Aquinas wrote his *Commentary* to preserve the general framework of Aristotelian metaphysics – as he understood that system. – But as the same time, it is clear that Aquinas has transformed that system by the introduction of *esse*.

We have now reached the point where we can legitimately attempt to reconstruct the metaphysics operative in Aquinas' *Commentary*. By the investigations of the last two chapters, we have seen how Aquinas opposes his work to those of his predecessors. His work, like theirs, is built around a few theories on the formal object and on the movement from universal science to first science. Since this is the case, we have no reason to divide any given paragraph of Thomas' work from any other paragraph. The *Commentary* in its entirety was written in the light of a theory of metaphysics, and as an entirety was directed against the works of Avicenna, Averroes, and Albert.

Nevertheless, it remains that Aquinas' *Commentary* does not present Aristotle's system exactly as the latter saw it. For Aristotle, *esse* was not a metaphysical principle. However, Aquinas saw in Aristotle a proof for God. And as shall become clear in the following chapter, the proof for God in Aquinas' *Commentary* is built on a theory of *esse* as a metaphysical principle of substances. Hence, Aquinas' introduction of *esse* has some of the appearances of a *Deus ex machina*: to show the legitimacy of Aristotle's theory that beings are caused, beings are explained by Aquinas as *esse*-essence composites, participating in *Esse*. But these are facts that are to be discussed in the following chapter.

CHAPTER VI

THE METHOD OF METAPHYSICS

The reconstruction of the metaphysics operative in Aquinas' *Commentary* is naturally dependent upon the successful completion of a search for the statements on method contained in that work. Some of Aquinas' thoughts on method or procedure will be explicit, but more will be implied by the expositions of individual topics. It is with such statements, both explicit and implicit, that we are interested in the present chapter. From them, we must attempt to reconstruct the metaphysics which guided Aquinas in the composition of the *Commentary*.

In treating Aquinas' method we come to the heart of the study of his conception of metaphysics. In discussing the method of metaphysics, we shall be dealing directly with the theories grounding Aquinas' thoughts on the objects of metaphysics and on the relations between the universal and the particular sciences. Thus before we begin the reconstruction of Aquinas' metaphysics let us recall briefly the basic doctrines already discussed, the doctrines which bear on the task at hand. Thomas' metaphysics, whatever it may be, must be conformable to these doctrines:

1) metaphysics studies being as such – that is, the ten categories of being as well as the essential attributes of being as such; to express the same doctrine in a different way: metaphysics studies the *communia* used by all the sciences;
2) it is through its study of the *communia* – through the study of sensible being as being – that metaphysics fulfills its role as the universal science; it is through the study of the *communia* that metaphysics learns of the existence of the immaterial being; and so it is through being the universal science that metaphysics can become the first science, the study of the first being.
3) in knowing sensible being as "being", metaphysics knows the essence of that sensible being.

Whatever we discover in the search for Aquinas' method of metaphysics it must be conformable to these three doctrines.

When discussing the method of the *Commentary*, it is important to distinguish clearly between two types of statements found in Aquinas' exposition. On one hand, there are the numerous explanations of the connection between the various books, and even of the connections between the parts of individual books. As a rule, the points concerning method made in this regard are of little use in an attempt to grasp the metaphysics at work in Aquinas' mind as he wrote. Generally speaking, these statements on method are little more than Aquinas' attempt to place a bit of order in what is undoubtedly a more or less incomplete collection of metaphysical discussions: Aristotle's *Metaphysics*.

Opposed to this type of statement, we have others which bear directly on the logical connection between the various key doctrines of metaphysics. For example, at times Aquinas indicates a relation between the concept of "being" and the proof for the existence of God. It is this type of statement that is of aid in an attempt to reconstruct the *Commentary*'s metaphysics; only statements such as these indicate the method to be followed in constructing the philolsophy of being of Aquinas' *Commentary*.

Unfortunately, Aquinas makes very few explicit pronouncements on the central issues of metaphysical method. In fact, there are but six: 1) metaphysics must study the *communia*; *In IV Meta.*, 1, 531; 2) metaphysics studies substances above all else: *In IV Meta.*, 1, 546; 3) in its initial stages, metaphysics proceeds by an investigation of predication: *In VII Meta.*, 3, 1308; 4) metaphysics proves the distinction of matter and form by such an investigation: *In VII Meta.*, 2, 1287; 5) God is proved as the cause of being; *In VI Meta.*, 1, 1164; 6) there are proofs for God based on participation: *In II Meta.*, 2, 296, on measure: *In X Meta.*, 3, 1973, and on finality: *In I Meta.*, 15, 233. Of course, there are other explicit statements on method, but they deal with doctrines less central to metaphysics, for example, the general theory of potency and act.

In addition to the six points on method just mentioned, others can be gleaned from a study of the relationship between doctrines, which would indicate a relative priority and posteriority of treatments. For example, since the proof of God's existence involves knowledge of the predication of "substance," obviously Aquinas would treat the predication of "substance" before he would treat the existence of God. By discovering hints such as these, we can go far in our attempt to uncover the metaphysics operative in Aquinas' mind as he wrote.

If we combine the information on method implied by Aquinas, with the six explicit statements on the procedure of metaphysics already mentioned, we obtain the following list:

1) Physics would be the first philosophy if there were no immaterial being.
2) The ancient philosophers intended to discuss all being by speaking of material being.
3) Metaphysics is the last of the sciences according to the order of discovery.
4) Metaphysics must use a logical method in its initial phases; that is, it must investigate the manner of predicating.
5) Extra-mental beings are divided into ten groups which correspond to the ten ways of predicating "being" *per se*.
6) Since nine of the *per se* predications of "being" refer to the tenth, where "being" means "substance," it is real substances that the metaphysician must principally study.
7) *Ens* expresses an *esse*-essence composition; the judgment: "Socrates est ens" expresses no understanding, or intelligibility, about Socrates not possessed in the act of understanding Socrates as *ens*; a judgment adds to apprehension only the knowledge of the truth, or of the rectitude of the apprehension.
8) By investigating the manner of predication, the metaphysician proves the distinction of matter and form in material substances.
9) The metaphysician must look for a first substance, in relation to which all other substances are measured.
10) The metaphysician can prove the existence of God by "participation," and can subsequently discuss some of the divine attributes.

When one reflects on the doctrines Thomas exposes concerning the seemingly numberless topics discussed in the *Commentary*, one begins to discover that the various doctrines group themselves around the ten hints and indications we have just enumerated. These ten thus appear as expressing among them the core of the metaphysics present in the *Commentary*, regardless of whose metaphysics that may prove to be. Without more ado, then, let us turn to the task of reconstructing Aquinas' metaphysics around these points.

1. THE *COMMENTARY'S* METAPHYSICS

A. The birth of metaphysics

In our investigation into the historical aspects of the *Commentary* in Chapter V, we noted that Aquinas presents in his work a being-centered philosophy which proves God as the cause of being; moreover, we discovered that this conception of Thomas is placed in opposition to the theories of Averroes and Albert. For both of these latter commentators, metaphysics can not prove the existence of a cause of being. Thus, Averroes reduces the supreme science to a study of God Himself, but of course, God understood as the prime mover of physics. Albert on the other hand conceives his science as the sutdy of created *esse simplex*; however, the knowledge of created being is at once knowledge of the Divine Light from which all creation flows. All three, Averroes, Albert, Aquinas, disgressed in their expositions of Book E, Z, and Λ to explain notions central to their conceptions of metaphysics. The fact that all three paused at the very same spot in their work cannot be explained as a coincidence. No, Albert was placing his opinion next to Averroes' so that all the world could choose between them; in like fashion, Aquinas opposes his thoughts to those of his two predecessors so that students of the *Metaphysics* could compare the three interpretations, and thus choose the correct one. Interestingly enough, the issue at stake was the relation, first between the universal science and the first sience, and second, between physics and metaphysics. In the face of the metaphysical pessimism of his predecessors, Aquinas explained his own, more optimistic view of Aristotle. God is proved to be the cause of being, he remarks; metaphysics takes a formal object which permits it to study all sensible beings as being, and consequently to rise to knowledge of the cause of the being of these sensible beings.[1] Thomas, however, although he thus touches on the relation of first and universal science, does not – to the best of my knowledge – ever explain openly why and when metaphysics is born. Nevertheless, his opposition to Averroes and to Albert is instructive. The latter pair of philosophers maintained that metaphysics begins where physics leaves off – with the discovery of the prime mover. This aspect of their writings is never mentioned by Aquinas. This fact, I believe, indicates that he did not object to their theories on the genesis of metaphysics; it indicates that Aquinas agreed with Albert and Averroes in seeing the discovery of the prime mover as the impluse which turns the philosopher into the metaphysician.

Yet even if we were unaware of this relation of Aquinas to his two predecessors, we could still conclude from his writing that metaphysics is born when one discovers a need to posit a first cause of motion; with this discovery, one realizes that the study of mobile being as such is not a study of reality in all that it means to be real. Let us note a few doctrines of Aquinas' *Commentary* which point to this conclusion.

1) Physics would be the first science if there were no immaterial being. On this point Thomas is quite explicit. Physics deals with "natural" objects, he writes, that is, with those in which one can discern a "nature" or a principle of motion and rest, etc. If all beings

[1] Cf. Chapter V, section 2, B with the footnotes of the subsection.

were natural in this sense, then physics would be first among the sciences. On the other hand, if there is an immaterial substance or being, then the study of this substance will be prior to physics, just as this immaterial substance will be prior to natural or material being.[1] In affirming this, Thomas appears to imply that if there were no immaterial beings, the highest or first science would study material beings *qua* material being, and not *qua* being (that is, as composed of *esse* and essence.).[2]

2) The ancient physicists, Thomas maintains, were quite justified in speaking physically of all things, granted their supposition that all beings are physical or material realities. Given their belief that all things were mobile substances, these early philosophers felt they knew what it meant "to be" when they grasped mobile being.[3] In other words, "being" or "that which is" was understood by them in some material sense, such as "that which consists of atoms of fire, water, and so on." The implication is clear, I believe: to realize that "to be" means something other than "to be composed of atoms," or "to be subject of motion," or something of this kind, it is required that one know there is at least one being that is not composed of atoms, or that is not subject to motion.

3) Metaphysics is the last of the sciences according to the order of discovery. As Thomas explains in an early part of his *Commentary*, when man sets himself the task of seeking out the natural properties and causes of things, the first properties and causes he discovers are properties and causes of a particular genus or species of being. After much effort, however, he will arrive at a knowledge of the universal

[1] "...si non est aliqua alia substantia praeter ens quae consistunt secundum naturam, de quibus est physica, physica erit prima scientia. Sed, si est aliqua substantia immobilis, ista erit prior substantia naturali; et per consequens philosophia considerans huiusmodi substantiam, erit philosophia prima". *In VI Meta.*, 1, 1170; *In III Meta.*, 6, 398. Cf. *In XI Meta.*, 7, 2267. Also, *Ibid.*, 2261–62.

[2] One of the Aristotelian *dubitationes* was: "whether or not there is any substance other then sensible ones". In commenting on this, Thomas notes that metaphysics is generally thought to be about some separated substance: nevertheless, "...si nihil est praeter sensibilia, tunc sola sensibilia sunt entia. Et cum sapientia sit scientia entium, sequitur quod sapientia sit circa sola sensibilia...". *In XI Meta.*, 2, 2175. Thomas does not mention the formal object metaphysics would use in this case, but since "being" would be sensible being, it appears the formal object would be "sensible being".

[3] "...quidam tamen naturalium de his [primis principiis] se intromiserunt; *et hoc non sine ratione.* Antiqui enim non opinabantur aliqua substantiam esse praeter substantiam corpoream mobilem, de qua physicus tractat. Et ideo creditum est, quod soli determinent de tota natura, *et per consequens de ente*; et ita etiam de primis principiis quae sunt simul consideranda cum ente". *In IV Meta.*, 5, 593; Italics added. Thomas' words "et hoc non sine ratione" refer to the study of the first principles; obviously, Thomas would also apply them to the belief of the naturalists that they were treating being as such. Cf. *Ibid.*, 12, 681–82.

intrinsic causes of all being; still later and after more effort, he arrives at knowledge of the immaterial beings. Since metaphysics is the knowledge of both the universal intrinsic causes and the immaterial substances, metaphysics is the last of the sciences to be learned by man.[1]

On the basis of this doctrine, the question arises: what is the act of knowledge, the act of discovery, that turns man from an interest in some particular genus of being to an interest in all being *qua* being? If we link the text just noted with those referred to previously in this section, we must admit that man can begin metaphysics, or the study of being as such, when he discovers the existence of an immaterial being. Hence our problem: which science will discover the existence of this being?

4) Physics, not mathematics nor any other science, is to be credited with the discovery of an immaterial being. Many texts must be synthesized to arrive at this conclusion. First of all, as Thomas notes, there are only two speculative sciences other than metaphysics; these are physics and mathematics.[2] Besides these three speculative sciences there are many others called *factivae* or mechanical arts, and *activae* or moral sciences; these last two types of science, however, are interested in *ordering* either what is external to man, or immanent to him –

[1] "Sed quantum ad investigationem naturalium proprietatum et causarum, prius sunt nota minus communia; et quod per causas particulares, quae sunt unius generis vel speciei, pervenimus in causas universales. Ea autem quae sunt universalia in causando, sunt posterius nota quo ad nos, licet sint prius nota secundum naturam... Facienda est etiam vis in hoc quod maxime universalia non dicit simpliciter esse difficillima, sed "fere". Illa enim quae sunt a materia penitus separata secundum esse, sicut substantiae immateriales, sunt magis difficilia nobis ad cognoscendum, quam etiam universalia: et ideo ista scientia, quae sapientia dicitur, quamvis sit prima in dignitate, est tamen ultima in addiscendo". *In I Meta.*, 2, 46. This text, which appears to prove this point, is not as clear when seen in its context. There two difficulties arise. 1) In the parts of paragraph 46 which have been omitted, Thomas speaks of the most universal predicate "being"; this idea is man's first concept, he says. Yet in the context of the *Commentary*, that idea of "being" is the idea of metaphysical being. How then can Thomas say that metaphysics is the last science man knows? (Cf. *In IV Meta.*, 6, 605; *In XI Meta.*, 4, 2210; for the first idea as the metaphysical idea of "being"). 2) The texts of footnotes 1–3, p. 243 tended to prove that we discover the existence of an immaterial being before we arrive at the idea of "being" as something other than "material being". Yet in the text from par. 46 just quoted, Thomas says that it is easier to know ideas like "being", the *universalia in praedicando* than the first, extrinsic causes of everything, the immaterial substances. – Thomas' meaning, and the reconciliation of these seemingly contradictory ideas, appears from the coordination of many different texts. This coordination can not be made in a footnote; in fact the present section of this chapter is devoted to such a coordination. For the moment suffice it to mention in brief some steps in the construction of metaphysics; first, man's first idea is that of "being"; engrossed as he is with material things, man doesn't realize that matter need not be in the content of that idea; second, he discovers the unmoved mover; in the attempt to find a concept common to that mover and the moved, he comes upon the content of "being"; thirdly, in his attempt to view the world as "being" he arrives at knowledge of the immaterial unmoved mover as the first cause of being.

[2] Cf. *In I Meta.*, 3, 53; *In VI Meta.*, 1, 1152–65.

their interest, thus, does not lead them to the discovery of anything.[1] Of the two sciences which are left, mathematics could never claim to discover an immaterial being, for its interest lies with material objects: it considers what must exist in matter, yet it considers it only under the aspect of an immaterial formality. Mathematics, thus, never seeks to go beyond the material world in a search for an efficient or final cause.[2] Hence, we are left with physics: it is from this science that man must receive the impulse to think metaphysically – it is the physician who must tell us of the existence of an immaterial being. But how?

5) The physician discovers the immaterial being in his search for the "why?" of physical things. Here our procedure must imitate that of the successful detective. After the fashion of a philosophical Hercule Poirot note the clues. a) The physician must consider mobile being.[3] b) His procedure in considering mobile is being styled *via motus*; that is, we must suppose, after the fashion of the writer of Aristotle's *Physics*, the physicist observes motion to see what such a fact demands.[4] c) In his consideration of mobile being, a point will come when he asks, for example: "why is this bronze object round?" By this question he will wonder why it is round rather than rectangular or triangular. He in no case expects the answer: "Because it has the form of roundness instead of the form of some other shape." To give him such an answer would be non-sensical, since he knew, before he asked the question, that the bronze object had the form of roundness. It is precisely because he knew the roundness of the object that he could ask for the "why" of its roundness. In any question we begin from a certain bit of knowledge and, by our question, show our desire to discover something as yet unknown. This was true for our question: "Why is this bronze object round?"; it is true as well for: "Why is man a man?." Since, then, we know the formal cause before we ask the question (we know "roundness" or "manness" in our two examples), we are asking for some other cause: we want to know why "roundness" or "manness" happen to be in these globs of matter rather than "rec-

[1] Cf. *In VI Meta.*, 1, 1152; *In XI Meta.*, 7, 2253.

[2] For the doctrine of mathematics as a science studying the material world from an immaterial point of view, cf. *In VI Meta.*, 1, 1161; 1163; *In XI Meta.*, 7, 2256–58; 2260–61. That mathematics is not interested in final causality, see *In III Meta.*, 4, 375; where there is no interest in final causality, there can be no interest in an agent, since when there is no end in view, there is no reason to speak of efficient causality.

[3] Cf. *In XI Meta.*, 7, 2264; *In I Meta.*, 12, 182 and 200.

[4] The procedure of physics is called *via motus*, cf. *In VII Meta.*, 2, 1287; *Ibid.*, 11, 1526. For an example of the use of this method, observe the physician's study of change: *Ibid.*, 1285–86.

tangularity" or "dogness." The answer is, understandably: "Because an agent put these forms in these instances of matter rather than any other forms." The question asks, then, for the identity of the agent and for his goal.[1] d) There cannot be an infinite regress in the case of such causes. Although thus far we have spoken of the possession of a form (of "humanity" for example), we wanted to know the cause of the *motion* which resulted in the possession of this form. Thus when we asked: "Why is this man a man?," and we answered: "Because an agent put this form in this matter to attain a goal," we wanted to discover the efficient and final causes of the motion. Since there cannot be an infinite regress in the efficient and final causes of the material universe, our question did not introduce us into a possibly infinite series of further questions. That is to say, although we can ask for the efficient cause of this agent's "putting" a form, we cannot then ask for that agent's cause and so *ad infinitum*. As Thomas explains, if we have an infinite series of agents, then they are all *media*, or intermediary causes. There is, he says, an absolute need of a first cause – a cause which is not an intermediary between some other cause and an end.[2]

Thus it appears that the physician can discover the existence of the first efficient and final causes of motion – if he cares to ask "why?" in the sense explained. Since this first cause is not caused it would not itself be moving; if it were moving, we would again ask "why?" and so on. And since what does not move is immobile or immaterial, the physician has discovered an immaterial first mover.

Given these many and varied texts of the *Commentary*, it would appear correct to understand the birth of metaphysics to be synonymous with the discovery of the immaterial mover.[3] If we grant this,

[1] Thomas speaks of this matter in two places; the more complete treatment is *In VII Meta., lectio* 17, but is naturally too long to be quoted. Cf. *In VIII Meta.*, 5, 1759. – See the discussion of these points in Ch. V, section 1, C.

[2] *In II Meta.*, 3, 303–304. Cf. *In XII Meta.*, 4, 2474.

[3] One might prefer to maintain that from faith we know there is a God, and that there are angels, all of which are immaterial. Thus we know that "being" does not necessarily say something about matter. One might certainly put forth this claim, but there is nothing in the *Commentary* implying that Thomas begins his metaphysics by an appeal to faith. One might note that, as was pointed out in Chapter V, section 2, B. Thomas is quite insistent that metaphysics proves the existence of God and does not accept the prime mover of physics as a subject to be described. It is difficult to understand why he should emphasize the ability of proving God as the cause of being, if he were to assume from faith the existence of the immaterial God as the starting point of his metaphysics. Why get excited over the possibility of proving the existence of a cause of being if one assumes the cause exists? That would be the same as insisting on the right to think in a vicious circle. – For an author who would agree that physics must prove the prime mover before metaphysics can adopt the formal object of "being", understood as a concept which does not explicitly express materiality, cf. GEIGER, "Abstraction et Séparation d'après Saint Thomas...", pp. 24–26.

what is the first step in the construction of metaphysics? Before attempting to answer this, let us discover what procedure or method we are to use. As Thomas says: "...prius oportet quaerere modum scientiae, quam ipsam scientiam."[1]

B. Metaphysics and the investigation of predication

This present sub-division of this section on Thomas' method touches one of the essential aspects of the conception of metaphysics present in the *Commentary*. Fortunately on this point Aquinas is rather more explicit than he is on many other topics. For instance:

...haec scientia [id est, metaphysica] habet quandam affinitatem cum Logica propter utriusque communitatem. Et ideo modus logicus huic scientiae proprius est, et ab eo convenienter incipit. Magis autem logice [Aristoteles] dicit se de eo quod quid est dicturum, inquantum investigat quid sit quod quid erat esse ex modo praedicandi. Hoc enim ad logicum proprie pertinet.[2]

Metaphysics and logic have some "community" – they share something most likely – and thus they have a certain kinship. Because of this kinship, metaphysics begins by using a logical method: it investigates the manner of predicating employed when we predicate essences.

Thomas was not content to affirm this doctrine but once. In an earlier lesson, Thomas speaks of a doctrine which is proved by the *metaphysician*"... per viam praedicationis, quae est propria Logicae, quam in quarto huius dicit affinem esse huic scientiae."[3] This kinship, he notes, is treated in Book IV by Aristotle. Thus Thomas twice affirms the kinship of logic and metaphysics, and states that metaphysics uses a logical method, at least in its initial stages.

In a still earlier passage, this time in Book III, Thomas refers once more to the affinity of logic and metaphysics. The context of the passage is the discussion entitled: "Why does metaphysics doubt about all things.?"[4] Thomas assigns two reasons for this procedure and then gives a third one which, he notes, is that of Averroes:

Tertiam [rationem] assignat Averroes dicens hoc esse propter affinitatem huius scientiae ad logicam, quae tangitur infra in quarto. Et ideo dialecticam disputationem posuit quasi partes principales huius scientiae.[5]

[1] *In III Meta.*, 2, 346.
[2] *In VII Meta.*, 3, 1308.
[3] *Ibid.*, 2, 1287.
[4] Little, if anything, has been made here of the idea of metaphysics as beginning with an investigation into the truth of our knowledge, for the passage from Book III does not speak of truth in the epistemological sense, as for example, Mgr. Noël believed it did. Cf. A. MANSION, "Universalis dubitatio de veritate...", p. 541. The passage from Aquinas made so much of by Mgr. Noël: *In III Meta.*, 1, 338.
[5] *In III Meta.*, 1, 345.

One should note, of course, that Aquinas neither agrees nor disagrees with Averroes' reasoning. In addition, the reference to Book IV has nothing to do with Averroes' idea, but is Aquinas' indication that Aristotle treats the kinship of metaphysics and logic in Book IV.

The kinship of metaphysics and logic is treated, as Thomas has predicted, in Book IV – in lesson four. Here Thomas discusses the various arguments given by Aristotle to prove that metaphysics studies the essential attributes of being; these are the attributes which, as was noted in Chapters IV–V, Thomas calls *communia*. It must be to the second of these arguments that Thomas referred when he noted Book IV contained information on the kinship of metaphysics and logic. The passage is important as it shows clearly how metaphysics is to proceed.

The argument in brief is as follows. The dialectician (or logician) and the sophist ape the metaphysician ("induunt figuram eamdem philosopho"); moreover, they have a quasi-resemblance to him: "quasi similitudinem cum eo habentes." Since, therefore, the dialectician and the sophist treat the *communia*, the metaphysician must do the same.[1]

This argument is anything but profound. Nevertheless, the explanation of the similarity as well as the differences between the three men reveals the argument to be much more than mere words. The dialectician, Thomas writes (and we can omit the discussion of the sophist). has this in common with the philosopher or metaphysician: he studies everything insofar as each thing is contained under being and the essential attributes of being.[2]

On the other hand, the metaphysician and the dialectician differ insofar as the former knows his object with certitude while the latter has only probable knowledge or opinion. This difference in the quality of their knowledge flows from the different starting points: that of the metaphysician is certain principles from which he demonstrates, that of the dialectician is probable principles.[3]

Thomas traces this difference back even further, right to its very source: to the difference in the subject matters of logic and of metaphysics. The metaphysician, who studies *ens naturae* – real being, uses principles of reality ("ex principiis ipsius") in his attempt to prove certain things about the *communia*. The dialectician or logician, studying *ens rationis*, considers these same *communia*; he too tries to

[1] *In IV Meta.*, 4, 572.
[2] *Ibid.*, 573.
[3] *Ibid.*, 574.

prove certain things about the *communia*. Unlike the metaphysician, however, the logician bases his proofs on the principles of *ens rationis*; thus, using principles extrinsic to the *communia*, he is more properly said "to try" to prove something, rather, than "to prove" something about *communia*.[1]

The logician's method of studying the *communia* is, then, a matter of applying the principles of *ens rationis* to them. These principles, Aquinas explains, are the intelligible intentions such as genus, species, and so on, which are attributes of things only insofar as things are known by the mind. It would be, then, through considering the *communia* as genus or as species, and so on that one would treat them in logic.

The exact characteristics of the metaphysician's consideration of the *communia* are not discussed as clearly as those of the logician's investigation. Thomas merely says that the principles used would be the principles of the philosopher, probably therefore, principles of *ens naturae*. Aquinas does not elaborate by identifying these principles.[2]

The difference between the logician's and the metaphysician's investigations becomes clearer, however, if one reflects on the principles used by the logician: he uses principles of objects as known. That is, the logician is using those characteristics of our knowledge – genus, species, specific difference, etc. – these characteristics which can rightly be termed the "form" of objects as known. It is this type of "formal" principle which the logician can use and which the metaphysician would not.[3] Thus, although the metaphysician is to follow a logical procedure, namely: investigate predication, he will not use the same principles as those employed by the logician – he will not investigate reality as a formal logician would. Such an investigation would yield only probable knowledge of the real – hardly the goal of metaphysical investigation. The principles used by the metaphysician, on the other hand, are opposed to the logical principles; the metaphysician uses principles intrinsic to real beings when he studies the *communia*, Aquinas implies. Here then again, Thomas is teaching that the metaphysician studies *communia*. As we noted in Chapter IV the investigation of being and its essential attributes was seen to involve a study of the content of these *communia*; and in Chapter V, Aquinas' explanation of metaphysics as the study of sensible beings as "beings"

[1] *Ibid.*.

[2] *Ibid.*.

[3] Cf. *In VII Meta.*, 9, 1462–63, where Thomas criticizes an attempt to affirm an identity between the form of our thought and the form of the real.

was seen to mean that metaphysics studies the content of the *communia*.[1] Hence we have come upon nearly the same doctrine in three different contexts. The present contexts, however, the comparison of logic and metaphysics, makes the point that metaphysics studies the *communia* by using the principles of real things.

Does this mean that metaphysics is to form the *communia* (or perhaps: clarify their content) through a consideration of real beings? If we said "yes," would we not be contradicting the doctrine given when Aquinas explained how "the study of sensible being as being" was identical to "the study of the content of the *communia*"? This latter doctrine gave as the material object, the content of the *communia*; an affirmative answer to the present question, however, would mean that real sensible beings fall into the material object. But this would destroy Thomas' attempt to explain that metaphysics, unlike physics and mathematics, studies being as separated in *ratio* and in *esse*. This problem is not as acute as it may seem; a negative answer to the question posed above does not do violence to Aquinas' words. In fact, in the context of the comparison of logic and metaphysics, Thomas notes that metaphysics studies the same thing logic does: the *communia*. To study the *communia* in the light of the intrinsic principles of reality was opposed to the logician's study which uses principles extrinsic to things. One can always understand this doctrine thus, as far as it concerns metaphysics: one studies the content of *communia* insofar as they express the intrinsic principles of real beings.

Such an interpretation has the considerable merit of rendering more meaningful, and of synthesizing many aspects of the *Commentary's* doctrine: 1) Metaphysics is to investigate predication but not as a logician does. 2) To study sensible beings as being is to study the content of the *communia*. A study of predication would involve a study of something such as: "John is a man." The logician would note that "man" is a species, but that type of thinking does not interest Aquinas' metaphysician. Hence, if the metaphysician investigates the content of the predicate "man," he must do it in this manner. "What does 'man' tell us about John – not insofar as John is a man, but insofar as he is a being?" That type of question is: 1) an investigation of predication, 2) yet not a logical investigation. Thus, it is 3) an investigation into the content of a predicate; 4) yet not into the predicate as the expression of sensible being, 5) but rather as the expression of being. And thus, it

[1] Cf. Ch. IV, section 1, B; Ch. V, section 3, B.

is an investigation 6) into the *commune ens*, insofar as *ens*, expressed in "man," expresses the real principles of the sensible being.

There is, of course, a jump in our argument that is not established by any evidence. "Man," we said, expresses *ens*; "man" expresses John as an *ens*. Up to the present, I have presented no evidence for such a theory, at least no explicit evidence. But if we turn now to an investigation into Aquinas' explanation of the use of the name "being" in *In V Meta.*, *lectio* 9, we shall see that concepts such as "man" do express *ens*. Thus our investigation into this exposition of "being" will both justify our synthesis in the present sub-section, as well as carry out the first step in Aquinas' metaphysics – the investigation into predication, the investigation into the content of *communia*.[1]

Thus, Aquinas attributes to metaphysics the necessity of beginning by using a logical method – the investigation of predication. Aquinas, we have discovered, gives as his reason, the fact that logic and metaphysics have a common subject: everything. Moreover, both of them are interested in everything as every thing "comes together in being and its essential attributes." Quite obviously, all of reality "comes together" only in a concept (or concepts) – in a *commune*. If one wishes to study the manner in which things are included within or under a concept, one has two paths open: 1) one can study the concept (or concepts) insofar as it expresses the reality of each thing; this would be a science of the real – metaphysics; 2) but one can also study the concept (or concepts) insofar as the mind has a certain way of relating one concept to another; this would be a science of the rules of predicating – (formal) logic. But how does one study a concept as expressing the reality of things? By looking at its use, or course: by looking at its predication. Thus metaphysics is to begin by an examination of the predication of those concepts which express all the real as real.

Yet here we must be very careful. We must not believe that "predication" is an activity identical with "judgment." Quite the contrary, as Aquinas notes, a predication such as: "S is P" is an expression of an

[1] I have often wondered whether Aquinas' three references to metaphysics as kin to logic and as using the logical investigation of predication were not revised after the exposition of Book IV, treating the kinship, was completed. That is, in *In VII Meta.*, 2, 1287; *Ibid.*, 3, 1308; *In III Meta.*, 1, 346, Aquinas refers to this kinship as justifying the metaphysical study of predication; perhaps he intended to re-write *In IV Meta.*, 4, 572–77, and to include in the new exposition a mention of this metaphysical use of the logical investigation. Somehow, the exposition in Book IV does not live up to the expectations generated by the doctrines of the other books. There is, however, no apparent way to verify such an hypothesis. (For the various dates of the books, see Chapter I, section 2, p. 21). For a study of Book IV, *lectio* 4 which takes the lesson entirely out of context, and so fails to see what Aquinas wishes to say, cf. MARITAIN, *A Preface to Metaphysics...*, pp. 38–42.

act of apprehension. A judgment states the truth of a prior act of knowledge. A judgment expresses the correctness of conceiving "S" as "P." And the conception of "S" as "P" is expressed in apprehension as: "S" is "P."

Thus we see Aquinas explain that an act of knowledge is everything it should be if it involves a similarity of the object being known.[1] Accordingly, even an act of sense knowledge can be "everything it should be" or perfect; however, sense is incapable of recognizing that it is perfect.[2] So it is that in intellectual activity, writes Aquinas, we must distinguish two activities; first, there is the possession of the similarity of objects; and secondly, there is the activity of knowing that our knowledge conforms to the object known. Thus we distinguish apprehension, or the "being similar to the object known," and judgment, or the "knowing that the intellect is similar to the object known."[3] Let us suppose for a moment that we conceive Socrates as "mortal, rational animal," or as "man." Our intellect would possess the similarity of the manness of Socrates. Yet we do not, on the strength of this possession, know whether our knowledge of Socrates as "man" is true or false.[4] To know the truth, or the correctness of conceiving Socrates as "man," one must reflect upon the idea "man," thus recognizing that "man" is an adequate expression of Socrates.[5]

It is a simple matter to read between these lines and grasp Aquinas' theory of intentionality. To apprehend Socrates as "man" is not some sort of mechanical action. It is not to bring down a curtain between, on the one side, an intellect engrossed in the contemplation of an idea "man," and on the other, the real Socrates. No, to apprehend Socrates as "man" is to be engrossed in a simple, direct, intellectual gaze at Socrates; "man" is precisely the form this gaze takes. One can choose

[1] "...cum qualibet cognitio perficiatur per hoc quod similitudo rei cognitae est in cognoscente; sicut perfectio rei cognitae consistit in hoc quod habet talem formam per quam est res talis, ita perfectio cognitionis consistit in hoc, quod habet similitudinem formae praedictae". *In VI Meta.*, 4, 1234.

[2] "Licet autem in cognitione sensitiva possit esse similitudo rei cognitae, non tamen rationem huius similitudinis cognoscere ad sensum pertinet, sed solum ad intellectum. Et ideo, licet sensus de sensibili possit esse verus, tamen sensus veritatem non cognoscit...". *Ibid.*, 1235.

[3] "Intellectus autem habet apud se similitudinem rei intellectae, secundum quod rationes incomplexorum concipit; non tamen propter hoc ipsam similitudimen diiudicat, sed solum cum componit vel dividit". *Ibid.*, 1236.

[4] "Cum enim intellectus concipit hoc quod est animal rationale mortale, apud se similitudinem hominis habet; sed non propter hoc cognoscit se hanc similitudinem habere, quod non iudicat hominem esse animal rationale et mortale". *Ibid.*.

[5] "...in hac sola secunda operatione intellectus est veritas et falsitas, secundum quam non solum intellectus habet similitudinem rei intellectae, sed etiam super ipsam similitudinem reflectitur, cognoscendo et diiudicando ipsam". *Ibid.*.

to look at Socrates as "animal," "fat," "man," "mouth to feed," etc. Each of these different ways of looking can be called the "form" of the intellectual gaze. Thus to apprehend Socrates as "man" is to see "man" in Socrates. The meaning, the content of such an activity can be expressed as: "Socrates-man." The judgment, as Aquinas has remarked, is merely the result of an activity of reflecting on the "Socrates-man"; the judgment expresses whether or not "Socrates-man" is the correct content of a gaze in the direction of Socrates.

Although it is very true that Aquinas speaks of judgment as "composing and dividing," we must not understand him as meaning something such as this: the judgment tells us that "Socrates is, or exists as a man."[1] We must not thus understand Aquinas, for he has taken pains to explain that he does not mean this. In an expression such as: "John is blind" – we consider this expression here as the ordinary form of our judgments – the verb "is" denotes the correctness of conceiving "John" as an "object incapable of sight."[2] On the other hand, when we say: "Socrates is white," we can mean two things; first, taking this statement as the normal manner of expressing a judgment, we can express in this way that it is true to conceive Socrates as a white being. Thus instead of giving: "Socrates is white" as the expression of a judgment we should express a judgment as: "It is true to conceive Socrates as a white being."[3] (Often when we are questioned about something which we have imagined, we employ this type of circumlocution; for example, "John is bald, that is, as I have imagined him"). Secondly, "Socrates is white" can be taken also as an expression of a really existent being. Here "is" does not express the truth of a manner of conceiving Socrates; rather it refers directly to the extra mental reality. In this sense, "Socrates is white" is an expression of apprehension. Here, "Socrates is white" is like the "Socrates-man" already discussed. "Socrates-man" and "Socrates-white" are the content of an intellectual gaze at Socrates.

[1] Cf. the texts already quoted, which far from implying that the judgment consists in a joining of two concepts, explicitly note that prior to the judgment man must possess the "similitudo" of the external object being understood. Cf. also: *In IX Meta.*, 11, 1896–1903. For an example of a presentation of the view I believe to be erroneous, cf. G. P. Phelan, "Verum Sequitur Esse Rerum", *Mediaeval Studies*, I, 1939, pp. 11–22.

[2] In his exposition of Book Δ 7, Aquinas explains how "is" can express 1) an extra-mental reality, as well as 2) the truth of a proposition. In the first use of "is", Aquinas is referring to the predicate *ens*; he refers to the content which *ens* expresses about an object. In the quote below, the second use of "is" is explained: "Dicitur, enim, quod caecitas est secundo modo, ex eo quod vera est propositio, qua dicitur aliquid esse caecum..." *In V Meta.*, 9, 896.

[3] "Cum enim dicimus aliquid esse, significamus propositionem esse veram. Et cum dicimus non esse, significamus non esse veram; et hoc sive in affirmando, sive in negando. In affirmando quidem, sicut dicimus quod Socrates est albus, quia hoc verum est. In negando vero, t Socrates non est albus, quia hoc est verum, scilicet ipsum esse non album". *Ibid.*, 895.

Moreover, there is no difference between "Socrates-white" and "Socrates is white." To gaze at Socrates as "white" is to gaze at Socrates as "is white." Thus, the expression of an apprehension is: "Socrates is white"; and the expression of a judgment is: "It is true to conceive of Socrates as white."[1]

Accordingly, we can conclude that a proposition such as :"Socrates is white" is the expression of apprehension. Hence to study the use of predicates such as "white" is to study them insofar as they appear in apprehension; to study the use of predicates involves, then no study of judgments.[2]

[1] This doctrine is gleaned from numerous aspects of Aquinas' works. He is very careful to distinguish between a) uses of "is" which express something about extra-mental reality, and b) uses which express something about the mind. Those uses of "is" expressing something about the mind are the judgmental uses of "is"; here, Aquinas refers to the "is" of a judgment, the "is" considered as having a role to play in communicating some knowledge – an expression of truth or falsity. (*In V Meta.*, 9, 895–96). When "is" expresses something about reality, it is not therefore the "is" of a judgment. Rather this "is" is part of a predicate such as "man" in so far as "man" is attached to an object such as John. (*Ibid.*, 889–94; esp. 893). This "attachment of 'man' or of 'is man' to a predicate" is not something taking place in a judgment; it must rather be prior to a judgment. If this is not Thomas' theory, the exposition of *In V Meta.*, *lectio* 9 makes no sense.

[2] In a recent article M. Breton has attempted to rediscover the metaphysical core of the two texts in which Aquinas deduces the ten categories (*In III Phys.*, 5, 332; *In V Meta.*, 9, 889–92). Cf. S. Breton, "La déduction thomiste des catégories", *Revue Philosophique de Louvain*, LX, 1962, pp. 5–32. M. Breton's article is not an historical exegesis (pp. 5–6, footnote 2). Yet he makes many affirmations about Aquinas' work, among the more important of which are several quite opposed to the present study. For example, the heart of Thomas' deduction of the categories in Book V of the *Commentary* is said to be a general theory of judgment (p. 5). One can consider the judgment as uniting a subject and a predicate; here the judgment (*compositio*) overcomes the intellect's weakness by constructing a unitary complex mental work; thus the intellect "moves toward" an act of simple understanding (p. 7; no textual proof is offered of these doctrines). – When he speaks of a *compositio* in these words, I wonder whether M. Breton is not unconsciously sliding from logic to psychology: from a fact of formal logic, namely the composition of a subject and predicate, he concludes that the intellect does not *understand* the subject through the concept predicated, but rather knows this union of subject and predicate through a judgment. To attribute a growth in man's knowledge (other than a growth in knowledge of logical truth) to a cause other than the activity of understanding is not to follow the doctrine of Aquinas. As P. Hoenen, S. J. has explained: For Thomas, every judgment is preceded by "...la représentation pure, mais composée, qui précède le jugement, une synthèse caractéristique de (dans le jugement qui suivra) sujet et prédicat... chez lui aussi, la composition des concepts du sujet et du prédicat précède le jugement, se trouve déjà dans les données sensibles, concrets; chez lui aussi "juger" est tout autre chose que "se représenter", c'est le résultat d'un retour critique sur la représentation composée..." Cf. P. Hoenen, S. J., *La théorie du jugement d'après St. Thomas d'Aquin*, edition altera, Analecta Gregoriana, Vol. XXXIX, Series Fac. Phil. section A (n. 3), Rome, 1953, pp. 69–70. – M. Breton's characterization of the judgment as a psychological linking of two concepts sets the tone for his explanation of the deduction of the categories. Thus despite the great philosophical interest of his article, its tone, its spirit can not be taken as true to Aquinas' thought. For an exposition of Aquinas' thought on knowledge which agrees with that of P. Hoenen, cf. B. Lonergan, S. J., "The Concept of *Verbum* in the Writings of St. Thomas Aquinas", *Theological Studies*, VII, 1946, pp. 349–92; VIII, 1947, pp. 35–79 and 404–44; X, 1949, pp. 3–40 and 359–93. See also the other writings of Fr. Lonergan mentioned in our bibliography. In general both the writings of Fr. Lonergan and P. Hoenen are reconciliable with the view expressed throughout the present study with one

Such an emphasis on apprehension, to the exclusion of judgment, is readily understandable if we recall that Aquinas has rejected the Avicennian theory of *esse*. Metaphysics must study all things in the light of the here-and-now aspects grasped by judgment, explained Avicenna. No, says Aquinas; the judgment expressess only the truth of our knowledge. Aquinas implied, of course, that the intellect, between apprehension and judgment, sees the extra-mental here-and-now of the objects apprehended. Yet that "here-and-now" is not *esse* as we shall shortly discover.

C. The investigation of predication: the concept of "being"

Metaphysics must examine the *communia* to see how they express reality, we have remarked. Yet a metaphysician who has not yet completed his science, does not know the identity of the *communia*. The *communia* are not a group of concepts, *all* of which are used by all the intellectual disciplines. Some of the *communia, ens,* for example, are used by all the sciences; others, such as efficient and final causes are not used by all.[1] Moreover, the *communia* express the metaphysician's knowledge of reality. Hence, at the beginning of his science, the metaphysician cannot be certain which concepts are to be called *communia,* nor what the content of any of them might be. Accordingly, the problem arises: how is the metaphysician to identify the concepts he must study? This problem is more apparent than real. Since the fledging metaphysician would not know the identity of the *communia,* Aquinas' statements about metaphysics as the study of *communia* are to be taken as statements of someone in full possession of his science. It is thus more important for us to reflect on the doctrines concerning the birth of metaphysics than on those concerning the study of *communia.* As will appear, in the construction of his science, bit by bit, the metaphysician will discover concepts which later will be used by other sciences, and which become thus *communia.*

Metaphysics is born, Aquinas implied, when physics discovers the existence of the immaterial prime mover. A new science comes into existence because the physical point of view is incapable of investigating the newly discovered immaterial mover. Accordingly, the meta-

exception; they both admit that for Aquinas metaphysical *esse* is know in a judgment. That theory is one I am unable to admit as a theory proper to the mature Aquinas: cf. Ch. VII below. An author with whom the present work agrees: R. GARRIGOU-LAGRANGE, O. P., "Notre premier jugement d'existence selon saint Thomas d'Aquin", *Studia Mediaevalia in honorem admodum Rev. Patris R. J. Martin, O. P.*, De Tempel Brugis, 1948, pp. 289–302.

[1] Mathematics, for example, has no interest in these two causes; cf. *In III Meta.*, 4, 375.

physician must first of all discover what point of view he is to take: how to look at all beings insofar as they are whatever they are? The most obvious way – in fact, the only way – is to look at the concept which expresses all beings as such. Yet the metaphysician does not yet realize what concept this is. If fact, he knows only one thing: there is an immaterial first mover, and hence the concept which expresses beings as such must not express all beings as if they were material; yet at the same time, this concept must not positively exclude materiality, for some beings are material. Hence, the metaphysician is forced to initiate a search for such a concept. Moreover, there appears to be but one possible procedure: examine our concepts to discover how they express realities *independent of their expression of the materiality* of the realities we know. When we discover how a concept expresses reality, we shall be in position to state the content of a concept expressing both material and immaterial reality. The only method of dicovering how a concept expresses reality is to examine our concepts in their very act of expressing reality – in predication, that is, in understanding or in apprehension.[1]

Hence, the metaphysician does not begin by looking at *communia*. Rather, he examines all predication, all uses of concepts in apprehension; through such an investigation, the metaphysician will realize what a concept must express if it is to express reality as such.

We can turn now to an examination of Aquinas' investigation of predication. (Earlier, in our study of the necessity of using a logical method, we noted as the correct interpretation: metaphysics studies the content of *communia*. This study would be carried through, it was argued, if a study of John as "man" could be a study of the manner in which "man" expresses the being (*ens*) of John. Insofar as we discover this point, our interpretation of "metaphysics must use a logical method" will be found consistent with all the other doctrines of the *Commentary* exposed in this chapter). *In V Meta., lectio* 9 gives us the

[1] This is one of the points that divides the present exposition from most others current at the present time. The *Commentary* demands that the metaphysician instigate a search for the concept expressing reality as such. Moreover, the doctrine of *In IV Meta., lectio* 2 demands that somehow the concept *ens* be considered as a strict concept, as was argued in Chapter IV. Thus *ens*, as a concept, is not an expression only of essence. For the opposite view see authors such as: GILSON, *Being and Some Philosophers*, on practically every page; OWENS, "The Accidental and Essential Characteristics of Being...", pp. 5–8; C. FABRO, *La nozione metafisica di partecipazione secondo S. Tommaso d'Aquino*, Soc. editrice internaz., Torino, 1950, 2 edit., pp. 137–44. All these authors are guilty of serious assumptions, and all quite frequently make very central affirmations without giving any textual justification for their work. P. Fabro is particularly guilty of this with his theories of "formal thought", "real thought", "intensive abstraction", and so on.

initial elements of a logical analysis resulting in the discovery of the formal object of metaphysics, and the concepts principally expressing that point of view; these elements are complemented by doctrines scattered throughout the *Commentary*. At times the logical character of the investigations is obscured by the method of exposition chosen by Aquinas – a paragraph by paragraph discussion of the topics presented by Aristotle. Thus, we find no step by step development of the investigation of predication in Aquinas' *Commentary*; we have rather a mixture of conclusions, steps to reach the conclusions, and declarations of intention – everything is there in the *Commentary*, but not in any orderly fashion. Hence, in our exposition, because we are interested in the gradual unfolding of metaphysics, we can not be content with a point by point discussion of Aquinas' words. Rather, we must carry through the examination of predication which seems to be demanded by the expositions of *ens* given here and there by Aquinas.

As metaphysicians our search is for a concept which will express both the immaterial, immobile mover and the material things around us. Let us line up all the possible types of propositions with an eye to discovering, first of all, what concepts are similar to the concept we want to form.

Propositions can be of the form: "John is a man" or "Socrates is a goat." Or again: "John is a musician." Or: "John is white." Then too, there is: "John is next to me," and so on. Since we want to discover a concept which not only has the proper *form*, but *reveals*, or tells us about, reality, it is not enough to discern different forms of propositions. Since, that is, our examination is not one of formal logic, but is rather an example of material logic, we must know what propositions tell us of reality. Our first question is then: What part of a proposition primarily communicates something about reality? The subject? The predicate? The word "is"? In "John is a man," obviously the subject "John" refers to reality; if it did not in fact, then the proposition as a whole would not do so. The predicate "man," if taken alone, certainly does not refer to reality, but because it is said of, or applied to "John," it does refer to reality and thus communicates something about it. This leaves "is": what does it do? At first it might appear that "is" accomplishes nothing. Some languages don't have such a verb, but should say simply: "John man." Hungarian, for example. Be that as it may, we are working with English (and even more important, with Aquinas' Latin). And in English (and in Latin) we say: "John is a man," and not: "John man." If we notice the relationship of: "John walks" and "John is walking" we get a clue to the task of "is." Obviously, we say or convey the same thing about "John" by both "walks" and "is walking." Hence, "is" seems to be a part of the predicate. Thus, it appears that a proposition refers to reality, and thus communicates something about it, if the subject refers to reality; the predicate of the form "is X," because it is applied to the subject, expresses what we wish to communicate about some reality.[1]

Now the metaphysician wants to form a concept which can be predicated of material beings and of the immaterial mover; this concept, when predicated, must be of the form "is X." To see what that concept must be, let us examine more closely predicates of the form "is X."

Sometimes we say: "John is white"; at other times: "John is a man"; and so on. An examination of these predicates should reveal many things, although it is impossible to ascribe any order to the appearances of these aspects. For the sake of convenience let us note first the distinction between propositions expressing what *happens to be* the case, and those expressing what *must be* the case. For example, when we say: "John is white," we could mean: "John happens to be white, but *qua* man he is not necessarily white," On the other hand, "John is colored" tells us something necessary about "John," for John as a material being, must have some color. This same distinction between propositions expressing what is necessary and what just happens to be, can be seen as a distinction between propositions which cannot be a conclusion through a syllogism beginning from an universal proposition, and those which can be so deduced. For example, a proposition such as: "John is colored" can be concluded syllogistically from: "All material things

[1] One will look in vain for an explicit text in the *Commentary* to prove that a proposition communicates something about reality whenever the speaker refers to reality by the subject. However, the closing lines of paragraph 896 would imply this: by admitting that the "is" of "Socrates is" can be taken (*accipiatur*) as referring to the truth of the statement, or (and) to the substance of Socrates, Thomas implies that the speaker can at least limit the primary goal of the statement by the reference given the subject. That "is" is a part of the predicate and not a verbal copula in an existential proposition, see *Ibid.*, 893: "Quia vero quaedam pradicantur, in quibus manifeste non apponitur hoc verbum Est, ne credatur quod illae praedicationes non pertineant ad praedicationem entis, ut cum dicitur, homo ambulat, ideo consequenter hoc removet, dicens quod in omnibus huiusmodi praedicationibus significatur aliquid esse. Verbum enim quodlibet resolvitur in hoc verbum Est, et participium" – That predicates of the form "is X" reveal something about reality in existential propositions, see *Ibid.*, 889–90, where Thomas divides reality in ten classes on the basis of the ten types of predicates; for example: "Primo distinguit ens, quod est extra animam, per decem praedicamenta" (889). "...Ens contrahatur ad diversa genera secundum diversum modum praedicandi, qui consequitur diversum modum essendi; quia "quoties ens dicitur", idest quot modis aliquid praedicatur, "toties esse significatur", idest tot modis significatur aliquid esse. Et propter hoc ea in quae dividitur ens primo, dicuntur esse praedicamenta, quia distinguuntur secundum modum diversum praedicandi" (890). "Unde patet quod quot modis praedicatio fit, tot modis ens dicitur." *Ibid.*, 893. Cf. *In XI Meta.*, 9, 2290; *In X Meta.*, 3, 1982.

are colored"; on the other hand: "John is white" cannot be deduced from any universal proposition.[1]

It is among the propositions that can be concluded from some universal proposition that we must search in our investigation of predicates. The reason for this is simple: the metaphysician, in seeking a concept which fits all things, is not interested in what things de facto happen to be (at least that is not his first interest); he seeks rather to construct a science, to know the necessity in things.

When he turns to investigate the propositions which express some necessary aspect of things, he discovers a great number of different types. In some, it appears, we express the *that which is* of the subject; in others, the how of the subject; in still others, the posture of the subject, and so on. From a listing of the different types, ten necessary aspects of material beings are revealed to us; that is, we discover that we have ten different ways of using predicates that express the real to us, and hence the real can be of ten different types. 1) Every material being is *that which it is*: "John is a man"; 2) every such being has certain *how*'s: "John is red"; 3) or *how much*'s: "John is five feet tall"; 4) *relations*: "John is next to me"; 5) is *in a place*: "John is in Africa"; 6) is *in time*: "John is here today"; 7) has *posture*: "John is standing"; 8) has a *having*: "John is clothed"; 9) is *acting*: "John is speaking"; and 10) is *being acted on*: "John is being hit."[2] (Here then is Aquinas' way of saying: every concept reveals the *particular* being of something. To discover how the concept expresses the *being*, we must remove the aspect of *particularity*).

Thus, there are ten types of these propositions; there are ten types of predicates which express some necessary aspect of material things. A further, very brief examination will reveal to us both an independence

[1] The propositions which express what merely happens to be the case are called *per accidens* propositions. "Ostendit quot modi dicitur ens per accidens; et dicit, quod tribus; quorum unus est, quando accidens praedicatur de accidente, ut cum dicitur, justus est musicus. Secundus, cum accidens praedicatur de subiecto, ut cum dicitur, homo est musicus. Tertius cum subiectum praedicatur de accidente, ut cum dicitur musicus est homo... In omnibus enim his, Esse, nihil aliud significat quam accidere". *In V Meta.*, 9, 886–87; cf. *Ibid.*, 888; *In VI Meta.*, 4, 1242. The propositions expressing what must be the case concerning an individual object – those which can be concluded from an universal proposition about a class, granted the individual object is a member of the class – these are what are called *per se* propositions.

[2] "...Ea in quae dividitur ens primo, dicuntur esse praedicamenta, quia distinguuntur secundum diversum modum praedicandi. Quia igitur eorum quae praedicantur, quaedam significant quid, idest substantiam, quaedam quale, quaedam quantum, et sic de aliis; oportet quod unicuique modo praedicandi, esse significet idem; ut cum dicitur homo est animal, esse significat substantiam. Cum autem dicitur, homo est albus, significat qualitatem, et sic de aliis." *In V Meta.*, 9, 890. Paragraphs 891–92 give the caracteristics of each of the ten different types.

of the first type of predicate (the type which expresses "that which is") as regards the other nine, and a dependence of the last nine on the "that which is" type. A predicate of the type "that which is" expresses something about the object represented by the subject insofar as that object is independent of everything else: by expressing what the object is, this type of predicate ignores all else but the object itself. Thus, "is a man" sets John apart from whatever is around him, in that it shows him to be standing on his own in the particular way that a man does. On the other hand, the nine remaining types of predicates always involve a reference to the "that which is" of the subject. "John is white": "is white" expresses a certain manner in which a "that which is" can be; but "is white" itself does not express a "that which is," as if John were what he is by being "white"; John and "white" are not identical as John and "man" are. Obviously the most important of these different types of predicates is the type "that which is."[1]

It is quite evident then that the metaphysician, in seeking a concept common to material beings and to the immobile being, is first of all trying to discover a concept of the type "that which is." Since this type of concept is the most basic of all, then this one at least must be implied in the two propositions: "A material being is," and "The immobile being is." Moreover, since objects which stand on their own are commonly called "things," or "substances," or "beings," our common concept is"" can be lengthened to read: "is a thing," "is a substance," or "is a being."[2] The metaphysician's question is then: What is the content of the predicate "is a thing," when this predicate expresses the "that which is" common both to material beings and the immaterial mover?

As soon as he asks the question, the metaphysician sees he already partially has the answer – or ,if he is a pessimist, that there is no answer. For as soon as he asks this question, the metaphysician realizes that one thing he can know about the "that which is" of an immaterial thing is

[1] "Sicut si aliquis definiat album, oportet quod dicat rationem hominis albi; quia oportet quod in definitione accidentis ponatur subiectum". *In VII Meta.*, 3, 1320. "Substantia enim quae habet quidditatem absolutam, non dependet in sua quidditate ex alio. Accidens autem dependet a subiecto, licet subiectum non sit de essentia accidentis; ...et propter hoc oportet quod subiectum in accidentis definitione ponatur...". *Ibid.*, 4, 1352. Cf. *In IV Meta.*, 1, 539; *In V Meta.*, 9, 894; *In VII Meta.*, 2, 1289.

[2] Thomas, in *In V Meta.*, 9, 896, speaks of the "is" of: "Socrates is" as a substantial predicate. He is referring to the type of predicate exemplified by "is a man", which is called in *Ibid.*, 890–91 a "substantial predicate", a predicate revealing the substance of an object. The "is" of: "Man is an animal" signifies substance. Hence it appears that the sentence: "The immobile mover is" can be lenghthened to read: "The immobile mover ia s substance", or: "The immobile mover is a being".

that an immaterial is a "that which is".[1] However, this knowledge is not sufficient; the metaphysician must continue his study.

The concept expressing both immaterial and material thing is, then, the concept "that which is". The metaphysician can now begin the construction of his science for he has now gained admittance to it: he has a point of view and he knows what that point of view is – "that which is." The next step is to begin the investigation of both immaterial and material beings from the point of view of "that which is." What does it mean for a material or an immaterial being to be a "that which is."? The most important answer to this question will be: "that which is" is composed of *esse* and essence.[2]

In running through the various types of propositions when he was separating those referring to reality in a necessary manner, the metaphysician could not but be struck by the strange type of proposition such as: "Humanity is in some way connected with objects such as John." Does this proposition say anything about reality? This is a question one would have to ask. As noted earlier, a proposition refers to reality if its subject does. "Humanity" does not directly refer to a reality as "John" would. Furthermore, "humanity" is not a concept expressing an object after the fashion of a "that which is"; nor is it like the other nine types of concepts – the *how*'s, the *how much*'s, etc. These latter concepts, even though they do not express a "that which is," at

[1] The question might arise whether the metaphysician actually knows that the immobile being has a "that which is". Can we really say that the metaphysician knows more than that the immobile being, as any other being, must be thought of as having a "that which is"? Do we know that our concept, in the case of the immobile being, reflects reality? Thomas certainly felt that if one knows the necessity of thinking of a being as a "that which is", one knows as well that a being is a "that which is." We have already remarked that Thomas, by the examination of the ways of predicating, knows that our use of predicates follows on ways of existing. Thus, since the metaphysician knows there exists an immaterial being, and since the most basic thing we say of anything is that it is a "that which is," Thomas does not hesitate to assume that all beings will be like the beings we know directly: that is, if to-be, as found in material beings, is to be a "that which is," then the same must hold true for anything which exists.

[2] In the actual construction of metaphysics, it is impossible to state the order in which one would discover the distinction of matter and form, and the distinction of *esse* and essence; the order of discovery would depend on what aspect of the use of substantial predicates one *happened* to notice first. If one happened to note first that the predicate can bear the burden of expressing all the formality of an object, then the first discovery is that of the matter-form composition of material beings. On the other hand, one's attention might chance, first of all, to be attracted by the fact that "man" can be predicated whereas "humanity" can not; thus one discovers *esse*-essence composition. Despite the impossibility of determining the order of discovery, one realizes, however, that it is the composition of *esse*-essence that must be discussed first in a logical development of metaphysics: the metaphysician's first task is to construct a concept common both to material and immaterial being; since such a concept is a substantial one, it will express both material and immaterial beings as *esse*-essence composites. It is only following the construction of the common concept of "being" that the metaphysician turns back to material being and then discusses their matter-form composition.

least they involve a reference to a "that which is". No, "humanity" is not a concept like the others we have noted, for it does not seem to involve in any way a "that which is". The strangest thing about "humanity" – or "whiteness," too – is that it pretends to be a "that which is"; that is, the mind conceives it as a *something*, as if it were an independent object after the fashion of "man" or "tree". Yet despite its pretentions of being a "that which is", there is nothing under heaven which seems to deserve the name "humanity"; we can never say: "X is humanity".[1]

Thus, although: "John is a man" is true, as is: "Humanity has something to do with John", we cannot say: "John is humanity". This impossibility of predicating "humanity" in a proposition revealing the "that which is" of John is quite enlightening, however. What is the difference between "humanity" and "man"? "Man" is a concept expressing "that which is", but "humanity" seems to reveal only "that by which a thing is a man". Does this not mean that the proposition: "John is humanity" is wrong because John is more than a "that by which"? Obiously this is so, for our concept "man" expresses much more about John than "humanity". A concept such as "man" expresses a "that which is" and not only a "that by which". However, "man" seems to include "humanity" in some way: "man" obviously expresses "that by which a man is a man" plus something else. But what is this something else, this other aspect? We must return to "humanity" and "man".

When we try to look at John both as "man" and as "humanity", we get an inkling of this "other aspect" present in "man". That object standing before us – can we understand it as "humanity"? No, because "humanity" is somehow incomplete, it does not express a totality, an unity, but only a part; an aspect of something. "Humanity" is something such as the part or principle of "man" which insures that "man" expresses John as a rational animal rather than as a rock, or as a horse. Thus, "humanity" appears to express all in virtue of which, or by which

[1] Humanity "...accipitur ut principium formale eius, quod est quod quid erat esse..." *In VII Meta.*, 5, 1378. "Humanitas... importat tantum principia essentialia hominis..." *Ibid.*, 1379. What Aquinas says of concepts such as "whiteness" and "white", "snubness" and "snub," can be applied to "humanity" and "man," if one makes allowances for the fact that "white" and "snub" are accidents or physical parts of substance, rather than substance itself. Thus: "Albedo enim etsi significet accidens, non tamen per modum accidentis, sed per modum substantiae. Unde nullo modo consignificat subiectum." *In V Meta.*, 9, 894. "...Oportet quod subiectum in accidentis definitione ponatur, quandoque quidem in recto, quandoque vero in obliquo. In recto quidem, quando accidens significatur ut accidens in concretione ad subiectum... Quando vero accidens significatur per modum substantiae in abstracto, tunc subiectum ponitur in definitione eius in obliquo, ut differentia; sicut dicitur, simitas est concavitas nasi." *In VII Meta.*, 4, 1352–53.

an object can be conceived as an intelligent animal. The "by which" again! Obviously John is more than this principle in virtue of which we are able to understand him as "man". John somehow is thought of as "having this by which"; he is conceived as living it, as acting it out. Much as Hamlet is a reality when Shakespeare's lines are acted out, so is John a "man" when we conceive "humanity" as exercised, or had, or lived out. Thus the unknown other aspect or other thing involved in "man" appears to be something like an "act of exercising humanity." "Man" as a "that which is" is "having humanity." In all such cases the "act of exercising" is always in proportion to the "that by which." The concept "thing," or "being," or "substance" (as commonly understood, or course) expresses then both "that by which" and the "act of exercising".[1] And so the metaphysician, through an examination of

[1] As was pointed out in the last footnote "humanity" expresses the formal or essential principles of the *quod quid erat esse*. – It will be objected that the reason why "humanity" can not be predicated of "John" is *only* because "humanity" expresses the essential principles of "man" and positively excludes the accidents, among which is individuating matter; since "humanity" excludes something which "John" is, namely his material parts, we can not say: "John is humanity." Thus, it will be argued, we are wrong in what we have noted as the root of the impossibility of predicating "humanity" of an object such as John. – The answer to this is somewhat long; fundamentally it amounts to this: the objection, even though founded on certain texts of the *Commentary*, is incorrect because first, it removes Thomas' words from the historical context of Platonism, and second, it overlooks an important doctrine expressed elsewhere in the *Commentary*. (The same objection could be made by quoting texts from *De ente et essentia*; such an argument does not interest us here as we are dealing with the Thomas of the *Commentary*.) First, the historical context of Thomas' treatment of concepts such as "humanity": the passages are especially *In VII Meta.*, 5, 1356–80; *In VIII Meta.*, 3, 1703–14. The key to the discussion in Book VII is in paragraphs 1362 – 63 where it is evident that Thomas (and Aristotle before him) is dealing with Platonic subsistent ideas – the essences which are separated from the objects whose essences they are. As paragraph 1367 expresses it, material things are what they are because they participate in the ideas. As was earlier explained by Thomas, there ideas are forms (cf. *In I Meta.*, 10, 153); and in material things which participate in the ideas, one must distinguish the nature (or the idea) and signate matter (or the principle of individuation) (cf. *Ibid.*, 155; *In III Meta.*, 3, 360.) In the discussion ot *In VII Meta.*, *lectio* 5, Thomas explains Aristotle's arguments against Plato in paragraphs 1356–77, and then he adds a note on his own in paragraphs 1378–80. It is in this addition of Thomas that one finds the information on "humanity" and "man" which is often interpretated thus: the only difference between the two concepts is that the first positively rules out accidents, while the second leaves open the possibility of accidents. But this interpretation makes Thomas say more than the context permits. Thomas, as he says, wishes to *add something which will show the reasonableness of the previous refutation of Platonism.* For Plato, the idea of "man" would not contain individual matter; it would be something like "humanity" which excludes this individual matter. But the Platonic idea of "man" and our idea of "humanity" would be similar in that neither could be predicated of an individual. The individual man, for Plato, was the form (idea) plus the individual matter; for us, the individual man is not "humanity" (the formal principles of a thing) plus the individual matter; since "humanity" *positively* excludes individual matter in its context. – Thomas does not, in these paragraphs, state or imply that he has given either *all* the differences between "humanity" and "man," nor *totally* explained why "humanity" can not be predicated of an individual. Thomas merely wishes to show that Aristotle was right in rejecting Plato's theory; to do this he adds something, as he says. Thomas' addition amounts to this: "We do not predicate "humanity" of man because "humanity" positively excludes individual matter; since Plato's idea of "man"

our modes of predicating, is led to posit in the concept of "being", not only a "that by which", but a proportionate "act of exercising" as well. Consequently, both material and immaterial beings are to be regarded as "that which is" where this expresses both "that by which" and an "act of exercising".

Does this mean we have reached a distinction of essence and *esse*? Does "that which is" *thus* express the composition of essence and *esse* in beings? No it does not, Aquinas would most emphatically insist. "Man" expresses more about John than "humanity" does; "man" can be said to express John as "having humanity", as "living out humanity".[1] Thus, "man" expresses the reality which John is, the reality identical to John. So "man" expresss the *ens* which John is. If then, "man" expresses "having humanity", where "having" is *esse* and "humanity" is essence, then *ens* would express *habens essentiam*. But this is definitely not the case with *ens*: it does not express *habens essentiam*.

A further consideration of "man" and "humanity" will reveal that

would exclude matter, it could not be legitimately predicated of a man, despite the fact that Plato thought it could be." Thus, Thomas seems to say, if Plato wants to continue holding his theory (that ideas express only the form of material things), then we will have to admit that "man" is not predicable of Socrates – a conclusion which would not at all suit Plato. – The second text in which Thomas treats again Plato's ideas is *In VIII Meta.*, 3, 1703–14. This text is often joined with the text we just studied (*In VII Meta.*, 5, 1356–80). Because the text of Book VIII explains that Socrates is not the same as "man" because Socrates has individual matter not contained in the specific term "man"; because this text also explains that if Socrates were not a material being, then the term expressing only the form of man would be identical with Socrates; because of these doctrines, one sees in this text from Book VIII a confirmation of the theory read into Book VII's teaching; "humanity," like the specific name of an immaterial being, positively excludes accidents such as matter; the term "man" differs from "humanity" because "man" is an open concept, it neither includes nor excludes the individual matter. What Thomas says in Book VIII is usually correctly interpretated, although it can easily be taken out of its context which then renders the doctrine slightly false. In Book VIII Thomas is dealing with the Platonic theory that specific names such as "man" express only form, even though the names are said of material objects. After answering this theory, Thomas wishes to illustrate the principles from which he took the wherewithal to refute Plato. The fundamental principle was that there is a difference between the specific essence of a material thing and the thing whose essence it is. (*Ibid.*, 1709). This is to be understood in the sense that the individualized essence of a material thing is not the specific essence alone, but has individual matter not included in the specific essence (*Ibid.*, 1710). Thus it is that specific names of material things do not express an essence which is identical with the thing defined. Thomas' theory is to be interpretated in relation to Platonism we have already noted several times. Just because Thomas says that, if "man" signified only a form, then there would be no difference between the specific essence "man" and John, it does not follow that a name like "humanity" can not be predicated *only* because of some relation to matter. Cf. *In VII Meta.*, 11, 1536. For interpretations opposed to the present one, cf. FOREST, *La structure métaphysique du concret...*, pp. 74–75; 93; 95–96. G. VAN RIET, "La notion centrale du réalisme thomiste: l'abstraction," *Problèmes d'épistémologie*, Nauwelaerts, Louvain-Paris, 1960, pp. 11–12; 14–16; 19–21; 34.

[1] "Humanitas... non est omnino idem cum homine, quia importat tantum princ ia essentialia hominis, et exclusionem omnium accidentium. Est enim humanitas, qua homo est homo... Hoc autem ipsum quod est homo, est quod habet principia essentialia..." *In VII Meta.*, 5, 1379.

it is the human intellect that has formed "humanity" as the means of expressing all that is important in "man." Originally, we conceive John as "that", as "something"; we conceive him as "rational animal," a "be-souled intelligent glob of matter", or something else of this type. It is only later when we are faced with some necessity of expressing the difference between John and something such as a tree, that we form a concept such as "humanity". To express how John is different from a tree, we formalize all the intelligibility of "man" in the concept "humanity". "Humanity" does not express just the form or soul included in "man"; quite the contrary, the material aspects of "man" are included. Yet these material aspects are expressed "formally": the mind pretends to gather together, to extract, all the intelligibility from "man" and to express this in "humanity."[1]

It is only consequent to the formation of "humanity" that the intellect comes to regard "man" as *habens humanitatem.* Thus, to see in "man" the content "that which has humanity" is a more sophisticated activity than the understanding of "man" as "rational animal." It was yet a still more complex, more unusual activity that was involved in our comparison of "man" and "humanity," the comparison which enabled us to conclude to the presence of two elements in "man": "act of exercising" and "that by which." Hence, we must remember that these two elements, expressed by *habens humanitatem,* are parts of "man" *due to our slightly unusual fooling around with concepts.* "Man," as that concept is spontaneously formed by us, does not express "that which has humanity." Let us not then argue from our sophisticated view of "man" to a division in John. The natural way of conceiving John is "rational animal" and not *habens humanitatem.*[2] In fact, we must recognize that "that which has humanity" is more an expression of the concept "man" than an expression of John. *Habens humanitatem* is conceived as sharing – as participating – in *humanitas.*[3] It would be quite incorrect to argue from the sophisticated intricacies of our concepts to a real distinction in John between a "that by which" and an

[1] "Humanitas enim non respondetur quaerenti quid est homo... Sed tamen humanitas accipitur ut principium formale eius, quod est quod quid erat esse..." *Ibid.*, 1378.

[2] We do not discover Aquinas speaking of "natural" and "sophisticated" acts of knowledge, but: "Sciendum est... quod quod quid est esse est id quod definitio significat... Non igitur est quod quid est esse hominis humanitas quae de homine non praedicatur, sed animal rationale mortale." *Ibid.*, For the relation between the abstraction of "form from sensible matter" and "universal from particular," see *In VIII Meta.*, 1, 1683; *In VI Meta.*, 1, 1157 and 1161.

[3] "Humanity" participates, or is conceived as participating in nothing: "Quod enim totaliter est aliquid, non participat illud, sed est per essentiam idem illi." *In I Meta.*, 10, 154. Yet "man," or *habens humanitatem,* is conceived as participating in "humanity": "Quod vero non totaliter est aliquid habens aliquid aliud adiunctum, proprie participare dicitur." *Ibid.*.

"act of excercising". Yet our investigation into the difference between "man" and "humanity" has not been completely valueless. For we do know that the reality of John is somehow identical to *his* humanity. Everything we express by "humanity" is somehow found in John. John appears to be shot through and through with humanity, even though we have no right to think that humanity is a part or principle of John. Let us pursue this line of thought.

John is certainly not anything other than the man who stands before us. The being (the reality) of John and the manhood (the humanity) of John are one and the same. Whether we say: "John is a man", or "John is a that which is," we are referring to the same object. But are we knowing the same thing about that object in these two cases? Do "is a that which is" and "is a man" express the same content of knowledge? Since the reality of John is his manhood, if "being" ("that which is") expresses John's reality, and if "man" expresses John's humanity, then obviously, "being" and "man" have the same content whenever we *use* them to express our knowledge of John. (That is, once we know what "John" is as "being" and as "man," then to know John as "being" and as "man" are the same).[1] Thus for John, "is" and "is a man" are the same; "to be" and "to be a man" are identical for John. Accordingly, as expressions of John, "man" and "being" are, respectively, "that which has humanity" and "that which has to be". Hence, whereas "man" is expressed as *habens humanitatem*, "being" is expressed as *habens esse* – certainly not as *habens essentiam.*[2]

[1] The knowledge we have when we know John as "man" or as "being" would be the same, once we have grasped what *ens* means. Aquinas says this in slightly different words when he explains that the metaphysician, by knowing an object as *ens*, would know the particular essence of that object: "Et quia eiusdem est considerare de ente inquantum est ens, "et de eo quod quid est," idest de quidditate rei... ideo etiam aliae scientiae particulares "nullam mentionem", idest determinationem faciunt de eo "quod quid est," idest de quidditate rei, et de definitione, quae ipsam significat. Sed "ex hoc," idest est ipso quod quid est ad alia procedunt..." *In VI Meta.*, 1, 1148. Cf. *In XI Meta.*, 7, 2249–51. For a slightly different theory see L. B. Geiger, O. P., "De l'unité de l'être," *Revue des Sciences Philosophiques et Théologiques*, XXXIII, 1949, pp. 3–14. P. Geiger appears to base his exposition on the distinction between "reference" and "meaning"; both the concepts "man" and *ens* would refer to everything in a concrete object. Yet both concepts do not have the same meaning. Through the concept "man," we seize the quiddity, the object as ordained to a special sort of activity. But through the concept *ens*, we see a determined manner of exercising the act of being (pp. 7–8). – If P. Geiger is correct, then how can an examination of concepts such as "man" permit us to realize the meaning of the concept by which we express reality as such? Cf. *In V Meta., lectio* 9.

[2] "To be" and "to be a man" are the same for John. This would follow from the text noted in the preceding footnote as well as from the discussion of the predication of *ens*, *unum*, and *res*: cf. *In IV Meta., lectio* 2; cf. Chapter IV, section 1, B. This identity of "to be" and "to be a man" in any given man is ultimately the reason why Aquinas could say :"...unumquodque habet esse per suam quidditatem." *In VI Meta.*, 1, 1148. "Cum enim Socrates incipit esse homo, dicitur simpliciter quod incipit esse. Unde patet quod esse hominem significat esse

Thus we say that in the case of a real thing such as John, the thing can be expressed as "that which has to be" and as "that which has a particular kind of to be". For John, "to be" is a "type of to be". The type of thing which John is impregnates his "to be", or vice versa. One can not separate John's "to be" and John's "to be a man". (One should note that here we are not arguing from the content of our concept to separate principles in John as we would be doing were we to admit that the content of man – *habens humanitatem* – meant there were two principles in John. Here we are merely using our concept insofar as it represents the content of the intellectual grasp of John as a being. We are no so much arguing "from the content of man" as "looking through the content".)

"To be John" is John's "to be". *Vivere viventibus est esse*, it has been said. A "to be" for any given object is, then, a particular "to be". "To be" is, moreover, the "actualizing" which always takes on a particular mode according to the particular object whose "to be" it is. That actualizing which is continuously going on in John is his "to be". Naturally this actualizing is the one proper to a man, and to this particular man named John. Thus when we express John as "being", as "that which is", we express him as "that which has actualizing" – *habens esse*. And when we say: "John is a man", we express "that which has an actualizing of the type proper to humanity".[1] "To be", then, is the "actualizing". In John's case, "John's actualizing" is the best we can do to express the particularity of his "to be" known when we know him as *habens esse*.

We can get an even better grasp of the "actuality" or "actualizing"

simpliciter." *In VII Meta.*, 1, 1256. This same doctrine is expressed by the very familiar: "Vivere viventibus est esse." – I have the impression that P. Fabro would regard *vivere* as a type of *esse essentiae* in potency to an *esse existentiae*; cf. FABRO, *La nozione metafisica di partecipazione*..., pp. 189–96; esp. the distinction on p. 192 between 1) the first notion of being: *ens commune*; 2) the later notions of being, namely, a) the being which means an essence that exists, and b) the being wich means an act of essence, and c) the being *ut verum*; and 3) the still later notion of being which is the synthesis of every formality. It is notion number 3) that is the foundation of the metaphysical notion of participation; hence, one concludes, this is what Aquinas would have referred to as *esse*. But then what is notion number 2, b? Is it something like *vivere* which is still in potency to being number 3? – That *ens* means *habens esse* is clear: "Nam ens dicitur quasi esse habens, hoc autem solum est substantia, quae subsistit." *In XII Meta.*, 1, 2419. This *quasi* offers a problem. Does it mean *ens* almost has an *esse*? This would appear to be the position of those who accept the predicate *ens* as an expression only of essence; cf. GILSON, *The Christian Philosophy of St. Thomas Aquinas*, pp. 40–41; 448, note 30. This theory, however, has no basis if one admits that Aquinas rejects the Avicennian judgment of *esse*. – Does *quasi* mean that *ens* has an "almost *esse*"? Obviously not, for although Aquinas might have said that an accident has a "quasi esse," he could not have said as much of substance. Hence, the *quasi* must modify *dicitur*: *Ens* is almost defined (said) as *habens esse*.

[1] "To be" has been replaced by "actualizing". The footnotes on p. 266 justify this.

expressed in our concept of "man" or "being" if we distinguish between a concept such as "man" considered as an universal, and the same concept used in knowing John. The two concepts are actually the same; their content does not vary in the two cases. However, when we define the universal concept, we pay no attention to the actuality expressed in the concept. We pay attention only to what are usually termed the "intelligible notes." To define is, thus to give the intelligible notes as if we had conceived them as *merely possible*. At the moment, we are not defining the universal concept "man"; rather we are using the concept "man" in knowing John. If we reflect a moment on such knowledge, if we reflect on: "John is a man," it becomes evident that "is a man" does not express something John might *possibly* be, but something he *actually* is conceived to be. Thus even thought we initially said "man" is *habens humanitatem*, it is better to say "man," as used in knowing John, is *habens actualem humanitatem*. Or even: *habens esse-hominem*. When we know John as "man", when we apprehend him as "man", we conceive him as an actual rational animal, and not as a merely possible. (To conceive John as a merely possible man would be to do something which experience tells us is impossible).

Have we reached the distinction of *esse* and essence yet? No, Aquinas would say once again, although it has been hovering in the background for pages. The *esse* of John is the *esse-hominem*, the "humanity-actualizing". Because we conceive John as "that which has an humanity-actualizing" whenever we say: "John is a man", we must ask ourselves more definitely what it is in John that we refer to by all the elements of our concept: "that which has humanity-actualizing". We have already noted that the two elements "that which has" and "humanity" are elements of our concept "man" in relation to "humanity". But what exactly do we refer to in John by "that"? And what relation does the "that" have to John's humanity?

Of course, "that" refers to the whole of John. The "which has" merely expresses the necessity of seeing that John is not purely and simply what it means to be humanity: John is not what humanity would be, if it existed. Nevertheless, John is impregnated by humanity, for we understand John as "actual man"; we understand him as if humanity were somehow actually found in every aspect of him. Thus it is that we cannot say that we are understanding John as "actual man" plus "definite bones and blood and flesh". Moreover, we do not understand the blood and bones as if they were something covered over with humanity or with "actual man". Rather, in our thought, "actual man" is far from being superimposed on definite bones and blood and flesh. Just the contrary, to say: "John is a man" is to entertain an intellectual gaze directed toward a definite pile of bones and blood; yet the content of that gaze, expressed as "actual man" involves blood and bones, which are not definite. Moreover, that content involves blood and bones and flesh impregnated by actual man, by actually exercised humanity. All the intelligibility of "actual humanity" is understood as penetrating blood and bones. Thus although we understand John as "that which has actualized humanity", this amounts to "actual material object impregnated with humanity".[1]

The point of this discussion is simple. The difference between "humanity" and "man" would have supplied Aquinas with the knowledge that *ens* is *habens esse*, or "that which has actuality". Then, when Aquinas looked closer at the "that which has", he realized even more clearly that it is an element of his thought. The "that which has" does not refer to the determinate matter of John, nor to anything else in John. Thus, Aquinas sees that he is gazing at a determinate material object as "actual matter impregnated with humanity".

Thus John, when understood as *ens*, is understood as impregnated with *esse* or actuality. Since John is a man, he is seen as impregnated with *esse-hominem* or with actual humanity. It is at this point that the *esse*-essence distinction can appear.

Up to this point we have been careful to examine the manner in which our concepts could express the reality of an object; thus, we have investigated predication – the use of concepts in apprehension. But to discover the distinction of *esse* and essence, we must turn for a moment to a judgment.

When we conceive John as "man," when we understand him as impregnated with actual humanity, we still do not know whether or not John exists.[2] When the intellect carries through its pre-judgmental reflection which follows the understanding, it grasps the here-and-now existence of the "man" in John: prior to expressing the truth of our apprehension of John, the intellect sees the here-and-now of the actual humanity which impregnates the entirety of John. Thus the intellect becomes fully aware that in understanding John as "man", it saw the

[1] This is implied by the texts on "humanity" given in footnotes 1, p. 262–3, p. 265.

[2] "Et sicut nulla scientia particularis determinat quod quid est, ita etiam nulla earum dicit de genere subiecto, circa quod versatur, est, aut non est. Et hoc rationabiliter accidit; quia eiusdem scientiae est determinare quaestionum an est, et manifestare quid est. Oportet enim quod quid est accipere ut medium ad ostendendum an est." *In VI Meta.*, 1, 1151. Cf. Ch. V, section 1, C above.

lack of any necessity of existing. For the first time, the intellect becomes aware that the following is the case: although we conceive John as "actual material object impregnated with humanity", we do not understand any necessity of John's actually being as we have conceived him to be. Even more, we do not know any such necessity, precisely because we do not find it in what it means to be a man.[1] At this moment we are in a position to distinguish between *esse* and essence.

Our knowledge of John as "man" expresses a) all that it *means* to be a man; yet in all that this means, there is no necessity of being actual – of actually being. Moreover, our knowledge of John as "man" expresses all that it means to be a man b) as if it were *actual* – not as if it were non-actualized, whatever that might mean! Thus in the content of our knowledge of man we can distinguish "whatever it means to be a man" and the "actuality of all that man means." In other words, "man" does not only express all that we need to *understand*, to *define* John as "man". "Man" expresses more than this; "man" expresses as well the *actuality* of all that "man" means.

"Actuality" is what we have often referred to as "to be", as *esse*. "All that *man* means" is essence. These two aspects are present in the concepts which express a "that which is".[2]

[1] In so far as we discover in being as such some need of a cause of *esse*, we must see that the *esse* is not part of the essence of anything around us. This is discussed further in sub-section E below, where the proof of God is treated.

[2] "Esse enim rei quamvis sit aliud ab eius essentia..." *In IV Meta.*, 2, 558; "...unum et ens, quod predicantur per se et non secundum accidens de substantia cuiuslibet rei." *Ibid.*, 554. As was argued at the end of Ch. IV, section 1, *ens* expresses the essence as real, or as related to *esse*. As was noted in footnote 2, p. 266, I do not think P. Fabro would agree to interpret the essence-*esse* distinction as one between "all that means it to be an X" (essence) and "the actuality of all that it means to be an X" (*esse*). Nor do we believe our explanation conforms with that of F. Van Steenberghen, "Le problème de l'existence de Dieu dans le "De ente et essentia" de saint Thomas d'Aquin," *Mélanges J. De Ghellinck, S. J.*, Tome II: *Moyen-âge, époques moderne et contemporaine*, Duculot, Gembloux, 1951, pp. 837–47; not with that of L. De Raeymaeker, "L'idée inspiratrice de la métaphysique thomiste," *Aquinas*, III, 1960, pp. 61–82. M. Van Steenberghen explains, for example, that there is a three-fold distinction to make in *esse*: 1) the "is" of a judgment of existence; 2) the *esse* which is a principle of composed being; and 3) the *esse* which is the transcendantal perfection common to all beings, intrinsic to them, and expressed in all substantial concepts. (pp. 841; 846). I do not think the *Commentary* would admit the distinction between *esse*'s 2 and 3, for as Aquinas writes: "...unumquodque habet esse per suam quidditatem." *In VI Meta.*, 1, 1148: this is *esse* in the sense of a metaphysical principle. "Cum enim incipit esse homo, dicitur simpliciter quod incipit esse. Unde patet quod esse hominem significat esse simpliciter." *In VII Meta.*, 1, 1256: what is the *esse* of *esse hominem*? It is the *esse* of: "Nam ens dicitur quasi esse habens, hoc autem solum est substantia, quae subsistit." *In XII Meta.*, 1, 2419. And this *esse* is the substantial principle. We find in the *Commentary* no use of *esse* in the sense of a transcendental perfection that is expressed in every concept of a substance, and which is not the metaphysical principle of the substance. Hence, I do not see how one can distinguish between the *esse* expressed in the concept of a substance, and the metaphysical principle, except as an idea is opposed to an extra-mental object.

Yet much earlier we had spoken of "man" as *habens humanitatem*, and of *ens* as *habens esse. Ens*, we noted, is "that which has actuality". The *esse* or actuality is conceived as impregnating the determined matter of an object. Thus "to be man" impregnates the matter of John. Now, however, we see that in any *habens esse* there is a distinction to be made: in John, for example, there is all that an *esse-hominem* means (emphasis on *means*), and then there is the very actuality of it. Hence, any time we gaze at a determinate material object as *habens esse*, we must realize that we *conceive* of the determinate object as a) impregnated with the *meaning* of its particular *esse*, and b) as *actually* impregnated. Thus, to understand anything as *ens* is to understand it as "actual essence" which needn't be actual. But what needn't be actual and yet is actual, receives its actualization from without; when we understand that in the essence there is no necessity to be actual, we know that the extra-mental essence receives its actuality from without. Accordingly our understanding of *ens* is no longer "actual essence," but "actualized essence."[1] Thus, the extra-mental *ens* is known to be composed of *esse* and essence.

"Actualized essence" expresses the causal effect worked on essence by *esse*. Thus, "actualized" is not an expression of *esse*. Rather, "actualized essence" is an expression of *ens*. "Actualized" emphasizes the result *esse* works on essence. It is the *esse* which imparts perfection or actuality which is had by an object, by an *ens*. Thus, *ens*, or "actualized essence", or "that which is", all express a composite of essence and *esse*; yet they do not express the composing principles as the separate words *esse* and "essence" can. *Ens*, and its synonyms, express the result of the union of *esse* and essence *Ens* expresses "actualized essence".[2]

[1] Aquinas would have said: "that which is" participates in actuality because it itself is not actuality: "Quod vero non totaliter est aliquid habens aliquid aliud adiunctum, proprie participare dicitur." *In I Meta.*, 10, 154. Because a real *ens* is participating in *esse* or actuality, does it follow that the concept *ens* must express this participation? If the concept *ens* expresses an object as it is in itself, then the concept should express the participating character of the object. Thus, *ens* has been defined as "actualized essence". The passive element in "actualized essence" expresses a "reception of actuality from something else"; thus this passive element expresses *ens* as sharing or participating. – In speaking of the relation Plato set between ideas and sensible beings, Aquinas gave a doctrine which appears to apply to the present situation: "...nomen quod per participationem praedicatur, dicitur per respectum ad illud quod praedicatur per se, quod non est pura aequivocatio, sed multiplicitas analogiae. Si autem essent omnino aequivoca a casu idea et substantia sensibilis, sequeretur quod per unum non potest cognosci aliud, sicut aequivoca non se invicem notificant." *Ibid.*, 14, 224.

[2] It is interesting to note the attacks directed by Dietrich of Freiberg against Aquinas' distinction of essence and *esse* as found in the *De ente*. Apparently Dietrich arrived in Paris in 1276 and wrote, after that date, as follows: we can not understand an essence such as "man" without understanding it as "actually existing man"; the fact that we do not understand whether or not any given man actually exists does not alter the fact that our understanding

All this Aquinas discovers by his logical analysis of predication. In brief, the principle steps of the investigation were as follows:

1) Some concepts used, or formed, in apprehension express what an object is, but need not be; others express what an object need be.
2) There are ten different categories, or ways to express what a material object need be. Of these ten, nine refer to, or include a reference to, the type expressing the "that which is" of a subject.
3) Accordingly, our concept of material and immaterial being must be "that which is"; an example of such a concept would be "man" or "that which has humanity"; here "humanity" is conceived as a "that by which".
4) *Habens humanitatem* gives the general form of "that which is" concepts. As regards these concepts in which we include the "having" aspect, we must be aware that our way of expressing all the perfection of an object is to formalize that perfection in an abstraction, e.g. "humanity"; only after the expression of such an abstraction, we relate concepts such as "man" and "humanity". The former concept is then defined as "having humanity". Thus we begin to conceive the object such as John as "sharing" in a formality which he, John, is not.
5) Yet the reality of an object such as John is identical to his humanity. The "to be" of John is "to be a man", even "to be John". Thus it is more correct to say we think of John as "having an actual humanity" than as "having humanity". *Ens* therefore is *habens esse*.
6) Thus "to be" is always a particular type of "to be" in the world around us. "To be" or "actualizing" always takes on a particular mode according to the particular object whose "to be" it is.
7) Whenever we conceive a thing, we conceive not just the "intelligible notes" of a possible essence, but the intelligible notes as actual. "Man" is *habens actualem humanitatem*.
8) When we know John as "that which has actual humanity", "that" refers to the whole of John. "Which has" shows the necessity of stating that John is not what humanity would be if it existed. Rather, John is saturated with humanity.
9) Because we think of John as "actual humanity" or "actual man" when he needn't be actual, we realize the distinction between a) all that it means to be John (essence) and b) the actuality of all that John means (*esse*).
10) Because *esse* is had when it need not be had, the essence is actualized. *Ens* means "actualized essence".

Even more briefly, this entire examination of predication can be explained as the successive realization of the content of "man" and *ens*. Thus one would proceed through the following stages of understanding:

1) *ens* – "that which is"
2) "*man*" – "that which has humanity"
3) *ens* – *habens esse* and not *habens essentiam*

is expressed as "actually existing man"; consequently, the distinction between understanding what man is, and judging that he is, reveals only that we have two ways of knowing, not that a given man is composed of *esse* and essence. (Cf. A. MAURER, C. S. B., "The *De Quidditatibus Entium* of Dietrich of Freiberg and its Criticism of Thomistic Metaphysics," *Mediaeval Studies*, XVIII, 1956, pp. 173–74.) When Aquinas rejects Avicennian *esse* in the *Commentary*, he de facto agrees with Dietrich. However, Aquinas retains his own theory of *esse*, the actuality of an essence.

4) *"man"* – "that which has actual humanity"
5) *"man"* – "actual object impregnated by humanity"
6) *ens* – "object impregnated by actuality" or "actual object"
7) *"man"* – "actual essence"; yet John needn't exist.
8) *ens* – "actualized essence"

Thus the end of the first stage in our investigation of predication has been reached. This rather unusual procedure has been followed because Aquinas indicated that metaphysics begins with such an investigation. Although he didn't say that metaphysics was to look for the concept of *ens*, the concept expressing the formal object of this science, that would seem to be the logical reason for the investigation. But I need not rest my case solely on the strength of this "logical reason," for the following aspects are also to be found in Aquinas' *Commentary*: 1) the presentation of the concept "that which is", as well as the concepts expressing the other nine categories, is given in the context of a logical investigation of predication; 2) most of the statements about *esse* or *ens* or essence are given in terms of an analysis of predication; 3) the final form of the concept of *ens* is a passive one; that is, the very understanding of that concept involves that we admit the dependence of *ens* on Pure *Esse*.

Although I believe my procedure has been justified, it would be ridiculous to characterize this discovery of "actualized essence" as that of Aquinas. One can never know exactly how Aquinas would have proceeded, that is, what would have been the content of each successive step in the investigation. His statements, his conclusions, only hint at the procedure to follow. The most that can be said of the reconstruction in this sub-section is that it is in the spirit of Aquinas' metaphysics.

We have reached the end of the first stage of our study of predication. Yet there are additional stages to be noted, which even if less important nevertheless reveal the importance Aquinas attached to the logical method in metaphysics.

In running through the various types of propositions which reveal something about reality, we noted there were only ten types. Nine of the ten referred to the proposition expressing the "that which is" of the subject. It does not take very much thought before the metaphysician realizes that it is the materiality of some beings which ultimately makes possible the nine types of concepts, and the nine types of propositions, which refer to the "that which is". Thus these nine types will never be used in relation to the immaterial being. Is there any reason then to study these nine types? If metaphysics were only the study of all beings from the point of view of "that which is", then of course there would be no reason to study these nine types of concepts and propositions. But, metaphysics is the study of *ens qua ens*, of all reality as such, and hence of the nine types of reality expressed by the nine types of concepts which refer to the "that which is" type. Thus we must be interested in the reality expressed by the concepts "quality", "quantity", and so on.

But as the metaphysician soon realizes, there is very little to say about these concepts, and so about the nine types of accidents revealed by these nine types of propositions. For example, one can say that the concept of "quality" expresses the "how" of a subject, e.g.: "John is white".[1] The concept of "white" expresses a "how", to be sure, but the "how" of a substance, of a "that which is".[2] The concept of "whiteness", moreover, can not be predicated of John because in "whiteness" the "how" is conceived as if it were something.[3] Thus, although concepts such as "man" and "humanity" are related as a whole and a part, concepts such as "white" and "whiteness" are related as two manners of conceiving the same "how". By looking at the knowledge: "John is white", and at the impossible: "John is whiteness", we realize that "white" expresses our knowledge of the "how" of "that which is". We can distinguish *esse* and essence in our "white", but the *esse* and essence belong to the "that which is" which supports "white". Hence, there is no reason to consider this "how" as anything but a modification of an *esse*-essence composite.[4] Other than these few items of doctrine, the

[1] *In VII Meta., lectio* 1 offers abundant proof that when we know substance, (when we know the meaning of "is a being" where "substance" can replace "being") then we sufficiently know the nature of being. For example: "In prima ostendit quod ad determinandum de ente, prout in decem praedicamenta dividitur, oportet determinare de sola substantia... Illud quod est primum inter entia quasi ens simpliciter et non secundum quid, sufficienter demonstrat naturam entis; sed substantia est huiusmodi; ergo sufficit ad cognoscendum naturam entis determinare de substantia." *In VII Meta.*, 1, 1246. When we study the other predicaments we learn only something such as: quality is a "quale quid"; we learn only something about the disposition of a substance: "Illud enim est primum secundum cognitionem, quod est magis notus et magis manifestat rem. Res autem unaquaeque magis noscitur, quando scitur eius substantia, quam quando scitur eius quantitas aut qualitas. Tunc enim putamus nos maxime scire singula, quando noscitur quid est homo aut ignis, magis quam quando cognoscimus quale est aut quantum, aut ubi, aut secundum aliquod aliud praedicamentum. Quare enim de ipsis, quae sunt in praedicamentis accidentium, tunc scimus quid est ipsum quale, scimus qualitatem, et quando scimus quid est ipsum quantum, scimus quantitatem. Sicut enim alia praedicamenta non habent esse nisi per hoc quod insunt substantiae, ita non habent cognosci nisi inquantum participant aliquid de modo cognitionis substantiae, quae est cognoscere quid est." *Ibid.*, 1259.

[2] "Licet autem modus essendi accidentium non sit ut per se sint, sed solum ut insint, intellectus tamen potest ea per se intelligere, cum sit natus dividere ea quae secundum naturam coniuncta sunt. Et ideo nomina abstracta accidentium significant entia quae quidem inhaerent, licet non significent ea per modum inhaerentium." *Ibid.*, 1254.

[3] *Ibid.*, Also: "Albedo enim etsi significet accidens, non tamen per modum accidentis, sed per modum substantiae. Unde nullo modo consignificat subiectum." *In V Meta.*, 9, 894.

[4] "Et quia accidentia non videntur entia prout secundum se significantur, sed solum prout significantur in concretione ad substantiam, palem est quod singula aliorum entium sunt entia propter substantiam." *In VII Meta.*, 1, 1256.

metaphysician can say nothing about the concepts expressing "how". Nor can he speak of the other eight types of categories which refer to "that which is" except to say something parallel to what has been said of the "how" type.

At this point we know rather well how concepts express reality. There are ten different types, we can say, but nine of them refer to the tenth or the "that which is" kind. Most important of all, we see now that both material beings and the immaterial first mover can be called "that which is". Moreover, we know that the content of this idea is "actualized essence". Thus we imply the distinction of *esse* and essence in substantial beings: *esse* is the actuality of essence. But are we ready to apply that concept of "actualized essence" to the first mover? In a sense we are, for we know now that to think of "actualized essence" is not to conceive materiality; this is so, even if we can think of a material John as "actualized essence". However, before we continue this line of thought any further, we must note another aspect of material beings which has been revealed by our study of predication: namely, the distinction of matter and form in material objects.[1]

D. The investigation of predication: the distinction of matter and form

(In this sub-section we need not reconstruct Thomas' exposition after the fashion of our study of "being"; that is, we need not reconstruct the analyses Thomas would have made if, etc... We need not reconstruct this analysis, because Thomas has left us his personal one.)

When the metaphysician begins his investigation of the substance of material beings, to follow Thomas' words, he considers material beings

[1] At this point it should be clear why the separation is not invoked as a privileged instrument in the formation of the concept of "being". The *Commentary* indicates that "being" is man's first concept, and that the movement in metaphysics consists in discovering what it is in reality that corresponds to "being". Thus our initial concept of being is expressable as "that which is", just as the final concept of being is. However, at the beginning of man's intellectual life, he is inclined to have no suspicions that matter is not one of the principles of being as such. According to the *Commentary*, when the physician discovers the existence of an immaterial mover, then the philosopher knows that matter can not be one of the principles of being as such. Yet in all this movement, he does not need to change the verbal expression of "being"; "being" can always be expressed as "that which is". It is evident that the separation or negative judgment is used by the philosopher in his attempt to discover what it is in reality that corresponds to "being". The most important use, moreover, is the judgment that all being need not be material; hence, "to be is not to be material". Yet by this judgment the philosopher is not "forming" a concept that he had not already "abstracted". To abstract is to gaze at an object with a meaning-full gaze. That is, there is some content to this gaze. And in the case of "being", the content of the gaze is "that which is". From the very first step in his intellectual life, man gazes at objects as "that which is"; but it is only much later that he comes to realize that, in a that which is, one must distinguish *esse* and essence. The theory of separation is judged unacceptable, because it tends to hide the fact that man abstracts "being" as naturally as he breathes.

as able to serve as logical substances – as subjects of predication. This is not the end of the investigation, Thomas notes; it is but the beginning, for a logical definition of substance gives us a common characteristic of things which will serve as a means to know the real principles of material beings.[1] This logical definition to which Thomas refers can be expressed: "that which is not predicated of a subject, and of which everything else is predicated".[2] Thus, to be a logical substance is to be the subject of predication. Thomas' method is, then, to examine material beings insofar as they can serve as the subject of predication.

As we have already noted, predication is of two types. First of all, there is what is called *per se* predication; the proposition expressing this type contains a predicate which states what *must be* the case concerning an individual object – the proposition is one which can be concluded from an universal proposition about a class. We have already examined this type of proposition, our examination resulting in the discovery of the *esse*-essence composition of beings. There is nothing more to be gained from direct examination of such propositions, unless we realize that a *per se* proposition where the predicate is a substantial one (a "that which is"), can be turned into a *per accidens* one. *Per accidens* propositions, it will be remembered, express what merely happens to be the case. An example of a *per accidens* predication would be: "The man is musical";[3] an exampe of a *per se* proposition when the predicate is a substantial one: "John is a man".[4]

A *per accidens* proposition can be characterized in yet another way; one might say that the definitions of the subject and the predicate are totally different. For example, in: "The man is musical", "man" is defined as "rational animal", whereas "musical" (that is, "musical thing") is defined as "that which has something to do with music". Thus, not only men can be musical, but also instruments, theaters, sheets of paper, and so on.[5]

[1] "...nunc dictum est quid sit substantia "solum typo," idest dictum est solum in universali, quod substantia est illud, quod non dicitur de subiecto, sed de quo dicuntur alia; sed oportet non solum ita cognoscere substantiam et alias res, scilicet per definitionem universalem et logicam: hoc enim non est sufficiens ad cognoscendum naturam rei, quia hoc ipsum quod assignatur pro definitione tale, est manifestum. Non enim huiusmodi definitione tanguntur principia rei, ex quibus cognitio rei dependet; sed tangitur aliqua communis conditio rei per quam talis notificatio datur." *In VII Meta.*, 2, 1280.

[2] *Ibid.*, Cf. also *Ibid.*. 1273.

[3] Cf. *In V Meta.*, 9, 886.

[4] Cf. *Ibid.*, 891, where the example is: "Socrates est animal."

[5] Thomas speaks of propositions where the subject and the predicate have different definitions: cf. *In VII Meta.*, 2, 1287–90. Although he does not say these are *per accidens* propositions, he would have said that given the doctrine of *In V Meta.*, *lectio* 9.

"John is a man": beginning with such a *per se* proposition, let us transform it into a *per accidens* one such as: "The man is musical". For example, "John is a man": first, we must transliterate this as: "this animal is a man". Now it is evident that the subject is included in the definition of the predicate. If we remove all formality from the subject and place it all in the predicate, we would have a proposition where the definitions of subject and predicate are different. Here, "John is a man", would read: "This is a man", or "This materialed thing is a man". Here the subject "this materialed thing" must be understood as if it were an *esse*-essence composite, where the essence is pure matter. The predicate "man" has as definition something such as "rational, sentient, loving being". Thus in: "This materialed thing is a man", all the emphasis is placed on the predicate. Thus, Thomas says, the ultimate subject – of a proposition, and likewise in the real thing – if considered in what it is essentially, is neither substance, nor quantity, nor any of the other ten categories of being. Thus matter – the ultimate subject – is totally distinct from form in the essence of material beings.[1]

One will have noticed a difference between the exposition of the investigation that yields the notion of "being", and the exposition, just finished, that leads us to the recognition that matter is not form. Whereas in the former we spoke often of the manner in which we "intellectually gaze" at a material being, in the exposition yielding the distinction of matter and form we spoke only of propositions. The ultimate cause of this difference is that the latter exposition has been carried through on paper by Aquinas, while the more important discovery of "being" was only hinted at. The exposition of the discovery of matter and form can be easily transposed into more psychological terms. For example, we can imagine a comparison to be made between John and other objects. If we compare John and a dog, we begin from the knowledge that they are both "animal". Now, if we "intellectually gaze" at John, the animal, and at Fido, the animal, with the goal of determining how they

[1] "Attamen diversitatem materiae ab omnibus formis non probat Philosophus per viam motus, quae quidem probatio est per viam naturalis Philosophiae, sed per viam praedicationis, quae est propria Logicae... Dicit ergo, quod oportet aliquid esse, de quo omnia praedicata praedicentur; ita tamen quod sit diversum esse illi subiecto de quo praedicantur, et unicuique eorum quae de "ipso praedicantur", idest diversa quidditas et essentia... Sicut enim haec est vera: homo est albus, non autem haec: homo est albedo, vel: humanitas est albedo, ita haec est vera: hoc materiatum est homo, non autem haec: materia est homo, vel materia est humanitas. Ipsa ergo concretiva, sive denominative praedicatio ostendit, quod sicut substantia est aliud per essentiam ab accidentibus, ita per essentiam aliud est materia a formis substantialibus. Quare sequetur quod illud quod est ultimum subiectum per se loquendo, "neque est quid", idest substantia, neque quantitas, neque aliquid aliud quod sit in aliquo gendere entium." *In VII Meta.*, 2, 1287–89.

are different, we will ultimately arrive at a stage where the "gaze" directed toward John is filled with "having humanity" and that directed toward Fido is filled with "having dogness." Then we can compare John and an oak tree. We begin by knowing that both objects are living. Thus, we first must "gaze" at John, the living being, and at the tree, the living being. Gradually we will come to the realization that John is "having animality" while the tree is "having treeness" or having plantness". If we continue comparing John with less perfect beings, we will ultimately arrive at the stage where we are comparing him with a lump of rock, for example, or a lump of coal. What is the difference between the two beings? Our gaze toward John is filled with "having living, sentient, rational animality," while the lump of coal is viewed as "having coalness." In other words, we are "gazing" at the bare un-*formed*, materiality of John and of the lump of coal. Every bit of form possessed by the two objects is expressed in the concepts corresponding to the content of our "gazes" at the objects. Then, following the realization that John is an individual material object in front of us, as well as the realization that George is another object viewable in the same manner as John – following the realization that "man" is not found in one object, we conclude that there is more to John and to George than the formality of "man". There must be something, completely other than form, which limits form. There must be prime matter. Thus Aquinas' exposition is easily transformable into the same type of expressions used in the investigation of "being."

What we have done, following Thomas' lead, is to have concluded the distinction of matter and form through an investigation with its roots in the logical definition of substance. Thus we understand why Aquinas wrote: *Definitione logica substantiae* "...aliqua communis conditio rei per quam talis notificatio [id est notificatio naturae rei materialis] datur".[1] It should be evident as well, that we have but continued the investigation of material being as "that which is".

When we began our investigation of predication, we hoped to find something of help in our search for knowledge common to all beings – material and immaterial. The distinction of matter and form has not aided us in that search. Yet since metaphysics must be the legislator of other sciences – since it must clarify the concepts used by them (the *communia*) – this discussion has not been in vain: because of it metaphysics legislates that each science, dealing with any material substance,

[1] Cf. *Ibid.*, 1280.

must distinguish a formal and a material element when it illustrates through sense knowledge the essence of its subject.

It is interesting to compare the discussion given above with Aquinas' refutation of the theory which would see in the logical parts of a definition a reference to real parts of material beings. This comparison illustrates a most important aspect of Aquinas' "probatio... per viam praedicationis";[1] this proof is not from *formal* logic, but rather is based on a study of *material* logic.

In a definition there is room to speak of a genus, a species, and a specific difference. When we say, for example: "Man is a rational animal" we can speak of the species "rational animal", the genus "animal", or the specific difference "rational".

Even though we have predicated "rational animal" of man, we should not conclude to the reality of two parts in men: one corresponding to "animal" (genus) and the other to "rational" (difference). The reason is, of course, because both "rational" and "animal" are predicated of man. If they were parts of man, parts combining to form man, then of course they could not be predicated of man.[2] (Note that we have looked to the contents of the concepts, and to the validity of the predication of these contents – this is material logic.)

On the other hand, Thomas writes, the parts of the definition – genus and difference – can be said to refer to the parts of real men: that is, "animal" is taken from a certain part of man, as is "rational", although each is taken from a different part. These parts are matter and form, respectively. Matter is that which must be informed, it is the subject of the form. "Animal", in its role in the definition, resembles matter; the genus of "animal" must be determined to a certain species by the difference "rational". Thus "rational" resembles the form, which gives the ultimate perfection to matter.[3]

If Thomas were interested in proving a distinction between real parts of beings solely because of a certain form present in his thought, that is,

[1] Cf. *Ibid.*, 1287.

[2] "Quod autem hic dicitur, quod sicut se habet definitio ad rem, ita se habet pars definitionis ad partem rei, videtur habere dubitationem. Definitio enim est idem rei. Unde videtur sequi quod partes definitionis sint idem partibus rei; quod patet esse falsum. Nam partes definitionis praedicantur de definitio, sicut de homine, animal et rationale; nulla autem pars inegralis praedicatur de toto." *In VII Meta.*, 9, 1462.

[3] "Sed dicendum est, quod partes definitionis significant partes rei, inquantum a partibus rei sumuntur partes definitionis; non ita quod partes definitionis sint partes rei. Non enim animal est pars hominis, neque rationale; sed animal sumitur ab una parte, et rationale ab alia. Animal enim est quod habet naturam sensitivam, rationale vero quod habet rationem. Natura autem sensitiva est ut materialis respectu rationis. Et inde est quod genus sumitur a materia, differentia a forma, species autem a forma et materia simul. Nam homo est, quod habet rationem in natura sensitiva." *Ibid.*, 1463. Also: *In VIII Meta.*, 2, 1696–98.

because of a form of his thought, then real men would be said to be composed of "animal" and "rational." But because Thomas is interested not only in the form of his thought, but in the *validity* of its *content* as well, he is involved in material logic. Thus when he realized the validity of: "This materialed thing is man", and "That materialed thing is animal", and "This materialed thing is rational", he realized that the content of his thought corresponded to the composition in reality. On the other hand, he realized that the validity of predicating of "man" the contents "animal" and "rational" meant that real men are animal and rational, and that thus animal and rational are not real parts of men. (And because he recognized the validity of saying: "John is a man", where "man" is "actualized essence", the expression of an essence which need not be actualized, Aquinas realized the *esse*-essence composition of reality.)[1]

[1] Thomas' thought is well expressed in *In IX Meta.*, 11, 1896–98; e.g.: "Oportet enim veritatem et falsitatem quae est in oratione vel opinione, reduci ad dispositionem rei sicut ad causam. Cum autem intellectus compositionem format, accipit duo, quorum unum se habet ut formale respectu alterius: unde accipit id ut in alio existens, propter quod praedicata tenentur formaliter. Et ideo, si talis operatio intellectus ad rem debeat reduci sicut ad causam, oportet quod in compositis substantiis ipsa compositio formae ad materiam, aut eius quod se habet per modum formae et materiae, vel etiam compositio accidentis ad subiectum, respondeat quasi fundamentum et causa veritatis, compositioni, quam intellectus interius format et exprimit voce. Sicut cum dico, Socrates est homo, veritas huius enunciationis causatur ex compositione formae humanae ad materiam individualem, per quam Socrates est hic homo: et cum dico, homo est albus, causa veritatis est compositio albedinis ad subiectum: et similiter est in aliis. Et idem patet in divisione." *In IX Meta.*, 11, 1898. It is interesting to note in this connection, that M. A. Forest does not mention the proof based on the manner of predicating when he discusses Thomas' thought on the distinction of matter and form. Even though M. Forest cites lesson two of Thomas' exposition in Book VII three times, he never mentions the paragraphs discussing the logical proof (paragraphs 1287–90). Cf. Forest, *La structure métaphysique du concret...*, pp. 100; 211–12. One can not help but wonder over this omission. Perhaps M. Forest considered this exposition to be merely a resumé of Aristotle's thought. If M. Forest had read the *Metaphysics* of Aristotle in conjunction with Thomas' words, he would have noted 1) many additions and developments made by Thomas and 2) that Aristotle ends by rejecting the entire discussion of the logical investigation, whereas Thomas does not. – Another indication that M. Forest, for some reason, failed to grasp the spirit of Thomas' metaphysics appears in the exposition given in pp. 72–97 of his work. There M. Forest in discussing the relation between universal concepts and extra-mental beings, notes Thomas' undying opposition to the Platonistic doctrine which tended to identify the parts of a definition with the parts of the reality defined. Avicebron, from his *Fons vitae*, proposed this doctrine, thus asserting that realities are really composed of genus and specific difference. For Thomas, as M. Forest points out, such a doctrine is nonsense. M. Forest does not note, and such a discussion would have provided the perfect background for noting, that the transposition made by Avicebron and rejected by Thomas is one involving only formal logic. Avicebron, and Thomas in refuting him, made no mention of what material logic could tell us about the real. However, in the discussion of *In VII Meta.*, *lectio* 2, Thomas is clear what material logic (although he doesn't call it by name) can teach us: namely, the distinction of matter and form. – The only work we know which has followed Thomas' proof of matter and form based on predication is C. Hart, *Thomistic Metaphysics. An Inquiry into the Act of Existing*, Prentice-Hall, Englewood Cliffs, 1959, p. 122.

E. The discovery of the existence of God

Before we discuss the proofs for the existence of God, it is best to listen to Aquinas' words on the characteristics of our knowledge of the Divine Being.[1] Whenever we ask a question, notes Thomas, we seek some new knowledge about something already known. For example, we can ask: "Why does it thunder?" In such a case we have heard a lot of noise, and we ask for the formal cause of the noise: hence the answer to our question is: "sounds in the clouds."[2]

Or we could ask for the efficient and final causes, the reason why a certain form is present in matter. For example: "Why are these stones and pieces of wood a house?" The answer desired: "Because John placed them in such a way that they are a house"; and "He did this to have a place to live."[3]

In all these questions we have either known a material result (thunder) and asked for the formal cause of that being (of the resulting being); or else we have seen a matter-form composite, recognized its form, and consequently asked for its agent and its end. But what of the beings that have no matter? As Aquinas says: such substances are either totally known by us, or else totally ignored.[4] Thus when we turn to a study of immaterial substances we must proceed differently than in our sciences of material beings. In regard to material substances, we begin our sciences with a *doctrina*, with knowledge of the essence of our subject which serves as the middle term of our demonstrative syllogisms. But we do not have that *doctrina*, that knowledge of the essences of immaterial beings.[5]

[1] As we have already noted, one discovers the existence of the extrinsic causes of beings after the discovery of the intrinsic causes. Cf. *In I Meta.*, 2, 46; see footnote 1, p. 244.

[2] "...In omnibus quaestionibus quaeritur aliquid de aliquo, sicut de materiae causa, quae est formalis..." *In VII Meta.*, 17, 1669. "In quaerendo autem propter quid de aliquo, aliquando quaeritur causa, quae est forma in materia. Unde cum quaeritur, propter quid tonat? Respondetur quia sonitus fit in nubibus: hic enim constat quod aliud de alio est quod quaeritur. Est enim sonitus in nubibus, vel tonitruum in aëre." *Ibid.*, 1656.

[3] "...In omnibus quaestionibus quaeritur aliquid de aliquo, sicut de materiae causa, quae est formalis vel vcausa formae in materia, ut finis et agens..." *Ibid.*, 1669. "Aliquando autem quaeritur causa ipsius formae in materia quae est efficiens vel finis, ut cum quaerimus propter quid haec, scilicet lapides et lateres, sunt domus? In ista enim quaestione... [quaeritur] propter quid huiusmodi sunt domus." *Ibid.*, 1657.

[4] "...Ex quo in omnibus quaestionibus quaeritur aliquid de aliquo... palam est, quod in substantiis simplicibus, quae non sunt compositae ex materia et forma, non est aliqua quaestio. In omni enim quaestione, ...oportet aliquid esse notum, et aliquid quaeri quod ignoramus. Tales autem substantiae, vel totae cognoscuntur, vel totae ignorantur, ut in nono infra dicetur. Unde non est in eis quaestio." *Ibid.*, 1669. Cf. also *In IX Meta.*, 11, 1901–1909.

[5] "Et propter hoc de eis [idest, de substantiis simplicibus] etiam non potest esse doctrina, sicut est in scientiis speculativis. Nam doctrina est generatio scientiae; scientia autem fit in nobis per hoc quod scimus propter quid. Syllogismi enim demonstrativi facientis scire, medium est propter quid est." *In VII Meta.*, 17, 1670.

Thus before we even begin our study of immaterial beings, we appear to be defeated. Yet this is not really true, for in our consideration of immaterial beings we can receive some help from the *doctrina* which played a role in the genesis of the sciences of material beings. However, since we use the *doctrina* of the physical world, we force ourselves to adopt a new style of questioning the world if we wish to achieve knowledge of non-physical beings.[1] Since this *doctrina* consists of knowledge of the essences of physical things, we can no longer pose questions asking for the formal cause of individual physical beings, nor questions asking for the agent and final causes of these beings as individual beings.[2] Rather, Thomas writes, we must prove the existence of immaterial beings as the final and efficient causes of there being any formal, material, final or efficient causes within the physical world; that is, we must discover the causes of "why there is being rather than nothing in the world around us." In this sense, then, a new style of questioning the world is had.[3]

It should be noted, however, that these theories about a new "style of questioning the world" are the theories elaborated by Aquinas, the philosopher who already knows where his science of metaphysics is going; this is Aquinas the teacher who is speaking, and Aquinas who is striving to construct a metaphysics. Nevertheless, his words can be of value to us.

If we have been correct thus far in our reconstruction of Thomas' metaphysics, then the theories on the new style of questioning involved

[1] "Sed ne videatur consideratio talium [immaterialium] sbustantiarum omnino aliena esse a physica doctrina ideo subiungit, quod alter est modus quaestionis talium." *Ibid.*, 1671.

[2] Because we have a *doctrina* relating to the physical world, it does not follow that we can not ask for the formal cause of *some* physical being. Because we know, for example, that "life" is "a principle of immanent activity in an organic body," it does not follow that we can not ask why (that is, seek the formal cause why) a *certain* material thing performs a certain action; the answer to this "why" will be that the thing possesses a certain type of substantial form or a soul. Yet in asking such a question we will not learn of the existence of God.

[3] "In cognitione enim harum substantiarum non pervenimus nisi ex substantiis sensibilibus, quarum substantiae simplices sunt quodammodo causae. Et ideo utimur substantiis sensibilibus ut notis, et per eas quaerimus substantias simplices. Sicut Philosophus infra, per motum investigat substantias immateriales moventes. Et ideo in doctrinis et quaestionibus de talibus, utimur effectibus, quasi medio ad investigandum substantias simplices, quarum quidditates ignoramus. Et etiam patet, quod illae substantiae comparantur ad istas in via doctrinae, sicut formae et aliae causae ad materiam. Sicut enim quaerimus in substantiis materialibus formam, finem et agentem ut causas materiae; ita quaerimus substantias simplices ut causas substantiarum materialium." *In VII Meta.*, 17, 1671. "Primae autem substantiae non cognoscuntur a nobis ut sciamus de eis quod quid est... et sic in earum cognitione non habet locum causa formalis. Sed quamvis ipsae sint immobiles secundum seipsas, sunt tamen causa motus aliorum per modum finis; et ideo ad hanc scientiam, inquantum est considerativa primarum substantiarum, praecipue pertinet considerare causam finalem, et etiam aliqualiter causam moventem." *In III Meta.*, 4, 384.

in our knowledge of immaterial substances should be interpreted thus: in our search for a concept common to both material and immaterial beings, and in our further reflection on material substance with an eye to learning something more about being as such, we are suddenly struck by a certain lack in material substances – we are suddenly struck by the fact that the immaterial being, discovered by the physician as first cause of movement, is itself – or else another immaterial being is – the uncaused cause of being, the necessary source of all. The question facing us, then, concerns the lack of intelligibility found in the world of our experience: what is it?

In the first place, there is more than a suggestion in Aquinas' words that the lack of intelligibility is not a lack in one material substance alone, but in all of them as a whole:

> Et etiam patet, quod illae substantiae immateriales comparantur ad istas materiales in via doctrinae, sicut formae et aliae causae ad materiam. Sicut enim quaerimus in substantiis materialibus formam, finem et agentem ut causas materiae; ita quaerimus substantias simplices ut causas substantiarum materialium.[1]

Just as the final, formal, and efficient causes are what we search for in our study of material substance where we are looking for the causes of individual globs of matter as we find them, so a similar statement holds for the study of immaterial substances: taking the material world as we find it, we seek the causes, which, while extrinsic to it, make it as a whole whatever it is.

Granted then that the lack of the intelligibility is a lack shared by the material universe as a whole, in what does it consist? As Thomas himself says, we want to study the causes and principle of whatever is called *ens per se* – of "that which is" and we want to study these causes and principles insofar as they are the causes of the "that which is" of things.[2] Some of these causes and principles have already been studied: we know now that every being has an essence as well as an *esse*, distinct from and proportionate to its essence. Perhaps this question could arise: What are the efficient and final causes of the essence-*esse* composite of material thing? Or better still: Taking the universe of material, composed beings as a whole, is there anything in the fact of *esse*-essence

[1] *In VII Meta.*, 17, 1671.

[2] "Unde si determinetur sufficienter illud genus entis quod continetur sub praedicamento, manifestum erit et de ente per accidens, et de ente vero. Et propter hoc huiusmodi entia praetermittuntur. Sed perscrutandae sunt causae et principia ipsius entis per se dicti, inquantum est ens." *In VI Meta.*, 4, 1244.

composition which requires us first, to look for a cause, and second, to look outside the universe of material, composed beings?[1]

It appears worthwhile to consider a paragraph which tempts one to say that Aquinas would answer "yes"; there is a paragraph which appears to assert the presence of "something" in the fact of *esse*-essence composition of material beings (or of beings as we know them) which requires the existence of a cause of that composition. That "something" appears to be the very fact of *esse*-essence composition. Thomas appears to affirm: whenever one finds composition, one finds the effect of a cause – "... necesse est ut omnia composita et participantia, reducantur in ea, quae sunt per essentiam, sicut in causas."[2] Since we know that material substances are composed of *esse* and essence, this general statement given by Thomas seems to mean that there must be a substance which is not composed, whose essence is its *esse*. This doctrine of Thomas is readily understandable if we reflect on the fact that we know (and we *must* know) material being as "actualized essence": the essence of material being is *actual* even though it needn't be; hence its essence is *actualized*.

There is a difficulty connected with the use of this principle, however. It is indeed true that Aquinas gives it as the reason which necessitates that we assign to the heavenly bodies a cause of *esse* as well as a cause of motion.[3] But even granting this meaning given to the principle in its context, Thomas applies it in a way which proves only that there is some immaterial being which is the cause of the generation of material beings. That is, Thomas appears to prove, not that the cause is an immaterial being whose *esse* is its essence, but only that the matter-form composition of material beings must have as cause an immaterial being. Let us examine Thomas' words; immediately following the principle quoted above, he writes:

Omnia autem corporalia sunt entia in actu, inquantum participant aliquas formas. Unde necesse est substantiam separatam, quae est forma per suam essentiam, corporalis substantiae principum esse.[4]

[1] It seems worthwhile to note that we ask if there is any lack of intelligibility in the universe of *material* beings. That is, we neglect for the moment the immobile being whom the physician discovered. We are busily engaged in an examination of the material universe in an effort to discover what we can say about all beings, about the immobile and the mobile. We will soon discover that the immobile mover does not have the same *esse*-essence relation we found in material beings, and which we predicated of the immobile mover, at least by implication, when we first acquired our concept of "being."

[2] *In II Meta.*, 2, 296.

[3] The heavenly bodies "... habent causam non solum quantum ad suum moveri, ut quidam opinati sunt, sed etiam quantum ad suum esse, ut hic Philosophus expresse dicit. Et hoc est necessarium: quia necesse est ut omnia composita et participantia, reducantur in ea, quae sunt per essentiam sicut in causas." *Ibid.*, 295–6.

[4] *Ibid.*, 296.

By "corporalia," there is little doubt that Aquinas refers both to heavenly and to earthly beings.[1] Because these beings have composed essences, they must be caused by some being which is "forma per suam essentiam." Does such a being have an *esse* distinct from its essence, in the sense that the *esse* is said to be proportionate to, and the actuality of, the essence after the fashion of the beings of our experience? Unless the term "forma per suam essentiam" means a being whose essence is its *esse*, then Thomas does not seem to prove here the existence of a being whose essence is its *esse*. Perhaps also in the section preceding the statement of the principle (concerning the reduction of composed beings to simple ones), Thomas by *causa essendi* meant only a cause of the presence of forms in matter. Moreover, do we know that this principle of the reduction of composed beings is absolutely universal, that is, that it applies to all compositions? Unless it is, we have not found in this section a statement leading us to see in the *esse*-essence composition a lack which requires a cause outside the universe of material beings.

Beginning with the last two aspects of the problem, there does not seem to be any doubt that the principle should be taken in the sense of a universal reference to all composition – that of *esse*-essence as well as that of matter and form. The *causa essendi* seems to be meant as a cause of every aspect of things. Otherwise there would be no need for Thomas to write: "...habent causam non solum quantum ad suum moveri, ut quidam opinati sunt, sed etiam quantum ad suum esse..."[2] Nor would there be any reason for Thomas to refer to Aristotle as explicitly teaching that all things have a cause of *esse*: "...habent causam ...quantum ad suum esse, ut hic Philosophus expresse dicit."[3] This reference, when related to a later one referring to Aristotle's God as the cause of *esse*, makes it clear that Thomas intends at least to imply here the reduction of *esse*-essence composition to pure *esse*.[4]

Yet if we emphasize Thomas' words: "...substantiam separatam, quae est forma per suam essentiam, corporalis substantiae principium esse," our conclusion is less helpful. "Forma per suam essentiam": a

[1] Cf. *Ibid.*, 295 where heavenly bodies are mentioned; a fortiori this principle of the reduction of composed to simple beings would hold for earth-bound bodies.

[2] *Ibid.*.

[3] *Ibid.*.

[4] "Unde sunt causae entium secundum quod sunt entia, quae inquiruntur in prima philosophia... Ex hoc autem apparet manifeste falsitas opinionis illorum, qui posuerunt Aristotelem sensisse, quod Deus non sit causa substantiae caeli, sed solum motus eius." *In VI Meta.*, 1, 1164. – P. Geiger would agree that *In II Meta.*, 2, 296 is a statement that the composition of *esse* and essence in creatures is to be reduced to the Pure *Esse*. Cf. GEIGER, *La participation dans la philosophie de St .Thomas d'Aquin*, pp. 26–27, footnote 3. However P. Geiger would not agree with the understanding of this proof for God as set forth here. See footnote 1, p. 287.

similar term is sometimes used by Aquinas: "ens per seipsam." Are these two identical? It would seem so. When he speaks of "ens per seipsam," Thomas is speaking of a substance which has no matter; thus a "being through itself" is opposed to a "being through its form," that is, to a being composed of matter and form.[1] The term "forma per essentiam" would seem to be a term applicable to an "ens per seipsam": both terms express the fact that the being is immaterial. It is in this sense that Thomas in Book VIII argues against Plato that when the specific name of a substance expresses form alone, then the substance in question is its essence, that is, the specific essence and the individualized essence are identical.[2] Thus a proof of the existence of a "forma per suam essentiam" is not a proof of God, it is not a proof of a being whose essence is its *esse*.

Thus we are faced with this problem: 1) although Thomas gives a principle which appears to be universal, and 2) although it is given as the justification for assigning a *causa essendi* to heavenly bodies, 3) Thomas concludes to nothing more than a being whose essence is its form.

Thus depending on whether one wishes to emphasize 1) the "forma per suam essentiam," or 2) the "causa essendi" and the reference to Aristotle, one has 1) no proof for God and a principle which is not universal, or 2) a proof for God, and an universal principle. In the light of the polemic between Aquinas and the God-centered philosophers which has been underlined in the past chapters, it appears necessary to emphasize the reference to Aristotle; consequently, we feel obliged to accept the second alternative mentioned above: Aquinas has given a proof for God based on the necessity of reducing all composition to something simple.

How do I think Aquinas would have justified such a necessity, such an universal principle? The answer lies in what has been said regarding our knowledge of *esse* and essence. It was pointed out that we understood John as "man," as a thing impregnated with "actualized humanity," with "actualized human essence." As Thomas would be quick to insist: Since there is truth in the knowledge expressed as: "John is a

[1] "Omnis autem substantia vel est ens per seipsam, si sit forma tantum; vel si sit composita ex materia et forma, est ens per suam formam...." *In III Meta.*, 4, 384.

[2] "Sic igitur patet, quod si nomen speciei significet formam tantum, cuiuslibet rei idem est quod quid erat esse et esse suum; sicut homo erit quod quid est esse suum, et equus, et omnia huiusmodi. Si autem nomina speciei significant compositum ex materia et forma, tunc non idem erit rebus quod quid erat esse earum." *In VIII Meta.*, 3, 1711. In this context *esse suum* does not refer to the metaphysical principle of the being, but to the individualized essence of the being: *esse suum* is the *ens*.

man," where "man" is "actualized human essence," it follows that John is not a being who is actual by right. Rather he has received his actuality from some other being: his *esse* comes from a source outside himself. Ultimately, we must trace received *esse* back to some being whose essence is *esse*. In other words, because we understand all things around us as "actualized essence", we must look for some unactualized, but actual being, an *esse* or an actuality which has not been actualized.[1] Hence we are not forced to justify this proof of Pure Actuality or Pure *Esse* by justifying the principle of "participation," the principle of the reduction of the composed and participating to the simple. Rather, because we discovered that the basic reality of things, the "that which is," has an aspect of passivity, we must posit the existence of a purely active being. *Esse*, as seen in the beings of our experience, is the actuality of essence. In this respect, there is no passivity in *esse*; all the passivity is in essence. Yet *ens*, "that which is," is not "actuality"; rather, "that which is" is "actualized essence" – *ens* is passive precisely insofar as an actual essence need not be actual. What is actual although it need not be actual is actualized. Thus *ens* is dependent in its totality on something other than itself. One might be tempted to demand that we justify the principle: "What is actual, when it need not be actual, has been actualized." Such a principle cannot, of course, be justified. It is a law of thought which must be accepted if one is to think at all: if we know an object as actualized, that is, as passive, and if our knowledge

[1] This is the sense in which one must understand "analogy" of the name "being" in so far as there is an analogical use of "being" of both God and creatures. Thus, Aquinas would mean the following when he speaks of "predication *per participationem*" and "predication *essentialiter*": "being" is predicated of the essence of a creature, for here we are dealing with *per se* predication as opposed to *per accidens*; yet "being" as a *per se* predicate used of a creature expresses that the creature only has *esse*, not that it is *esse*; hence men must look for something whose essence is *esse*; when said of something whose essence is *esse*, that is of God, "being" becomes a *per se* predicate that is used *essentialiter*: to express His essence as *esse*, not as having *esse*. – P. Fabro would agree with this view that "analogy between God and creatures" is a mere expression of the mechanics of the use of "being" in the proof of God as Pure *Esse*; cf. FABRO, *La nozione metafisica di partecipazione secondo S. Tommaso d'Aquino*, p. 189, footnote 2. However, that is as far as agreement between his position and ours extends, for there is nothing in common between his view of metaphysical "being" and that presented here. There is not sufficient space in a note to discuss the reasons for disagreeing with P. Fabro. It must be sufficient to note that 1) I do not think he has proved his view on "being" (pp. 189–96) and 2) the present work is based on the *Commentary*, a work which P. Fabro has almost entirely overlooked. – P. Geiger would not at all agree with this present proof of God. The composition of *esse* and essence is dependent on the logically prior composition of matter and form; it is because an essence is limited, that it limits *esse*, he says. Hence the real problem is prior to the limitation of *esse* by essence. Cf. GEIGER, *La participation dans la philosophie de S. Thomas d'Aquin*, p. 26, footnote 3; pp. 213–17; esp. p. 198, footnote 2. – More will be said on this problem in the general conclusion below. For a criticism of P. Geiger's views, see: C. FABRO, *Participation et causalité selon S. Thomas d'Aquin*, Nauwelaerts, Louvain-Paris, 1961, pp. 63–73.

is true, then we are forced to maintain the existence of some totally actual or active being, a being in no sense passive. (It hardly seems necessary to point out that this proof through "participation" is a proof through "causality." Since "being" is conceived as "actualized essence," it is conceived as *dependent* on the source of its actualization.)

This proof of Pure *Esse* can be described as based on an intellectual necessity of seeking the pure X in virtue of which material beings are said to have X. The same "intellectual necessity" appears in another exposition of the *Commentary*. (And hence, I am inclined to believe that the proof of the *causa essendi* discussed above, the proof based on the necessity of reducing the composed and participating to the simple, was a proof of God.) In this new exposition, although it is not a new proof but rather the same proof in different terms, Aquinas notes that all objects falling into the genus of substance are called "substance"; the fact that we call things "substance" forces us to seek that thing which is first called "substance".[1]

Thomas' treatment of this necessity is straight-forward. He gives the example of colors: we are forced to seek the first color, the color which measures all other colors. This first color is "white," Thomas notes, and all other colors are colors insofar as they are related to "white."[2] In the same way there is some substance which is first among all substances. Moreover, it is necessary that we seek that first substance.[3]

But need this first substance be that whose essence is "to be"? Aquinas it appears, would answer affirmatively. For what we are seeking

[1] The general principle is enunciated as follows:"...oportet quod unum similiter se habeat in omnibus generbius, quia ens et unum aequaliter de omnibus generibus praedicantur. In omnibus autem generibus quaeritur aliquid quod est unum, quasi ipsa unitas non sit ipsa natura quae dicitur una; sicut patet in qualitatibus et in omnibus generibus non est sufficiens dicere, quod hoc ipsum quod est unum, sit natura ipsius quod unum dicitur; sed oportet quarere quid est quod est unum, et ens." *In X Meta.*, 3, 1967. The application to the predication of "substance" is had in paragraph 1973: cf. footnote 3. See also*In III Meta.*, 10, 465.

[2] "Et quod in qualitatibus et in quantitatibus oporteat quaerere quid est quod dicitur unum, manifestat per exempla. Et primo in coloribus. Quarerimus enim aliquid quod est unum, sicut album quod est primum inter colores. Unde si in quolibet genere est unum id quod est primum, oportet quod album sit unum in genere colorum, et quasi mensura aliorum colorum; quia unusquisque color tanto perfectior est, quanto magis accedit ad album. Et quod album sit primum in coloribus, ostendit, quia colores medii generantur ex albo et nigro, et ita sunt posteriores. Nigrum etiam est posterius albo, quia est privatio albi, sicut tenebrae privatio lucis. Non autem sic est intelligendum, quod nigrum sit pura privatio, sicut tenebrae; cum nigrum sit species coloris, et per consequens natura coloris in eo servetur; sed quia in nigro est minimum de luce, quae facit colores. Et sic comparatur ad album, sicut defectus lucis ad lucem." *In X Meta.*, 3, 1968.

[3] "Et sicut in coloribus cum dicimus unum, quaerimus aliquem colorem qui dicatur unum; sic in substantia cum dicimus, unum, necesse est quaerere aliquam substantiam, de qua dicatur ipsum unum. Et hoc primo et principaliter dicitur de eo, quod est primum in substantiis (quod inquiret inferius); et per consequens de aliis generibus etc." *Ibid.*, 1973.

when we seek the first substance is the measure of all other substances.[1] And since for us to be called "substance" is to be said to be a composite of *esse* and essence, to be said to be "actualized essence," the substance which measures us will have to be the measure of the actualizing, or actualization. And just as the standard measure of one meter is not *said to be itself* one meter long – if it were so said, then it itself would be measured – so too the first substance cannot be thought of as "actualized", but it can only be thought of as the measure of the "actualization of others." Just as a piece of string, by coming closer and closer to equal the measure of a meter's length, is said to be a meter long, so beings, by coming closer and closer to their measure of being, can be said to be, or to be actual. If we consider a piece of string which is found to be one meter long as having an essence (one meter long) and an *esse* (to be measured, or *to be found* to be of a certain length), then it is obvious that the standard measure of one meter doesn't receive the *esse* as the string did: the measure is not itself *measured*; but the standard measure is *by right* what other things (like string) are not until they are measured (until they receive their *esse*). In the same way, the substance which measures other substances is *by right* what they can only receive; the substance which measures is *by right* an essence; it cannot become or be made that essence. Other substances are not their essences by right; but they must be made to be them, to have them. Thus the first substance must be said to be by right what it is, or its essence is actuality or *esse*.[2] Thus once again we are faced with this "intellectual necessity" to

[1] *Ibid.*, 1968.

[2] In the preceding footnotes texts have been refered to in which *unum* and *mensura* have been connected. In the lesson preceding the one from which these texts were taken, Aquinas wrote: "...ratio unius sit indivisibile esse; id autum quod est aliquo modo indivisibile in quolibet genere sit mensura; maxime dicetur in hoc quod est esse primam mensuram cuiuslibet generis." *In X Meta.*, 2, 1938. Thus the greatest substance is the measure of other substances: "... de ratione unius est, quod sit mensura. Et hoc maxime proprium est in quantitate; deinde in qualitate, *et in aliis generibus*; quia id quod est mensura, debet esse indivisibile, aut secundum quantitatem, aut secundum qualitatem. Et ita sequitur, quod unum sit indivisibiliter, sicut unitas quae est principium numeri, *aut"secundum quid" idest inquantem est unum, ut dictum est in aliis mensuris.*" *Ibid.*, 1960; Italics added. How then is the greatest substance *unum*? First, how is any substance *unum*? A substance is *unum* if it is "...quod est in se indivisum." *In IV Meta.*, 2, 553. Moreover, it is the essence which is called *unum per se*: "Substantia [in context, essence] enim cuiuslibet rei est unum per se..." *Ibid.*, 554. Of course, it is the essence as *real*, as part of a being which is called *unum:* cf. the discussion of the predication of *unum* in *Ibid.*, 550–60. To be *unum* requires that all division be foreign to the essence as "had": "Sed unum quod cum ente convertitur importat privationem divisionis formalis quae fit per opposita, cuius prima radix est oppositio affirmationis et negationis." *Ibid.*, 3, 566. Thus any substance is *unum* in so far as its *real essence* is undivided; of course, any essence is undivided *if* and *insofar as* it is "had": "Est enim unum *ens* indivisum." *Ibid.*, 2, 553 Italics added. The greatest substance, the greatest *unum*, is the measure of the *unum* of any other substance; cf. *In X Meta.*, 2, 1938 and 1960. The greatest subs tance, in its essence, must have the least formal division of any substance. But no substan ce has

seek that in virtue of which our thoughts about the beings around us are thinkable; that is, we are forced to seek that which enables us to continue to think the real as intelligible – we are forced to seek that which enables us to have a consistent view of the universe.[1]

These two expositions given by Aquinas (the principle regarding the reduction of composed beings to a simple being; the search for the first substance which measures other substances) do not, I believe, constitute two proofs for God; rather both involve the reduction of *ens* understood as "actualized essence" to *Ens* understood as "actual essence." There

any formal division in its essence. Thus we are forced to say, not that this greatest substance *has* the least formal division, but (as the measure of the lack of formal division) simply *measures* the lack of formal division. Since all formal division is reduced to the opposition of the first affirmative judgment and the first negative one, all lack of formal division is reduced to the first affirmative judgment, the first principle. But if the greatest substance measures all lack of formal division, it measures also the first affirmative judgment. Its essence must be "that which is" or *ens*. It necessarily is.

[1] This reduction of the proof by "measure" to that by "participation" will not be universally accepted. Cf. G. Isaye, S. J. , *La théorie de a mesure et l'existence d'un maximum selon Saint Thomas*, Archives de Philosophie, XVI, Cahier I, Beauchesne, Paris, 1940, pp. 115–25; 131; 78–83. P. Isaye claims that Aquinas' *quarta via* is actually a catch-all proof for God; several separate proofs are mixed together as if Aquinas did not want to lose any of several different insights, but was unable to synthesize them. In this context the proof of the *maximum* in being is a proof of a measure of being; Aquinas saw that his intellect strives toward such a being; hence He exists. – P. Isayes's work is impressive; even if one discounts those sections which attribute to Aquinas P. Maréchal's views on knowing, in particular the knowledge of metaphysical *esse* in a judgment, there yet remains an abundance of evidence that Thomas at one time taught a proof for God based on the need of a measure of being. But is P. Isaye correct in divorcing "causality" and "participation" from a "proof through measure" (pp. 10–16)? His treatment is much too summary, only 6 pages, and so, leaves one unconvinced. In addition, certain texts noted in the course of some of the discussions make one wonder whether Aquinas' theory of measure did not undergo an evolution parallel to those in regard to *esse*. E. g. *In I Sent.*, d. 8, q. 4, a. 2, ad 3*um*: here Aquinas goes from substances around us to a Pure *Esse* which is the measure of other substances; however, this is not a proof of God. Rather, His existence was assumed and, because a measure of substance is needed, Aquinas decides that God is that measure. In later works, e.g. *Sum. theol. I*, 2, 3 in the *quarta via*, measure is clearly used to prove the *maximum* in being. The same is true of the *Com. in Meta.*: Aquinas says we are to look for the first substance which measures all others. But as we have noted in the text above, the proof through "measure" seems to be identified with the proof through "participation" or through "causality". Because of these differences between Aquinas' attitude in the *Sentences* and in the post-*De hebdomadibus* works, I feel that it may be possible that Aquinas has changed in his view of "measure" just as he did in his view on *esse*. At the beginning of his career, *esse* is the here and now brute fact of existence, whereas in, and after the *Commentary on the De hebdomadibus*, *esse* becomes act, actuality, perfection. In like manner, Aquinas' proof of God based on the distinction of essence and *esse* becomes one through participation after the *De hebdomadibus*, whereas it was one based on causality prior to that work. Perhaps the Aquinas of *esse* as act, and as participating in Pure *Esse*, is a philosopher of measure as well, where measure is conceived as the standard needed to explain the measurement found in the world. The measurement will be that of *ens*. Hence, proofs through "participation," through "causality," and through "measure" would be identical: *ens*, as participating, shares in Pure *Esse*; *ens*, as caused, is dependent on Pure *Ens*; *ens*, as measured, is determined to a certain degree of perfection by a Pure Measure. And all three aspects – participating, dependent, determined or measured – would be expressed in the concept *ens* in so far as creatures are understood as passive in respect to the perfection they have received.

are, however, two other proofs Thomas gives of God: the first is a proof of an intelligent exemplary and ordering cause, the second of an efficient cause; this latter proof, however, is radically identical with the physician's proof of the first mover.

The proof of God as the intelligent exemplar and ordainer of all things is given by Aquinas in the midst of a refutation of Plato's theory of ideas as subsistent. Natural things – material beings acting according to the laws inscribed in their very being – tend naturally to generate things similar to themselves. Such a "tending" – such natural intentionality – must be reduced to some primary directing intelligence. This intelligence is thus said to direct all natural (therefore, material) things to their end according to reasons known primarily by the first intelligence.[1]

Before we began the discussion of the proof of God based on the principle of the reduction of composed beings to a simple one, we noted that any proof for God must be rooted in the *doctrina physica*, in the knowledge we have of material things. The lack of intelligibility in material things was sufficiently noted in the first proof (through "participation" or "measure"). In the second proof – that of God as the intelligent exemplar and ordaining cause – it should also be evident that our *doctrina physica* involved knowledge of beings in whose essences or natures were inscribed certain goals. To explain the presence of the tendency toward these goals, it is necessary to posit the existence, outside the world of natural beings, of an intelligent exemplar and ordering cause. Such a cause is not, *qua* directing intelligence, the author of *esse*. But given the fact that being is totally caused by the first substance, this first substance, must be the intelligent ordering cause, for the total creation of being must be according to an intelligent plan.

At first sight, there appears to be a third proof for God in the *Commentary*, a proof from motion. This proof is, however, only a restatement of the physician's proof of the first mover, joined to an affirmation of the necessity of positing a cause of *esse* in the event that the proof of the first mover is not absolutely certain. Thomas' formulation of this proof is: "if X, then Y; but if not X, then Z, and so Y", where "X"

[1] "Sciendum autem quod illa ratio, etsi destruat exemplaria separata a Platone posita, non tamen removet divinam scientiam esse rerum omnium exemplarem. Cum enim res naturaliter intendant similitudines in res generatas inducere, oportet quod ista intentio ad aliquod principium dirigens reducatur, quod est in finem ordinans unumquodque. Et hoc non potest esse nisi intellectus cuius sit cognoscere finem et proportionem rerum in finem. Et sic ista similitudo effectum ad causas naturales reducitur, sicut in primum principium, in intellectum aliquem. Non autem oportet quod in aliquas alias formas separatas: quia ad similitudinem praedicatam sufficit praedicta directio in finem, dua virtutes naturales diriguntur a primo intellectu." *In I Meta.*, 15, 233.

represents the eternal character of the world; "Z", creation in time; and "Y" the first agent cause.

The world, Thomas says, is perhaps eternal, perhaps not.[1] Let us assume it is eternal: "If X". In this case, there has always been motion and time.[2] Such motion is possible only if there is, not only eternally moved substances but as well a mover-substance, also eternal or mover in act.[3] Thus the essence of this substance is act; its essence is to be.[4] Moreover, as pure act, this substance is totally separated from matter.[5]

This argument, based on the assumption that motion is eternal, is not absolutely necessary or certain, for there is no absolutely demonstrative proof that motion – and so the world – is eternal.[6] Thus, let us assume

[1] "Et praetermissis aliis rationibus quae hic non tangit, manifestum est quod ratio quam hic posuit ad probandum sempiternitatem temporis, non est demonstrativa. Non enim, si ponimus tempus quandoque incepisse, oportet ponere prius nisi quid imaginatum. Sicut cum dicimus quod extra caelum non est corpus, quod dicimus extra, non est nisi quid imaginatum. Sicut igitur extra caelum non oportet ponere locum, quamvis extra videatur locum significare, ita non est necessarium quod tempus sit prius quam incipiat vel postquam desinet, licet prius et posterius videantur tempus significare. Sed quamvis rationes probantes sempiternitatem motus et temporis non sint demonstrativae et ex necessitate concludentes..." *In XII Meta.*, 5, 2498.

[2] "Videtur igitur necesse esse quod tempus sit sempiternum. Et si tempus est continum et sempiternum, necesse est quod motus sit continuus et sempiternus: quia motus et tempus, aut sunt idem, ut quidem posuerunt, aut tempus est aliqua passio motus, ut rei veritas habet. Est enim tempus numerus motus...Sed tamen non est accipiendum de omni motu, quod possit esse sempiternus et continuus. Non enim hoc potest esse verum nisi de motus locali Et inter motus locales, solum de motu circulari..." *Ibid.*, 2490–91.

[3] "...ad sempiternitatem motus sustinendam, necesse est ponere substantiam sempiternam semper moventem vel agentem; dicens, quod cum necesse sit, si motus est sempiternus, quod sit substantia motiva et effectiva sempiterna, ulterius oportet quod sit movens et agens in actu semper; quia si esset "motiva aut effectiva," idest potens movere et efficere notum, et non agens in actu, sequeretur quod non esset motus in actu. Non enim est necessarium, si habeat potentiam movendi, quod moveat id actu: contingit enim id quod habet potentiam agendi non agere; et ita motus non erit sempiternus. Ad hoc igitur, quod motus sit sempiternus, necesse est ponere alquam substantiam sempiternam moventem et agentem in actu." *Ibid.*, 2492.

[4] "...neque est sufficiens ad sempiternitatem motus, si substantia sempiterna agat, sed tamen secundum suam substantiam sit in potentia; sicut si ponamus principia esse ignem aut aquam secundum positionem antiquorum naturalium: non enim poterit esse motus sempiternus. Si enim sit tale movens, in cuius substantia admiscetur potentia, contingit id non esse. Quia quod est in potentia contingit non esse. Et per consequens continget quod motus non sit, et sic motus non erit ex necessitate, et sempiternus. Relinquitur ergo, quod oportet esse aliquod primum principium motus tale cuius substantia non sit in potentia, sed sit actus tantum." *Ibid.*, 2494.

[5] "..ex praedictis sequitur, quod huiusmodi substantias, quae sunt principia motus sempiterni, oportet esse sine materia. Nam materia est in potentia. Oportet igitur eas esse sempiternas, si aliquid aliud est sempiternus, utpote motus et tempus. Et sic sequitur quod sint in actu." *Ibid.*, 2495.

[6] "...rationes probantes sempiternitatem motus et temporis non sint demonstrativae et ex necessitate concludentes..." *Ibid.*, 2499. One should connect this doctrine with that of *In VIII Phys., lectio* 2. There Aquinas notes that Aristotle, in talking about the eternity of motion, only wishes to prove that in motion, even if it is eternal, some first mover is required. As evidence that Aristotle does not wish to deny creation even though speaking of

the world is not eternal: "If not X"; in this case we can still posit the existence of a first efficient cause. In this case, we shall not argue from motion, but from the world's necessity of being "produced in *esse*" – totally created. There must be some pre-existing, transcendant substance which caused the world to exist (whether the world be eternal or not); otherwise there would be no world at all. If this pre-existing substance were not itself eternal, then it too must have been created by some other substance. To avoid regress *ad infinitum*, as avoid it we must, it is necessary to posit an eternal substance, an eternal, immaterial substance. This eternal substance, as the total cause of being, is totally being: its essence must be to exist.[1] Thus whether or not the world is eternal, we must posit the existence of an immaterial, eternal act or substance whose essence is to be.

This argument to a first efficient cause on the hypothesis of the eternity of motion is similar to the physician's proof of an immaterial, first mover: "If X, then Y", would seem to be the physician's proof. The argument to the first cause of being, given against the background of the hypothesis that the world is not eternal, appears to be the proof through participation or measure. Only the terms of the proof from participation have been expressed differently in Book XII due to the differences of context no doubt. This argument to an efficient cause of being could be given by the metaphysician in a slightly different form than the one we have noted as the *Commentary's* expression of it. To be absolutely complete the metaphysician could write: If motion is eternal, there is a first efficient cause of it; but since the beings who are moved have received being, therefore there is also a first efficient cause of being. If motion is not eternal, there must still be a first efficient cause of it; moreover, for the reason given above, there must be a first efficient cause of being as well. A consequent examination would reveal that the first cause of motion is the cause of being.[2]

Aquinas, I feel, did not bother to give the form mentioned above to his argument since first, he was following Aristotle's exposition, and

eternal motion, Aquinas points to the proof of God as the cause of being in Book II of the *Metaphysics*. Cf. JOLIVET, *Essai sur les rapports entre la pensée grecque et la pensée chrétienne*, pp. 14–17.

[1] "...ea quae hic probantur de sempiternitate et immaterialitate primae substantiae, ex necessitate sequuntur. Quia si non fuerit mundus aeternus, necesse est quod fuerit productus in esse ab aliquo praeexistente. Et si hoc non sit aeternum, oportet iterum quod sit productum ab aliquo. Et cum hoc non possit procedere in infinitum, ut supra in secundo probatum est, necesse est ponere aliquam substantiam sempiternam, in cuius substantia non sit potentia, et per consequens immaterialem." *In XII Meta.*, 5,2499.

[2] This argument could be reconstructed from texts in the *Commentary*; in particular, those texts already quoted from *In XII Meta.*, *lectio* 5 would be utilized.

two, he had already proved in Books II and X that there must be a first being, the source or actuality, the measure of other substances. Here in Book XII, he was following Aristotle's proof of a first cause of motion, and thus adds the conditioned "if" to Aristotle's "X, then Y"; and in addition completes Aristotle with "but if not X, then Z, and so Y".

Thus Aquinas has given us two proofs in his *Commentary*; there must be the source of *esse*, in which we participate, and by which we are measured; there must be as well – or better, there must be for another reason – an intelligent exemplar.[1] Such a source of *esse* is a being whose essence is *esse*. But this contradicts the very starting point of our metaphysics; that is, it contradicts the discovery of the concept "that which is" or "is a being", which we applied to all beings, the first mover as well as the beings of our experience. "That which is" was found to express the result of a composition of essence and *esse*, where the *esse* is different from, and proportionate to the essence: "actualized essence". Thus several questions arise: Do we understand what the concept "that which is" means when applied to the immobile, first substance? Do we in any way have a concept common both to this first substance and to material substances? It is time once more to return to an analysis of our concepts.[2]

[1] Metaphysics, as Aquinas observes, must study the final cause: cf. *In I Meta.*, 2, 51. Such a study does not constitute a proof for God, but rather is dependent on one. First, one must discover the existence of some being which can be the goal or final cause of other beings; such a being is proved to exist as first substance, or as exemplary and ordaining cause, or as efficient cause. After such proofs it can be studied as final cause: cf. *In XII Meta.*, 7, 2519–34.

[2] It has been said that metaphysics really only begins its main work when it discovers God: "in the light of that principle (idest: God) it may *commence* its main work of dealing with its subject in terms of creation, conservation, and occurrence." J. Owens, C. Ss. R, "Review of: *Metaphysics and the Existence of God*, by Thomas C. O' Brien, O. P. ...," *The New Scholasticism*, XXXVI, 1962, p. 251. – That metaphysics is not completed once God is discovered to exist, that is rather evident. Yet to say that metapyhsics "commences its main work" at that stage seems slightly exaggerated. As Aquinas maintains, metaphysics can only be "first" because it is universal. Because it discovers a need for a first cause through the study of beings as such, metaphysics can study the first being. It is certainly true that metaphysics must explain being as such in terms of its final and efficient causes. Yet if the study of being as such which precedes the proof of God concludes to the *esse*-essence composition, and realizes that *esse* is received, then the problem of creation and conservation have already been partially, if not almost totally, studied. Fr. Owens however does not agree that one proves God through participation, a legitimate theory, granted his view of *esse* as known in a judgment. Such an *esse* can hardly be understood as anything other than a brute fact which needed to be caused; such an *esse* is not understood as "participating" Cf. J. Owens, C. Ss. R., "The Causal Proposition – Principle or Consclusion?", *The Modern Schoolman*, XXVII, 1955, pp. 159–71; 257–70; 323–39. – An interesting recent work by Fr. Owens indicates in some detail his understanding of a metaphysics drawing its inspiration from Aquinas; cf. *An Elementary Christian Metaphysics*, Bruce, Milwaukee, 1963.

F. The human attempt to speak of God

In company with Aquinas, we have just affirmed the existence of a being whose essence is *esse*. Do we know what we are talking about when we speak of Pure *Esse*? Let us examine several aspects of the problem before we attempt to answer.

First, the pure being, whose essence and *esse* are identical, is said to be "substance". That is, we say: "The pure being is", and by "is" under stand "is a substance". Here "substance" must mean that this being's *esse* is its essence: "actual essence". When material things are called "substance", the word "substance" must have a different meaning. As "substance", material things are understood as being the result of *esse*-essence composition: "actualized essence".

Secondly, Thomas says that heavenly bodies and material bodies are both called "substance", but that they fall under the same genus only logically and not ontologically.[1] If this is so, then surely would not Thomas say that God, material beings, and heavenly bodies are all called "substance" only in a logical use of the word "substance"? Since Thomas does say that all these substances are considered by metaphysics insofar as they fall under the genus of *ens per se*, must he not understand this as a logical genus?[2] Let us pause a moment to consider this point.

A definition expresses the genus and the specific difference proper to the subject to which the definition is applied. For example, "John (or man) is a rational animal". "Animal" expresses the genus to which John (or man) belongs: it expresses what John (or man) has in common with certain other things – they are all living, sentient, material substances. Thus, a generic term is taken from that aspect of a being, which aspect could and must (ideally) be further perfected in one of several ways through a qualitative addition: "animal" could be perfected by adding "rational" or "irrational", thus giving us "man" or "unthinking animal". Thus a generic term is like matter to which form (the

[1] "Corruptibile autem et incorruptibile dividunt per se ens: quia corruptibile autem est quod potest non esse, incorruptibile autem quod non potest non esse. Unde, cum ens non sit genus, non mirum si corruptibile et incorruptibile non conveniant in aliquo uno genere." *In X Meta.*, 12, 2145. "Nam corruptibilium et incorruptibilium non potest esse materia una. Genus autem physice loquendo, a materia sumitur. Unde supra dictum est, quod ea quae non communicant in materia, sunt genere diversa. Logice autem loquendo, nihil prohibet quod conveniant in genere, inquantum conveniunt in una communi ratione, vel substantiae, vel qualitatis, vel alicuius huiusmodi." *Ibid.*, 2142.

[2] "Et veritas est, quod haec scienta est de omnibus substantiis, licet de quibusdam principalius, scilicet de substantiis separatis, inquantum omnes conveniunt in uno genere, quod est ens per se." *In XI Meta.*, 1, 2153.

specific difference) must be added to arrive at the composite (the definition).[1]

In the case of heavenly and material bodies, what could Thomas discover as their common or material aspect? Nothing, as he says himself. For they cannot both be said to be "bodies," where "body" is taken as an univocal term; the heavenly body, as Thomas often remarked, is incorruptible, and hence its matter is different from ours. Thus we qua bodies have nothing in common with heavenly bodies also qua bodies. And so we have no common genus *physice* (that is, ontologically) *loquendo*.[2]

Yet heavenly bodies and material bodies do fall into a common logical genus – that of "substance". To be a member of the *logical genus of substance* is not the same as being a *logical substance,* we must note. "John is a man." The predicate "is a man" falls into the logical genus of substance, but "John "does not. "John", the word in the proposition, is a logical substance: "John" cannot be predicated of anything else, and yet other words can be predicated of it.[3] As Thomas mentions, the logical genus of substance reveals to us the substantial matter of existing.[4] This substantial manner of existing is had, of course, by particular substances,[5] and these particular substances, used as the subject of propositions, become logical substances.

Thus if we say: "Mars is a heavenly body", Aquinas would analyze our proposition as follows. The particular substance Mars functions in your proposition as a logical substance. Even more, by the predicate "is a heavenly body", you have expressed (albeit imperfectly) the essence of Mars. Insofar as you consider this predicate "is a heavenly body", you are working with one instance of the logical genus of substance: you are considering a word which must be predicated according to certain rules, etc. Moreover, "is a heavenly body" can be compared with other instances of the logical genus of substance, say for example

[1] "...cum in definitione unum comparetur ad aliud ut actus ad materiam, quidam definientes res per materiam tantum insufficienter definiunt. Sicut definientes per domum caementum et lapides et ligna, quae sunt materia domus; quae talis definitio non notificat domum in actu, sed in potentia. Qui vero dicunt, quod domus est coopertura pecuniarum et corporum, dicunt formam domus sed non materiam. Qui vero dicunt utrumque, definiuntur compositam substantiam. Et ideo eorum definitio est perfecta ratio. Ratio vera, quae sumitur ex differentiis, pertinet ad formam. Quae vero ex partibus intrinsecis, pertinet ad materiam." *In VIII Meta.*, 2, 1700. Cf. *In VII Meta.*, 9, 1463.

[2] Cf. *In X Meta.*, 12, 2142–45: see above, footnote 1, p. 295.

[3] Cf. *In VII Meta.*, 2, 1280.

[4] Cf. *In V Meta.*, 9, 890.

[5] Cf. *Ibid.*, 981; *Ibid.*, 10, 898.

with "is a goat". "Thus is a goat" and "is a heavenly body" fall within the same genus – a logical one to be sure.

Moreover, it will be remembered that an examination of predicates of the type "is a goat" and "is a heavenly body" – predicates used in one manner of *per se* predication – revealed that particular substances are composed of essence and *esse*. Thus, when we say: "Socrates is a (particular) substance", and "Mars is a (particular) substance", we have a very complex understanding of Socrates and Mars as 1) things; 2) resulting from the composition of *esse* and essence; 3) where the *esse* is a "to be in itself and not in another" and is proportioned to essence. But what type of word is "substance" in these cases? Is it a generic term? And if so, is it ontological or logical?

Thomas would answer that "is a substance" is an alternative form of "is a being". "Being" and thus "substance" are not generic terms because there is nothing outside them which could serve as a specific difference.[1] Thus "being" and "substance" express everything in, say Mars or Socrates.[2] Hence "substance" as said of Mars has no meaning common to "substance" as said of Socrates, since Mars and Socrates are ontologically different beings. Thus "substance" in these cases is not a genus.

We asked shortly above: Since heavenly bodies and material bodies fall under the same genus logically, do not God, heavenly bodies, and material bodies fall also under a logical genus? The answer, we must now say, is: "No". Heavenly and material bodies fall under the logical genus of substance because we have substantial predicates for both types of bodies. Moreover, the fact that we do have these substantial predicates enabled us to form the predicate "is a substance"; by this

[1] "Nulla enim differentia paritcipat actu genus; quia differentia sumitur a forma, genus autum a materia... Nulla autem posset differentia sumi, de cuius intellectu non esse unum et ens. Unde unum et ens non possunt habere aliquas differentias. Et ita non possunt esse genera, cum omne genus habeat differentias." *In XI Meta.*, 1, 2169.

[2] That "being" expresses everything in the object to which it is applied is not stated explicitly by Aquinas; it is obviously implied in the following text however. "Quanto aliquae scientiae sunt priores naturaliter, tanto sunt certiores: quod ex hoc patet, quia illae scientiae, quae dicuntur ex additione ad alias, sunt minus certae scientiis quae pauciora in sua consideratione comprehendunt ut arithmetica certior est geometria, nam ea quae sunt in geometria, sunt ex additione ad ea quae sunt in arithmetica... Scientiae particulares sunt posteriores secundum nartuam universalibus scientiis, quia subiecta earum addunt ad subiecta scientiarum universalium: sicut patet, quod ens mobile de quo est naturalis philosophia, addit supra ens simpliciter, de quo est metaphysica, et supra ens quantum de quo est mathematica: ergo scientia illa quae est de ente, et maxime universalibus, est certissima. Nec illud est contrarium, quia dicitur esse ex paucioribus, cum supra dictum sit, quod *sciat omnia. Nam universale quidem comprehendit pauciora in actu, sed plura in potentia.* Et tanto aliqua scientia est certior, quanto ad sui subiecti considerationem pauciora actu consideranda requiruntur." *In I Meta.*, 2, 47. Italics added.

predicate we expressed the fact that both heavenly and material beings are conceived as resulting from the composition of *esse* and essence: "actualized essence". To turn now to God: we do not have a substantial predicate proper to Him, although we could form a type of substitute substantial predicate on the basis of our proofs of His existence. Thus: "God is X", where "is X" means "immaterial, eternal being whose essence is *esse*". As soon as we admit that God's *esse* and essence are identical, we make this substitute substantial predicate a different type than all the other substantial predicates we have: "actual essence". We were able to place Mars and John under the logical genus of substance because our substantial predicates for them (and "is a substance" as said of them) have the same form: "actualized essence". Thus by the construction of our substitute substantial predicate for God, we placed Him outside that logical genus of substance which we had earlier studied and so come to understand.

But can we not say that we have two logical genera of substance? God alone falls under genus number two, while all composed beings fall under genus number one? In answer, let us enumerate the ways of saying "substance".

There appear to be five distinctly different ways: 1) *Logical* substance: insofar as any real object is represented by a name or word serving as the subject of a proposition, that name is a logical substance; it can be predicated of nothing and all else is predicated of it. 2) *Logical genus of composed substance*: insofar as we have for certain real objects substantial predicates emphasizing an *esse*-essence composition ("actualized essence"), these objects are gathered under the logical genus of composed substances. 3) *Ontological, composed substance*: all real beings which fall under the logical genus of composed substances in virtue of the substantial predicates we have for them, can be considered as falling under the class of those beings composed of essence and *esse*; the difference between the logical genus of substance and this ontological class, is that the former is directly a class of predicates, and indirectly of the beings to whom the predicates are applicable, whereas the ontological class is directly, and only, a class of real beings. Of these three classes, God can fall only under the first, logical substance; we can say of God: "God is". It would seem however that a fourth and fifth class of substance must be enumerated. 4) *Logical genus of simple substance*: insofar as we have for God a substitute substantial predicate expressing the lack of an *esse*-essence composition, we have a logical genus different from the logical genus of composed substance. 5) *Ontological, simple substance*: God, the only being for whom we have a substantial predicate falling under the logical genus of simple substance, would constitute the ontological class of simple substance.

Now although we see the necessity of affirming a being such as the one we have placed in the ontological class of simple substance, do we actually have a substantial predicate by which we express (that is, understand) the lack of an *esse*-essence composition? Do we really have a logical genus of simple substance? What that question amounts to is this: Can we understand *esse* and essence as anything other than incomplete principles, totally separate from one another? Thomas' answer would appear to be a categorical "no":

> Inde est quod prima rerum principia non definimus nisi per negationes posteriorum; sicut dicimus quod punctum est, cuius pars non est; et Deum cognoscimus per negationes, inquantum dicimus Deum incorporeum esse, immobilem, infinitum.[1]

Thus Thomas would appear to deny the legitimacy of our fourth use of "substance": there may be a possibility for some intellect of forming a logical genus of simple substance, but man does not have that intellect!

Now we can return to the question with which we began this subsection: Do we know what we mean when we affirm the existence of a being whose essence is *esse*? The answer: We do not know (or understand) *what* we mean by: "God is" where "is" would be a substantial predicate, although we understand *why* we must say: "God is".

Thus it follows that the word "substance" can be applied to God and *understood* in only one way: "God is a logical substance". We can say: "God is in the logical genus of simple substance", and "God is in the ontological class of simple substance" – but we do not have the predicate which puts God in the logical genus, nor do we understand anything about the ontological class in which He must fit.

Can we now form a concept of "common being", a concept common to God and to the other beings? If so, then, this concept could not be a common substantial predicate: the substantial predicate "being" expresses for us: "actualized essence"; it implies a relation of *esse* and essence; it implies that these are two distinct principles. Since God's essence and *esse* are in themselves unlike the principles of other beings, there can be no substantial predicate expressing at once both God's and creatures' being.

Yet it will be remarked, and quite correctly so, that one can conceive both created being and God *qua* "that in which *esse* and essence can be identified." Since this would not be a substantial concept, since it

[1] *In X Meta.*, 4, 1990

would not express the "that which is" either of God or of creatures, it would not express all the depth of each being to which it is applied. Thus, although one could form such a concept, it would not be a metaphysical concept, for as Thomas notes, the metaphysical concept of "being" expresses everything about the objects to which it is applied.[1]

If follows, thus, that the metaphycisian never advances beyond the content of his first idea of "is a substance", of "is a being". His wisdom consists in learning the limits of the extension of his concept.

Notwithstanding the non-metaphysical character of the idea of being as "that in which *esse* and essence can be identified", it is worthwhile to investigate it a bit. Our proof leading to the jugdment: "God is", consisted in noting certain lacunae in the intelligibility of the beings around us; those lacks required that we negate of God an *esse*-essence composition (insofar as composition means that *esse* and essence are distinct principles). Now if we try to form a concept which expresses "that in which *esse* and essence can be identified", we must form that concept by taking "being" as we know it, as expressing the result of the composition of *esse* and essence, and by negating the element of causal effect produced by *esse*. The content of our new idea of "being" includes this negation of the causal effect of *esse*. What we have conceived is an idea not of real, extra-mental beings, but of a negative judgment. Thus the immediate object known through our new idea is a mental act; only indirectly – through the negative judgment – does our act refer to the original idea of "being" and to our proof of God, and only through these acts does it refer to extra -mental beings. It is no wonder that this new idea is not metaphysical, removed as it is from any activity in which man knows extra-mental reality.

Moreover, it is interesting to note that the entire positive content of this new idea of "being" is no different from that our original idea of "being", where *esse* and essence are denoted as distinct from one another. For just as the positive content of the proofs for God consists in the positive content of our ideas of the world, so the positive content of the new idea, built on the proofs for God, consists in the same positive content. Thus we never really form a new idea if "new" means new positive content. The idea is "new" only in the sense that it is a different act of the mind.

But can we not form the concept of "actual essence"? This concept would signify, it seems, only the positive aspects of *ens*, without including the passive aspect of composed beings. God is "actual essence"; we are examples of "actualized essence". The question is, naturally,

[1] *In Meta I.*, 2, 47: see footnote 2, p. 297.

reduceable to: What would we understand by "actual essence"? The content of this idea would be identical to that of "actuality", some would say. However, that seems to be a bit inaccurate. "Actuality" must always be the "actuality of something", and that is the way we understand it. "To be" is always "to be man" in the world around us, or else it is "to be horse", or something of this kind. It might be objected that we can understand "to be" in the sense of the common element in "to be horse" and "to be man". Now there is one sense in which it is very true that we can understand this common "to be". It is the essential "to be" to which one refers when one speaks of the Porphyrian tree: "being" is either "material" or "immaterial"; "material being" is "animate" or "inanimate", and so on. Such a "being", such a "to be" is the *least intelligible of the possible intelligible notes of a meaning.* But opposed to this essential "to be", this "to be" of meaning, Aquinas distinguishes a non-essential "to be", the "to be" known as metaphysical *esse.* In a non-essential order, we have this "to be" which is the actuality of an essence. It is this latter "to be", this non-essential "to be" which cannot be grasped except as the actuality of some essence. Hence, if we try to conceive "actual essence", where essence means some definite essence, e.g. "horse" or "man", then we have a concept of a creature. If we replace "essence" of "actual essence" with the essential "to be", then we have a concept of the least perfect being God could create; in such an idea, the non-essential "to be" (the actuality) is proportionate to the essential "to be" of the Porphyrian tree – the weakest essence possible. Hence, it appears that any attempt to construct a common concept of "actual essence" is bound to failure.

If all this is correct, then what did Aquinas mean when he said that metaphysics studies all substances as falling under the genus of *ens per se*?[1] The meaning involves a reference to the problem which evoked Thomas' words on the problem of the identity of the substances metaphysics studies – all substances or only certain kinds?[2] When Aquinas answers that we study all substances as falling into the genus *ens per se,* he can be understood as referring to the fact that we begin with material beings *qua* being *per se,* posit the separate, pure being as the cause of being, and then say what we can about this pure being.[3] Of course, in this explanation, the word "genus" loses its strict meaning. There is a second way of understanding Aquinas' words, however; yet this way,

[1] *In XI Meta.* 1, 2153; *In XII Meta.*, 2, 2427.
[2] *In XI Meta.*, 1, 2152.
[3] *In IV Meta.*, 2, 563.

like the first, makes a mockery of the word "genus". Here *ens per se* is to be understood as *esse per se*: "to be on one's own, and because of one's self". A creature can be said to fall under this genus insofar as it has received this "to be one's own and because of one's self"; God is said to fall within this genus because He is the "to be on one's own and because of one's self". Thus *esse per se* is said "in a participative way" of creatures, but "essentially" of God.[1] Yet we do not really have a concept of *esse per se*, for there is no difference between *esse per se* and "substantial actuality".

So much for what we meant when we affirmed the existence of an eternal and immaterial substance, and for our attempts to reconstruct new ideas on the basis of that affirmation. Is there anything we can say about the attributes of God? Thomas, following Aristotle's lead, does not hesitate to speak of God as "intellect", as "good", and so on.[2] The individual attributes of God need not interest us here. Rather, it is the procedure that should hold our attention. Two rules appear in the *Commentary*. One of these we have already noted: in any predication made of God, there must be included a negation. We know God's essence only by denying the things which are connected with our world of composed beings, *qua* material beings.[3] A second rule appears involved as well: "Propter quod autem unumquodque tale, et illud magis". The untranslatable latin *dictum*; let us simply say, in lieu of a translation: if there is any thing in creatures which is "divine" and "noble", it is found in the highest degree in God.[4]

G. Conclusions

This attempt to reconstruct Aquinas' metaphysics has followed a long andsome what torturous path. We began with certain statements

[1] For the notion of *per participationem* predication, cf. *In I Meta.*, 14, 224.

[2] *In XII Meta.*, 8, 2536–44.

[3] "...ea quae sunt priora secundum naturam et magis nota, sunt posterior et minus nota quod ad nos, eo quod rerum notitiam per sensum accipimus. Composita autem et confusa prius cadunt in sensu...Et inde est, quod composita prius cadunt in nostram cognitionem. Simpliciora autem quae sunt priora et notiora secundum naturam, cadunt in cognitonem nostram per posterius. Inde est quod prima rerum principia non definimus nisi per negationem posteriorem...et Deum cognoscimus per negationes, inquantum dicimus Deum incorporeum esse, immobilem, infinitum." *In X Meta.*, 4, 1990.

[4] "Propter quod autem unumquodque tale, et illud magis. Et ideo sequitur quod quicquid divinum et nobile, sicut est intelligere et declectari, invenitur in intellectu attingente, multo magis invenitur in intelligibili primo quod attingitur. Et ideo consideratio eiusdem, et delectabilissima est et optima. Huiusmodi autem primum intelligibile dicitur Deus.Cum igitur delectatio, quam nos habemus intelligendo, sit optima, quamvis eam non possimus habere nisi modico tempore, si Deus semper eam habet, sicut nos quandoque, mirabilis est eius felicitas. Sed adhuc mirabilior, si eam habet potiorem semper, quam nos modico tempore." *In XII Meta.*, 8, 2543.

indicating that physics must discover the existence of an unmoved mover. Following an investigation of some remarks on the need of following a logical study of predication, we noted that that study was not to be conducted after the fashion of formal logic; instead, we were to try to discover how the real principles of things are expressed by our concepts. Our next step, the investigation of the different types of predicates, led us to the discovery of the concept "being", the expression of the result of the *esse*-essence composition of finite beings; "actualized essence". Following this, we discovered the distinction of matter and form through the use of the same mode of investigation. Then, noting the lack of the intelligibility in our knowledge of material things, we concluded that there must be a simple, transcendent being, whose essence is *esse*. Finally, we averted to the fact that, in affirming: "God is", we are aware that we must make this affirmation, although we do not really understand what it is that we affirm.

In this very brief discussion of Aquinas' metaphysics, we omitted many elements, e.g. the defense of first principles; the discussion of the *communia* other than *ens*; the proof of the divine attributes. But as far as the general lines of Thomas' thought are concerned, they have been sufficiently sketched.

2. AQUINAS AND ARISTOTLE

A. Aristotle on the birth of metaphysics

As has been argued, Aquinas taught that the birth of metaphysics coincides with the physician's discovery of the existence of the unmoved mover. In other words, if one thinks all beings are material, then highest science will take the point of view of "material being", or "mobile being", or something similar to these.

In presenting this doctrine, Thomas seems not to have deviated from the teaching of Aristotle. Thomas, we noted, explained that physics would be the first science if there were immaterial being; in saying this, Thomas' exposition is merely a repetition of a text of the Aristotelian *Metaphysics*.[1] Moreover, for Thomas, there is no stigmata to be attached to the ancient natural philosophers who, ignorant of the existence of non-sensible being, thought they were treating being as such in treating natural being; we must not blame these philosophers for, given their

[1] "Now if natural substances are the first of existing things, physics must be the first of sciences; but if there is another entity and substance, separable and unmovable, the knowledge of it must be different and prior to physics and universal because it is prior." K 7, 1064a9–14. Cf. also E 1, 1026a29–31. For Thomas: *In XI Meta.*, 7, 2267; *In VI Meta.*, 1, 1170.

beliefs, their conduct was justified. Again, such an exposition by Thomas adds nothing to the Aristotelian text.[1] And finally, metaphysics as explained by Thomas is the final science to be learned by man – again an Aristotelian doctrine, albeit that Thomas' exposition is better organized, more thorough than Aristotle's.[2]

On the basis of these three points in Aquinas' *Commentary*, we raised the question: What for Thomas would be the act of knowledge that turns man into a metaphysician? To answer this question we were forced to synthesize numerous texts of the *Commentary*. Although the texts discussed do not differ substantially from Aristotle's statements,[3] we need not look to the statements of Aristotle's *Metaphysics* to discover that it is the physician who discovers the immobile being, nor to discover the exact workings of the proof; in the case of Aristotle's physician, the discovery of the prime mover is quite clearly elaborated in Book VIII of the *Physics*.[4] Given Aristotle's views that metaphysics is the ultimate science to be learned by man, it appears that metaphysics is born when the physician discovers the existence of the unmoved, first mover.

The ultimate explanation of the "retarded birth" of metaphysics is found in the Aristotelian theory of the necessity of exercising the intellect bit by bit before one can be accustomed to the highest grade of intelligibility. Thus man must pass through the stage of intellectual knowledge known as physics before reaching the metaphysical stage.[5] This theory is not surprising; the metaphysician who wishes to know all reality from a determined viewpoint – qua being – must elucidate his point of view. For Aristotle, this elucidation was to be accomplished

[1] "Unde nullus secundum partem intendencium presumit dicere aliquid de ipsis, si sint vera aut non, – neque geometra neque arismeticus, set phisicorum quidam hoc merito agentes; soli enim opinati sunt deque tota natura intendere et de ente." Γ 3, 1005a29–33: *Vetus*, p. 307, l. 26–30. Aquinas: *In IV Meta.*, 5, 593.

[2] "Fere autem et difficillima hec cognoscere hominibus, maxime universalia, sunt, longissime enim a sensibilibus." A 2, 982a23–25: *Vetus*, p. 258, l. 6–8. Aquinas : *In I Meta.*, 2, 46. – M. Aubenque would not agree that Aquinas' interpretation was correct, for Aristotle, it is maintained, would consider the "logically prior" science to be also *de facto* first in our order of learning. Cf. Aubenque, *Le Problème de l'être chez Aristote...*, pp. 45–68; e.g. p. 67. The difficulties of M. Aubenque's work can not be discussed here; it is sufficient to note that Aquinas could have legitimately understood Aristotle as he did.

[3] Compare the texts given above in footnotes 1, p. 243–2, p. 246 with the corresponding passages of Aristotle.

[4] For a good summary of the argument, see W. D. Ross, *Aristotle's Physics. A Revised Text with Introduction and Commentary*, Clarendon Press, Oxford, 1936, pp. 85–94.

[5] For a discussion of this necessity of exercising the intellect, see G. Verbeke, Démarches de la réflexion métaphysique chaez Aristote," in *Aristote et les problèmes de méthode. Communications présentées au Symposium Aristotelicum tenu à Louvain du 24 août au Ier eptembre 1960*, Nauwelaerts, Louvain-Paris, 1961, pp. 108–109.

by means of apories.[1] For example in the fourth aporie, he puzzles whether there are immaterial substances as well as material ones (B 1, 995b 13–18); aporie three asks if one science can study all substances (B 1, 995b 10–13). As has been pointed out by Fr. Owens, the books containing the apories would have been presented to Aristotle's pupils; a discussion would have taken place, for example, on apories three and four; following this discussion the pupils would have received the next section from Aristotle and discussed it, and so on.[2] Thus Aristotle's students would have realized the breadth of the formal and material objects before going on to study the science as such. Not only does Fr. Owens say that Aristotle's pupils would have understood through these discussions that they were going to study all being from the point of view of being, but Aquinas as well held some such opinion.[3] Thomas, in treating the third and fourth apories, mentions first that these are problems concerning the material object of the study.[4] When he comments on Aristotle's lists of pro's and con's for aporie three, Thomas makes the point: this question will be solved in Book IV when Aristotle shows that the first science – the study of *ens qua ens* – must treat all substances under the common *ratio* or meaning of substance.[5] Thomas by thus introducing the formal object of metaphysics ("qua ens") into a discussion of whether all substances are treated, illustrates the doctrine that the formal object of a science depends somewhat upon the identity of the material object. For Thomas, then, this aporie showed that Aristotle was interested in discovering the formal object of the science, or better: in showing to his students the breadth of the formal object.[6] When Thomas treats Aristotle's fourth aporie, however, he treats it totally as an historical problem solved in Books XII–XIV.[7]

Accordingly, the doctrine of Aquinas' *Commentary* can be declared

[1] *Ibid.*, p. 115, footnote 16.

[2] OWENS, *The Doctrine of Being...*, pp. 28–29.

[3] References to Fr. Owens' theory should not be construed as a belief that he is right in his overall view of Aristotle. He is quoted only to illustrate a possible way of understanding Aristotle, a way which is substantially that adopted by Aquinas: "De talibus ergo quae ab aliis imperfecte dicta sunt, dictum est prius. Iterum in tertio libro recapitulabimus de istis quaecumque circa hoc potest aliquis dubitare ad unam partem vel ad aliam. Ex talibus enim dubitationibus forsitan investigabimus aliquid utile ad dubitationes, quas posterius per totam scientiam prosequi et determinare oportet." *In I Meta.*, 17, 272. See also: *In III Meta.*, 1, 339 and 343.

[4] *In III Meta.*, 2, 346, 348, 350.

[5] *Ibid.*, 6, 389. That it is possible to view Aristotle's book IV as speaking of common *ratio* of substance, cf. MERLAN, *From Platonism to Neoplatonism*, pp. 161–69.

[6] The same concern with the formal object is manifested in the treatment accorded this aporie in *In XI Meta.*, 1, 2153.

[7] *In III Meta.*, 7, 403–404.

Aristotelian. Thomas was not deviating from Aristotle's position when he implies that one must discover the prime mover before one realized that being is not identical to material being. The apories, Thomas would have remarked, and this is a legitimate position to take, represent a literary or pedagogical device through which one introduces the student to the formal object; aporie three is especially important in this regard: "Are we to study all substances?" In the answer to this aporie, Thomas would note, Aristotle points out that the point of view taken (being) permits us to study all substances. Thus, this aporie presents the student with a problem solvable only through the discovery of the formal object; only if the student discovers that he will look at objects as "being", will he see that he is thus looking at all objects. Thomas' doctrine is, then, the *intentio Aristotelis*.

B. The necessity of using a logical method

Our second point in the reconstruction of Aquinas' metaphysics concerned the logical method to be used by the metaphysician. Twice Aquinas explicitly stated that the metaphysician begins his investigation with an examination of predication, that is, with a logical study. The reason behind this procedure lies in the "community" obtaining between metaphysics and logic.[1] The kinship or family resemblance of logic and metaphysics is treated by Aristotle in Book IV, Thomas noted.[2] An examination of Book IV disclosed Thomas' exposition in the context of lesson four. The logician (or dialectician) studies all being and all the essential attributes of being; in this he resembles the metaphysician.[3] The manners in which these two men study beings distinguish them, however; the logician uses the principles extrinsic to the *communia* in his attempt to prove something about the *communia*, whereas the metaphysician uses principles taken from real beings in his proofs.[4] This means, we concluded, that in Thomas' eyes, the metaphysician must study the *communia* to discover how they express the principles of being.

Our question now concerns the Aristotelian character of this exposition: *verba* or *intentio Aristotelis*? The evidence inclines one to answer that we have in Thomas' words the *intentio* of Aristotle. Let us note several comparisons of Aristotle and Aquinas.

Judging from Thomas' exposition of *In VII Meta.*, 3, 1308, Thomas

[1] *In VII Meta.*, 3, 1308. *Ibid.*, 2, 1287.
[2] *Ibid.*, 2, 1287. *In III Meta.*, 1, 345.
[3] *In IV Meta.*, 4, 573.
[4] *Ibid.*, 574.

read in the *Moerbecana* something about a logical investigation of quiddity; Thomas writes: "...Primum dicemus de eo quod quid erat esse quaedam logice ... Magis autem logice dicit [Aristoteles] se de eo quod quid est dicturum, inquantum investigat quid est quod quid erat esse ex modo praedicandi". Thus it would appear that Thomas' Latin version announced a projected logical investigation;[1] one can doubt, of course, whether the *logica investigatio* was explained by some phrase such as: "an investigation of the quiddity based on our manner of predicating concepts." Yet Thomas had but to look at the general tenor of many of the expositions in Book Z to conclude that Aristotle was engaged in an investigation of predicates: for example, chapter 4, to which corresponds Thomas' *In VII Meta., lectio 3* and part of *lectio 4*. In chapter 4, Aristotle, after his declaration of intention, begins to speak of essence as what must be said or predicated in virtue of itself.[2] Thomas appears to have a correct translation for he writes: "Hoc autem primo sciendum est de eo quod quid erat esse, quod oportet quod praedicetur secundum se."[3] Thus we can consider Thomas to be giving the *verba Aristotelis* when he speaks of Aristotle in his commentary on Book Z 4, as beginning by a logical investigation, by an analysis of predication.[4]

But Thomas says more in *In VII Meta.*, 3, 1308 than Aristotle in Z 4, 1029b 11–16. Aristotle merely remarks that he will give some "linguistic remarks" about essence; Thomas in addition states that we begin with a logical study since it is proper to metaphysics to use the method of logic, and the propriety of this usage is based on the kinship of logic and metaphysics. At first sight we appear fortunate for Thomas seems to justify his addition: Aristotle has said (in Book Γ) that logic and metaphysics have a kinship because of their "sharing" – thus the method of logic is proper to metaphysics. Turning to the section to which Thomas refers, we note that it occupies only a few lines: Γ 2, 1004b 17–26. Yet we find that Thomas' exposition fills nearly one page of the Cathala-Spiazzi edition. Evidently Thomas attached importance to this section of Aristotle; not only does he expose the doctrine carefully, but refers

[1] The Greek also gives the impression of a projected logical investigation: "And first let us make some linguistic remarks about it, that is, essence." Z 4, 1029b13.

[2] Z 4, 1029b13–14.

[3] *In VII Meta.*, 3, 1309.

[4] At the moment there is no reason to discuss the earlier passage where Aquinas explains Aristotle's method as a logical investigation: *In VII Meta.*, 2, 1277. This passage is obviously logical. (Cf. Ross, *Aristotle's Metaphysics...*, Vol. II, p. 165.) It will be considered when we note the proof of the distinction of matter and form.

to it three times in connection with the "affinity" of logic and metaphysics: *In VII Meta.*, 2, 1287; 3, 1308; *In III Meta.*, 1, 345.[1]

How does Aquinas' exposition of these lines from Γ 2 compare with Aristotle's doctrine? Has Thomas added anything? In general, the basic lines of Aristotle' sdoctrine appear in Thomas' work. 1) The fact that dialecticians and sophists imitate the philosophers by treating all things, would illustrate the claim that the philosopher treats all things; 2) all three sciences – sophistic, dialectic and philosophy – study all beings; yet 3) philosophy differs from sophistic by a difference of goals;[2] 4) thus sophistic is an appearance of knowledge, and dialectic engages in an attempt to know, while philosophy busies itself with true knowing.[3] To this doctrine Thomas had made additions, especially in points 3) and 4). These additions, however, would appear to be based on the *intentio Aristotelis.* In paragraph 574, which treats points 3) and 4) Thomas explains the difference between dialectic and metaphysics. The difference lies in their difference of "power": "Differunt autem abinvicem. Philosophus [sic.] quidem a dialectico secundum potestatem." The philosopher – the metaphysician – treats the concepts called *communia* by demonstrating from principles intrinsic to real beings, and hence from principles expressed in the *communia,* for the *communia* express beings as real; thus he is certain in his knowledge. On the other hand, the dialectician procedes from principles of the common concepts considered as objects of knowledge – from principles of *ens rationis*; thus he has only opinion.[4] These additions can hardly be considered unjustified. First of all, Thomas found this Aristotelian comparison of the dialectician, the sophist, and the philosopher in the context of an explanation that metaphysics treats the essential attributes of being as such as well as all substances and accidents *qua* being.[5] This meant, among other things as Aristotle explains, that metaphysics has to study

[1] The reference in paragraph 1308 has merely: "Sicut enim supra dictum est...," but this refers to paragraph 1287 of the preceding lesson, and this in turn to Book IV.

[2] "Signum autem est: – dialectici enim et sophiste eandem quidem induunt figuram philosofo, sophistica enim visa solum sapiencia est, dialectici autem disputant de omnibus, commune autem omnibus est ipsum ens; disputant autem de his, videlicet propter id quod philosophie sunt ipsa propria. Circa quidem enim illud genus convertitur sophistica et dialectica cum phylosofia, set differt hac quidem modo potencie, alia vero vite voluptate. Est autem dialectica temptativa de quibus philosofia est cognoscitiva; sophistica vero apparet quidem non autem est." Γ 2, 1004b17–26: *Vetus*, p. 306, l. 11–20.

[3] Whereas the *Vetus* says dialectic is "temptativa", and philosophy "cognoscitiva", the Greek has dialectic as "critical", and philosophy as "claims to know".

[4] There are other, small additions in paragraphs 576–77, but these are of no interest to us at the moment.

[5] Γ 1–2.

the concepts expressing these common attributes of beings.[1] Secondly, Aristotle's comparison of the sophist, the dialectician, and the philosopher was given as a "sign", as an "indication" that metaphysics treats all these common concepts.[2] Thus there is little wonder that Thomas sees this passage as ultimately linked to the different ways of studying *communia*. But why should Thomas explain these different ways of studying *communia* as ultimately based on the difference in principles used, or what is the same thing, in the point of view taken? Did he find this doctrine in Aristotle? Aristotle's doctrine is too well known to merit explanation. Aristotle most certainly maintained that one studies reality by using the principles of reality as premisses to demonstrate; and that one studies logical being through the use of notions such as "genus", "species", and so on. The fact that Thomas made the addition noted shows his desire to reveal the meaning implicit in this passage; he adds nothing not contained in the *intentio Aristotelis*.

However it is another question to ask why Aquinas went into such detail on this point in Book IV. If one found there a clear explanation of *why*, or even a statement *that*, "metaphysics should investigate predication", then one could easily say that Aquinas took such pains because he wished to prepare for his exposition in Book VII. However in the passage from Book IV one finds no such explanation; yet it is certainly implied. One does have, for example, the statement that the metaphysician, as the dialectician, must study all things as "coming together in being and in the essential attributes of being".[3] Is not such "being" the idea of "being"? If it is not the idea, then how do things "come together in it"?[4] Moreover when he opposes the "being" of the logician (dialectician) to that of the metaphysician, Aquinas contrasts *ens rationis* and *ens naturae*. Both the logician and the metaphysician study all things, because all things are *ens naturae* and can become *ens rationis*. Both men are to decide, each in his own way, what must be considered about the essential attributes.[5] And it is not understood in this discussion that

[1] Γ 2, 1004a31–b 1.

[2] Γ 2, 1004b17. *In IV Meta.*, 4, 572.

[3] "Conveniunt autem in hoc, quod dialectici est considerare de omnibus. Hoc autem esse non posset, nisi consideraret omnia secundum quod in aliquo uno conveniunt: quia unius scientiae unum subiectum est, et unius artis una est materia, circa quam operatur. Cum igitur omnes res non conveniant nisi in ente, manifestum est quod dialecticae materia est ens, et ea quae sunt entis, de quibus etiam philosophus considerat." *In IV Meta.*, 4, 573.

[4] There is only one alternative to saying the "being" spoken of is the idea. That would be to identify it as God. As "coming together in being," we come together in 1) an idea, and 2) in the real being which is God, in so far as we participate in His *Esse*.

[5] "...ens est duplex: ens scilicet rationis et ens naturae. Ens autem rationis dicitur proprie de illis intentionibus, quas ratio adinvenit in rebus consideratis; sicut intentio generis

the metaphysician investigates how the *communia* express the reality of things?

Necessitas authem huius scientiae quae speculatur ens et per se accidentia entis, ex hoc apparet, quia huiusmodi non debent ignota remanere, cum ex eis aliorum dependeat cognitio; sicut ex cognitione communium dependet cognitio rerum propriarium.[1]

Thus, it appears legitimate to see in the exposition of this passage from Book IV a preparation for the treatment – through a logical investigation – of being in Book VII. (Of course, the puzzling question remains: why was Thomas not more explicit?)[2]

What then is our conclusion regarding Aquinas' exposition of Aristotle's thought on this point? Insofar as Thomas explains that metaphysics is to begin with an investigation of predication (*In VII Meta.*, 2, 1287; 3, 1308), we have the *verba Aristotelis*. Insofar as Thomas discusses Aristotle's comparison of dialectic and metaphysics (*In IV Meta.* 4, 572–77), he presents the *intentio Aristotelis*. Finally, insofar as Thomas justifies the use of a logical investigation on the basis of the kinship of metaphysics and logic what do we have? On the basis of our investigations, we can certainly see that this doctrine is not the *verba Aristotelis*. Yet we should not be tempted to view it is as a meaning Thomas adds *ex sensu proprio*. And this for the following reasons. 1) For Aristotle, physics discovers the unmoved mover, and so metaphysics is born.[3] 2) The metaphysician knows therefore, as Aristotle would say, that his point of view is not expressed in a concept defined through the use of matter.[4] 3) Besides noticing these two points of doctrine, Thomas saw the logical investigation of substance given by Aristotle in Book Z (VII). (As will be clear if one reflects on the expositions of Averroes and Albert, but especially the former, Thomas' attention would have been drawn to the logical character of Aristotle's work by his predecessors, if he did not recognize it on his own).[5] 4) It appears not too hypothetical to maintain that Thomas would have asked "why" Aristotle

...Et huiusmodi, scilicet ens rationis, est proprie subiectum logicae. Huiusmodi autem intentiones intelligibiles, entibus naturae aequiparantur, eo quod omnia entia naturae sub consideratione rationis cadunt. Et ideo subiectum logicae ad omnia se extendit, de quibus ens naturae praedicatur... Philosophus igitur ex principiis ipsius procedit ad probandum ea quae sunt consideranda circa huiusmodi communia accidentia entis. Dialectus autem procedit ad ea consideranda ex intentionibus rationis, quae sunt extranea a natura rerum." *In IV Meta.*, 4, 574.

1 *Ibid.*, 1, 531.

2 Cf. the hypothesis of footnote 1, p. 251.

3 Cf. the preceding sub-section.

4 See the discussion of the definitions of the three sciences in E 1, 1025b29 sqq.

5 Cf. Ch. II, sections 2–3 above.

carried through an investigation of predication, and that he would have answered: "From formal logic we would know only that our concepts of all beings, of immaterial as well as of material, must be either generic, or specific, or proper names, and so on; but Aristotle is using logic to arrive at knowledge of things themselves, and so must be using material logic – a study of what the content of concepts reveals about reality. In fact, this was the only method open to Aristotle, for one cannot sense immaterial beings. Yet if we know they exist, and there is an unmoved mover, they must be whatever material substances *qua* substance are seen to be. What we must do – what Aristotle does – is to look at our use of concepts, at predication, and observe the characteristics of material substances as substances, that is, independent of the materiality and its consequences expressed by our concepts." In other words, Thomas would observe that Aristotle, by investigating predication, had chosen the only possible way of discovering what our concepts reveal of reality, apart from reality's material aspects. Thomas appears to have gone through a mental process similar to this, for he links the "sharing" of logic and metaphysics to the logical investigation as proper to metaphysics: because both disciplines study all beings (as either *ens rationis* or *ens naturae*) the investigation of predication is proper to metaphysics. The "missing link" can be only: "since man does not sense immaterial beings, he can construct a concept of beings *qua* being only by examining the characteristics common to valid knowledge as knowledge."[1] Thus, Thomas' statement that the method properly used by metaphysics in its initial stages is the investigation of predication represents the *intentio Aristotelis*.[2]

C. The investigation of predication: the concept of "being":

Thus far in our comparison of Aristotle and Aquinas, we have found that the latter expresses doctrines which can be characterized as the "intentions" of Aristotle; when we compare their statements on the concept of "being", the situation changes somewhat.

In our examination of Aquinas' thought on the investigation leading to the discovery of the content of "being" (actualized essence), we were forced to synthesize many widely scattered passages. Several of these passages, however, were found in lesson nine of Book V; in fact this

[1] We speak here of the characteristics of "knowledge as knowledge," that is, of knowledge as the expression of realities. We are not referring to the relationship between various acts of knowledge, for that is as matter of formal logic.

[2] For a somewhat similar view of Aristotle's method, see Ross, *Aristotle's Metaphysics...*, Vol. I, p. lxxvii; Allan, *Aristote le philosophe*, p. 109.

lesson set the tone for our entire reconstruction. Thus in this sub-section let us first examine Aristotle's Δ 7, the passage forming the base of Thomas' Book V, lesson nine.

Aristotle opens the discussion by dividing the predication of "being": "Ens dicitur hoc quidem secundum accidens, illud vero secundum se", we read in the *Media*, the basis of Thomas' exposition in this lesson.[1] The *secundum accidens* predication is then subdivided into three classes; it is important only to note that in all classes, "being" means "happens": " 'Hoc enim esse hoc' significat accidere huic hoc."[2]

When Aristotle turns to the *secundum se* use of "being", he remarks that things said to be in this way are all those things which signify the figures of predication. If we ask: "What signifies a figure of predication?" it is hard to answer other than: "Concepts". But of course we would not say that the concepts *exist secundum se.* The inexactness of our language is corrected by the reason Aristotle gives: As many ways as something is predicated, in that many ways "being" is signified. Thus Aristotle means that an investigation of predication will reveal the various categories; moreover, each category signifies or reveals a manner of "being".[3]

Turning then to an investigation of these *per se* or *secundum se* types of predicates, Aristotle lists only eight: some signify *quid est*, others *quale, quantum, ad aliquid, facere, pati, ubi, quando.*[4] Corresponding to each of these categories, he notes, is a meaning of "being".[5] Thus Aristotle has first, divided *secundum accidens* from *secundum se* predication; and secondly, divided *secundum se* predication into eight types according to the eight types of predicates; and finally, stated that here are eight meanings of "being" – and eight types of extra mental being, for he implies the validity of our knowledge.

As if he were giving something logically presumed in the divisions just made, Aristotle explains that the word "is" is part of the predicate;

[1] Δ 7, 1017a7. The *Media* version of Δ 7 is found on pp. 233–34 of the edition of B. Geyer.

[2] Δ 7, 1017a12–13, *Media.*

[3] Δ 7, 1017a22–24. There is a slight difference between the *Media* and the Greek texts. Whereas the *Media* directs us to concepts, the Greek directs us to the figures of predication. Thus the *Media* does not presuppose the figures of predication; rather, it indicates that we are to move from an examination of concepts to the discovery of the figures. On the other hand, the Greek would seem to presuppose the figures; look at them, it seems to say, and see how many types of concepts there are. Thus the *Media* indicates that it is the metaphysician who is to discover the categories in his attempt to construct the concept which expresses reality as such. And that is the doctrine presented by Aquinas in *In V Meta.*, 9, 889–90.

[4] Δ 7, 1017a24–27, Cf. Ross, *Aristotle*, p. 166.

[5] Δ 7, 1017a27.

there is no difference between: "The man convalesces", and "The man is convalescing".[1]

This addition is followed by the remark that there are two more types of *secundum se* predication of "being". First, "is" can signify the truth of a proposition. For example: "Socrates is a musician" – here "is" is our way of saying that one is correct in thinking of "musical" as a quality of Socrates.[2] In a second type of *secundum se* predication of "being", "is" can be "potentially is" or "actually is".[3]

Thus Aristotle has given four ways of predicating "being" (*ens*). 1) The word "is" can mean "happens". 2) The word "is" is part of the predicate; there are eight types of predicates such as "is X"; to these correspond the figures of predication as well as eight types of being. 3) The word "is" can signify the truth of a proposition. 4) The word "is" can signify actual or potential being. The implication is that the last two significations of "is" can be had by any of the eight types of predicates which reveal the eight types of reality, as by any of the predicates which reveal what merely "happens" to be the case.

If we turn now to the question of conformity between Thomas' exposition and Aristotle's, we must concede that Thomas has added only details, and a reference to Avicenna (and, unless one knows ahead of time what he means: confusion). The detail is important, however, for Thomas mentions the substantial, metaphysical *esse* had by substantial beings. Moreover, Thomas mentions that the concept *ens* is superior to any other concept just as the concept "animal" is superior to the concept "man".[4] Thus for Thomas, "being", as a substantial predicate of the form "is X", would express completely, though potentially, everything expressed by a less general, substantial concept.[5] By the reference to Avicenna, Thomas makes the following point: a concept of an accident includes a reference to substance.[6]

One looks in vain, of course, for a doctrine of metaphysical *esse* in Aristotle: an existing thing is a substance (*ousia*) and one cannot dis-

[1] Δ 7, 1017a27–30.

[2] Δ 7, 1017,31–35.

[3] Δ 7, 1017a35–b9. For a general discussion of Δ 7, cf. Ross, *Aristotle's Metaphysics...*, Vol. I, pp. 306–309. G. E. M. Anscombe and P. T. Geach, *Three Philosophers. (Aristotle, Aquinas, Frege)*, Blackwell, Oxford, 1961, pp. 11–39.

[4] "Esse vero quod in sui natura unaquaeque res habet, est substantiale. Et ideo, cum dicitur, Socrates est, si ille Est primo modo accipiatur, est de praedicato substantiali. Nam ens est superius ad unumquodque entium, sicut animal ad hominem." *In V Meta.*, 9, 896. In the context, *est* is to be understood as the predicate or concept "is a being".

[5] Also Aquinas adds two categories to Aristotle's list of eight, but the additional two are given by Aristotle in *Cat.*, 4, 1b25–5, 2a19.

[6] *In V Meta.*, 9, 894.

tinguish between the substance of a thing and its species (*ousia*), except insofar as the substance has accidents which are not part of the species.[1] Thus, the doctrine of *esse* as a substantial principle is definitely not something Aristotle even hinted at.

On the other hand, Aristotle would seem to have held that the concept "being" – as a substantial predicate – expresses completely though potentially everything expressed by a less general substantial concept. At least Thomas indicates that he understands Aristotle to have given this doctrine. For example, in commenting on the general characteristics attributed to the wise man by the common consent of the "man on the street", Thomas introduces the notion of "being", as the most general of all concepts, and justifies this doctrine by a reference to the *Physics*.[2] Immediately following this paragraph, Thomas explains that the concept of "being" expresses all aspects of everything.[3] As we noted in the preceding chapter, these doctrines can be considered as *verba Aristotelis*.[4]

A third addition, made in connection with a reference to Avicenna, is in reality a doctrine which can be called "new" only in reference to Book Δ 7, for Aristotle explains in many other places that every definition of an accident must refer to substance.[5] Thus, Thomas by making this point in his exposition is only insuring that the dependence of the nine categories of accidents on the category of substance is correctly understood: Thomas safeguards Aristotelian truth against the Avicennian interpretation by placing the former in its total context. Thomas is giving thus the *verba Aristotelis* and nothing more.

Lesson nine of Thomas' commentary on Book Δ is then almost identical to the corresponding passage in Aristotle. Yet that "almost identical" makes all the difference in the world: Thomas has replaced the essential being of Aristotelian metaphysics with an *esse*-essence composite being. Thus, Thomas' exposition is actually a transformation of Aristotle. Insofar as Thomas' writing contains an investigation of predication which leads to the discovery of the category of being, of the concept of

[1] Thus, both Aristotle and Aquinas could admit that a concrete thing is its essence. But when Aquinas teaches this doctrine, he implies that we must distinguish between the essence and ist actuality. That is, we must distinguish the meaning of the "to be" of the object, and the actuality of that meaning-full "to be". Thus, for Aquinas, a concrete thing is the actuazlied essence (plus the accidental determinations); yet an "actualized essence" is composed of two principles, composed of actuality and of essence.

[2] *In I Meta.*, 2, 46.

[3] *Ibid.*, 47.

[4] Cf. Ch. V, section 4,A; section 5.

[5] E. g. Z 4, 1029b22 – 1030a2; 5, 1031a1–4. Thomas introduces this doctrine into his *In V Meta.*, 9, 894.

"substance", Thomas is true to the Aristotelian Δ 7; but insofar as the category of "substance" is a category containing concepts expressing an *esse*-essence composite, Thomas' thought is not that of Aristotle.

In the preceding section which dealt with the reconstruction of Aquinas' metaphysics, an attempt was made to explain how Aquinas would expose the discovery of the distinction of *esse* and essence. Accomplishing this was like putting together a jig-saw puzzle. We had Thomas' starting point: an investigation of predication. We had Thomas' goal: the discovery of the *esse*-essence composition of beings. And we had various doctrines to help us pass from the initial investigation to the goal: the results of *In V Meta., lectio* 9; "ens dicitur quasi esse habens"; "humanity" expresses the essence of a man; "man" expresses "having humanity"; and so on. Such doctrines implied both the distinction of *esse* and essence, as well as the manner of conceiving an *esse*-essence composite. In our present study of Aristotle, let us look to the doctrines corresponding to those of Thomas such as "ens dicitur quasi esse habens".

1) Take for example *In XII Meta.*, 1, 2419. It appears rather difficult not to see here an affirmation of "being" as a concept involving the expression of *esse*: "Nam ens dicitur quasi esse habens..."[1] Thomas has introduced this phrase into a proof that metaphysics considers substances and not accidents. Aristotle wrote: "The subject of our inquiry is substance; ...At the same time these latter [id est, the accidents] are not even being in the full sense, but are qualities and movements of it..."[2] It is in the context of explaining why accidents are not being in the full sense that Thomas presents the "full sense" of being: "Nam ens dicitur quasi esse habens, hoc autem solum est substantia, quae subsistit. Accidentia autem dicuntur entia, non quia sunt, sed quia magis ipsis aliquid est..." In this case Thomas has completely transformed Aristotle's essential being into an *esse*-essence composite.

2) Another interesting exposition is that of *In VI Meta.*, 4, 1241. The context is Aristotle's declaration that the being which is expressed by the "is" of a judgment is not studied in metaphysics. Aristotle writes:

> But since the combination and the separation are in thought and not in the things, and that which is in this sense is a different sort of "being" from the things that are in the full sense (for the thought attaches or removes either the subject's "what" or its having a certain quality...), that which is accidentally and that which is in the sense of being true must be dismissed.[3]

[1] M. Gilson would disagree, as was mentioned in footnote 2, p. 266.

[2] Λ 1, 1069a18; a21–23.

[3] E 4, 1027b29–34.

Aristotle says here that the intellect "attaches or removes" a "what"; Thomas says that the intellect finds in things a composition which makes a thing one thing, and that the intellect expresses this composition as a "that which is":

...compositio et divisio, ...est in mente, et non in rebus. Invenitur siquidem et in rebus aliqua compositio; sed talis compositio efficit unam rem, quam intellectus recipit ut unum simplici conceptione. Sed illa compositio vel divisio, qua intellectus coniungit vel dividit sua concepta, est tantum in intellectu, non in rebus... Et ideo illud, quod est ita ens sicut verum in tali compositione consistens, est alterum ab his quae proprie sunt entia, ...quarum unaquaeque est "aut quod quid est," id est substantia, aut quale, aut quantum"...[1]

Aristotle and Aquinas hardly say the same thing in these parallel passages. For Aristotle, the mind can either conceive the "what" of an object, or judge something about the object. For Thomas, the mind can conceive what makes an object one thing, or else judge something about that object. And what makes a thing one? Although he implies a plurality of component principles in a thing, Aquinas doesn't identify either the nature or the number of these principles. Yet if we recall what he said previously about the *esse*-essence composition which makes a thing a substance, we realize that once again Aquinas has gone beyond the meaning of Aristotle's texts.

There is but one way to understand what Aquinas is trying to say in his *Commentary*, and that is to place each text in the total context of the work. Such an effort reveals that for Aquinas, "being" means more than essence. "Being" or "that which is" means "esse habens", or "habens quidditatem". And this is not at all what one finds in the *Metaphysics* of Aristotle.

D. The investigation of predication: the distinction of matter and form

Aquinas' metaphysics have thus far been revealed to be anything but the mere repetition of Aristotle's words. Thomas, we have seen, gives the intention of Aristotle when he speaks of the genesis of metaphysics, and of its initial method. Yet these doctrines, expressing the *intentio* of the Greek, are transformed by Aquinas insofar as Aristotle's "being" has given place to the "*esse*-essence composite" of Aquinas. In that step of metaphysics we have now to discuss, the distinction of matter and form, it is extremely difficult to determine whether Aquinas believed he was giving the words or merely the intention of Aristotle.

After Aquinas had discovered the distinction of *esse* and essence within his concept of "being," he spelled out another doctrine discover-

[1] *In VI Meta.*, 4, 1241.

ed by the same logical investigation as produced the knowledge of the concept of "being". The ability of material substances to serve as logical subjects (subjects of predication) was brought to the fore as Thomas asked: What does it mean for a material being to be capable of serving as the subject of a proposition?[1] (This question had already been partially answered through the study of *per se* predication: a material substance, like all substances, must be composed of *esse* and essence. But there is still the *per accidens* predication to investigate: a material substance, e.g. Socrates, can be the logical substance of *per accidens* predication). An investigation into the propositions concerning a material substance, propositions where the definitions of the subject and of the predicate are different, can lead us to realize the distinction of matter and form in the essence of such substances.

As we have remarked, it is difficult to determine exactly Thomas' intention in *In VII Meta.*, 2, 1280; 1287–89, where he treats this matter. Obviously much of the doctrine he exposes in the context of this lesson is not the thought contained in Aristotle's Z 3. Let us compare Aristotle with Aquinas closely to discover exactly what the latter has done.

Aristotle	*Aquinas*
1) There are four uses of "substance," but of these, the one most commonly thought to be correct is the use of "substance" to denote a substratum. Let us examine that. (Z 3, 1028b33–1029a2)	1) Practically the same as Aristotle except that he notes that the substratum called "substance" is first substance. (*In VII Meta.*, 2, 1270–75)
2) There are three items which seem to fit the definition of substratum: "what is not predicated of anything else, and of which all things are predicated"; those three are matter, form and the composite of them. (1029a2–5)	2) First subject, or the logical subject of predication can be divided into matter, form, and the composite. (*Ibid.* 1276–77
3) Form is obviously prior to matter and to the composite of matter and form. What we have said about substratum in the definition given immediately above is not sufficient. (1029a5–10)	3) Form is indeed prior to the other two items. Aristotle has not sufficiently explained "substance" in the sense of a logical subject by these few lines; what he had thus far said – the definition of logical subject – is only a condition to help us know what first substance really is in itself. (*Ibid.*, 1278–1280)
4) This definition of "substance" is not sufficient, for it would follow that matter is especially substance. Proof: take away length, breadth, potency, quality, take away all these and we	4) Aristotle now begins to explain the theory of some forgotten philosopher who, because he was unable to distinguish between substantial and accidental form, concluded that matter was the

[1] *In VII Meta.*, 2, 1280.

have matter left. Such attributes belong to matter; hence it is matter that is above all substance. (1029a10–19)

substance par excellence. (*Ibid.*, 1281–84)

5) By matter I mean what is outside all the categories. Thus if we adopt the point of view expressed in the definition above (point 2), matter is substance. (1029a20–27)

5) Since the opinion just given followed on ignorance of the true nature of matter, Aristotle begins to explain what it really is. He chooses first to follow the physical method, the study of motion, for that method is the only source of adequate knowledge of matter. Yet it is also possible to give a metaphysical proof based on an investigation of predication. In this examination, one is interested in "denominative predication", that is, the quiddities of the subject and of the precdiate are completely different. E.g.: "This man is white". If we put all the formality in the precdiate, we have: "This materialed thing is a white man". Hence we know that there must be some ultimate, unformed X which receives forms; there must be prime matter. (*Ibid.*, 1285–90)

6) It is impossible to accept the conclusion that matter is most especially substance. A substance has "separability" and "thisness" while matter has neither. (1029a27–30)

6) Now Aristotle speaks of the philosopher mentioned in point 4). He notes that a substance must be "separable" and a "this". Obviously neither of these belongs to matter, but rather to the composite of matter and form. (*Ibid.*, 1291–93).

Interestingly enough, despite the changes Aquinas has made in this chapter of Aristotle's Book Z, the goal of the chapter has not been affected. Aristotle starts out with the possibility that if we seek what fits the logical definition of substance, we might arrive at knowledge of what substance is. But it turns out that it is matter which most perfectly fulfills the definition of a logical subject: "what is not predicated of anything else, and of which all else is predicated". Hence we must begin anew, examining now either form or the composites of matter and form; but it's better not to examine the composite, both because it is posterior to matter and form, and because we know already what a composite is. Hence, we shall examine form.[1] And we shall study it in the essences of material substances.[2] Thus Aristotle's goal was to illustrate how wrong it would be to examine what most perfectly fits the

[1] Z 3, 1029a30–b2.
[2] Z 3, 1029b1–12.

definition of a logical subject; his goal was to show concretely, by a series of eliminations, that we must study the form of sensible beings.

Now in the long run, Aquinas' exposition does not deviate from this goal of Aristotle. It is true that he sees a metaphysical proof of the distinction of matter and form where there was none, that is, where there was none if we judge from the role lines 1029a20–27 were primarily intended to fulfill (point 5 above). Yet in the end, Thomas' work rejoins that of Aristotle's; after proving the distinction of matter and form, Aquinas remarks that we know matter through knowing form, and so it is form that we want to study.[1] Moreover, knowledge of the composite of matter and form is posterior to knowledge of matter and knowledge of form; hence we do not want to study the composite first of all.[2] Thus we are left with only form as a proper subject of our investigations. Where shall we study it? Certainly not in immaterial substances, but in material ones.[3] Thus, Aquinas concludes, we choose to study the quiddity of sensible substances.[4]

If, then, Thomas' metaphysical proof for the distinction of matter and form does not affect the main point of the chapter of Book Z, can we see any reason why Thomas introduced it? That is, should we characterize this proof as *verba*, as *intentio Aristotelis*, or as a Thomist meaning?

As an aid in formulating our answer, let us consider several points: 1) ambiguity in Z; 2) a passage from Z 13 which seemingly contradicts Z 3; and finally 3) certain aspects of Thomas' *Commentary* and of his conception of metaphysics.

1) There is quite definitely a good deal of Z 3 which is none too clear. Aristotle opens with a definition of substratum in terms of predication; then he reveals the impossible consequences: matter is substance. Yet when he explains what he means by matter, Aristotle comes to using the definition of substratum he says is misleading. Substratum is that which is predicated of nothing, and of which all else is predicated. And when Aristotle explains prime matter, he seems to say it is that which is predicated of nothing, and of which substance is predicated; although he does not say "all else" is predicated of matter, he does say that all concepts of accident are predicated of substance and substance of matter.[5] In other words, even though he set up the argument for matter

[1] *In VII Meta.*, 2, 1296.
[2] *Ibid.*, 1295.
[3] *Ibid.*, 1297–98.
[4] *Ibid.*, 1299.
[5] Note the following:"By matter I mean that which in itself is neither a particular thing

as substance and then refuted it, his refutation does not do its job. One wonders whether Thomas was not taken aback by this passage.

2) A passage of Z 13 contradicts Z 3, if the latter chapter is understood as denying that matter is substance. In Z 13, 1038b2–6 Aristotle makes these following points. We are investigating substance, and have seen that substratum, essence, composite of substratum and essence, and universal are all called "substance". We have already treated the first two of these four. Regarding substratum, we have said "that it underlies in two senses, either being a "this" which is the way in which an animal underlies its attributes, or as the matter underlies the complete reality".[1] It would appear that Aristotle contradicts our reading of Z 3. But what did Thomas think of this passage in Z 13? He could most certainly read here that matter is called "substance", and this is exactly how he did interpret it.[2] Moreover, Aquinas understood this Aristotelian passage in relation to the earlier chapter 3; this division of Z 13, he wrote, is actually the same as the four-fold division of substance given in the beginning of Book VII.[3] Thus Thomas appears to have believed Aristotle did indeed hold the theory that matter is substance in some sense.[4] Accordingly, one can with some probability conclude: because of Aristotle's statements in Z 13, Thomas felt that Z 13 must be interpreted as an affirmation of the validity of saying "Matter is substance".

Thus it appears quite probable that Aquinas exposed Z 3 in the light of Z 13. He might have thought there was something "impossible" about affirming matter to be substance, unless it was possible to interpret Aristotle's Z 3 as accepting matter as substance. Perhaps A 3 proved helpful in this regard, as there Aristotle had discussed the theories of the ancient philosophers who saw the substance of everything as matter: perhaps their problem was touched in Z 3 – they did not know how to distinguish substantial form from accidental.[5] Thus Aquinas could explain Aristotle's criticism of "matter is substance", as a criticism of "only matter is substance". And so he could understand

nor of a certain quantity nor assigned to any other of the categories by which being is determined. For there is something of which each of these is predicated, whose being is different from that of each of the predicates (for the predicates other than substance are predicated of substance, while substance is predicated of matter)." Z 3, 1029a 20–30.

[1] Z 13, 1038b1–6.

[2] *In VII Meta.*, 13, 1566.

[3] *Ibid.*, 1567.

[4] When Aquinas explains how matter has been treated, he does not refer back to lesson 2 (or to Aristotle's Ch. 3) however; rather he implies that it is lessons 7–12 he has in mind. He, does not, though, imply that the proof that matter is substance was not treated in lesson 2. Cf. *In VII Meta.*, 13, 1568.

[5] *Ibid.*, 2, 1284.

why Aristotle begins in Z 3 to disclose the nature of matter, and to do this by using a principle very similar to the one which caused the trouble for the ancient philosophers. As Thomas says: "Because these philosophers erred through their ignorance of matter, Aristotle shows how matter needs form to constitute a substance."[1] Thus first of all the physical proof of the distinction of matter and form is given (par. 1285–86), and then the metaphysical proof (par. 1287–90). Moreover, the metaphysical proof is based on the proper use of the original definition of substance (par. 1280–98). In paragraph 1280 Thomas mentions that this logical definition of substance is the instrument through whose use we come to know the principles of things; in paragraphs 1287–1289 he uses this principle, slightly modified, or better still: properly understood. The ancient philosopher implicitly argued this way: if we take the proposition "Socrates is", we can gradually remove the qualifications or determinations from the subject and place them in the predicate; thus we could say "Socrates is white, cold, tall, etc." and eventually we would have nothing left in the subject but the substance, that is: fire, air, earth, and water – only matter is left. (par. 1282–84). To this argument of the ancient philosopher, Aquinas adds this precision: whenever we take a qualification, a determination, from the subject and place it in the predicate, it is a formal aspect of our knowledge that we are thus transforming to the predicate. Thus in "Socrates is black", we have placed our knowledge of his color into the predicate; finally, when we say: "Socrates is black, fat, tall man", what is left in the subject? Nothing except that which is neither *quid*, nor *quale*, nor *quantum*, nor anything else; there is left only that of which everything else is predicated: there is only prime matter left in the subject. (par. 1287–89). What is important is Thomas' decision to see the impossibility of affirming: "Matter is substance", as meaning: "It is impossible to say that substance is only matter". We cannot know whether this was Thomas' *Deus ex machina*, that is, his manner of integrating an otherwise senseless passage (Z 3, 1029a10–29), or whether he actually felt Aristotle was referring to these early philosophers mentioned previously in Book A. Yet we can say, and we must, that because Thomas read in Aristotle's Z 3 a metaphysical proof of the distinction of matter and form, his action was to some extent dependent on the affirmation of Z 13: "matter is substance". Thus we are inclined to see in Z 3 an example of how Thomas gives the *intentio Aristotelis*.

3) There are still other aspects of the problem, aspects arising from

[1] *Ibid.*, 1285.

the *Commentary* as a whole. We noted in Chapter IV the importance of the concepts *ens, unum,* and *res*: the metaphysician desires to clarify these concepts until he can see each and everything through them; in Chapter V we brought out the doctrine of metaphysics as the study of the *communia,* those common notions to be used in all scientific knowledge. Thus we should not be surprised at what Thomas says and does in Book V; he notes that that book studies the notions common to all things; and in explaining the various notions, Thomas always at least implies the primary meaning of every name, and the relations of other, secondary meanings to the primary one.[1] Nor should we be surprised when Thomas begins the study of the essence of sensible substances by saying that first we study essence logically – through logical common meanings – and only afterward do we study the real principles of sensible substances.[2] In other words, the metaphysician must clarify his concept of *ens* until he can see any being through it. He takes a step in this direction when he distinguishes *esse* and essence in that concept; he takes a further step when he works with *ens* as applicable to material beings and distinguishes not only *esse* and essence, but within essence matter and form. Moreover, if the metaphysician does not prove the distinction of matter and form, but takes it from the physician, then he is not actually clarifying the concept of being as applied to sensible substances.[3] There are still further aspects of the *Commentary.* For example, if the metaphysician is to study concepts, he is obviously close to logic; it would naturally be wise for Thomas to say this clearly. Thus we find a clear affirmation: 1) that matter and form are shown to be distinct by the metaphysician, 2) who discovers this by investigation of predication; 3) moreover, this method is proper to logic; 4) yet metaphysics and logic are kin (and so, it is implied, metaphysics can use this logical method).[4] Again, it is clearly stated that 1) metaphysics begins its study of the essence of sensible substances by a logical investigation; 2) this method is proper to metaphysics because metaphysics and logic have a "community", a sharing of the same object.[5]

Now all these aspects of the *Commentary,* excepting of course the notion of "being" as an *esse*-essence composite, are Aristotelian:

[1] E. g. *In V Meta.*, 5, 824 for the meaning of *natura.* For the analysis of the name *ens* its primary and secondary meanings, see: R. Mc Inerny, "Notes on being and Predication," *Laval Phiosophique et Théologique*, XV, 1959, pp. 236–74.

[2] *In VII Meta.*, 3, 1306.

[3] On the *communia* and the metaphysical study of them, see: *In XI Meta.*, 1, 2146; *In VI Meta.*, 1, 1147–51, and 1165; *In VI Meta.*, 1, 531.

[4] *In VII Meta.*, 2, 1287.

[5] *Ibid.*, 3, 1308.

intentiones at least. Hence, from these considerations of the total context of the *Commentary*, it appears necessary to conclude that the presentation of the metaphysical distinction between matter and form was part of Aquinas' explanation of the *intentio Aristotelis*.[1]

E. The discovery of the existence of God

In exposing Thomas' thoughts on the proof of the existence of God, we made three important points. 1) We noted that scientific knowledge is concerned with a *doctrina*, that is with knowledge of the essence of its material object. A science can ask for the formal, efficient, or final causes of its material object as known in the *doctrina*. If a science proves the existence of an immaterial being, this proof must involve a search for the efficient and final causes of the object the science expressed in its *doctrina*. Yet this search for the efficient and final causes, resulting in the discovery of God, must be different from all other "searches" for other efficient and final causes. These latter investigations must seek the particular efficient or final causes of particular material beings. The former search – the metaphysical one – is an attempt to find the particular efficient and final causes of there being any formal, final, or efficient causality within the universe we can know. Thus the causes found by the metaphysician logically (that is, in the order of sciences) and ontologically ground the particular causes of an individual being. 2) Our second point was to ask the question: But psychologically speaking, what is it that makes the metaphysician search for the causes which will logically and ontologically ground the particular causality found within the material world? In other words, what is the lack of intelligibility within the material world, which when recognized, drives the metaphysician to the discovery of an immaterial being? This lack of intelligibility, it was argued, resides in the *esse*-essence composition of material beings. Because we must think of things as "actualized essence," we must think of them as having received their *esse*'s from a being which is identical to its *esse*. As was argued, this proof seems implied in Thomas' declaration: "All composed and participating things must be reduced to a simple thing". Moreover, we felt this same proof was involved in Thomas' discussion of measure in Book X. There he wrote that we are forced to look for some being which will be the measure of substance. 3) A second proof of God was found in the need of an intelli-

[1] Perhaps the anti-Averroistic aspect of Aquinas' *Commentary* explains in part the attempt to integrate the study of matter into metaphysics. For Averroes' theories, see Ch. II, section 2 above.

gent ordaining cause for the intelligent activity present in the world. And finally we discussed the proof from motion given in Book XII; this proof, it was explained, is not a new one, but represents Aquinas' method of insuring that the statement "God is" is acceptable to all. Thus Aquinas notes a possible fault with Aristotle's physical proof from the eternity of motion, but at the same instant, he insists that we can prove God as the cause of being even if the physical proof rests on the false theory of the eternity of motion. We must now compare these three aspects of Aquinas' thought with the writings of Aristotle.

First of all, what is Aristotle's mind on the *doctrina* of metaphysics and on the manner of studying this *doctrina*? Our reconstruction of Thomas' thought revolved around *In VII Meta., lectio* 17. In the corresponding Aristotelian passage (Z 17) we find much the same doctrine as given in Thomas' work; as usual, Thomas' exposition is more orderly and leaves little unsaid. Summarized briefly, Aristotle's thought is as follows: 1) whenever we ask "why?", although we know something already about an object, we are seeking to discover something else, something new.[1] 2) "why?" can be a question seeking the end, or even the first mover;[2] 3) however, in other cases, by asking "why?" we seek the formal cause "by reason of which the matter is some definite thing";[3] 4) whenever we are studying simple beings, we cannot ask "why?" in this same way – "our attitude towards such things is other than that of inquiry".[4] In this last statement Aristotle is of course thinking in terms of the immateriality of simple things: because they have no matter, we cannot first know something about them, e.g. that they exist as such or such things, and then ask for their form.[5]

Such a doctrine is identical to that exposed by Aquinas, if we except the latter's developments and additions. For example, Aristotle was content to note that our study of simple things must employ a different method than our study of material things; to this Thomas adds a note on the method used in investigating simple beings. As Thomas explains, we begin with material beings as effects, and from a consideration of these rise to a knowledge of their causes, the simple beings; thus the knowledge of material beings is a "quasi middle term" in our search for the immaterial beings. This method, Thomas writes, is the one Aristotle is found to practice in Book XII when he discusses the first

1 Z 17, 1041a10–28.
2 Z 17, 1041a28–30. Cf. Ross, *Aristotle*, pp. 172–73.
3 Z 17, 1041a32–b9.
4 Z 17, 1041b9–11.
5 Z 17, 1041b4–9. Also: Θ 10, 1051b17–33.

cause of all motion.[1] That Thomas' additions are in fact Aristotelian seems sufficiently obvious.[2] The only difference between Thomas and Aristotle seems to be one between an explicit doctrine on method (Thomas) and the practice of that method (Aristotle).

Our second point in the reconstruction of Aquinas' metaphysical search for God was to seek the psychological origin of this search. As we noted, Thomas insists on reducing composed or participating beings to simple ones.[3] Although there was a difficulty involved in the interpretation of this principle, we concluded that we must see in it a necessity of positing a substance whose essence is *esse*.[4] Thus, we saw in the knowledge of the participated character of material beings, the psychological origin of the search for the cause of being.

At first glance, this emphasis on "participation" appears anything but Aristotelian. However, let us note certain passages of Books α and I. α 1, 993b19–31 of the *Arabica* clearly notes this principle: "Et unumquodque principiorum proprio est causa eorum, secundum quod sunt aliae res, quae conveniunt in nomine, et intentione, verbi gratia ignis in fine caliditatis".[5] This principle deals with effects similar to their cause insofar as they belong to the same species: e.g. all hot objects as hot, are in the same species as fire, the cause of their heat. In what follows the statement of this principle, Aristotle appears to emphasize only that the cause of any quality will possess that quality to the highest degree; he forgets that his principle implies that the cause and effect belong to the same species: "Conveniunt in nomine, et intentione". In this spirit, lines b25–31 emphasize that the cause of being and the cause of truth will be the most true and (by implication) most being; thus the principles of heavenly bodies are the most true and the most being.[6] Aristotle thus implies that the first cause of being is in the same species as all other being; but on the other hand, since being is neither

[1] *In VII Meta.*, 17, 1671; cf. Λ6, 1071b3–26.

[2] Cf. *In III Meta.*, 4, 384 with its reference to Book IX (probably to Θ 10, 1051b17–33; cf. *In IX Meta.*, 11, 1901–1903). See also: VERBEKE, "Démarches de la réflexion métaphysique chez Aristote," esp. pp. 121–29.

[3] *In II Meta.*, 2, 296.

[4] This text was connected with the search for the "measuring *unum*" of *In X Meta.*, 3, 1967–73.

[5] For the *Arabica* see: fol. 29v,L. On the use of this text by Thomas, both in the *Commentary* and his other works, see DE COUESNONGLE, "La causality du *maximum*...," pp. 434–44; 658–80.

[6] "Ex quo oportet ut maxime verum sit illud, quod est causa veritatis rerum, quae sunt post. Et ideo necesse est ut principia rerum, quae sunt semper, sint semper et in fine veritatis. Quia non sunt vera in aliquo tempore, et in alio non: neque habent causam in essendo vera in eo quod sunt vera, sed illa sunt causa in hoc aliarum rerum. Quapropter necesse est ut dispositio cuiuslibet rei in esse, sit sua dispositio in rei veritate." *Arabica*, fol. 29v,L–M.

species nor genus, he obviously is not implying this at all, regardless of what the immediate context may force one to believe.[1] But what Aristotle does at least explicitly say is this: everything has a cause of its being, a cause which is the most being.

Thomas' exposition follows Aristotle very closely. Thomas noted the following points. 1) The principle given in lines b23–25 deals with causes and effects falling in the same species.[2] 2) Aristotle mentions this limited principle by way of contrast to the relation between the cause of being and truth, and the effects of this cause.[3] 3) Thus Aristotle expressly says that the heavenly bodies have a cause of *esse* and not only of motion.[4] Thus far Thomas' attitude is surely one of interest in the *verba* of his Greek master. But in paragraph 296, Thomas writes: "Et hoc est necessarium: quia necesse est ut..."; thus, here Thomas sets out to explain the reason which forces one to write as Aristotle did. As Thomas explains it, this reason is the necessity of reducing all composed and participating things to a simple one. But did Thomas find this reason in Aristotle? That is, does Thomas give this reason as the motive behind Aristotle's statement? Or is this reason intended to show the truth of Aristotle's statement?[5]

Aristotelian metaphysics – as it is, untainted by Thomist *esse* – demands to be completed by the theory of participation. Aristotle in Book I 2, 1053b24–1054a13 deals with the search for the first substance, the measure of all other substances. Moreover, the universal science is to become the first science; the implication is that one rises from the study of being as such to the cause of being.[6] And finally, in Book α Aristotle speaks of a cause of being. Now the prime mover of Book Λ is not the cause of being; *qua* mover, he is not the ultimate reason why material substances are found to exist, or to be substances. But since Aristotle indicates that we must look for the first substance which measures others, which makes others to be substances, well, as Thomas could say: "Manifestly, it's incorrect to say that Aristotle does not prove the existence of a cause of being."[7] Yet, why does Aristotle set forth the

[1] Cf. B 3, 998b22–27; K 1, 1059b31–34.
[2] *In II Meta.*, 2, 292–93.
[3] *Ibid.*, 293–95.
[4] *Ibid.*, 295
[5] Although Aristotle's remarks on participation are not especially favorable, these remarks belong in a well-defined anti-Platonic context; see: A 9, 992a28–29; 991a20–22; *Vetus*, p. 279, l. 8–9; p. 276, l. 35-p. 277, 1. 2. Hence, from the presence of these remarks, one can not argue that Aristotle would be against participation in every sense of the word.
[6] Cf. the discussions of Ch. IV, section 3, B and Ch. V, section 1, A.
[7] Cf. *In VI Meta.*, 1, 1164.

necessity of looking for a first substance which measures others, if he doesn't find one? And why does Aristotle, although stating in Book α that everything needs a cause of its being, nevertheless fail to look for that cause? These are unanswerable questions from a medieval point of view.[1]

Thus Thomas was faced with some serious difficulties in composing his *Commentary*. First, Aristotle on three occasions at least implies the need to discover a cause of being (Books α, E, and I); yet he doesn't succeed in reaching anything higher than a first mover. Thomas' method of resolving these difficulties we know: the introduction of *esse*.

Hence, the introduction into Book II of the principle of the reduction of the participating or composed to the participated or simple is, at best, Aquinas' method of co-ordinating Aristotle's scattered remarks. By the inclusion of that principle, Thomas implies the theory of *esse*, to be sure; Aquinas says in effect, that because Aristotle has a proof for God, it is not impossible that it was based on a participation in *esse*. It does not seem possible for us to say more than this; it does not appear clear that Aquinas introduced participation, or *esse*, because he wished to expose his own personal thought.

The third and final point we must study in Aristotle concerns his thought first, on the proof of God as the directing intelligence, and secondly, his proof of the prime mover. As regards the first point, the proof of an ordaining intellect, Aquinas does not present this doctrine as a proof. That is to say, there is no evidence that his presentation is the presentation of a proof for God. Just the opposite, the doctrine offered as a "proof" is given by Thomas in a type of aside: Don't think that what Aristotle has said destroys the doctrine of God's directive intelligence, he remarked; quite the contrary, the fact of intellectually directed activity in our world demands that there be a supreme organizer of the entire affair.[2] Yet despite the informality of the context, what Thomas said here can be construed as a proof for God. Of course, there is nothing in the corresponding Aristotelian passage that even remotely resembles this paragraph of Aquinas, nor does looking to the context of the entire *Metaphysics* yield anything comparable. Thomas' disgression is, then, his own way of making certain that Aristotle is not misunderstood in matters of moment.

[1] Yet they were real questions for Aquinas, who maintained that Aristotle does indicate there must be a cause of being, just as there is one of truth. Aquinas' opposition to Averroes and Albert on this point, cf. Ch. IV, section 2, B, reveals his certitude that Aristotle did indicate the necessity of a cause of being.

[2] *In I Meta.*, 15, 233.

In discussing Aquinas' proofs of God, we remarked that he indicates Aristotle's famous proof of the prime mover in Book Λ to be only conditionally correct. The correct proof should be expressed thus: "If X, then Y, but if not X, then Z, and so Y". In this formula, "X" represents the eternal character of the world; "Z", creation in time; and "Y" the first agent cause. Thus Thomas notes that if the world, and so motion, is eternal, then we need a first mover to explain motion. However, the arguments for the eternity of the world are not absolutely convincing, or necessary; hence if the world is not eternal, but temporal, then we still need a first agent – a cause of being as well as motion.[1] It is only too obvious that Λ 6 gives only a proof of the prime mover of motion in an eternal world: "If X, then Y". Thus in Thomas' *Commentary* on this passage we have one of the rare examples where Thomas openly goes beyond Aristotle's thought.

The elements Aquinas has added to Aristotle's passage are two in number: first, it is only with probability that we can prove the eternity of the world, and second, eternal or not the world is created. (The context of the *Commentary* demands that this proof be related to the proof of the cause of being; hence, we know that the world, whether temporal or not, has been caused by the Pure Being). The first of these additions is of course not an Aristotelian doctrine. The second, however, was indeed a doctrine of Aristotle, as Thomas twice affirmed: *In II Meta.*, 2, 295; *In VI Meta.*, 1, 1164. Thus if one looks at the conclusion proposed by Aquinas – there is a first cause of being – this is an *intentio* of Aristotle.

F. The human attempt to speak of God

At the end of our reconstruction of Aquinas' theories on the science of being as such, we asked what it meant to say: "God is". Our investigation was of necessity detailed. Suffice it to recall here that for Thomas, one can speak of five uses of "substance"; 1) *Logical substance*: insofar as any real object is represented by a name or words serving as the subject of a proposition, that name is a logical substance; it can be predicated of nothing else and all else is predicable of it. Thomas speaks of this substance in several places, e.g. *In VII Meta.*, 2, 1280. In the corresponding Aristotelian passage this same notion of "substance" appears; it is not referred to as "logical", however.[2] 2) *Logical genus of composed substances*: insofar as we have for certain real objects sub-

[1] *In XII Meta.*, 5, 2490–99. Cf.. Ross, *Aristotle*, pp. 179–80.
[2] Z 3, 1028b33–1029a2; cf. also Δ 8, 1017b10–14.

stantial predicates stating an *esse*-essence composition, those objects are gathered under the logical genus of composed substances. Thomas speaks of this logical genus in many places; for example, the discussion of "being" in *In V Meta*,. 9, 891; the discussion of the difference between corruptible and incorruptible beings in *In X Meta.*, 12, 2142 and 2145; the discussion of *idem in genere* in *In V Meta.*, 22, 1124–27. On this point, Aristotle's doctrine is not found outwardly to differ from Thomas'.[1] Thomas' work is at first glance merely more explicit; he notes, for example in paragraph 2142, that two things not in the same genus in the sense of having a common element in their definition are nevertheless in the same logical genus or category of substance. Yet there is a real difference between Aristotle and Thomas. Of course both agree in admitting the existence of the logical genus of substance insofar as we have substantial concepts for substance. Yet they would obviously disagree on the content of these concepts; for Aristotle these concepts express essence alone, while for Thomas, they express the *esse* proper to the essence as well. Thus for Thomas, the logical genus of substance is a logical genus of *composed* substances: composed in their being. And so we understand that the substantial concept for God is not found in this class. However, for Aristotle the concept expressing the substance of God fits into the logical genus of substance.[2] 3) *Ontological composed substances*: for Aquinas, all real beings of our world which fall under the logical genus of composed substances in virtue of their substantial predicates, can be considered as falling under the class of "beings composed of *esse* and essence". Naturally, Aristotle would not have understood this class, for what is *esse*? 4) *Logical genus of simple substance*: insofar as Thomas has for God a substantial predicate expressing the lack of *esse*-essence composition, he has a concept found in this class; of course, he has no such concept, although he would have to admit the possibility of this class for someone who could understand God. However, Aristotle would have no way of distinguishing between this class of predicates and the second use of substance mentioned above, the logical genus of composed substances. 5) *Ontological simple substance*: in Thomas' eyes, this group comprises only one being God, for He is the only being for which it is possible to be pure *esse*. Again, this class is non-sensical for Aristotle.

Thus for Thomas there are five uses of "substance," and for Aristotle

[1] Δ 7, 1017a24–27; I 10, 1059a8–10 and 12–14; Δ 28, 1024b9–16.
[2] Δ 28, 1024b9–16.

only three.[1] The fact that Thomas has extended Aristotle's three uses to five is explainable by the introduction of *esse*. Since Aquinas explained created beings as composed of *esse* and essence, two distinct principles, it was necessary to construct two new uses of "substance" to speak of God and of the concept representing Him. Again, then, Thomas has given here his own thought, not that of Aristotle.

There is one more item to mention in regard to man's method of talking about God: the general rules to be followed. We noted that Aquinas gave two: 1) in any predication of God, one must include a negation (*In X Meta.*, 4, 1990); 2) if there is any perfection in creatures, the same perfection is had in the highest way by God, provided of course that the perfection is not in any way intrinsically related to matter (*In XII Meta.*, 8, 2543). As regards the first rule, it appears legitimate to consider it as Aristotelian although it does not appear in the passage corresponding to Thomas' reference to it.[2] In these lines Aristotle speaks of *unum*, noting that it derives both its name and meaning from its contrary; thus "indivisible" is defined in terms of a negation of "divisible". Aristotle's reason is important: we must think in these patterns, he writes, "because of the conditions of perception". And when he comments on these lines, Thomas refers once again to the first book of the *Physics*. Just as he did in *In I Meta.*, 2, 45–46 (the discussion of our knowledge of being), Thomas refers to Aristotle's doctrine that knowledge of simple beings is the last knowledge man achieves. Thus it is, Thomas explains, that we define God by noting that He is foreign to the composition of creatures. Although Aristotle perhaps did not express this doctrine as clearly as Thomas, it is most certainly the legitimate outcome of his thought, if not his actual meaning; e.g. Z 3, 1029b3–12 where he discusses the necessity of starting our search for knowledge of being by inspecting the essences of sensible substances. On this point then, Thomas' thought appears to be at least the intention of Aristotle.

The second rule affirms that God must possess in a more perfect way anything perfect possessed by creatures. Thomas' words (*In XII Meta.*, 8, 2543) again surpass the exact content of the Aristotelian passage: Λ 7, 1072b22–26. Yet the immediate context of these lines appears to Thomas as Aristotle's description of the "condition" of God;[3] and this is not a bad characterization of what Aristotle is doing. Thus, since Aristotle

[1] It is understood of course that we do not attempt to speak of the divisions of "substance" explicitly given by Aristotle and by Aquinas in Books Δ, Z, and H.

[2] I 3, 1054a26–29.

[3] *In XII Meta.*, 7, 2519.

writes: "...God is always in that good state in which we sometimes are...",[1] it does not seem incorrect for Thomas to note that whatever is perfect in our world must be even more perfect in God. Thomas' doctrine is, accordingly, little more than a formulation of the method Aristotle follows: the *intentio Aristotelis*.

G. Conclusion: Aquinas' Commentary on the Metaphysics

We have reached the point where we can determine with certainty at least one important aspect of the nature of Aquinas' massive volume. *The Commentary on the Metaphysics* most definitely does not present the metaphysical doctrines proper to that famous and learned Greek thinker who was Aristotle. Quite the contrary, Aquinas' work is a personal synthesis of metaphysics, based on Aristotle's work to be sure, yet totally cast in the perspective of *esse*. Yet the very fact of this difference between the two does not permit us to conclude that Aquinas was conscious of the transformation he wrought by introducing *esse* as a co-principle of substance.

The *Commentary's* metaphysics quite obviously is based on Aristotle's system. Aristotle began his metaphysics because he had discovered an immaterial first mover. In his attempt to construct a viewpoint sufficiently broad to enable his study to be the "universal science", Aristotle put his logic to work: what do we mean when we say, "John is a man", or "Peter is big"? The examination led to the concept of "being", and through "being" to "substance" and its nine satellite-concepts. "Substance", however, was a vague, fluid concept, used alternately to refer to matter, to form, and to the composite of matter and form. Step by step, Aristotle investigates each of these uses until he arrives at a well-defined knowledge of essential form as "substance". From such knowledge, he could have concluded that his immaterial prime mover, a form, has every right to be called "substance", and presumably he did so conclude. Still more, since he knew the necessity both of finding a substance capable of measuring others, and of searching for a cause of being, he must have wondered whether his first mover was not just that – the cause of all being, the ultimate explanation of the reality of substances. Unfortunately, his genius failed him, and he had to be content with a mere repetition of the physical proof of God. Such was Aristotle as we can see him from our perspective.

Aquinas, however, appears to have seen Aristotle a bit differently. He recognized Aristotle's desire to find a cause of being, a measure of

[1] Λ 7, 1072b24–25.

substance. Because some writers had denied this point, Aquinas insisted emphatically on it; even more, to make still clearer the importance of the proof of God in Aristotle's system, Aquinas indicated the basis of the proof (*esse*-essence composition), and the mechanics of the proof (participation). That such is indeed what Aquinas was about in the *Commentary* cannot be denied on the grounds of any evidence we have discovered.

3. CONCLUSION OF PART TWO

In the introduction we proposed carrying through a comparative study of Aquinas' *Commentary*, Aristotle's *Metaphysics*, and the corresponding *Commentaries* of Averroes, Albert, and Avicenna. It was noted that such a comparison would show first, that Aquinas' work is a unit, written in the light of one conception of metaphysics; moreover, that it is directed against the interpretations of Avicenna, Averroes, and Albert; and finally, that the Aristotelian metaphysics presented by Aquinas is not identical to the system operative in Aristotle's work own. These three facts have been sufficiently illustrated.

In the chapter devoted to the analysis of these philosophers, I have constantly maintained that one finds no evidence within the *Commentary* itself which permits us to conclude that Aquinas was conscious of the transformation he was working in Aristotelianism. Undoubtedly to some this attitude will appear excessively cautious. Nevertheless, I believed it the proper one to adopt, given the numerous examples of Aquinas' quite evident desire to be true to the text of Aristotle he was exposing. Time and again we have noted his method of referring to the immediate context, or to other books of the *Metaphysics*, or even to other works of Aristotle, references made in an attempt to clarify the meaning of the passage being exposed. Is not this the method of one who wishes to present the thought of the author examined, in this case of Aristotle?

A brief summary of the results of our studies in Chapters IV–VI will illustrate Aquinas' efforts to present the true Aristotelian metaphysics in the face of the incorrect systems exposed by the three earlier commentators.

In Chapter IV, the discussion of the object of metaphysics revealed that Aquinas had explained Aristotle's Book Γ in the light of the *esse*-essence composition of the beings of man's world. Aristotle set forth a theory of a science interested in all substances and accidents, but primarily in substances. In Aristotle's eyes, the unity of metaphysics rest-ed upon the analogical predication of "being": everything called "being" was to studied. However, since all predication of "being" involved a reference to substance, the metaphysician's concern lay first of all with the use of "being" in the case of "substance". Yet the metaphysician had other concerns as well – the essential attributes of being such as "unity". These attributes were used by all sciences, but only metaphysics could study them. When Aquinas commented on these theories, he made two additions: *res* and *esse*. "Thing" or *res* was introduced on the same footing with "unity": both were to be studied and for the same reasons. The addition of *esse* was much more important, however, for *esse* was said to be every bit as much a principle of substance as essence is. The concept "being" (*ens*), Aquinas remarked, is a name taken from the *esse* of things; yet "being" does not mean *esse*, for "being" expresses the essence of any substance to which it is applied. This meant, we decided, that "being" expresses essence as related to *esse*. Now this exposition of Thomas Aquinas was quite evidently not the same as that of Aristotle. Yet, one has no reason to suspect that Aquinas felt this theory to be anything but Aristotle's. There is no positive evidence in favor of such a hypothesis.

Prior to Aquinas' time, Avicenna too had exposed an Aristotelian metaphysics of sorts. However, there was not too much kinship between Avicenna and his sources in Aristotle. Aristotle had distinguished between "what it is" and "that it is" in his *Posterior Analytics*. Yet he felt that "that it is", or the knowledge of "that it is", indicated nothing about the composition of being as being. Avicenna, on the other hand, made the opposition between apprehension of "what it is" and judgment "that it is" into a difference in the knowledge of real principles of substances. *Esse* or existence became for Avicenna the formal object of metaphysics; it was through the adoption of the "here and now" as the metaphysical viewpoint, that Avicenna succeeded in drawing so many varied objects into the unity of his universal study.

Averroes knew this system of Avicenna and did not even attempt to hide his disgust for it. Avicennian *esse* is the truth of a proposition, Averroes explained. To know *esse* is not to know a principle of things, but is rather only a here-and-now accident of the man knowing the *de facto* reality of things. In thus criticizing Avicenna, Averroes was being nothing but Aristotelian. However, on other points of doctrine, Averroes' insights were not so acute. Metaphysics, he wrote, studies only the final and formal causes of substances.

Both Averroes and Avicenna were known by Albert. Albert too disagreed with Avicennian *esse*. The word *esse* was totally acceptable to Albert, but the Avicennian meaning given that term was anything but sacred. Hence Albert too substituted for the Avicennian meaning one of his own: *esse simplex*, he wrote, expresses the first creature of God, it expresses that without which a thing would not be. It expresses, in short, the form of the whole. Just as Albert did not approve of Avicennian *esse*, so too he disliked Averroes' theory of the causes to be studied by the metaphysician. In Albert's exposition, metaphysics, although primarily investigating the formal and final causes of substance, must look as well at the material and efficient causes. The point of this doctrine, as we concluded, is that to know being or substance, one studied the form of substance, as well as its goal; yet too, one must study all the substances of which "being" can be predicated – both material substances and the first being, the first efficient cause of motion, must be examined. Hence, for Albert all four causes were to be investigated. The crucial point of these theses was touched upon by Albert when he explained that to know a being one had to know its form.

When Aquinas wrote his *Commentary*, he had before him, at least in a figurative sense, the works of Avicenna, Averroes, and Albert. Avicenna's interpretation of the *Metaphysics* was wrong, Aquinas said, because it was based on the theory of *esse*. In Aquinas' eyes the objections advanced by Averroes against this theory were quite justified: the *esse* known in a judgment was the truth of an apprehension. Hence Avicennian *esse* was put aside. By removing this un-Aristotelian note, Aquinas illustrated a desire to expose one of the most basic theories proper to the metaphysical system of Aristotle.

Just as Aquinas attacked Avicenna, so too he appears to have attacked both Averroes and Albert. Parallel to Averroes' exposition of metaphysics as interested solely in the formal and final causes, Aquinas remarked that all four causes must be studied. Albert had earlier written in a passage also parallel to that of Averroes, and so to that of Aquinas, that metaphysics was to know all four causes. The reason given by Albert for the interest in formal causality is important: knowledge of the formal cause is the most perfect knowledge of an object. (In the light of the studies in Chapter V–VI, this doctrine is seen to be an expression of Albert's Neoplatonism). Aquinas had something to say in regard to this theory of Albert. We do not know the formal cause of immaterial beings, he noted. Hence in metaphysics one does not seek to know the formal cause on the grounds that such knowledge is the best way to know an object; rather, one seeks to discover the formal cause because the form is the source of perfection.

The meaning of these various anti-Avicenna, or anti-Averroes, or anti-Albert doctrines should not be minimized. The very fact of these oppositions, revealed in Chapter IV, evidences Aquinas' desire to expose the Aristotelian system.

This conclusion, as well as these oppositions between Aquinas and his predecessors, were encountered again in the investigations carried through in Chapter V. There it was explained how Aristotle wished metaphysics to prepare for the use of other sciences the concepts which can be called "communia". These concepts were nothing more than expressions of the metaphysician's knowledge of beings as being. Somewhere in the examination of being as such, that is somewhere in the universal science – or again, somewhere in his attempt to clarify these *communia*, the Aristotelian metaphysician discovered the need to posit a first cause of being .

When Avicenna explained the procedure of his metaphysics, much the same general outline appears: from a study of beings as "here and now", the metaphysician arrives at a realization that God, the Pure, Necessary Being, exists. Yet Avicenna's thought is essentially different from Aristotle's, for it revolves around the viewpoint of "here-and-now existence".

For Averroes, the entire Avicennian system was rather worthless. The true nature of metaphysics is this, Averroes would have said: one studies the prime mover of physics, one must describe this immobile mover. If one succeeds in this, if one can explain the prime mover as the formal and final cause of substance, then one knows all beings as being. Quite obviously this is a reversal of the Aristotelian perspective. For Aristotle the universal science becomes first science; for Averroes the movement is just the opposite.

Albert apparently saw this incorrect aspect of Averroes' exposition, yet one finds no doctrine directly attacking it. However Albert did clearly maintain that *esse simplex*, the form of the whole, is the object of metaphysics. The study of such a *prima creatura Dei* was for Albert universal science, the study of being as such. In his solution of the problem of moving from the universal to the first science, Albert drew on two sources: on Averroes and on Neoplatonism. Like Averroes, Albert explains that God is proved to be the first mover by the physician. Implying that the metaphysician, on the other hand, cannot prove that the prime mover is also the cause of being, Albert is content to present a Neoplatonic doctrine as the basis of man's ability to discover something about God. All creatures flow from their divine source, wrote Albert, and thus bear within themselves the stamp of the divine intellect. Hence, if one knows the form of things as real (the *esse simplex*), one will know something about God. Thus metaphysics is the first science.

When he read such theories as those of Avicenna, Averroes, and Albert, Aquinas was obviously not at all satisfied. Avicennian *esse* he rejected as unfitting to serve as a formal object. Both Averroes and Albert too fell under his disapproval when he categorically stated, not once, but twice, that Aristotle did prove God to be the ultimate cause of being. A complementary theory of Averroes and Albert was also rejected. Averroes, because he conceived metaphysics to be the study of God, noted that all metaphysical definitions were in terms of God. Albert, because of his Neoplatonic theory of creatures as images or mirrors of God's light, could repeat the same doctrine of a science studying all objects defined in terms of God. Precisely at the point in his *Commentary* which parallels the expositions of Albert and Averroes regarding definition in terms of God, Aquinas explained the truly Aristotelian view of metaphysics. Metaphysics studies God as the cause of being, but as well it studies sensible beings *qua* being. That is, it studies the concepts which can express any sensible being as being.

This doctrine of metaphysics as the study of the common concepts, as well as Aquinas' opposition to the other three commentators, becomes more meaningful in the light of our investigations in Chapter VI. There we noted that Aquinas carries through a logical study, an examination of our use of concepts. Through such a study, he discovered how a concept expresses the reality of any being. Thus he realized how "being" expresses reality; thus he learned that "being", when used to know any sensible object, expresses the actualized essence which is the reality of the being in question.

Moreover, because the metaphysical concept of "being" expresses the total reality of any being, one can understand the relation Aquinas posits between metaphysics and the other sciences. The metaphysician is the legislator, the organizer, the ruler of other intellectual disciplines. Once the metaphysician views an object as "being", for example, once he knows the horse standing in his garden as "being", then he has grasped the reality of that horse. He has grasped the actualized essence of that horse. He has directed his attention to the mass of hide and muscle and bones which is the horse, and he has understood the "actualized essence" incarnated in that mass. He has thus known the horse as "being". Since the being of the horse is nothing separate from the *actual* to-be-a-horse, the metaphysician has grasped the essence of the horse in all its specific reality. Thus it is that the supreme science can decide, consequent to such knowledge, that someone had better get busy and study, now in an explicit way, all the particularity of that being called "horse". Hence it is that the lower sciences are dependent on metaphysics.

Avicenna, Averroes, and Albert all spoke of this same doctrine of the relations of metaphysics to the lower sciences, but none of them presented the same view as Thomas. Avicenna, by studying all beings as "here and now", could come up with nothing more interesting than the general characteristics of what it means to be here and now. Averroes and Albert, on the other hand, knew only the form of substance; they knew only that form is the cause of substance. None of these theories was acceptable to Aquinas. By his opposition to Avicennian *esse*, by his rejection of the Averroestic and Albertian doctrines on the relation of first and universal philosophy, Aquinas made it quite clear that Aristotle's metaphysics of being was not the same as those theories of the three earlier commentators.

It is this continual opposition of Aquinas to his three predecessors, as well as his evident attempts to discover and clarify the meaning of the *Metaphysics*' many discussions, that incline one to view the *Commentary* as, at least, the result of Aquinas' desire to present the metaphysical system of Aristotle. It is not at all impossible that the *Commentary* is much more than this; it is not impossible that Aquinas accepted the entirety of the system he attributes to Aristotle. Thus, it is not impossible that the *Commentary* present a doctrine which is simultaneously the metaphysics Aquinas judged to be proper to Aristotle, and the system Aquinas accepted as his own. However, the examination of the *Commentary* proper, that is, divided from the *Prooemium*, presents no evidence in favor of such a conclusion.

Yet the *Prooemium* was composed for a purpose. Evidently, we are expected to read it as an introduction to what follows, or if not that, at least we are invited to compare it with what follows in the *Commentary* proper. Hence, as a final word, let us make such a comparison.

One will remember that the *Prooemium* spoke of metaphysics as the study of *ens commune*, the concept common to all other knowledge; in addition, there was mention of other concepts; we called them the "retinue of concepts following *ens*" – these too were assumed by all other knowledge. Now in the *Commentary* we have discovered the same type of study attributed to metaphysics: it has to investigate the concept of "being" and all the essential attributes of being, the *communia*; one of the roles of metaphysics was said to be the clarification of the *communia* with a view to their use by other, lower sciences.

In the *Prooemium, ens commune* was said to have as its causes, God and the other separate substances. In the *Commentary* proper, we noted that metaphysics studies the most common concept "being" and in virtue of this concept rises to knowledge of the existence of a first cause of being. (In our study of the *Commentary* proper no mention was made of a plurality of separate substances, although they are often referred to by Aquinas; e.g. *In XII Meta., lectiones* 6–7.)

Moreover, in the *Prooemium* we noted the presence of a difficulty. *Ens commune* was said to have God as its cause; and yet *ens commune* was said, in one of Aquinas' more cryptic sentences, to be predicated of many things, among them God. In the *Commentary* itself, we have the same difficulty. The reality corresponding to our initial idea of "being" has God as its cause; yet when we tried to form new concepts of "being", one which expresses God, and one which expresses both God and creatures, we really could not surpass our initial idea of "being"; we could not form any new concept with a new positive content. We could only tell ourselves that God is not like we are. Thus "being", whether taken in its meaning had at the end of metaphysics, or in its meaning had at the close of the study of the predication of "being", is exactly the same concept. Thus, the "being" of the *Commentary*'s many discussions is said of God, and yet God is the cause of the *ens commune* of creatures!

Then too the *Prooemium* presented a picture of metaphysics as the last science in the *via resolutionis*. This meant, we discovered, that all concepts and all causes are reduced respectively to the concept of "being" and to God. This reduction was noted to be a dynamic thing, somehow involved in the construction of the science of metaphysics. But as well, *via resolutionis* expresses a type of static relation between "being" and all other concepts; whatever the other concepts express is somehow contained in "being". Now the same idea was present in the body of the *Commentary*, though not in terms of "resolution". "Being" or "substance" is the most common of concepts; it is found by an examination of the use of other concepts: the reduction or resolution as a dynamic process. Moreover, "being" in the nine senses of accident is reduced to "being" as "substance" since "substance" enters into the definition of the accidents. Then too, all causes can be seen as reduced to the first cause of being, to God; as we mentioned, the metaphysician posits God as the reason why there is any causality at all in the world. Yet the resolution of all other concepts to "being" has its static side as well. As the *Commentary* explains, the metaphysician determines the scope and legitimacy of other sciences, thanks to his ability to know the essence of any object through his concept "being".

The *Prooemium* is seen to be exactly what it pretends to be – an introduction to what follows it in the volume entitled *Commentary on the Metaphysics.* True, many problems are left untouched in the *Prooemium,* for example, the mechanics of the proof of the first cause of being. But would the *Prooemium* be a *prooemium* if it exposed the workings of such a proof? Hence the similarity of the *Prooemium* with what follows it is a proof that the *Commentary* is Thomas' exposition of Thomas' thought.

This comparison of the *Prooemium* and the *Commentary* indicates that the metaphysical system expressed in the latter work was one Aquinas accepted as his own. This conclusion is further strengthened by two facts brought out in Chapter III where we studied the *Prooemium.* In the first place, we noted an opposition between Aquinas' *Prooemium* and those of Avicenna and Albert; Aquinas' metaphysics, as found in the *Prooemium,* appears to have been placed in opposition to the theories of the two earlier commentators. And secondly, the comparison of Aristotle's Book A, cc. 1–2 and Aquinas' *Prooemium* revealed the latter's thought to be basically Aristotelian: metaphysics knows the intrinsic causes of all things by forming the most universal of all concepts; and metaphysics studies God the first cause of being, because knowledge of God flows from knowledge of being as such. Since the *Prooemium* represents Aquinas' thought, and since this thought appears to agree with that exposed in the *Commentary*; since both the *Prooemium* and the *Commentary* were opposed to the theories of Albert and Avicenna on the basic points of metaphysics; and since the metaphysics of the *Prooemium* and of the *Commentary* both adhere to the fundamental framework of Aristotelian metaphysics, does it not seem very probable that both in the *Prooemium* and in the *Commentary* Aquinas exposed a philosophy he accepted as his own?

However, it is evident that this proof needs to be strengthened by further study of Aquinas' independent or non-commentary works. Nevertheless, it remains valid to conclude that the *Commentary* presents the Aristotelian metaphysics accepted by Aquinas. This conclusion is somewhat weak when compared to the other four, seemingly certain, facts our study has disclosed: Aquinas' *Commentary* is a unit, written in the light of one conception of metaphysics; it was directed against the interpretations of Avicenna, Averroes, and Albert; the Aristotelian metaphysics presented by Aquinas is not identical to the system operative in Aristotle's *Metaphysics*; yet Aquinas apparently believed he was exposing the true thought of Aristotle.

PART THREE

The *Commentary* has been written in the light of one system of metaphysics. This system we have seen operative in many passages of Aquinas' exposition. However the necessity of studying the *Commentary*'s metaphysics in the context of textual analysis may possibly have obscured some or other of the central theories of that system. Hence in this final division of our study, I propose to view that metaphysics, both independent of all analysis of texts, but as well in a synthetic fashion. Thus in Chapter VII I shall attempt to relate the most important doctrines of the *Commentary* to the basic theory which gives them their meaning.

As noted in Part Two, the metaphysics exposed by Aquinas apparently represents his understanding of Aristotle. But at the same time, as the theories of the *Prooemium* show with some probability, the metaphysics operative in the *Conmmetary* is the system Aquinas accepted as his own. This identity of Aquinas' view of Aristotle and Aquinas' own system will be indirectly treated in the following Chapter. Although we shall deal primarily with the central doctrine of the *Commentary*, together with some related issues, in footnotes I shall indicate texts from Aquinas' independent or non-commentary writings where the same doctrines are found; to these references, I have added others to the passages of the Thomist *Commentaries* containing the same doctrines. Thus even though these references are limited in number, they tend to bolster the conclusion achieved previously by the comparison of the *Prooemium* with the *Commentary*, namely that the Aristotelian system exposed by Aquinas was the metaphysics Aquinas accepted as his own.

In Chapter VII, as in Chapters IV–VI, when reference is made to "Aquinas' theory", I am speaking of a doctrine found in the *Commentary*. Accordingly, the use of that expression indicates only the source of the doctrine discussed and does not beg the question whether Aquinas accepted it as his own.

CHAPTER VII

THE BASIC INSIGHT OF AQUINAS' *COMMENTARY*

It has long been a truism to state that metaphysical thought revolves around a conception of "being." In fact, since the notion of "being" was first cast into the spotlight by Parmenides, man has taken it for granted that to talk metaphysically is to speak against the background of suppositions regarding the meaning given to "being." Hence, if we note that a theory of "being" is the doctrine central and operative in all the expositions of Aquinas' *Commentary*, we are not saying anything not already obvious.

However, it would be another matter to attempt to explain what Aquinas understood by "being". If one does that, one will grasp the *Commentary*'s metaphysics at its origin. Hence, as a synthetic conclusion to all the detailed analysis of the preceding chapters, nothing could be more appropriate than to trace the most important points of the *Commentary*'s metaphysics back to Aquinas' notion of "being", for only by such a reduction can one understand fully what Aquinas meant by "being" in his exposition of the Aristotelian *Metaphysics*.

In the first place, if one is to understand Aquinas' "being", one must grasp the complexity of meanings communicable by that most familiar of words – "is". Just as important, however, is the necessity of seeing clearly what Aquinas means by the intellectual act of knowing. After these problems are resolved, one can realize to a great extent what Aquinas means by "being"; one can see why that term expresses an act of understanding and not a judgment.

When one reaches this knowledge of "being", there will yet remain certain implications which one will not explicitly recognize without reflecting on Aquinas' conception of metaphysics as the lord of the sciences. Then too, a mention of the transcendentals and their relation to "being" renders still clearer, and more explicit, what one knows by knowing "being". And finally, it will be of help to trace Aquinas' teach-

phical knowledge of God back to its roots in "being". ...es, 1) the meaning of "is"; 2) intellectual knowledge; ... of "being"; 4) the transcendental concepts; 5) meta-... ...ling science; and 6) philosophical knowledge of God; ...es, central to the metaphysics revealed in our earlier anal-yses, are all bound together by the meaning of "being". And in this final chapter, it is these doctrines and their dependence on the theory of "being" that shall occupy our attention.

1. THE MEANING OF "IS"

Descartes insists repeatedly in his *Regulae ad directionem ingenii* that man is able to solve great problems, only if he is first patient enough to solve the small ones. I know of no better example of this truth than the necessity of understanding "is" before one attempts to solve the problem of "being".

On this point Aquinas is in whole-hearted agreement: unless we understand what is communicated by our use of the word "is", we cannot hope to correctly understand "being". This is true, Aquinas insists, because sometimes to say "is" is to say "being", and sometimes it is to say "true". (*In V Meta.*, 9, 889–96; *In VI Meta., lect. 4*; *In X Meta.*, 3, 1981–82).[1]

We cannot, of course, speak of "is" without speaking of knowledge, for the meaning of this word depends upon the object of knowledge expressed as "is".

Now "is", just as any verb, is capable of assuming many forms. "He *is* a bad man", we can say; yet also, "I *was* wrong", and "If it *were* so cold, why did the water not freeze?" Yet here we speak of "is" as a verb. As any verb, "is" can thus appear in many forms. But this is grammar, and Aquinas' philosophy has absolutely no interest in the grammatical aspect of verbs.

If, on the other hand, we consider "is" as a word, as a cognitive term, and no longer as a verb, then we do have a subject of philosophical import. As a verb "is" can turn up as "was", or "am", or "will be". But in these instances, "is" is considered as a verb, as a grammatical part of speech; it is considered only as governed by the temporal aspect we wish to express, by the mathematical number of subjects we wish to

[1] Note the following texts in which "is" of: "God is", "Christ is man", and "John is blind" are said to mean that prior conceptions of God, of Christ, and of John were true: *De ente*, c. 1, p. 2, l. 8–p.3, l. 7 (Edit. Roland–Gosselin); *Sum.theol.* I, 3, 5, ad *1um*; *Ibid.*, 48, 2, ad *2um*; *De pot.*, 7, 2, ad *1um*; *Quodl. IX*, q. 2, a. 2, ad *4um*; *Sum. theol. III*, 16, 9; and 17, 1

talk about, and so on. Now if "is" be taken in this sense, Aquinas would be quick to admit, it would have nothing directly to do with the problem of "being". But if we take "is" as a word, or as a cognitive term in a proposition, and here it is of no import whether we have "is", or "will be", or "was", or any other form, here "is" has much to do with "being". "Is", in the sense of a term appearing in a proposition, must be studied by the philosopher; such a study is called a logical one, but nevertheless is of paramount importance to the metaphysician. (*In VII Meta.*, 2, 1280 and 1287; *Ibid.*, 3, 1308; *In V Meta.*, 9, 889–94).

This difference between a grammarian and a logician can be reduced to the opposition between style and knowledge; or to say the same thing, grammar considers the conventions in terms of which we ought to express ourselves, while logic must see how we can express certain things. Grammar, for example, taught our parents to say: "It is I". But the people will not be ruled by Academies, and hence people as famous as Winston Churchill have once and for all established the right to proclaim: "It is me". Logic, however, is not subject to the whims of the majority. Thus, if we want to explain who walked on the tulips, we cannot say: "It was the fat". Logic will not stand for such arbitrariness; if one wishes to communicate the identity of something, one must mention a substantial essence, so says logic. Thus: "It was the fat *man* who stepped on the tulips".

A prime example of the philosopher's use of logic is had in metaphysics. The entire course of that science is determined by what the logician has to say of "is". But at this point an ambiguity appears in Aquinas' thought, an ambiguity clearly resolved by what since his time has come to be known as the distinction between material and formal logic.

As Aquinas explains logic, it would have as its domain the rules governing how we must speak if we wish to express something about an object; opposed to logic would be grammar, the study of the rules governing the style convention demands be present in what we say. Logic, notes Aquinas, studies the relations accruing to a content of thought precisely because this content is being thought. Yet there are two distinct types of relations which belong to any content of thought. On the one hand, a content can be related to another. Thus, in: "John is a man", the content "man" (or the word "man") is related to the content "John" (or the word "John"); this relation, called the relation of a species to a subject, would be the interest of what is now called formal logic. (*In IV Meta.*, 4, 574; *In IX Meta.*, 1, 1775). There is a second kind of relation which accrues to the content of thought, however. For example, in "John is a man", the content expressed by the proposition is related to the extra-mental, really here-and-now object called "John". This relation, just as much as that discussed by the formal logician, is one which exists only because someone has chosen to think. Moreover, this relation is the kind of relation it is, because someone wished to think about John as "man" rather than as "fat". Hence, this relation of thought to extra-mental reality is a matter of logic too; however, the study of such relations is now known by the name of material logic. (*In IV Meta.*, 4, 574; *Ibid.*, 17, 736; *In VII Meta.*, 2, 1280 and 1287).

The distinction between these two types of logical relations was never assigned by Aquinas to the study of different branches of logic. Nevertheless, Aquinas was never guilty of confusing them, and so it is quite proper to characterize some of his expositions as touching on matters of material logic, and others, as involving formal logic.[1]

Now both of these branches of logic deal with the rules governing how we must speak if we are to express something about an object; they have no interest in the style with which we speak, however. As far as the metaphysician is concerned, Aquinas' *Commentary* states that he ought to be interested in logic in so far as it reveals what we wish to say when we say "is"; yet the metaphysician's interest in logic never exceeds the limits of material logic.

"Is", the metaphysician may think, is never found except in what appears to be the expression of a judgment. Thus, statements such as: "Henry is a man", or "Henry is fat", both judgments, would be the only expressions of intellectual knowledge in which the word "is" appears. Thus, the use of "is" occurs only when the intellect affirms the here-and -now existence of an object. (For simplicity's sake, the metaphysician could speak thus only of affirmative statements; he would not therefore deny that "is not" can be used to deny here-and-now existence).

An apparent objection arises, when one recalls that one can say: "All men are mortal." Is this too an expression of a judgment? Many Thomists would experience little difficulty with this, for they would draw on their knowledge of the theories of universals and of abstraction, and hence point out how this statement is ultimately grounded on a judgment that a given existing man is indeed mortal. Hence, all universal

[1] Cf. the criticism directed against an improper use of formal logic in: *De sub. sep.*, c. 5, n. 62; *Sum con. gen. III*, c. 41.

statements involving a subject and predicate joined by "is" are actually traced back to a judgment of the existence of an individual.

The favorite expression for such judgments is "judgment of existence". The intellect is said to judge that a given object does indeed possess some formality, or is some formality. Such a judgment is always expressed, this is the usual docturne, by a statement of the form: "A is B". Most important of all, the adherents of this explanation will conclude that the "is" of such statements refers to existence; it means "exists".

Now it is quite obvious that such a theory is much too simple to do justice to the totality of man's statements. It was because he recognized this that Aquinas pointed out that the "is" of judgments never means exists. (*In VI Meta.*, 4, 1241; *In X Meta.*, 3, 1982; *In V Meta.*, 9, 895–96).[1]

If the "is" of all judgments means existence, wrote Aquinas, then obviously one could never say "Blindness is the privation of sight". If the "is" of such a statement means "exists", then blindness exists, and so our world is peopled not only with entities, but with privations or non-entities as well. Hence, Aquinas would not admit that the "is" of judgment means "exists". (*In V Meta.*, 9, 895).[2]

Some philosophers might feel that I, and Aquinas too, if I am correctly explaining his views, have failed to make a quite elementary distinction. To understand how we speak of "blindness", they remark, it is necessary to apply our statement to an individual whose eyes simply do not function. For example, "John's eyes do not function" or "John is blind". "Blindness is the privation of sight" would be thus an universal manner of characterizing individuals such as John. Hence, we can ultimately trace universal statements back to a judgment of existence.

Aquinas would not be satisfied with such a view. And he would an-

[1] The following passages can only be understood as implying that the "is" of a judgment means "true"; *In III Sent.*, d. 6, q. *unica*, a. 2, ad *2um*; *Sum. theol. I*, 3, 4, ad *2um*; Ibid., 3, 5, ad *1um; Ibid.*, 48, 2, ad *2um*; *De pot.*, 7, 2, ad *1um*; *Quodl. II*, q. 2, a. 2c; *Quodl. IX*, q. 2, a. 2, ad *4um*; *In I Peri herm.* 2, 2. The passage in the *Sentences* (above) gives the key to understanding the *dictum* of the early years: "Secunda operatio respicit esse." cf. *I Sent.*, d. 19, q. 5, a. 1, ad *7um*; even at this early stage of his career, Aquinas was aware that the "is" of a judgment demanded, or implied, an awareness of existence prior to the actual judgment. The mature Thomas certainly did not cease to hold this same theory. Interestingly enough, the mature, or post *De hebdomadibus* Aquinas refrains from using the *dictum* just mentioned. One can postulate his awareness of the following danger: if he says that the judgment "respicit esse", then 1) one will understand that *esse* as a brute fact of being here and now, and 2) one will not recognize that *ens*, the concept, expresses *esse* and essence, or an essence as participating in actuality; and if one does not realize this "participating" character of being, then one would have returned to the early Thomas. Given these dangers, Aquinas may have refrained from using the *dictum*.

[2] Cf. *De ente*, c. 1, p. 2, l. 8 –p. 3, l. 7 (édit. Roland-Gosselin): *Sum. theol. I*, 48, 2, ad *2um*.

swer it in this manner. If "John is blind" or "John's eyes do not function" is ultimately to judge about the existence, either of John, or of an accident in John, then you are affirming an existing privation of John. In other words, an eye is not an eye if it cannot see. Thus, in the case of a blind person, there is no eye. The blind John has only two globs of matter in the head cavities where two eyes should be. No matter what words we use to express this situation of blindness, the "B" of "A is B" must mean a privation, a non-entity. Hence, if the "is" of "A is B" means "exists", then to state something about John's blindness (a non-entity) is to affirm the existence of something which is not real.

Nor can the advocates of the existential "is" of judgments have recourse to the following argument: "John is blind" affirms the existence of two globs of matter in John's eye sockets. Fine, Aquinas would reply, but how then do we know that: "John has two globs of matter in his eye sockets" means: "John is blind"? The apparently overpowering answer comes back: When we hear any statement about the two globs of matter in John's eye sockets, we immediately compare this information with our knowledge that similar beings such as George have eyes instead of globs of matter; hence, we know John is blind. This answer would not convince Aquinas, however, for it still implies an existential "is" in the sense of an affirmation of the existence of a privation. If our minds did proceed in the manner just described by Aquinas' imaginary opponent, then the comparison of John's globs of matter to George's eyes would involve the recognition that John's globs should be eyes, but are not. Hence, we would know the *quia est* of an actual state of affairs outside our mind: we would know the truth of our knowledge that John's globs should be like George's eyes. To know this truth, Aquinas insists, we would have had to judge some such thing as: "John's globs are something which *should* function as George's eyes, although they do not so function". Does the "are" of this judgment mean "exists"? If so, we affirm the existence of a privation.

So it is that Aquinas insists that the "is" of a judgment never means "exists". The problem naturally arises, for it is a grave problem: how do we express existence, if not in the judgment? How do we know existence, if not in the judgment?

To this problem Aquinas answers with a distinction. "Is", he maintains, must be considered as used in the expression both of the first operation of the intellect – *intelligere* or understanding, and of the second operation – *judicare* or judgment. Moreover, a proposition such as: "John is a man" should be taken as the expression, not of judgment, but of understanding or *intelligere*. Thus, "A is B" is the form of the result of abstraction, not of judgment. It is quite true, however, that we spontaneously feel that "John is a man" expresses a judgment. Yet one should realize that the expression of judgment is more properly: "My understanding of John as 'man' is true". (*In IV Meta.*, 17, 736)[1] Aquinas habitually expresses this distinction between understanding (*intelligere*, abstraction, first intellectual operation) and judgment, by the distinction between *ens perfectum* and *ens ut verum*. (*In V Meta.*, 9, 889 and 895; *In VI Meta.*, *lect. 4*).[2]

Thus in Aquinas' mind, the "is" used in the expression of understanding is an expression of a real being. Of course, one can understand imaginary entities such as unicorns, rain storms that didn't take place, and so on. Yet it remains that one understands the object presented in imagination as an *ens perfectum*, as a perfect being, or simply as a being. On the other hand, when one judges, the "is" employed in such everyday statements as: "It is raining" or "John is a man", expresses the intellect's awareness of the truth of a prior understanding of the outside world, an understanding expressed as "is raining", "rain storm" or "John-man" or "John is a man".

It is important to pause here to note a very important aspect of judgment which is often overlooked. Judgment is not simply an automatic expression of something like: "John is a man". If we examine our knowledge, we shall see quite the contrary. Our first intellectual act must be an understanding of John as "man"; we must conceive "man" when we direct our attention to John. Following such an act of understanding, the intellect must reflect on what it has done; it must regard both the conception "man" or "John-man", and the sense data in the face of which "man" was conceived. The intellect must discover whether "man" was the proper concept to form in the face of the data to which one was attending. In some mysterious way, the intellect must also look to see whether or not the sense data corresponds to extra-mental reality. If the intellect discovers that "man" was the correct concept to form in the face of the sense data, and if the intellect learns that the data corresponds to extra-mental reality, then the intellect knows the truth of the understanding of John as "man". Consequently, the

[1] Aquinas never explicitly gives the doctrine I expose, but it is implied in the following: *De verit.*, 1, 9; *Sum. theol. I*, 16, 2; *Sum. con. gen. I*, c. 59; *Quodl. VIII*, q. 1, a. 1; *In I Peri herm.*, 3, 9; *In III De anima*, 11, 760–61.

[2] Aquinas' thought, if not his words, agrees with my exposition: *De ente*, c. 1, p. 2, l. 8 -p. 3, l. 17 (édit. Roland-Gosselin); *In I Sent.*, d. 8, q. 2, a. 2, ad *4um*; *Sum. theol. I*, 3, 4, ad *2um*.

intellect judges, in the sense of affirming, "It is true that John is man". Of course, we ordinarily say: "John is a man", but here "is" means "is true" or "is truly conceived". A judgment, thus, should be considered as a two-fold activity: 1) the reflection, and 2) the affirmation. It is very important to note that in the process of reflection, which follows apprehension and precedes the very affirmation, we must grasp 1) the conception, 2) the sense data as a legitimate source of the conception, and 3) the correspondance of the sense data to the extra-mental reality. If we are talking about, say John, then we must grasp John's existence. Thus, to affirm: "John is B", we must grasp his here and now reality prior to the actual affirmation. Accordingly, when we make a judgment which concerns a being, we do not know the existence of that being *by* or *in* affirming, but prior to it. Moreover, the "is" of the affirmed "A is B" means "is true" or "is truly". To say: "A is truly B", we must already know the existence of A, and as well, of B in A. One could say, to be sure, that we imply our knowledge of the existence of John as man, when we affirm that it is true to conceive John as "man"; yet the "is" of such an affirmation does not *mean* "exist".

In like manner, whenever the subject of the affirmed expression is an imaginary being, like "unicorn", or even something purely intelligible such as a circle, then in these cases too the pre-affirmation reflection must grasp the existence of the imaginary or intelligible being. Thus, before one affirms: "A unicorn is not human", or "A circle has equal radii", one must grasp the imaginary existence of the unicorn, or the intelligible existence of the circle. In other words, an act of the imagination or a thought of circle is something real; the intellect, through reflection, must know the exact type of existence had by the subject of a to-be-affirmed expression. Thus, the "is" of every affirmation, while meaning "is true", must imply prior knowledge of existence. (*In VI Meta.*, 4, 1234–36; *In IX Meta.*, 11, 1897; *In II Meta.*, 2, 298).[1]

Perhaps some might see no really important difference between the view of judgment just explained and the one previously characterized as incorrect. Immediately above, it was maintained that the "is" of any judgment means "is true", and that the very affirmation of something implies, but does not express, knowledge of some type of existence. Yet

[1] Cf. *Sum. theol. I*, 16, 2; *Ibid.*, 16, 1, ad *3um*; *Ibid.*, 48, 2; *De verit.*, 1, 9; *Ibid.*, 8, 14; *Sum. con. gen. I*, c. 55; *In I Peri herm.*, 3, 4; *Ibid.*, 5, 20–22; *In II Peri herm.*, 2, 2; *In I Post. anal.*, 2, 2–3; *Sum. theol II–II*, 173, 2; *Ibid.*, 175, 4. Note how man sees the union of two intelligibilities in the phantasm: *De verit.*, 11, 3; *Ibid.*, 12, 7; *De malo*, 16, 12, ad *7um*. Or note how he sees a composition present in thought: *In III Sent.*, d. 24, q. 1, a. 1, *sol.* 2. See also: *In III De anima*, *lectiones* 7–8.

earlier, in what was considered a view rejected by Aquinas, "is" was said to mean "exist". Is there really any important difference between these two theories? Aquinas would quite definitely feel there was an important difference, a difference best seen in terms of the analysis of the affirmation: "John is blind" or "It is true that John is blind."

We have already discussed the insolvable difficulties involved in the affirmation: "John is blind" if one holds that the "is" means "exists": no matter how one rephrases the statement, to judge that blindness has something to do with John would people the world with privations or non-entities, if the "is" of judgment means "exists". Now Aquinas' theory of judgment escapes all such lamentable conclusions. To judge: "John is blind", Aquinas would say, means that first, we have conceived (understood) John as having globs of matter fixed in his eye sockets, globs which do however have a function even if it is not that of seeing; second, that the intelllect, through reflection, grasps the existence of the globs of matter which function as any two globs of matter do; that is, the intellect grasps the existence of two pieces of matter with some function such as that of filling the two eye sockets; thirdly, the intellect recalls how it conceives the eyes of George which function as eyes; moreover, the intellect recalls that it is true to so conceive George; in other words, the intellect recalls its grasp of real, functioning eyes in the case of George; then, the intellect compares the proper or true ways of conceiving the eyes of George and the globs of matter of John; through such a comparison, the intellect becomes aware that it can *conceive* John's globs of matter as "privation of eyes such as those had by George". Thus, the affirmation: "John is blind" means "It is true to conceive John as having only globs of matter, where most other men have eyes", or even: "It is true to conceive John as lacking something had by most men". Thus, to affirm a difference between John and other men means one has grasped the existence of two realities, or better, of two parallel orders of realities. On the one hand, one grasps the existence of John and of the proper way of understanding him; one sees that it is true to think of John as rational animal; however, in the animal aspect of the knowledge of John, one knows only four senses, not five, for globs of matter are not eyes. Thus John is conceived in a positive manner, even if the perfection involved in that conception is less than that in the conception made regarding a person such as George. When we understand George, our thought is richer, for George has the perfection of sight. Now such an analysis of the affirmation: "John is blind" does not imply any prior grasp of an existing privation. One grasps the existence of John with four senses and two globs of matter, and the existence of George with five senses. Thus, in Aquinas' view, there is no necessity of affirming the positive existence of any lack of perfection. Yet in the view of those for whom the "is" of judgment means "exists", one is affirming the existence of privation, no matter how one rephrases the statement: "John is blind". Between *implying* the existence of John with four senses and *two globs of matter*, and *affirming* the existence of a *privation*, there is all the difference in the world. (*In IV Meta.*, 3, 565).

To avoid having a world of non-entities, Aquinas sharply distinguishes between the "is" of a judgment (*ens ut verum*), and the "is" of understanding (*ens perfectum*). The "is" which appears in the judgment implies prior knowledge of existence, but means "is true" and not "exists".

Now it is evident that a study of the "is" of the judgment was a study of material logic, for this branch of logic studies the relation of the content of thought to the object being known. In judgment, Aquinas would say, we know the conformity between an act of understanding and the object known. Thus, we know truth and falsity: "is" in a judgment means the understanding is properly related to the object known; "is not" would mean the understanding is not true, is not related to the object confronting our knowing faculties.

There remains one more "is" which the metaphysician must study through the use of material logic – the "is" of the act of understanding, the *ens perfectum*. Here again we have an investigation of material logic, for we study the relation of understanding to the object known. In the examination of the "is" of a judgment, we claimed that "is" expresses the fact of a relation of conformity of the understanding to the object understood. But in studying the "is" of understanding, we are not interested in the *fact* of a relation; we assume there is a relation, for after all we are dealing with an understanding – if there is no relation, there is no real understanding of the object confronting us. On the assumption of such a relation, we ask *how* the "is" of understanding is related to the object known. We want to discover how many ways "is" can be so related: how many ways, or what kinds of relation can be had? (*In V Meta.*, 9, 889–92).

In any act of understanding, Aquinas explains, we conceive the essential or necessary characteristics of the object understood. If we understand something such as John, we could, for example, conceive what it means to be "man", or "fat", or "father", and so on. However, it would not be correct to view the content of such conceptions as the essence "man", or the quasi-essence "fat". Rather, the content of our knowledge of John is "John-man", or "John-fat"; even more precisely, our knowledge is "John is man", or "John is fat". "Is", thus, is part of the content of "man" or "fat" through which John is known, or in terms of which John is known. (*In V Meta.*, 9, 893, 890, and 896).[1]

Thus, for Aquinas, the content of any act of understanding has the form "A is B". The "is" of these contents is not of course "exists". Rather, this "is" can be seen as representing one of the ten possible ways of understanding an object. Sometimes, for example, we understand John as a substance; in "John is a man", "is" denotes that we conceive John as a substance and "man" expresses what type of substance John is conceived to be. In other cases we could conceive John as qualified by an accident: "John is fat", "John is a father". The "is" of "is fat" expresses that we conceive John as having quantity; the "is" of "is a father" denotes that we conceive John as related to something. (*In V Meta.*, 9, 891–92; *In IV Meta.*, 1, 539–43).

There are, thus, ten possible ways in which "is" of the understanding is related to the object known. "Is" can express an object as a substance, or as having one of the nine accidents. Aquinas has spoken of such an "is" as "significative is", although perhaps it might be better to speak of it as the "referring is". There are, we can conclude, ten types of "referring is". (*In XI Meta.*, 3, 2197).

The study of these ten types of "referring is" can be styled as part of material logic. Such logic is concerned with the relations of knowledge to the object known, rather than with the interrelations of knowledge itself. All logic, we have also said, investigates what we wish to say from the point of view of how we should say it, if we wish to be understood properly. Such a description certainly characterizes the study of the "referring is"; for if we wish to express what a thing is, then we state it in a substantial way, and so on.

But why on earth should the metaphysician be interested in such a study? Such a question can be answered in two ways. The first of these was given in Chapter VI; the metaphysician must discover the possible ways of understanding objects, because he wants to find the concept to use in relation to material being and the immaterial mover, both considered as real. But this is the type of answer one gives before one has completed metaphysics. Aquinas, from the metaphysical knowledge he possesses after the construction of his science, could give a much different answer: he could start from the idea of "being" and show its

[1] Cf. *In I Peri herm.*, 5, 20–22.

relation to the "referring is" of our understanding. Thus, Aquinas' answer would run as follows. Whenever we understand a being (the object of our knowledge), our understanding contains an understanding of "being" (the content of our knowledge). In other words, in any expression of the content of understanding, "being" is found. No matter what may be the content of an act of understanding, that content involves "being."

This doctrine – "being" as the core of any concept, of any act of understanding – touches the root of Aquinas' metaphysical thinking. Moreover, this is the basic insight of Aquinas, with which this chapter is concerned. But before we can say any more about this point, we must turn to the theory of knowledge. Only when we realize what it means to understand, can we grasp how "being" is at the center of all understanding. And only then will we see why the metaphysician is interested in the types of "referring is".

2. INTELLECTUAL KNOWLEDGE

In discussing the meaning of "is", we necessarily presupposed a quite definite theory of intellectual knowledge. In all that we said, we presumed an opposition between the mental act of understanding and that of judging. We discussed somewhat the activity of judging by distinguishing two phases, the reflection and affirmation; yet the activity called "understanding" was left rather vague. It is time now to clarify what Aquinas means by understanding or *intelligere*.

In the *Summa Theologiae*, *intelligere* is characterized as *quasi intus legere*, almost like "to read what is within", or more poetically, "to read the inmost secrets of a thing". Much the same type of etymological analysis can be made of "understanding". The termination "ing" emphasizes, of course, the activity involved. Unfortunately, the word "understanding," unlike the Latin intelligere, does not bring out the fact that this activity is performed by a knower. Nevertheless, if we grant this fact, then "understanding" appears as the activity of entering into and seizing what supports or explains sensible knowledge. (*In IV Meta.*, 1, 531; *In VI Meta.*, 4, 1241; *In VII Meta.*, 17, 1667; *Prooemium*; *In I Meta.*, 1, 30; *Ibid.*, 2, 47; *In III Meta.*, 4, 384).[1]

However, it is very essential that we not mistake this view of knowledge as implying a return to the Kantian opposition between "thing in itself" and "appearance." When Aquinas speaks of understanding, he wishes to say, not that we grasp some superficial aspect of things,

[1] Cf. *Sum theol. I*, 86, 2 and 6; *In IV Sent.*, d. 50, q. 1, a. 4.

nor even something that is hidden at the center of things behind the appearances. Quite the contrary, to understand is to seize whatever a thing may be. In other words, it is to deny the legitimacy of the Kantian distinction.

It is understanding, conceived as the grasp of whatever a thing is, that supports or stands under sensible knowledge. In other words, intellectual knowledge supports – stands under – or explains sensible knowledge. Sense knowledge, Aquinas is found of repeating, attains only facts – the *quia est* as opposed to the *quid est* or the *propter quid est.* An eye sees colors; it knows the existence of colors. An ear hears sounds; like the eye, it knows that something exists, but not what that thing is. (*In I Meta.*, 1, 30). Now intellectual knowledge stands under such knowledge. Whereas the senses can know they are being acted on, that is, that they sense, only the intellect can comprehend or seize the reality of the object acting on the senses. Thus, understanding is the activity of entering into and grasping what explains sense knowledge. (*In VII Meta.*, 17, 1667).

At this point it is well to mention an improper view of knowledge. Let us suppose we know Socrates as a man. If asked *what* they know in such an act, some philosophers would say there are two possible answers. In the first place, one could emphasize the object being known; thus, they know Socrates. However, they would point out that one could also emphasize the idea formed, and so say one knows the essence "rational animal". Such an explanation of knowledge is one often proposed by the proponents of the existential "is" in judgment. Yet it is difficult to imagine a view of knowledge farther from that of Aquinas. If he could return today, and were asked about such a view of knowledge, Aquinas would undoubtedly remark that such a view has mistaken a metaphysical theory of knowledge for a psychological one. In a metaphysical view, knowledge is characterized as the successive actuation of potencies. Thus, a passive potency of sense is activated by an object, Socrates; other internal senses are in turn activated until, finally, the agent intellect impresses a species on the possible intellect. The species impressed in this case would be the essence of Socrates "rational animal". Now as a metaphysical view of knowledge, although an oversimplification, this theory is not unacceptable.[1] Yet there is no surer means of misunderstanding the *Commentary* than to approach it with

[1] Cf. *Sum. con. gen. II*, cc. 57, 73 and 76; *In II De anima, lectiones* 10–12; *In III De anima*, 2, 588–93; *Ibid., lectiones* 5–7; *De verit.*, 22, 5. ad *8um; In III Sent.*, d. 24, q. 2, a, 2. ad *1um*; *In III Sent.*, d. 14, a. 3, *sol.* 2c.

the impression that "knowledge as knowledge" is well characterized by such a metaphysical view of knowing. To speak of knowledge as knowledge is to forget all about metaphysical descriptions in terms of acts and potencies. This the reader of Aquinas often forgets but Aquinas himself was never guilty of mistaking a metaphysics of knowledge for a psychology of knowledge – for knowledge as knowledge.

In the *Commentary*, Aquinas constantly uses his psychological view of knowledge. Or to say the same, he regularly thinks of knowledge as knowledge. As already mentioned, understanding is the activity of grasping the "what" of an object sensed as "present" by the external senses. (This is not universally true as sometimes one understands what one has only imagined. But for simplicity's sake we speak only of understanding extra-mental reality.) Now to understand, psychologically speaking, is not to grasp either a possible or an universal essence. To understand Socrates as man is to grasp "man" in Socrates. It is to see how "man" penetrates completely every square inch of the material object who goes by the name "Socrates". It is to see how "man" makes that material object what it is: *the* man called Socrates. (*In VI Meta.*, 4, 1234–36; *Ibid.*, 1, 1147–49; *In V Meta.*, 9, 889–94).[1]

Now some might say this view is not that of Aquinas, for he holds that it is determinate matter, and not essence, which makes Socrates *the* man he is. Essence, they would insist, is not the principle of individuation. (*In VII Meta.*, 10, 1469).

However, it has not been said that essence is the individuating factor in Socrates. Essence, as Aquinas notes, is the source of all the perfection of an individual, if and because the essence is real or actual. That is, the *esse*, as entering into composition with essence, permits the essence to impart to a being whatever perfection it is able to impart. It is in this sense that the essence of Socrates makes him *the* man he is. It is because his essence is his, that he is *the* man he is. (*In III Meta.*, 4, 384; *In VIII Meta.*, 2, 1696; *Ibid.*, 3, 1710).[2]

Now what we see when we understand Socrates is the essence proper to Socrates as actually making Socrates *the* man he is. To approach this same doctrine, but from Socrates' point of view, there is no difference between the reality of Socrates and the humanity actually constituting Socrates. One cannot distinguish the individual being from the individualized essence, for the individualized essence is actually penetrating

[1] Cf. *Sum. theol. I*, 84, 7; *Ibid.*, 86, 1; *Ibid.*, 85, 1, ad 5*um*; *In Boeth. De Trin.*, q. 6, a. 2, ad 5*um*.

[2] *Sum. theol. I*, 75, 4; *Sum. con. gen. II*, c. 57; *De unit. intel.*, c. i, n. 175.

and actually constituting the man Socrates is, the reality Socrates is. Hence, to understand Socrates as man is to grasp the reality of Socrates as "man".

We have sufficiently characterized Aquinas' view on knowledge to return to our point of departure, the "referring is", of understanding. When one understands Socrates as "man", one grasps a particular instance of "substance". When one understands Fido as "dog", one grasps another. Now let us examine such acts of understanding from the point of view of material logic: how are these acts related to Fido and to Socrates? The answer is quite straight-forward: they express the what, the substance, the reality of Fido and Socrates. In like manner, to understand Socrates and Fido as "fat", or as "next to us", is to express respectively the "how much", and the "relation", of the reality of Socrates and Fido. But in thus characterizing what is expressed about Socrates through our understanding of him as "man" and as "fat", we note *how* the *reality* of Socrates is expressed. We note how the act of understanding is *related* to the *reality* of Socrates. Thus even though we understood the meaning of "Socrates is a man", when we asked for the relation of this understanding to Socrates, in answering that question, we mention only how this understanding expressed Socrates as *real*, and not as this particular reality which is a man. Thus, to examine an act of understanding in material logic is to discover how the act expresses the reality of a thing *qua* real, and not *qua* particular. (*In V Meta.*, 9, 889–94; *In XI Meta.*, 3, 2197).

The reason which Aquinas would give to explain this last mentioned fact is as follows. Socrates' being or reality is his individualized essence. Hence, to understand Socrates as "man" is to understand his reality as it is, as "man". Thus, if we do not ask *what* an act of understanding expresses about Socrates, but *how* it expresses him, then we receive the answer: as a "substance". When we ask what it expresses, we receive the answer "man", or what is the same, "substance which is man".

There is but one conclusion to draw from this discussion: when we express Socrates as "man", we actually express him as "substance which is man". In the case of Socrates, the substance is a human one; the substance is constituted a substance by an human essence. (*In VI Meta.*, 4, 1241; *Ibid.*, 1, 1147–49; *In V Meta.*, 9, 889–94).

It is in this sense that "being" or "substance" is at the heart of every act of understanding. When we express Socrates as "man", we mention the particular form substance has taken in this particular thing called

Socrates. Yet at the interior of "man" is "substance", completely penetrating every aspect of the content of that act of understanding Socrates which is expressed as "man".

Now it was precisely because he knew "being" was at the center of every understanding that Aquinas directs his fledgling metaphysician to examine all acts of understanding. In the case of knowing Socrates as "man", "substance" is at the heart of "man". Yet we can also know Socrates as "white"; here "being" in the sense of "quality" would be at the center of our understanding. (*In V Meta.*, 9, 889–94; *In XI Meta.*, 4, 2210; *Ibid.*, 5, 2211; *In I Meta.*, 2, 45–47; *In IV Meta.*, 6, 605)[1]

We have finished now our preparatory discussions. We have seen how "is" can mean either "true", or "being" in the sense of "substance" or "accident". Turning to knowledge, we discovered how the "is" which means "being" is at the center of every understanding. Hence, we see now why Aquinas values so highly the use of material logic. It remains now to speak more expressly of the idea of "being".

3. THE MEANING OF "BEING"

As Aquinas explains, there are ten ways in which an act of understanding can refer to the object known, and all ten express a different way to be real. To understand Socrates as "man", for example, is to express the reality of a substance. But Socrates can also be grasped as "fat", as "white", as "teacher of youth", and so on; by these three acts, we know the quantity, the quality, and the relation of Socrates. "To be white" expresses the "how" of Socrates' substance, just as "to be fat" and "to be teacher of youth" express respectively the quantity and the relation of his substance. All told, man is capable of expressing the basic reality of Socrates in ten ways. Thus, Aquinas remarks, the "is" of the understanding can be related to Socrates in ten manners: there are ten types of *esse significativum*. However, it is the same basic reality that is expressed, for Socrates has but one *esse effectivum*, and hence is but one substance. (*In XI Meta.*, 3, 2197; *In VI Meta.*, 4, 1236 and 1241; *In V Meta.*, 9, 889–94; *In IV Meta.*, 1, 539–43; *In VII Meta.*, 1, 1252–56).

Now this does not contradict what we noted above: there are ten ways to be real. Obviously, Socrates' quantity is real; so too is his quality of whiteness, his relation to his pupils as their teacher. But nevertheless, quantity, quality, relation, and the other six accidents are none

[1] Cf. *Sum. theol. I–II*, 94, 2, *De verit.*, 1, 1; *Ibid.*, 21, 1; *Sum theol. I*, 85, 3c, ad *1um*, and ad *2um*; *In I Phys.*, 1, 7–8.

of them the basic reality of Socrates. Only that basic reality, his substance, can stand on its own. Only his substance is real independent of all else. Quantity, quality, and the other accidents are only modifications – real ones, of course – of that substance. Hence, there is only one reality which is constituted a reality by *esse*: there is but one *esse effectivum* and, since that belongs to substance, only substance is real independent of all else.

Since this is so, our intellect understands the nine accidents as added, non-constitutive modifications of substance. As Aquinas maintains, our intellect is made for the real, and accordingly, its manner of knowing conforms to the real: ten ways of understanding the real, but all ten refer to the same basic reality of an object.

It was this characteristic of our knowledge that Aquinas explained in terms of analogy. The analogy of "being" is a logical theory which expresses the inter-relationship between our ten ways of understanding reality. When we understand Socrates as "man", we express his substance, we express that which exists by itself, free of any dependence on what is not. But when we understand Socrates as "white", "fat", or "teacher", we express those aspects which depend on his substance. Thus, the union of the ten ways of understanding Socrates is based on the union between the ten ways he is real.

Accordingly, the analogy of "being" is the logical schematization of the ten ways our knowledge is related to reality. But since the basic way is expressed by the "is" meaning "substance", metaphysics directs all its attention to grasping the substance of things. (*In IV Meta.*, 1, 534–47; *In VII Meta.*, 1, 1247).

"Being," we have mentioned, is at the core of every act of understanding. When we grasp Socrates as "man", we see his humanity actually constituting him to be a reality. "Substance", therefore, is to be expressed as "essence actually constituting the reality of a being". (*In VII Meta.*, 1, 1247).

Here, of course, Aquinas is using his psychological theory of knowledge. Knowledge of what "substance" means is acquired by looking to our understanding of Socrates as "man". We see "man" in him, actually penetrating all his matter, and actually constituting him as the real man he is. The meaning of "substance" is grasped when we see, not only what, but also how we have expressed the reality of Socrates by "man".

The meaning given to "substance" immediately above is not the final one, however. Aquinas knew well the necessity man has of jugding. As often as we do not know that our understanding is a true act of knowledge, we are forced to judge. Or to express the same necessity in other words, as long as in the content of our understanding we do not express any necessity on the part of the object of being real, then the pre-affirmation reflection must take place. We stand in absolute necessity of reflecting both on our understanding and on the source of that understanding (the phantasm, sense knowledge), before we can know that our knowledge is true, or that it actually refers to a real object. This necessity concerns the act of understanding Socrates as "substance", just as much as any other act of understanding him. Hence, Aquinas is quite aware, that although the substance of Socrates is the actual essence which constitutes the man he is, nevertheless there is no need that the actual essence constitute Socrates as real. Why is Socrates real? The answer is not found in his essence, that Aquinas sees when he realizes that to know Socrates as we understand him, is not yet to know that he is real. Hence, to be actual is something Socrates' essence has no title to. Obviously, Aquinas concludes, the actuality of the essence has been given to the essence: the "to be actually an essence" has been caused.

Thus, "substance" must be redefined. No longer shall we say it is "essence actually penetrating the matter of Socrates and constituting him what he is". Rather it is the "actualized essence penetrating and constituting Socrates as real". This sense of "substance" permits us to see how we express the basic reality of any object. (*In II Meta.*, 2, 295–96; *In VI Meta.*, 1, 1164; *In X Meta.*, 3, 1973).[1]

All this talk of actuality and of essence means we are working with the distinction of *esse* and essence. When we speak of "actualized essence", the final definition of "substance", we are dealing with the result of the *esse*-essence composition, we are dealing with a substantial *ens*. To some it may seem odd to speak of Socrates as an "actualized essence", for he is not an essence. Yet we have not said he is an essence. Rather, we have maintained that he is an actualized essence. Here it is imperative to keep in mind the nature of knowledge. To speak of Socrates as "substance" is to be engaged in the activity of understand-

[1] Aquinas' arguments to God from the fact of *esse*-essence composition underwent an evolution; in the early years, the proof was based on causality; after the *De hebdom.*, participation comes to the force. Thus, I feel, his idea of "Being" must have undergone a corresponding change: from "actual essence" to "actualized essence." Cf. *De ente*, c. 4, p, 35, l. 3–19 (édit. Roland -Gosselin); *In I Sent.*, d. 8, q. 5, a. 2; *In II Sent.*, d. 3, q. 1, a. 1c; *De verit.*, 8, 8; *In Boeth. De hebdom.*, *lect.* 2; *Sum con. gen. II*, c. 52; *De pot.*, 7, 2; *Sum. theol. I*, 44, 1; *Com. theol. I*, c. 68, n. 117. The later texts should be read in the light of *De pot.*, 3, 5, ad *1um* and *2um*; *Ibid.*, 1, 1c *fine*. The pre-*De hebdom.* texts should be read in the light of *De ente*, c. 4, p. 34, l. 7–15 (édit. Roland-Gosselin). See also the texts of footnote 1. p. 362.

ing a glob of matter capable of being sensed as present. Moreover, it is to see the essence which is actually penetrating that glob of matter and making it real. To understand Socrates in this way is not to say he is an essence. Just the contrary, it is to say he is not a pure or a possible essence. It is to say he is actually made what he is by an essence, which is quite a different thing than saying he is an essence.

Ens, in the sense of substance, is actualized essence. Here "actualized' cannot be separated from "essence". Rather, "actualized" must be understood as expressing the essence as 1) shot through and through by actuality, and 2) as possessing in itself no reason why it should be so shot through by actuality.[1] Actuality is, thus, seen as received by essence. Actuality is, in short, *esse*. "Actualized essence" expresses, thus an essence as having received *esse*; "actualized essence" expresses substance, or the effect of the *esse*-essence composition. (*In XI Meta.*, 3, 2197; *In IV Meta.*, 1, 539).

It is not an easy matter to characterize the difference between *esse*, or actuality, and essence, for here we are dealing with the stuff of all thought. Obviously, we cannot explain what is the stuff of thought by using thoughts, for they presuppose what we wish to explain. Scholastics have often said, *vivere est esse viventibus*.[2] In a way, that phrase isn't very helpful. In other words, it is impossible to think "man" without thinking "actual man." We cannot conceive what it means to be purely possible, and that is what essence as such would be. Any time we know Socrates as "man," we see an actual essence in him. Yet we need not think of him at all. Now the difference between "not thinking of Socrates as man" and "thinking of Socrates as man," is the closest we can come to understanding the difference between pure essence and actual essence in Socrates. The difference between the two is our only means of getting at the difference between possible essence and actual essence. The activity of thinking "man" resembles the *esse* of Socrates. The activity of thinking "man" is similar to the activity of *esse*. If we do not think, the meaning given to "man" in this activity of thinking is not; it is nothing. If *esse* is not had, Socrates is not actually being made real by essence.

Perhaps it is well to point out the difference between the actuality which is *esse* and the actuality of a brute fact. The actuality of *esse* has nothing to do with space and time, while the actuality of a brute fact is constituted by materiality present in space and time. It is well too to

[1] Cf. *De pot.*, 1, 1c *fine* and 2c; *Sum. con. gen. II*, c. 57; *Sum. theol. I*, 65, 3.
[2] E. g. *Sum. con. gen. II*, c. 57.

note that there is a difference between thinking of Socrates as "actualized essence" and intellectually knowing his here-and-now reality. An actualized essence, if it is material, is what can be known as here and now. Moreover, since the prime mover is known to be immaterial, we know he is not here and now; he is not a material being in space and time. Hence it is important to recognize that "actualized essence" is the basic reality, and so is prior to here-and-now. Here-and-now depends on matter. Thus, if an actualized essence is material, it is here and now. To be a brute fact – to be here-and-now – does not thus constitute a thing as real; to be a brute fact is an effect of being an actualized essence in a universe of matter.[1]

"Substance", thus, means "actualized essence constituting a thing". Such a concept is at the core of every concept we form. This fact explains why, in the *Commentary*, Aquinas uses material logic in an attempt to discover the ten ways of understanding reality. In all these ways, we espress the being of things. When we think of Socrates as "man", we grasp his most basic reality, his actualized human essence. (*In V Meta.*, 9, 889–94).

Yet it is often said that we are unable to abstract the concept of "being". To *form* it, we must use the judgment of separation. Thus, "being" is formed as the result of judgments such as: "To be is not to be material."[2]

Now if the separation is explained in these terms, if it is explained as the instrument of forming a concept which cannot be abstracted, then we have a doctrine open to two radically different interpretations, for there are two ways to understand "abstract" in this context. Do the advocates of this theory wish to say that "being" is not at the core of every concept? That is, do they wish to state that "being" is not contained within every act of understanding? If they mean this, then the *Commentary*, by implication, contradicts their view. However, there is a second way to understand this doctrine of separation. Here, the impossibility of abstracting "being" means that we are unable to abstract or take "being" out of concepts such as "man", unless we are aided by judgments. It is quite evident that we are unable to understand "being" unless we understand it in relation to Socrates or some

[1] Contrast the early works where *esse* was a brute fact: *De ente*, c. 3, p. 24, l. 1 –p. 25, l. 10 (édit. Roland-Gosselin); *Ibid.*, c. 4, p. 34, l, 7–15; and the *esse* of the post-*De hebdom.* works: *De pot.*, 7, 2; *Sum. theol. I*, 3, 4; *Ibid.*, 75, 5 ad 4*um*; *Quodl. XII*, 5, 1. In the last text mentioned, Aquinas is almost totally explicit in seeing the "here and now" as the effect of *esse*; an actual form has a here and now, but *esse* is given as "actuality".

[2] Cf. *In Boeth. De trin.*, q. 5, a. 3.

other object. Understood in relation to Socrates, "being" does not mean "actualized essence," but rather "actualized human essence." And since the objects understandable by man are all material and particular types, any direct act of understanding, at least in our adult life (cf. below), is one of grasping, not pure "being", but "particular being". Thus, in this second interpretation, the impossibility of abstracting "being" means that, without using a judgment, it is impossible to separate an understanding of "being" from our understanding of "particular being".

Thus, if one accepts the fact that "being" is had at the center of every abstraction, then the judgment of separation *forms* the concept of "being" only in so far as it enables us to see that the particularity of an essence is not intrinsic to the concept of "being" as such. When we understand Socrates as "being" – before we use any judgment of separation – it is "actualized human essence" we grasp. The separation can inform us that to be is not to be material. Hence, we can remove the materiality of Socrates from our understanding of him as "actualized human essence". But to exclude everything related to matter, we must exclude everything which *in our understanding* sets Socrates aside from other *men*, and from all other *material* objects such as dogs and trees. But to exclude such content from our understanding is to exclude the particularity of Socrates; it is to exclude his humanity, Fido's dogness, and so on. If we consider a judgment of separation as the instrument which tell us to exclude "materiality" from our acts of understanding the substances of material things, then the separation has indeed enabled us to form the concept of "being". And only in this sense would the *Commentary* admit the use of separation.

Thus, when we say that "being" is not abstracted, we mean only that this concept cannot be abstracted from our concepts of substances, concepts themselves formed by abstraction.

However, it is necessary to explain clearly what is meant by the negative judgment: "to be is not to be material." This negative judgment, this judgment which separates "materiality" from "to be", is formed on the basis of the proof of the immaterial first mover. (*In IV Meta.*, 5, 593). Now it is often objected that such a proof, and so such a judgment of separation, presupposes a notion of "being". But Aquinas would definitely not agree. The proof of the prime mover is basically the recognition of the impossibility of explaining the totality of motion by recourse to anything intrinsic to that totality. Moreover, the judgment: "There is a prime mover" means only that a prior conception, or act of understanding, is true. Let us outline the entire proof. First, the physician understands the extra-mental totality of motion in these terms: "a definite number of material bodies, all in motion, and whose motion is explained by three intrinsic principles, matter, form, and privation." Secondly, he judges that this understanding is correct. In Aquinas' terminology, the physician is said to know *quid est* ("a definite number, etc.") and *an est* (this conception is true). But the physician does not yet understand why bodies are moving: *propter quid est*. Hence the third step: he conceives the totality of moving bodies as dependent on an unmoved mover, who, to be considered as unmoved, must be thought to be immaterial. Fourthly, the physician judges that this conception is the true one to have; or in other words, he judges that he is correct in not seeing a material, unmoved mover in the totality of moving bodies, that he is correct in relating material, moved bodies to an unmoved, immaterial mover. Thus, "There is a first unmoved mover, an immaterial mover" means that it is true to conceive the moving world as dependent on something other than itself. In other words, the intellect understands that a determinate number of material bodies is in motion; the intellect does not see how it can understand such motion without understanding it as related to a first unmoved mover unless it sees him as immaterial, or as removed from all possibility of motion. Hence, the intellect, by the exegencies of its understanding, is forced to conceive motion as caused by an immaterial mover. This very act of understanding creates a new idea, if you will; it does not presuppose that man understands "to be" or "being" as other than "to be material". Rather, the desire to understand motion forces the intellect implicitly to create a new idea of "being" by positing a mover who cannot be material. Prior to our understanding that motion does not contain its own explanation, the physicist would have said that "being" is "material being". But when he sees that a universe of material bodies cannot cause motion unless they themselves are moving, he knows he has committed himself to accepting the judgment: "to be is not to be material". Thus, I do not think the proof of the prime mover presupposes a notion of "being" other than "material being" as present in the mind of the physicist; rather, the proof, based on the desire to understand, "creates" the notion of "being" as other than "material being": or better, it creates a new content for the notion of "being", which up to the time of the proof had meant "material being".

One can describe this creation of the notion or content of "being" as the result of the exegency of man's desire to know. (*In I Meta., lectiones*

1–2). Man is forced to posit this new content, because he is forced to understand not only what material things are *qua* moving, but also why they are moving. However, it does not follow that this creation of "being" implies that "being" is not found at the center of any act of understanding.

Yet even if an event such as the proof of the prime mover be needed before we can arrive at the meaning of "being", it does not follow that this concept is not the very first one we have. On this point, Aquinas is quite explicit and insistent: "being", he writes, is a concept man is obliged to form as soon as he begins to use his intellect. The understanding of an object as "being" is the initial step in the intellectual life. (*In I Meta.*, 2, 46).[1] Admittedly, understanding is a paradoxical activity when "being" must necessarily be the first of our thoughts, as well as one attained only after a great deal of speculation. But Aquinas was a great one for paradoxes! However, since to be paradoxical is not the same as to be unintelligible, Aquinas could have quite consistantly taught both the priority and the posteriority of our knowledge of "being".

Aquinas was found of asserting that for man to know "being" is the most natural thing in the world. When one begins to think, to understand, one must grasp the most essential aspects of reality; one must seize the most common aspect of reality. Thus, one's first thought cannot be that of "man", for this requires that one has already understood "animal". Nor could "animal" be the content of man's first understanding, for this supposes one knows "material being". Not even "material being" is first, for this must be preceded by the grasp of "being", "that which is". (*In I Meta.*, 2, 46; *In IV Meta.*, 6, 605; *In XI Meta.*, 4, 2210; *Ibid.*, 5, 2211).[2]

Now it should not be thought we are dealing here with the "being" of the Porphyrian tree. Such a "being" is the most general, most vague, most poor of all notions of essence. The "being" with which we deal here is not a notion of essence, but one of actualized essence. The difference between these two notions of "being" is rooted in the fact that the Porphyrian "being" has nothing to do with actual knowledge of reality. This latter "being" falls within the scope of the formal logician, who sets up a hierarchical arrangement of thought as known. That is, in discussing the relation of predicates to subjects of a proposition, the

[1] Cf. *De verit.*, 1, 1; *Ibid.*, 21, 1; *Sum. theol. I*, 85, 3c, ad *1um*, and ad *2um*; *In I Phys.*, 1, 7–8.

[2] Cf. *Sum. theol. I*, 85, 3; *In I Phys.*, 1, 7–8; *In I Post anal.*, 4, 43.

logician speaks of genus and species. A predicate related to a subject as its genus can be turned into a specific concept through the addition of a difference. Now it is in the hierarchical classification of generic and specific predicates that Porphyrian "being" appears. This "being" is the predicate to which must be added successive determinations, or specific differences, until one arrives at the species needed. Thus, to arrive at the specific term predicable of Socrates, to "being" one adds "material", "animate", "sentient", and "rational". Such "being" is not a concept used to know reality, however. It is purely a limit notion used by the formal logician in his discussion of species, genus, and difference.

Opposed to Porphyrian "being", there is the concept of "being" which is both man's initial concept and the concept grasped by the metaphysician. At the dawn of sentient life, a child is unable to distinguish between colors; although aware that it sees, a baby needs practice before he can distinguish green from blue, red from brown, and so on. Yet any child reveals that he is conscious that he sees. An analogous situation occurs in the initial acts of understanding. One understands an object, to be sure; one grasps an object as real. But it is only after one understands several men and several dogs, that one can differentiate between men and dogs. That this is so should not be surprising. The first act of understanding can be described as a grasp of an object as "that which is". In this grasp, the other is quite evidently known as an unity of activity, and if one knew the word at this stage of thought, an unity of meaning. Nor is this unity understood as possible, but on the contrary, as actual. Yet even though the first act of understanding is of "that which is", the intellect is incapable of explaining, at this stage, that the "that which is" is human rather than something else. Only when a person learns that "man" is used of Socrates, that "walking", "eating", "speaking" are said of man's activities, only then can one explain Socrates as anything other than a "that which is". Thus it is that the first act of understanding is of "being". One grasps initially an object as "that which is", although one does not see immediately what kind of thing "that which is" is in the case of any object so grasped. As well, one does not see what it is in the object that makes it a "that which is"; one does not yet understand that a thing must be a composition of essence and *esse* to be a "that which is". (*In I Meta.*, 2, 46).

Thus, man is indeed in a paradoxical situation. His initial apprehension yields "being". At this stage of his life, however, he has no interest in such a thought, for he has to learn to live. He cannot go through life mistaking a snarling dog for his mother. Hence his interest is totally absorbed in discovering the particularities of objects. He must grasp the differences between objects if he is ever to work with them. Unfortunately it becomes extremely difficult to distinguish between the basic manner of expressing, of understanding, any particular reality, and the expression of the particular type of reality had in any given instance.

Hence it is that Aquinas could say that the initial act of understanding is "being". Yet at the same time, he could insist that we understand "being" only after a great deal of intellectual effort.

4. THE TRANSCENDENTAL CONCEPTS

In the *Commentary* only three concepts are explicitly brought forth as enjoying universal application: "being", "one", and "thing". Yet a fourth concept, "true", appears to have all the attributes of a transcendental, for every reality is said to be "true" insofar as it is "being". In this section, however, we shall limit our primary considerations to the first three, for "true" appears to be a much less valuable concept.

In the second book, "true" is said to be a characteristic of all things: just as any object is being, so too it is true, Aquinas remarks, for an object is true insofar as it is being. In discussing the truth of being in this way, Aquinas reveals that "true" is every bit as much a transcendental as "being". (*In II Meta.*, 2, 294).[1] Yet when he deals with the *communia* in Book IV, Aquinas mentions only "being", "one", and "thing". These common concepts are those which are used by every science, and hence are quite important. (*In IV Meta.*, 1, 531; *Ibid.*, 4, 587; *Prooemium*).[2] "True", on the other hand, insofar as it expresses the relation of beings to an intellect, is not all important for the sciences other than metaphysics. Whereas the unity, the thinghood, and the being of a particular object (say of mobile being) must be explained by the science of that object (physics), the truth of an object is something completely presupposed by the science of that object. A particular science presupposes its object is true; it does not bother to say anything about the object's truth, however. Hence, we shall limit ourselves to a discussion of "being", "thing", and "one" in this section. In the following section, we shall note how particular sciences use these common concepts.

[1] Cf. *De verit.*, 1, 1; *Ibid.*, 21, 1.
[2] Cf. *In I Phys.*, 1, 4.

When we understand a substantial reality as "being", we grasp it as "actualized essence". But we can also grasp the nine accidents which can be had by material substances. However, in understanding any of these accidents, we imply an understanding of the substance, which is prior to our understanding of the accidents. For example, to know the whiteness of Socrates ("Socrates is white") is to imply our awareness of his substance ("Socrates is", or "Socrates is a substance"). (*In V Meta.*, 9, 894; *Ibid.*, 15, 980; *In VII Meta.*, 1, 1252–56).

In any of these nine possible ways of understanding accidents as "being", there is a reference of accidents to substance. But *qua* "being", a substance is referred to nothing else. Now the same is not the case when we deal with our understanding of objects expressed as "one", or as "thing". In these cases, an object is always related to some other substantial object.

Just as there are ten possible ways of understanding an object as "being" (substance and the nine accidents), so too are there ten ways of grasping reality as "one", and "thing". It is not only a substance that can be seized as "one" and "thing", but also quantity, quality, relation, and so on. (*In IV Meta.*, 2, 561–62).

When we understand the substance of Socrates as "one", we see the actualized essence as "undivided in itself and as divided from all others". But then, too, we can grasp Socrates' whiteness as "one", that is, as "undivided in itself and as separate from the color of any other substance". In these cases, we related Socrates' substance, and one of his qualities, to the substances, and colors, of other things. This comparison results in our understanding of his substance and his color as identical to themselves and as totally independent of all other substances and colors. (*In IV Meta.*, 2, 550–53 and 560–62).[1]

Our acts of understanding which are expressed by "thing" involve a similar activity of comparison. When we compare an actualized essence to the definition we can give of it, we understand the object as "thing". Thus, we grasp Socrates as "thing" insofar as we understand him as "actualized essence capable of being defined by man". In much the same way we speak of his whiteness as "capable of being defined by man". In other words, to understand a being as "thing" is to seize it as an object about which we can ask, and answer, the question *quid est*? (It should be clear that "thing" and "true" are not synonyms. Whereas "thing" expresses an object as capable of being submitted to intellectual scrutiny, "true" expresses beings as reflections of the Divine Intellect;

[1] Cf. *De verit.*, 1, 1; *Ibid.*, 21, 1.

an object is "true" insofar as its reality is as God desires). (*In IV Meta.*, 2, 550–53; *In I Meta.*, 12, 183).[1]

"Being", "thing", "one" – these are three transcendental notions. Insofar as a reality is an actualized essence, it can be known as such, as "being". In so far as it is an actualized essence, it can be seen as capable of undergoing intellectual scrutiny, for we can see how it can be known in answer to *quid est*? And finally, insofar as it is a being, an object can be known as undivided from itself and separated from all objects similar in being.

5. METAPHYSICS, THE RULING SCIENCE

It is the task of metaphysics to grasp the reality of all things under the formality of "real". By knowing an object as "real", as "being", the metaphysician will know it as an actualized essence. Moreover, because he knows material beings as "being", and because he knows this knowledge is true, he can direct a lower science to study material beings as such. (*In VI Meta.*, 1, 1147–51).

To be material, the metaphysician realizes, is to be an actualized essence composed of matter and form. However, to speak of a material object as "being" is to express it insofar as essence and *esse* constitute it. In this latter knowledge, nothing explicit is said about the materiality of the object. Insofar as the metaphysician expresses all the necessary characteristics of actualized essence, he has sufficiently expressed Socrates as "being". He has not, however, explicitly explained the motion of Socrates, the material activity of Socrates. Hence, a new study, an infra-metaphysical study of Socrates is needed, a study emphasizing his material aspects.

Now this may strike us as strange, for does not the metaphysical knowledge of Socrates as "being" express everything, absolutely everything expressable about Socrates? If we know Socrates as "being", do we not know whatever it is possible to know about him? Since Aquinas undoubtedly answers affirmatively, we are face to face with another of his paradoxes. Not only does Aquinas say our metaphysical knowledge of Socrates is an understanding of the total reality of Socrates, but as well, he maintains that metaphysical knowledge must be completed by the other sciences.

As is to be expected, the key to his paradox lies in our knowledge of "being". When speaking of our knowledge of "being", there is room for

[1] Cf. the texts of the preceding footnote. Also see: *In I Sent.*, d. 19, q. 5, a. 1, ad *2um* and ad *3um*.

an important distinction. On the one hand, we can consider the *knowledge* we have of Socrates or of any other individual reality when we know them as "being"; and on the other hand, we can speak of the *science* of metaphysics.

A science, Aquinas would say, is the habitual possession of the knowledge had about a definite number of beings considered under a given formality. Thus, Thomist physics is the permanent realization of what it means to know all existing material beings from the viewpoint of materiality or mobility. And metaphysics is the knowledge of all finite beings precisely as actualized essences. Insofar as a man is a philosophical physicist, or a metaphysician, he understands what it means for objects to be material, or real. But the point to emphasize is that, as a posesssor of these two sciences, one is not actually engaged in knowing any particular being. As a physicist, one knows what any material object is as moving or as capable of motion; as a metaphysician, one sees what is required to be real. But, as physicist, and as metaphysician, one does not actually grasp Socrates' capability of motion, nor his being; one is only capable of grasping Socrates as mobile and as being.

Opposed to this view of metaphysics and physics as sciences, there is the actual metaphysical knowledge of an object such as Socrates. To have actual metaphysical knowledge of Socrates is to see his concrete essence, his humanity, as actually penetrating the glob of matter named Socrates. To know him as "being" is, as well, to see that this actual essence, this humanity, need not be actual; it is to see an actualized essence constituting Socrates as the man he is. Quite obviously, when we have such knowledge of Socrates, we are aware that this essence is humanity, that it is composed of matter and form. (*In VI Meta.*, 1, 1147–48; *In III Meta.*, 4, 384). However, since we are acting as metaphysicians in knowing him thus, we are forbidden to explain to the world anything about his humanity as such. As metaphysicians, we *speak* in terms of a science; as metaphysicians, we can not speak at all of the psychological and physical activity of which Socrates *qua* man is capable. Rather, as metaphysicians, we must limit ourselves to those statements which can be part of the science of metaphysics. We must say nothing about Socrates that cannot be said about every other finite being. (*Prooemium*; *In IV Meta.*, 1, 531).[1]

[1] Compare *In IPhys.*, 1, 5 with *In III Meta.*, 4, 384. The latter text explains that all four causes are studied in metaphysics, while the former passage gives only three, omitting matter. But in the *Physics'* explanation, Aquinas is dealing with the science of metaphysics which demonstrates from three causes; in the passage of the *Metaphysics*, he speaks of the knowledge of the *Metaphysics*, he speaks of the knowledge one has of Socrates as "being." Thus, here is the opposition of metaphysics as science and as knowledge.

Sometimes when Aquinas speaks of the relation of metaphysics and the other sciences, he speaks in terms of sciences, not in terms of the activity of knowing individuals. In thus referring to metaphysics, he is not speaking of the activity of understanding Socrates as "being". Rather, he is speaking of the attempt to express our knowledge as a science. (In VI Meta., 1, 1149)[1] Thus it is that metaphysics must be completed by other sciences. So it is, that even if our knowledge of Socrates as "being" is an understanding of how he is constituted by his humanity, when we express that knowledge metaphysically – or in scientific, metaphysical terms – his humanity is not referred to as "humanity" but as "essence". Hence, after we have completedly expressed our knowledge of finite beings as "being", there is much that is only implicit in what we have said. We have explained the activity of actualized essences, but we have not explained the particular activity of particular types of essences. Hence, further explanations are needed, explanations which attempt to make explicit what we as metaphysicians only said implicitly. (*In I Meta.*, 2, 47).

And so metaphysics must be completed by other sciences. Naturally enough, these other sciences fall under the sway of metaphysics, for metaphysical knowledge of an object reaches implicitly every aspect of the object's reality, or simply: every aspect of the object. (*In VI Meta.*, 1, 1148).

As presented in the *Commentary*, Aquinas' metaphysician is an absolute dictator as concerns the scope, the interests of the other sciences. Nor is this surprising. When the metaphysician understands Socrates as "being". he sees the effect of the human essence in the constitution of a reality. Hence, he is in a position to tell physics, for example, what it must look for in studying man: the material activity whose source is actualized humanity. As we have pointed out, the metaphysician, by understanding Socrates as "being", has seen the constituting activity of Socrates' humanity, yet he is forbidden to speak of this humanity as such in the science of metaphysics. Hence, he is led to do the next best thing; he points out that Socrates is not only a being, but that he is both a material and a living one, and that thus he should be studied from the points of view of materiality and life. By pointing out these facts, the metaphysician establishes the legitimacy of physics and psychology. (*In VI Meta.*, 1, 1147–51).

It is interesting to note that Aquinas will not permit the physicist or the psychologist to discover the essence of man. The metaphysician,

[1] Cf. *In I Phys.*, *lectio* 4.

Aquinas writes, must see the humanity of a given man before he can point out the need to study man as mobile and as living. In other words, if the metaphysician does not see in Socrates or in some other man an actualized human essence, he will not be able to establish another science to study man from some particular point of view such as that of "living". Thus, because knowledge of the human essence precedes the establishment of the psychological study of man, psychology itself has nothing to do with the discovery of the human essence. As far as knowledge of the human essence is concerned, Aquinas remarks, psychology is reduced to describing it in terms of sense knowledge. (*In VI Meta.*, 1, 1148–49).

Thus far we have been dealing with the influence had on other sciences by the metaphysician's knowledge of "being". However, there are still the other two *communia* to consider, "one" and "thing". "One", the metaphysician claims, expresses an object as an actualized essence identical to itself and divided from all others; "thing", on the other hand, expresses an object as an actualized essence about which we can ask, and answer the question *quid est*? Both these bits of metaphysical knowledge have a role to play in guiding the sciences other than metaphysics. Yet in a certain sense, they have been at work in what we have already said. We have spoken several times of the metaphysician's manner of expressing his direct knowledge of Socrates' actualized humanity. In so doing, the metaphysician is supposing his knowledge of Socrates as "thing"; by expressing Socrates as "actualized essence which constitutes Socrates", he is presupposing that Socrates is a "thing", he presupposes the legitimacy of giving the *quid est* of Socrates. Moreover, when the metaphysician directs the physicist to describe, through the aid of sense knowledge, the essence of Socrates, he tells the physicist to have confidence in the fact that Socrates is a "thing". In other words, the metaphysician knows one can ask for the *quid est*? of Socrates, and that one can obtain an answer; the metaphysician knows this, and he expects the physician to trust him by simply going ahead in the search for the *quid est*? of Socrates as discoverable through a description in terms of sense knowledge.

Just as the metaphysician's knowledge of Socrates as "thing" influences the foundation of physics and of the other sciences, so too does his knowledge of Socrates as "one". Insofar as Socrates is understood as "one", he is not only seen as being constituted by an essence, but he is seen also to be constituted by an essence which is internally united, or to put it more bluntly, by an essence which is not what it is not. In other words, Socrates as "one" is well-defined in the sense of limited; he is constituted as having these essential perfections, but not those. Socrates is a man, he has all that it means to be a man; but because he has the essentials of man, he does not have those of an elephant or of a rose bush. All this is what the metaphysician wishes to say when he pronounces on the unity of Socrates.

It is not difficult to see how the metaphysical knowledge of Socrates as "one" has a role to play in establishing a science such as physics. The metaphysician, by understanding Socrates as "being", grasps the activity constituting human essence in its very constitution of Socrates as a reality. But insofar as the metaphysician sees oneness in Socrates, he explicitly adverts to the unity of Socrates' human essence in its constituting activity; he sees how matter is organized into an existing human being; he grasps the essence as limiting Socrates to human perfections, to the exclusion of all others such as those proper to elephants. It is because Socrates is thus understood by the metaphysician, that physics is directed to the study of a multitude of mobile beings, and not to some pantheistic whole. Let us explain. If the metaphysician did not see where the perfections of Socrates end, that is if he did not grasp the limits of the perfections both of Socrates and of the elephant Socrates is riding, then he would not be able to direct the physician's attention to the movement dependent on Socrates as distinct from that dependent on the elephant. In other words, if the physician is to describe Socrates' mobility as demanding matter, form, and privation, he must presuppose that certain motions are rooted in Socrates and that other movements have their source in the elephant. These presuppositions can be made by the physician, for the metaphysician had previously grasped the limits of the humanity of Socrates, he has seen the distinction between the influence of that humanity in the constitution of Socrates and the constituting activity of another essence in the elephant; he has seen how the elephant is the elephant, and how Socrates is Socrates. (*In IV Meta.*, 2, 550–53; and 560; *Ibid.*, 3, 566).

It is clear that we are not contradicting what has already been said about the knowledge of Socrates as "being". To know Socrates as "being" is to see the actualized human essence as constituting Socrates: attention is directed to the essence of Socrates alone. In knowing him as "being", we do not compare Socrates' essence with other actualized essences; here for example, we would not compare him with the elephant. When we do make such a comparison however, we have a new act of understanding, an act expressed as "one", an act in which we understand Socrates in relation to an elephant: "the actualized essence of Socrates is what it is and is not anything else." (*In X Meta.*, 4, 1996–98).[1]

Much of what we have thus far discussed concerning the relation between metaphysics and the other sciences is at times expressed by Aquinas in terms of "resolution". Thus, occasionally Aquinas speaks of resolution as a relationship between the concepts used in the various sciences. In this sense, resolution denotes the ability, or better, the necessity of understanding statements about reality made by a lower science as dependent upon logically prior metaphysical ones. (*Prooemium*; *In XI Meta.*, 5, 2211; *In IV Meta.*, 1, 531; *Ibid.*, 5, 591).[2] An example will make this point clear.

Suppose, for example, the psychologist argues from the presence of universal concepts to the immateriality of the soul. We might proceed in this fashion: We are aware that we know the meaning of "man" or of "body", for we are able to understand the common and essential characteristics found in all men or all bodies. We can call our awareness of such meanings the "possission of universal concepts". Thus our ability to know intellectually is not limited to the type of singular, individual knowledge we can have with our senses. For example, we might be able to see only a certain distance, and hence we can know visually only a very limited number of men in any one given act of sight. Yet our intellects are not chained down in this fashion, for we can know all existing men in a solitary act of understanding "man". This means, the psychologist concludes, that our intellects are not material but rather immaterial potencies. And because we sometimes think, and sometimes do not, thinking is an accidental, that is non-substantial activity of man; thinking is thus rooted in an ability, in a potency of man. Hence, we have an immaterial potency of knowledge, for the acts of this potency are not limited to here and now individual material objects. Thus, the source of this potency must be immaterial, and because this potency is specifically human, its immaterial source is the human form. Hence, the human soul is immaterial.[3] In this argument, the psychologist has presupposed metaphysical knowledge of human activity. He has presumed that the metaphysician has grasped in an

[1] Cf. *Sum. theol. I*, 11, 2, ad *4um*; *Ibid.*, 85, 8; *In III De anima*, 11, 758.

[2] On this type of resolution, cf. *De verit.*, 10, 8, ad *10um*; *Ibid.*, 11, 1, ad *13um*; *Ibid.*, 15, 2, ad *3um*; *In I Sent.*, d. 8, q. 1, a. 3, *Praeterea*; *In II Sent.*, d. 9, q. 1, a. 8, ad *1um*; *In III Sent.*, d. 23, q. 2, a. 2, *sol.* 1; *In Boeth. De Trin.*, q. 6, a. 1, ad *3am quaestionem, cor.*; *In Diony. De div. nom.*, C. IV, l. 7, n.. 375–76; *Sum. theol. I*, 79, 8c.

[3] Cf. *Sum. theol. I*, 75, 1, ad *2um*; and 2c.

individual man, first, a potency-act relation, and secondly, the relation of specific potencies to their source in the substantial form.

The psychologist, therefore, presupposes as true two theories: first, every act is proportionate to its potency; and second, all the activities proper to a species flow from the substantial form constituting a being in that species. Since these two theories are the expression of the metaphysician's knowledge of individuals as "being", the psychological discussion of the immateriality of the soul can be viewed as resolvable or reducible to metaphysical knowledge. (*In IV Meta.*, 5, 591).[1]

But there is yet a second sense of resolution in Aquinas' thought, which must not be confused with the one discussed above. Often resolution refers to a dynamic aspect of the construction of metaphysics. In this sense, Aquinas sometimes speaks of metaphysics as the last of the sciences in the *via resolutionis*. In thus using "resolution", Aquinas is thinking of the procedure we followed in Chapter VI, section 1, where we sketched the reconstruction of metaphysics. Metaphysics began, we noted, when the physicist discovered the existence of an immaterial first mover. In our search for the concept used to express both material beings and the immaterial being, we examined our other, our familiar concepts with the hope of discovering what a concept must express if it is to be the concept of reality as such. This search can be styled an attempt to trace back, or to resolve concepts of "particular being", to the concept of "being". In this sense, then, resolution denotes the method used in learning or in constructing metaphysics. (*Prooemium*; *In VII Meta.*, 3, 1308; *In V Meta.*, 9, 889; *In XI Meta.*, 5, 2211).[2]

Thus we reach the end of our discussion of metaphysics as the ruling science. The entire relationship of metaphysics to the lower sciences is rooted in what it means to know an object as "being", as "one", and as "thing". Only because the metaphysical knowledge of Socrates is a grasp of his actualized, determined human essence as actively constituting him, does the metaphysician possess rights over all other scientific studies of Socrates. Solely because of what he knows in grasping the reality of Socrates can the metaphysician establish and direct other sciences to a certain area of study.

[1] An example revealing the use of metaphysical principles is the *Sum. theol. I–II*. The entire book is nothing but an application of the notions of "potency", "form", and "act" first, to sense data about human activity, and secondly, in q. 109 sqq. to the data of revelation concerning grace inspired activity.

[2] Cf. *In I Phys.*, *lectio* 7. On this type of resolution, see: *In Boeth. De Trin.*, q. 6, a. 1, ad *3am quaestionem*, *cor.*; *In Diony. De div. nom.*, C. IV, l. 7, n. 375–76; *Sum. con. gen. III*, c. 41

6. PHILOSOPHICAL KNOWLEDGE OF GOD

To understand Socrates as "being" is to see him as constituted by an actual essence which need not be actual, by an actualized essence. Our knowledge of Socrates is thus a knowledge, not only of his human essence, but of his human essence as having something to which it has no claim – actuality. As already noted, the actuality of human essence in Socrates is not the actuality of the here and now, the spatial and temporal being. To be in space and time is not the same as to be real. On the contrary, a material being is in space and time because it is real. Thus, the actuality of Socrates' human essence causes Socrates to be here and now; it causes him to have the accidental perfections of living in a certain place at a certain time. (*In X Meta.*, 3, 1982).[1]

What, then, is this actuality had by human essence in Socrates? Whatever it is, it cannot be described, nor even defined, for actuality pertains to the very stuff of thought. No matter what words we use to describe the actuality of Socrates' essence, those very words presuppose what they wish to describe. To think is to think in terms of actuality. To explain, or to describe actuality is, then, to attempt to give an explanation of what must be at the center of any explanation. Hence, the closest we can come to grasping actuality is to direct our attention to the difference between thinking and not thinking. In our act of thinking or understanding Socrates as "man", the *meaning* of Socrates as "man" corresponds to essence, while the *very thinking* of that meaning corresponds to actuality. Thus, the best we can give in the way of a description of actuality or *esse* is this: Socrates' *esse* or actuality is similar to our grasp of what it means to have a real, intellectual fact of thinking that meaning.

When we understand any object as "being", we necessarily grasp the essence and the actuality of the object as mutually penetrating one another; however, as well, we understand the actuality as something that need not be penetrating the essence, for the essence has no claim to be real. Hence, to think Socrates as "being" is not to understand two principles of being as distinct one from another, although we understand Socrates as having two principles, but rather as the result of their composition: actualized essence.

Now it should be quite evident that in giving such a meaning to "being", Aquinas has ruled out all possibility of not proving God. However, he seems to have been quite aware of this situation and quite

[1] Cf. *Quodl.* XII, 5, 1.

content with it, for he insists on the necessity of reducing composed being to the *Esse* in which they participate. Actualized being, Aquinas would admit, makes it necessary to postulate a pure Actuality who has caused an actualized being to share in actuality. (*In II Meta.*, 2, 295–96).

The movement from actualized being to Actuality cannot be expressed except by the conjunction of two words: "participation" and "causality". This situation depends exclusively on the meaning Aquinas gives to the actuality found in the beings around us. Because actuality is other than the essence which need not be actual, quite clearly Aquinas must use the word "cause" to explain why an essence is *de facto* actual: because Socrates' essence is actually constituting him when it need not be, it was made to be so constituting him, it was caused to be an actually constituting principle. It is easy to see why the proof of pure Actuality is one involving causality, but how does participation enter in?

Participare is "to have a share", *partem capere.* A being, an *ens*, represents a limited amount of real perfection. It is *so* perfect, but no more: Socrates is more perfect than an elephant perhaps, but he is not the perfection represented by an elephant. But in these examples we are dealing primarily with the essential aspects of Socrates and the elephant. And essential perfection as such has nothing to do with proof of God as pure Actuality.

When we say that Socrates participates in actuality, we are not directly interested in his essence. Of course, indirectly, the human essence enters the picture, for we only know his essence has *received* actuality, that it is not actuality, when we realize that this essence is not to-be. But directly the essence of Socrates does not enter in the proof of pure Actuality; we do not compare his essence to any other essence, we are not interested in whether his essence, and so Socrates, lacks the essential perfections of elephants and rose bushes. In other words, all essence can do in the present problem is to inform us that there is a problem: Socrates' essence is not to-be. (A corollary of this fact is that we need not compare one being to another to arrive at God; we argue from one being only.)

Because we see that any given essence, say Socrates' humanity, is not actuality, we understand that it has received actuality. We understand Socrates' humanity as permeated by actuality. Thus we have two radically different orders of perfection: the order of essence, in this case: humanity, and the order of actuality, in this case: to-be-a-man. Hence, we have the distinction between an essential to-be-a-man, and an actual to-be-a-man. Now insofar as we consider Socrates as having this actual to-be-a-man, we consider him as having something which is not identical with the essential to-be-a-man. Even more, we understand the actuality in this case as limited to the particular actuality it is (to-be-a-man) because the essence is to-be-a-man. Thus, we are considering Socrates as having something which he need not have, which he is not.

"Having something which one is not" is no different from *partem capere* or participating. Hence, we say Socrates participates in pure Actuality, meaning by this, that he has somehow managed to gain a hold on what Actuality is.

Thus because Socrates has an actuality when he need not, he is caused by a prime agent cause. But because having an actuality when he need not is no different from having an actuality which he is not, we must also say that Socrates participates in Actuality.[1]

Thus God, known as pure Actuality, enters our metaphysics. Aquinas gives, by implication, "actualized essence" as the expression of the formal object of metaphysics. Quite obviously, God cannot be studied from that point of view. He can only be related to actualized essence as the cause of its actuality. Hence, if we wish to study actualized essence as such, we can only bring God in as a first cause. (*Prooemium*; *In VI Meta.*, 1, 1164). But metaphysics began as the study of being as such! It is true enough that metaphysics began by looking for a concept of being as such, a concept it decided was "actualized essence". But now that we have discovered a Being who is Actuality, mustn't we decide that being as such is "actual essence"? Mustn't we admit God as belonging to the material object of metaphysics.?

To these questions, we do not doubt that Aquinas answers with a resounding "no". In the first place, God is evidently part of metaphysics because He is the cause of the object studied in the universal science. Because metaphysics as universal science discovers the need of the cause of being as such, metaphysics becomes first science, or the study of God. But there is no indication in Aquinas' *Commentary* that after the discovery of God, one can change the limits of universal and first science so that they are identical. Quite the contrary, universal and first science remain quite distinct branches of metaphysics.[2]

[1] Cf. the texts of footnote 1, p. 360. Also: "...esse, quod rebus creatis inest, non potest intelligi nisi ut deductum ab esse divino..." *De Pot.*, 3, 5, ad *1um*.

[2] The following passages appear at first to deny this point, for they give *ens* or "substance" as common to material and to separated substances: *In III Meta.*, 6, 398; *In IV Meta.*, 4, 593; *In VI Meta.*, 1, 1170; *In VII Meta.*, 11, 1526; *In XI Meta.*, 7, 2263 and 2267; *In XII Meta.*, 2, 2427. However, the difficulty is only apparent since 1) these texts must be

But apart form this factual proof, there is a speculative reason we can give to show why God is not found under the material object. As Aquinas would say, to affirm "God is", or "Actuality is", is to affirm the truth of a prior conception, or of a prior act of understanding. This prior understanding is anything but a grasp of God as "being", for we have absolutely no inkling of what God's being might be. Hence, the affirmation "Actuality is" must be preceded by an understanding of creatures as dependent upon something which itself has not received actuality, something therefore which is non-creaturely. "Actuality is" can, then, be construed only to mean: "It is true to conceive actualized essence as dependent upon Actuality". Hence, the judgment "God is" implies no grasp of God's reality. That is to say, "God is" is not a statement in which we express some positive understanding of what God is. To say "God is" is not to know God's essence, quite evidently. Nor is it to know His *esse*, for His *esse* is His essence which we do not know. Thus, to affirm "God is" is not to understand God in any way, not even in the weakest, most partial way. Hence, if we can affirm "God is", and still be totally ignorant of His *esse* and His essence, it follows that we are totally ignorant of God, in the sense of completely failing to grasp His reality.[1]

But is this not agnosticism? Assuredly not. An agnostic is one, who although admitting knowledge of God's existence as valid, nevertheless refuses to accept the possibility of saying something meaningful about Him. In his *Commentary* Aquinas has given no such theory ."God is", Aquinas implies, insofar as we use it as the expression of a judgment, means: "It is true to conceive the beings of our experience as dependent on Pure Actuality". Thus, although we do not understand what God is, we are certain that the name "God" – the symbol "God" – *refers* to the source, to the cause, of actualized essences. In other words, "God" or "Pure *Esse*" or "Actuality" are synonymous terms which have a real reference. They refer to something, but we do not understand that something to which they refer. We only know that actualized essences

read in the light of the anti-Averroestic spirit in which they were written: "God can belong to the study of substance as such, because He is proved as its cause";2) *In XI Meta.*, 1, 2158–59 explains that the separated substances in question are the *primi motores*; these are studied as *ens*; however, "primi motores" and "prima causa entis" are not the same thing; see also: *In IX Meta.*, 1, 1770; 3) *In III Meta.*, 4, 384 explains we cannot know the forms of separated substance; hence a priori we cannot know God as *ens*. Hence it seems that the passages mentioned in the beginning of this footnote must be interpreted thus: God falls within the genus of substance, and hence within the metaphysical formal object, just as any cause of a genus falls within the genus; in other words, God is the cause of the substance of other things, and is not a substance in the same way other things are. Cf. *In II Meta.*, 2, 293.

[1] Cf. *Sum. theol. I*, 3, 4, ad *2um*; *Ibid.*, 13, 2c, and ad *2um*; *De pot.*, 7, 2, ad *1um*.

cannot be understood totally unless they are understood as depending, as related to something which is by right what they only happen to be, namely: real. It is because we understand that creatures have received their actuality, because we grasp them as "actualized", that we relate them to a Pure Actuality. When we thus see a creature as depending on Actuality, we do not understand this Actuality upon which it depends. Rather, we place the name "Actuality" in the forefront of thought as the signpost indicating the reality we must try to understand, if we are to continue to regard creatures as actualized essences, as dependent beings. But we can never understand the reality (ie: God) to which our names like "God", "*Esse*", or "Actuality" refer.

What we have been working with here is the distinction between "meaning" and "reference". If we understood the reality of God, then we would know the object - we would understand the object – to which "God" points or refers. If we understand that object, then "God" – *for us* – would be a meaningful term: it would express God as He is. But since we do not understand the reality of God, but only the reality of creatures, the only meaning we can give to "God" or to "*Esse*" is some meaning with actually fits only creatures. Since, however, we recognize that God must be other than creatures – Actuality instead of actualized essences – because we recognize this, we attempt to remove the creaturely aspect present in any idea such as "*Esse*". By a judgment, we tell ourselves that "*Esse*" is not an actuality which has been received by an essence, but is rather an actuality which is an essence. We do not understand what such an Actuality can be, but by this negative judgment, we point at God.

And so it is that to say: "God is" is not to understand anything about God. It is, rather, to understand our necessity of relating actualized essences to something else. It is our way of pointing from actualized essences to something else. Hence, we can say "God is" and be totally ignorant of the reality of God, in the sense that we do not understand Him, but only point *mutely* to Him.

Now such a theory is not agnostical, for Thomas can find many ways of pointing to God. For example, because he knows that the immateriality of the soul is the source of man's ability to know intellectually, God qua immaterial must be an intellect. Of course, he does not understand what God's intellect is; "divine intellect" is simply another signpost by which he points to the reality of God. There are many other terms such as "one", "true", etc., which he would realize are applicable to God, for the reality of creatures expressed by these terms is not intrinsically tied to imperfection. It is just the opposite: "one" or "true" express the reality of creatures as perfect, not as imperfect; "one" for example expresses the lack of division of being as such. Because Aquinas must point to God through the name "being" or "real", he knows he must also point to Him through the name "one". Yet he has no understanding of the perfection of God pointed at by "one".

And so it is that Aquinas can speak of God, which is something not at all acceptable to an agnostic. Thus, in spite of the fact that "God is" expresses nothing about God's reality, yet because it points to Him, because it states the truth of creatures' dependence on something else, Aquinas can construct a natural theology. Because of the distinction between the "reference" and the "meaning" of a term, Aquinas can speak of God, without knowing, that is, without understanding anything of the reality to which he refers. (*In XII Meta.*, 8, 2543).[1]

The fact that "God is" says nothing of God, but rather affirms the truth of a conception of creatures as unable to explain their own actuality, points up well the difference between philosophical knowledge of God, and the knowledge of Him resulting from faith. The only person who has experienced God, that is, who has known Him as He is, is God Himself. If He chooses to tell us that He exists, then we can accept this on the strength of His authority; thus we have knowledge that He exists, knowledge based on God's own experience of His reality. But in metaphysics, when we affirm "God is", we have no knowledge based either on our own, or on someone else's experience of God; rather, we have knowledge only of creatures' insufficiency to explain themselves. Thus, the difference between the "God is" of reason and the "God is" of faith is this: reason knows that it must think of creatures as related to some Supreme Reality; faith knows, not only that it must think a creature-God relation, but as well, it has been told of this relation by someone who has experienced it.

7. CONCLUSION

We have discussed many points in this chapter. Yet all our discussions have centered around the condept of "actualized essence". First, we remarked the various ways of considering "is". The most important way concerns the "is" found in the expression of an act of understanding, for as Aquinas remarks, it is this "is" which leads us to knowledge of "being".

[1] Cf. *Sum. theol. I*, 13, 2c and ad *2um* and ad *3um*; *In Boeth. De trin.*, q. 6, a, 3; *Sum. theol. I*, 88, 2, ad *2um*.

But to discover the meaning of the "is" of the understanding, and so of "being", one must realize what it means to understand. For example, to grasp Socrates as "man" is to see his human essence actually constituting him as the man he is. It is by looking at knowledge in this sense that one arrives at a comprehension of "being": "actualized essence which penetrates a thing, constituting it the thing it is".

From this view of "being", it is but a short step to "one" as "actualized essence undivided in itself and divided from all others", and to "thing" as "actualized essence capable of being defined by man". And on the basis of the metaphysical knowledge expressed by these three ideas, it was not difficult to see why metaphysics rules all other sciences.

Finally, we spoke of rational or philosophical knowledge of God. He is proved, we noted, simultaneously through causality and participation. Both of these paths to God are implied in the concept of "actualized essence". Yet the abrupt entrance of God on the philosophical scene does not induce any change in the concept expressing being as such. "Being" remains "actualized essence penetrating a thing, and constituting a thing the reality it is". And that is the fundamental insight found in the metaphysics operative in Aquinas' *Commentary*.

In the conclusion of Part Two, we gave a brief comparison of the *Commentary*'s metaphysics and the system we discovered in the *Prooemium*. The identity of these two philosophies, we understood as a probable sign that the Commentary presents simultaneously the metaphysics Aquinas considered as proper to Aristotle, and the metaphysics Aquinas accepted as his own. In the present chapter, we presented in the footnotes references to both the independent works and commentaries; the works thus referred to contain the same doctrine we have discovered in the *Commentary*. This is, we believe, another indication that Aquinas' *Commentary* presents the metaphysics which is simultaneously Aquinas' system, and the system Aquinas believed Aristotle's.

CONCLUSION

The preceding chapter developed the theme of "being" or "actualized essence" as the doctrine central to the metaphysics operative in Aquinas' *Commentary on the Metaphysics*. That key doctrine, and the metaphysical system built around it, were exposed by Aquinas in such a way that I feel justified in concluding that they represent Aristotle's thought as Aquinas understood that thought. However, whether or not this historical conclusion be true, independent of its truth or falsity, I believe it proper to see the metaphysics operative in the *Commentary* as most probably Aquinas' own system. And so, if I am correct on both points, it appears that the *Commentary* presents simultaneously Aquinas' understanding of Aristotle's system, and Aquinas' own personal philosophy of being, for these two metaphysics are identical.

On the basis of this identity, it is not at all difficult to understand why Aquinas should have decided to compose his exposition of the *Metaphysics*. In the intellectual circles of his era, Avicenna and Averroes were all the vogue. Their *Commentaries* were read, and to some extent were followed, for example, by Albert the Great. Consequently, in the field of metaphysics, the most widely studied works put forth a brand of Aristotelianism quite unfaithful to Aristotle himself – as Aquinas understood him. Aquinas would not have been content with this situation, for, after all, Aristotle's metaphysics was his metaphysics. Accordingly, we can surmise, Aquinas decided to present a correct exposition of Aristotle's thought, an exposition which took the form of the *Commentary*.

In the light of these facts and this surmise, we can better understand the theories of P. Chenu on the nature of Aquinas' *Commentaries*.[1] P. Chenu mentions the two-fold goal of a medieval commentator: to

[1] M.-D. Chenu, O. P., *Toward Understanding Saint Thomas*, Trans. by Landry and Hughes, Regnery, Chicago, 1964.

discover truth, and to disclose the true intention of an author. Since the "truth" is the case of the *Commentary on the Metaphysics* was the system proper both to Aquinas and to the Aristotle Aquinas understood, for Aquinas, "to discover truth" was "to disclose the true intention of the author". His complete acceptance of Aristotle explains why Aquinas searched so diligently for the *verba Aristotelis*. It explains too why he would turn to the *intentio Aristotelis* when the very words of Aristotle were not clear, or when they appeared to endanger either faith or some central doctrine of Aristotelianism. By this method Aquinas attempted to clarify the principles of Aristotle, to restore Aristotelianism to a more perfect harmony with its own principles. The identity between Aquinas' Aristotle and Aquinas' own system explains as well how Aquinas could introduce *esse* and participation as the corner-stone of his exposition. Aristotle, Aquinas knew, rose from a study of being as such to God, the cause of being. This Aquinas not only knew, but even more, with this he was in complete agreement. But how does Aristotle envisage this movement from beings to their Cause? He did not say. Yet it was imperative that Aquinas expose the working of the Aristotelian proof for God, if he were to counteract the influence of Averroes, Avicenna, and Albert. Thus it was that Aquinas inserted into his exposition doctrines not found in Aristotle, nor deducible from his principles. Such doctrines can be termed, to use the expression of P. Chenu, Thomist doctrines *ex sensu proprio*.[1] However, this expression does not imply that Aquinas was conscious of injecting new meaning into Aristotle. The *Commentary* itself gives no evidence that Aquinas was aware of the transformation he was effecting in Aristotelian metaphysics. Hence it is that Aquinas' *Commentary* presents what is *de facto* a Thomist metaphysics, and not the system proper to Aristotle.

Aquinas' exposition, because of its anti-Averroistic, anti-Albertian, and anti-Avicennian spirit, is seen to be a medieval's response to the intellectual problems of his time. Averroes and Albert had undermined the proof of God, while Avicenna had simply ruined Aristotelian (and Thomist) metaphysics. Hence, Aquinas exposed *the* philosophical "authority" in the field of metaphysics, making certain, by *esse* and participation, that the search for truth returned to what he felt was its proper direction, the direction indicated by Aristotle.

One of the most interesting aspects of the metaphysical system of the *Commentary* is its emphasis on the discovery of the concept of "being". It is, one can say, a metaphysics geared to the intellectual development

[1] *Ibid.*, pp. 206–14; 220–22; 153–55.

of man. It is one which takes into account the exigencies of a pupil: this metaphysics moves from the better known to us to the better known in itself. In other words, the *Commentary*'s metaphysics has in mind the student, the pupil who has still to attain the stage where he can view reality as such.

Thus Aquinas – for the *Commentary* most probably expresses his thought – would not have begun a metaphysics course without clearly underlining the point of intellectual development necessarily attained by all his students. They must already have completed a course in physics, a course which terminated in knowledge that there exists an immaterial first mover. This fact Aquinas would emphasize, for it is the awareness that "to be" is not "to be material" that turns man toward metaphysics.

But when he had called to his students' attention this fact of the difference between "being" and "material being", Aquinas would have turned directly to the problem of discovering the meaning of "being", for "being" expresses the only viewpoint from which one can study all reality as such. Aquinas' method has become familiar to us now: an investigation of predication with an eye to discovering our manner of understanding reality as such.

There is nothing esoteric about such a procedure. There is no demand, such as some metaphysicians make, that the student attempt to "intuite" the reality of things which "shines forth" through the particularities of objects – the "epiphany" of being. Nor are Aquinas' pupils asked to undertake the difficult task of grasping the "value" of being exemplified by man's inability to think away his own existence. Rather than directing attention to such demanding tasks, Aquinas simply asks his students to hunt for the order and the content found within our ordinary, everyday acts of understanding.

After he had led his pupils to a grasp of "being", Aquinas would not have instigated a search for God. For Aquinas, metaphysics does not set out to prove God after the fashion of one looking for hidden gold. Aquinas' metaphysician, far from looking for God, simply stumbles across Him. This Aquinas would have brought home to his students by pointing out how a radical dependence on Actuality is part and parcel of our notion of "being". Accordingly, Aquinas would say, there is a Pure Actuality.

In leading his students from an examination of predication to the existence of Pure *Esse*, Aquinas would have pushed them to the limits of human understanding. From acts of grasping "man" and "horse",

Aquinas would have moved his students to acts of pointing to Pure Actuality. In other words, Aquinas would have impressed on them the recognition of the impossibility of understanding God, apart from a super-natural vision. Thus, he would have explained the working of first philosophy – the systematic attempt to discover the words, the sign posts, which can truly be used to refer to the unique perfection of God.

This then is the metaphysics of Aquinas' *Commentary*. And so we reach the *finis operis*, the discovery of the nature of Aquinas' exposition of the Aristotelian *Metaphysics*. Of that exposition we can say: first, that it was written in the light of one metaphysical system; second, that it was directed against the interpretations of Aristotle given by Avicenna, Averroes, and Albert; third, that Aquinas attributed to Aristotle a metaphysics which is not that held by the ancient Greek; yet there are no indications that Aquinas was conscious of this error in his work; and fourth, that Aquinas accepted this metaphysics as his own.

BIBLIOGRAPHY

(I have included in this list, not only the books and articles referred to in my study, but also others found useful in my work.)

A. Primary Sources

ALBERT THE GREAT: *Metaphysica*, in *Opera Omnia*, T. XVI, Pars I and II, edit. B. Geyer, Aschendorf, Münster, 1960–64.

— *Physicorum Libri VIII*, in *Opera Omnia*, Vol. III, edit. A. Borgnet, Vivès, Paris, 1890.

AQUINAS, THOMAS: *Commentaria in Aristotelis Libros Peri Hermeneias et Posteriorum Analyticorum*, in *Opera omnia jussu Leonis XIII, P.M.*, T. I, Rome, 1882.

— *Commentaria in Libros Aristotelis De caelo et mundo, De generatione et corruptione, et Meteorologicorum*, in *Opera omnia jussu Leonis XIII, P.M.*, T. III, Rome, 1886.

— *Commentarium in octo Libros Physicorum Aristotelis*, in *Opera omnia jussu Leonis* XIII, *P.M.*, T. II, Rome, 1884.

— *Compendium theologiae ad fratrem Reginaldum*, in *Opuscula theologica*, T. I, edit. A. Verardo, Marietti, Turin, 1954.

— *Expositio super librum Boethii De Trinitate* ad fidem codicis autographi nec non ceterorum codicum manu scriptorum, edit. B. Decker, Brill, Leiden, 1955.

— *De aeternitate mundi contra murmurantes*, in *Opuscula Philosophica*, edit. R. Spiazzi, O.P., Marietti, Turin, 1954.

— *De natura accidentis*, in *Opuscula Philosophica*, edit. R. Spiazzi, O.P., (authenticity doubtful), Marietti, Turin, 1954.

— *De substantiis separatis seu de angelorum natura*, in *Opuscula Philosophica*, edit. R. Spiazzi, O.P., Marietti, Turin, 1954.

— *De unitate intellectus contra Averroistas*, edit. L.-W. Keeler, editio altera, Pont. Univ. Gregoriana, Romae, 1957.

— *In Libros de anima expositio*, edit. A. Pirotta, O.P., editio 3a, Marietti, Turin, 1948.

— *In duodecim libros Metaphysicorum Aristotelis expositio*, edit. R. Spiazzi, O.P., Marietti, Turin, 1950.

— *In librum Beati Dionysii De divinis nominibus expositio*, edit. Pera, Carmello, & Mazzantini, Marietti, Turin, 1950.

— *In librum Boetii De hebdomadibus expositio*, in *Opuscula theologica*, Vol. II, edit. M. Calcaterra, O.P., Marietti, Turin, 1954.

— *In librum De causis expositio*, edit. H. Saffrey, O.P., Nauwelaerts, Fribourg-Louvain, 1954.

— *Le "De ente et essentia" de S. Thomas d'Aquin*. Texte établi d'après les manuscripts parisiens. Introduction, Notes et Etudes historiques, par M.-D. Roland-Gosselin, O.P., Vrin, Paris, 1948.

— *Questiones disputatae*, Vol. I, edit. R. Spiazzi, O.P., editio 4a, 1953; Vol. II, edit. P. Bazzi and others, editio 9a, 1953; Marietti, Turin.
— *Quaestiones quodlibetales*, edit. R. Spiazzi, O.P., editio 8a, Marietti, Turin, 1949.
— *Scriptum super libros sententiarum*, Vol. I–II, edit. P. Mandonnet, O.P., 1929; Vol. III–IV, edit. M. Moos, 1933 and 1947; Lethielleux, Paris.
— *Summa contra gentiles*, in *Opera omnia jussu Leonis XIII, P.M.*, T. XIII–XV, Rome, 1920–30.
— *Summa theologiae*, in *Opera omnia jussu Leonis XIII, P.M.*, T. IV–XIII, Rome, 1896–1906.
ARISTOTLE: *Metaphysics:* Greek text in *Aristotle's Metaphysics. A Revised Text with Introduction and Commentary* by W. D. Ross, 2 vols., Clarendon Press, Oxford, 1958.
— *Metaphysics:* English translation in *The Works of Aristotle translated into English*, Vol. VIII, trans. by W. D. Ross, 2nd edition, Clarendon Press, Oxford, 1960.
— *Metaphysics: Versio Arabica in Aristotelis Opera Omnia*, Vol. VIII: *Metaphysicorum libri XIIII*, Junctas, Venetiis, 1562.
— *Metaphysics: Versio Media* in *Alberti Magni Opera Omnia*, T. XVI, Pars I–II, edit. B. Geyer, Aschendorf, Münster, 1960–64.
— *Metaphysics: Versio Vetus* in *Opera Hactenus inedita Rogeri Baconi*, Fasc. XI, edit. R. Steele, Clarendon Press, Oxford, 1932.
— *Physics:* Greek text in *Aristotle's Physics. A Revised Text with Introduction and Commentary* by W. D. Ross, Clarendon Press, Oxford, 1955.
— *Posterior analytics:* English translation in *The Works of Aristotle Translated into English*, Vol. I., trans. by G. R. G. Mure, Clarendon Press, Oxford, 1955.
— *Posterior analytics:* Greek text in *Aristotle's Prior and Posterior Analytics. A revised Text with Introduction and Commentary*, by W. D. Ross, Clarendon Press, Oxford, 1949.
AVERROES: *Commentarium in De coelo:* in *Aristotelis Opera Omnia*, Vol. V: *Aristotelis De Coelo, De Generatione et Corruptione, Meteorologicorum, De Plantis*, Junctas, Venetiis, 1562.
— *Commentarium in De physico auditu libri octo:* in *Aristotelis Opera Omnia*, Vol. IV: *Aristotelis de Physico auditu*, Junctas, Venetiis, 1562.
— *Commentarium in Metaphysicorum:* in *Aristotelis Opera Omnia*, Vol. VIII: *Metaphysicorum cum Averrois Cordubensis in eosdem commentariis*, Junctas, Venetiis, 1562.
— *Tahafut Al-Tahafut* (*The Incoherence of the Incoherence*). Translated from the Arabic with Introduction and Notes by S. Van den Bergh, 2 vols., University Press, Oxford, 1954.
AVICENNA: *De Coelo et Mundo* in *Opera Philosophica*, fol. 37r,A–fol. 42v,B, (Venise, 1508), Réimpression Bibl. S.J., Louvain, 1961.
— *Epistle of Definitions* in: A.-M. GOICHON, *Introduction à Avicenne. Son Epitre des définitions*, traductions avec notes, preface de M. A. Palacics, Desclée, Paris, 1923.
— *Logyca* in *Opera Philosophica*, fol. 2r,A–fol. 12v,B, (Venise, 1508), Réimpression Bibl. S.J., Louvain, 1961.
— *La Métaphysique du SHIFA'*, Traduction faite sur le texte arabe de la lithographie de Téhéran de 1303h par le P. M.-M. Anawati, O.P., Inst. d'Etudes Méd., Montréal, 1952.
— *Metaphysica sive prima philosophia*, (Venise, 1495), Réimpression Bibl. S.J., Louvain, 1961.
— *Metaphysices Compendium* ex arabo latinum reddidit et adnotationes adornavit N. Carame, Pont. Inst. Oriental. Stud., Romae, 1926.

— *Philosophia prima sive scientia divina* in *Opera Philosophica*, fol. 70r,A–fol. 109v,B, (Venise, 1508), Réimpression Bibl. S.J., Louvain, 1961.

B. Secondary Sources

— *Aristoteles Latinus, Pars Prior*, Corpus Philosophorum Medii Aevi, Roma, 1939.

AFNAN, S. M.: *Avicenna. His Life and Works*, Allen and Unwin, London, 1958.

ALLAN, D. J.: *Aristote le Philosophe*, traduit par C. Lefèvre, préface par A. Mansion, Nauwelaerts, Louvain-Paris, 1962.

ANSCOMBE, G. E. M. and GEACH, P. T.: *Three Philosophers* (*Aristotle, Aquinas, and Frege*), Blackwell, Oxford, 1961.

ARNOU, R.: *De quinque viis sancti Thomae ad demonstrandam Dei existentiam apud antiquos Graecos et Arabes et Judeos praeformatis vel adumbratis*, Pont. Univ. Greg., Romae, 1932.

AUBENQUE, P.: *Le problème de l'être chez Aristote. Essai sur problématique aristotélicienne*, Press Univ. de France, Paris, 1962.

— "Sur la notion aristotélicienne d'aporie" in *Aristote et les problèmes de méthode. Communications présentées au Symposium Aristotelicum tenu à Louvain du 24 août au 1er sept. 1960*, Nauwelaerts, Louvain-Paris, 1961, pp. 3–19.

BAISNEE, S. S. J.: "St. Thomas Aquinas' Proofs for the Existence of God Presented in their Chronological Order", *Philosophical Studies in Honor of Ignatius Smith*, Newman, Westminister, Md., 1952, pp. 29–64.

BALTHASAR, N.: *L'abstraction métaphysique et l'analogie des êtres dans l'être*, Warny, Louvain, 1935.

— *L'être et les principes métaphysiques*, Inst. Sup. de Phil., Louvain, 1914.

— *La méthode en métaphysique*, Inst. Sup. de Phil., Louvain, 1943.

— *Mon moi dans l'être*, Inst. Sup. de Phil., Louvain, 1946.

BEACH, J.: "Separate Entity as the Subject of Metaphysics", *The Thomist*, XX, 1957, pp. 75–95.

BERTOLA, E.: "Le traduzioni della opere filosofiche arabogiudaiche nei secoli XII e XIII" in *Scritti in onore de Prof. Mons. Francesco-Olgiati*, Vita e pensiero, Milano, 1960, pp. 1–35.

BRETON, S.: "La déduction thomiste des catégories", *Revue Philosophique de Louvain*, LX, 1962, pp. 5–32.

BROWNE, M.: "Circa intellectum et eius illuminationem apud S. Albertum Magnum", *Angelicum*, IX, 1932, pp. 187–202.

BURRELL, C. S. C., D.: "A Note on Analogy", *The New Scholasticism*, XXXV, 1962, pp. 225–32.

CHAHINE, O.: *Ontologie et théologie chez Avicenne*, Adrien-Maisonneuve, Paris, 1962.

CHARLIER, L.: "Les cinq voies de saint Thomas. Leur structure métaphysique", in *L'existence de Dieu*, Casterman, Tournai, 1961, pp. 181–227.

CHENU, O.P., M.-D.: *Introduction à l'étude de saint Thomas d'Aquin*, 2e édit., Vrin, Paris, 1954.

CLARKE, W. N.: "What is Really Real?", in *Progress in Philosophy*, edit. J. McWilliams, Sheed and Ward, Milwaukee, 1955, pp. 61–90.

COLLE, G.: "Commentaire" in *Aristote, La Métaphysique. Livre IV. Traductions et Commentaire*, Vrin, Paris, 1931.

CUNNINGHAM, F. A.: "Distinction According to St. Thomas", *The New Scholasticism*, XXXVI, 1962, pp. 279–312.

D'ALVERNY, M.-T.: "Avicenna Latinus", *Archives d'histoire doctrinale et littéraire du Moyen-Age*, XXVII, 1961, pp. 281–316.

— "Les traductions d'Avicenne (Moyen Age et Renaissance)", in *Avicenna nella*

Storia della Cultura Medioevale Relazioni e Discussione, (*15 Aprile 1955*), Accademia Nazionale dei Lincei, Roma, 1957, pp. 71–87.
— "Notes sur les traductions médiévales des oeuvres philosophiques d'Avicenne", *Archives d'histoire doctrinale et littéraire du Moyen Age*, XIX, 1953, pp. 337–58.
DECARIE, V.: *L'objet de la métaphysique selon Aristote*, Vrin, Paris, 1961.
DE COUESNONGLE, O.P., V.: "La causalité du *maximum*. I: L'utilisation par saint Thomas d'un passage d'Aristote, II: Pourquoi Saint Thomas a-t-il mal cité Aristote?", *Revue des Sciences Philosophiques et Théologiques*, XXXVIII, 1954, pp. 434–44, 658–80.
DE FINANCE, S.J., J.: *Etre et Agir dans la Philosophie de Saint Thomas*, 2e édit., Pont. Univ. Greg., Rome, 1960.
DE KONINCK, C.: "Metaphysics and the Interpretation of Words", *Laval Théologique et Philosophique*, XVII, 1961, pp. 23–34.
DEMAN, M.: "Remarques critiques de saint Thomas sur Aristote interprète de Platon", *Les sciences philosophiques et théologiques*, XXX, 1941–42, pp. 133–151.
DE RAEYMAEKER, L.: "Albert le Grand, Philosophe. Les lignes fondamentales de son système métaphysique", *Revue Néo-Scholastique de Philosophie*, XXXV, 1933, pp. 5–36.
— "L'idée inspiratrice de la métaphysique thomiste", *Aquinas*. III, 1960, pp. 61–82.
— *Metaphysica Generalis*, 2 vols., edit. altera, Warny, Louvain, 1935.
— *The Philosophy of Being. A Synthesis of Metaphysics*, trans. E. Ziegelmeyer, S.J., Herder, St. Louis, 1957.
— *Riflessioni su temi filosofici fondamentali*, Marzorati, Milano, 1957.
DE VAUX, C.: *Avicenne*, Alcan, Paris, 1900.
DE VAUX, R.: "La première entrée d'Averroès chez les Latins", *Revue des Sciences Philosophiques et Théologiques*, XXII, 1933, pp. 193–242.
DE VOGEL, C. J.: *Greek Philosophy. A Collection of Texts. Selected and Supplied with some Notes and Explanations*, Vol. II: *Aristotle, the Early Peripatetic School and the Early Academy*, Brill, Leiden, 1953.
— "La méthode d'Aristote en métaphysique d'après *Métaphysique* A 1–2", in *Aristote et les problèmes de méthode. Communications présentées au Symposium Aristotelicum tenu à Louvain du 24 août au 1er sept. 1960*, Nauwelaerts, Louvain-Paris, 1961, pp. 147–70.
DE WULF, M.: *History of Medieval Philosophy*, Vol. II: *The Thirteenth Century*, trans. E. Messenger, 3rd. Eng. edit. based on 6th French edit., Longmans, Green, and Co., London, 1938.
DHONDT, U.: "Science suprême et ontologie chez Aristote", Revue Philosophique de Louvain, LIX, 1961, pp. 5–30.
DONDAINE, O.P., A.: "Bulletin d'histoire", *Revue des Sciences Philosophiques et Theologiques.*, XXV, 1936, pp. 713–26.
— "Notes et Commentaires. Saint Thomas et les traductions latines des Métaphysiques d'Aristote", *Bulletin Thomiste*, X, 1933, pp. 199–213.
— *Secrétaires de S. Thomas*, 2 vols., Commissio Leonina, Rome, 1956.
DORRIE, H.: "Gedanken zur Methodik des Aristoteles in der Schrift περὶ ψυχῆς", in *Aristote et les problèmes de méthode. Communications présentées au Symposium Aristotelicum tenu à Louvain du 24 août au 1er sept. 1960*, Nauwelaerts, Louvain-Paris, 1961, pp. 223–44.
DUCHARME, L.: "'Esse' chez Saint Albert le Grand. Introduction à la métaphysique de ses premiers écrits", *Revue de l'Université d'Ottawa*, XXVII, 1957, pp. 209–52.

— *La constitution métaphysique de l'être crée dans les premières oeuvres de S. Albert le Grand,* unpublished doctoral thesis, Louvain, 1952.
DUCOIN, S.J., G.: "Saint Thomas commentateur d'Aristote" in *Histoire de la philosophie et métaphysique,* Desclee, Paris, pp. 85–107.
— "Saint Thomas commentateur d'Aristote. Etude sur le commentaire thomiste du livre Λ des Métaphysique d'Aristote", *Archives de Philosophie,* XX, 1957, pp. 78–117, 240–71, 392–445.
DUHEM, P.: *Le Système du monde. Histoire des doctrines cosmologiques de Platon à Copernic,* Vol. V, Hermann, Paris, 1917.
DUIN, J.: "Nouvelles précisions sur la chronologie du "Commentum in Metaphysicam" de S. Thomas", *Revue Philosophique de Louvain,* LIII, 1955, pp. 511–34.
ELDERS, S.V.D., L.: "Aristote et l'objet de la métaphysique", *Revue Philosophique de Louvain,* LX, 1962, pp. 165–83.
ESLICK, L.: "What is the Starting Point of Metaphysics?", *The Modern Schoolman,* XXXIV, May, 1957, pp. 247–63.
FABRO, C.P.S., C.: "Dall'essere di Aristotele allo 'esse' di Tommaso", in *Mélanges offerts à Etienne Gilson,* Vrin, Paris, pp. 227–47.
— "Intorno al fondamento della metafisica tomistica", *Aquinas,* III, 1960, pp. 83–135.
— *La nozione metafisica di partecipazione secondo S. Tommaso d'Aquino,* ediz. 2a., Soc. editrice internaz., Torino, 1950.
— *Participation et causalité selon S. Thomas d'Aquin,* Nauwelaerts, Louvain-Paris, 1961.
— "Un itinéraire de Saint Thomas. L'établissement de la distinction réelle entre essence et existence", *Revue de Philosophie,* IV, 1939, pp. 285–310.
FAY, C.: "Fr. Lonergan and the Participation School", *The New Scholasticism,* XXXIV, pp. 461–87.
FERNANDEZ, A.: "Scientiae et philosophia secundum Sanctum Albertum Magnum", *Angelicum,* XIII, 1936, pp. 24–59.
FESTUGIERE, A., "Notes sur les sources du commentaire de S. Thomas au livre XII des *Métaphysiques*", *Revue des Sciences Philosophiques et Théologiques,* XVIII, 1929, pp. 282–90, 657–63.
FLANNIGAN, C.S.J., Sister T.: "The Use of Analogy in the *Summa Contra Gentiles*", *The Modern Schoolman,* XXXV, Nov. 1957, pp. 21–37.
FOREST, A.: *La structure métaphysique du concret selon saint Thomas d'Aquin,* Avant-propos d'E. Gilson, 2e édit., Vrin, Paris, 1956.
GARDET, L.: *La pensée religieuse d'Avicenne* (*Ibn Sīnā*), Vrin, Paris, 1951.
GARRIGOU-LAGRANGE, O.P., R.: "Note premier jugement d'existence selon saint Thomas d'Aquin", in *Studia Mediaevalia in honorem admodum Rev. Patris R. J. Martin, O.P.,* De Tempel, Brugis, 1948, pp. 289–302.
— "Saint Thomas commentateur d'Aristote", *Dictionnaire de théologie catholique, Vol. XV,* col. 641–51.
GEIGER, O.P., L.-B.: "Abstraction et séparation d'après Saint Thomas, *In De Trinitate, q. 5, a. 3*", *Revue des Sciences Philosophiques et Théologiques,* XXXI, 1947, pp. 3–40.
— "De l'unité de l'être", *Revue des Sciences Philosophiques et Théologiques,* XXXIII, 1949, pp. 3–14.
— *La participation dans la philosophie de S. Thomas d'Aquin.* 2e édit., Vrin, Paris, 1953.
GEFFRE, C.-J.: "Théologie naturelle et révélation dans la connaissance du Dieu un", in *L'existence de Dieu,* Casterman, Tournai, 1961, pp. 297–317.
GEYER, B.: "Die Uebersetzungen der aristotelischen Metaphysik bei Albertus

Magnus und Thomas von Aquin", *Philosophisches Jahrbuch*, VII, 1917, pp. 392–407.

GIGNON, O.: "Methodische Probleme in der *Metaphysik* des Aristoteles", in *Aristote et les problèmes de méthode. Communications présentées au Symposium Aristotelicum tenu à Louvain du 24 août au 1er sept. 1960*, Nauwelaerts, Louvain-Paris, 1961, pp. 131–45.

GILSON, E.: "Avicenne et le point de départ de Duns Scot", *Archives d'histoire doctrinale et littéraire du Moyen Age*, II, 1927, pp. 89–149.

— *Being and Some Philosophers*, 2nd edit., Pont Inst. of Med. Studies, Toronto, 1952.

— *Elements of Christian Philosophy*, Doubleday, New York, 1960.

— *History of Christian Philosophy in the Middle Ages*, Random House, New York, 1955.

— "Les sources gréco-arabes de l'augustinisme avicennisant", *Archives d'histoire doctrinale et littéraire du Moyen Age*, IV, 1929, pp. 5–107.

— *The Christian Philosophy of St. Thomas Aquinas*, trans. L. Shook, C.S.B., Random House, New York, 1956.

GOICHON, A.-M.: "Introduction. L'évolution philosophique d'Ibn Sina", in *Ibn Sīnā (Avicienne). Livre des directives et remarques*, traduction avec introduction et notes, Vrin, Paris, 1951, pp. 1–74.

— *La distinction de l'essence et de l'existence d'après Ibn Sīnā (Avicienne)*, Desclée, Paris, 1937.

— *La philosophie d'Avicenne et son influence en Europe médiévale*, Forlong Lectures, 1940, 2e édit. revue, corrigée et augmentée, Adrien-Maisonneuve, Paris, 1951.

— "Notes" in *Introduction à Avicenne. Son Epitre des définitions*, traductions avec notes, préface de M. A. Palacics, Desclée, Paris, 1923.

GRABMANN, M.: *Circa historiam distinctionis essentiae et existentiae*, Excerptum ex *Acta Pontificio Academiae Romanae S. Thomae Aquinatis*, nova series, I, 1934.

— "Die Aristoteleskommentare des heiligen Thomas von Aquin", in *Mittelalterliches Geistesleben. Abhandlungen zur Geschichte der Scholastik und Mystik*, Band I, Hueber, München, 1926, pp. 266–313.

— *Die echten Schriften des hl. Thomas von Aquin. Auf Grund der alten Kataloge und der handschriftlichen Uberlieferung festgestellt.* Beiträge zur Geschichte der Philosophie und Theologie des Mittelalters, Band XXII, Heft 1, Aschendorf, Münster, 1920.

— "Die Lehre des heiligen Albertus Magnus vom Grunde der Vielheit der Dinge und der lateinische Averroismus", in *Mittelalterliches Geistesleben. Abhandlungen zur Geschichte der Scholastik und Mystik*, Band II, Hueber, München, 1936, pp. 287–312.

— "Die Schrift 'De ente et essentia' und die Seinsmetaphysik des heiligen Thomas von Aquin", in *Mittelalterliches Geistesleben. Abhandlungen zur Geschichte der Scholastik und Mystik*, Band I, Hueber, München, 1926, pp. 314–31.

— *Die Werke des hl. Thomas von Aquin. Eine literar-historische Untersuchung und Einführung*. Beiträge zur Geschichte der Philosophie und Theologie des Mittelalters, Band XXII, Heft. 1/2, 3. Auflage, Aschendorf, Münster, 1943.

— *Methoden und Hilfsmittel des Aristotelestudiums im Mittelalter*, Bayer. Akad. d. Wiss., München, 1939.

HANSEN, J.: "Zur Frage der anfangslosen und zeitlichen Schöpfung bei Albert dem Grossen", in *Studium Albertina Festschrift für B. Geyer zum 70. Geburtstage*. Beiträge zur Geschichte der Philosophie und Theologie des Mittelalters, Supplementband IV, Aschendorf, Münster, 1952, pp. 167–88.

HART, C.: *Thomistic Metaphysics. An Inquiry into the Act of Existing*, Prentice-Hall, Englewood Cliffs, N.J., 1959.

HAYEN S.J., A.: *La communication de l'être d'après saint Thomas d'Aquin*, Vol. I: *La métaphysique d'un théologien*, Desclée, Louvain-Paris, 1957.

— *L'intentionnel dans la philosophie de saint Thomas*, Desclée, Louvain-Paris, 1942.

HEATH, O.P., T.: "Saint Thomas and the Aristotelian *Metaphysics:* Some Observations", *The New Scholasticism*, XXXIV, 1961, pp. 438–60.

HENLE, S.J., R.: *Method in Metaphysics*, Marquette University Press, Milwaukee 1951.

— *Saint Thomas and Platonism. A Study of the Plato and Platonici Texts in the Writings of Saint Thomas*, Nijhoff, The Hague, 1956.

HOENEN, S.J., P.: *La théorie du jugement d'après St. Thomas d'Aquin*, Analecta Gregoriana XXXIX, editio altera, Pont. Univer. Gregoriana, Romae, 1953.

HORTEN, M.: *Die Hauptlehren des Averroes nach seiner Schrift: Die Widerlegung des Gazali*, aus dem arabischen Originale übersetzt und erläutert, Marcus, Bonn, 1913.

— *Die Metaphysik Avicennas*, enthaltend die Metaphysik, Theologie, Kosmologie und Ethik übersetzt und erläutert, (no publisher), Halle a.S., 1901.

— *Die Metaphysik des Averroes*, nach dem arabischen übersetzt und erläutert. Abhandlungen zur Philosophie und ihrer Geschichte XXXVI, (no publisher), Halle a. S., 1912.

ISAAC, O.P., J.: "Etudes critiques. No. 142. Mansion, A., 'Date de quelques...'", *Bulletin Thomiste*, XXIV–XXIX, 1947–52, pp. 174–76.

— "Saint Thomas interprète des oeuvres d'Aristote" in *Scholastica ratione historico-critica instaurando*, Acta congressus scholastici internationalis Romae anno sancto MCML celebrati, Pont. Athen. Antonianum, 1951, pp. 353–63.

ISAYE, S.J., G.: *La théorie de la Mesure et l'existence d'un maximum selon saint Thomas*, Archives de Philosophie XVI, Cahier I, Beauchesne, Paris, (no date).

JOLIVET, R.: *Essai sur les rapports entre la pensée grecque et la pensée chrétienne. Aristote et saint Thomas ou l'idée de création. Plotin et saint Augustin ou le problème du mal. Hellénisme et christianisme*, nouvelle édit., Vrin, Paris, 1955.

JANSSENS, E.: "Les premiers historiens de la vie de S. Thomas", *Revue Neo-Scholastique de Philosophie*, XVIII, 1924, pp. 201–14, 325–52, 452–76.

KAPPELI, T.: "Mitteilungen über Thomashandschriften in der Biblioteca Nazionale in Neapl. II. Ein Autograph des Metaphysikkommentars des hl. Thomas?", *Angelicum*, X, 1933, pp. 116–25.

KLUBERTANZ, S.J., G.: "Being and God According to Contemporary Scholastics" *The Modern Schoolman*, XXXII, 1954, pp. 1–17.

— *Introduction to the Philosophy of Being*, Appelton-Century-Crofts, New York, 1955.

— *St. Thomas Aquinas on Analogy. A Textual Analysis and Systematic Synthesis*, Loyola Univ. Press, Chicago, 1960.

— "St. Thomas on Learning Metaphysics", *Gregorianum*, XXXV, 1954, pp. 3–17.

— "The Teaching of Thomist Metaphysics", *Gregorianum*, XXXV, 1954, pp. 187–205.

LAUER, R.: "St. Albert and the Theory of Abstraction", *Thomist*, XVII, 1954, pp. 69–83.

LITTLE, S.J., A.: *The Platonic Heritage of Thomism*, Golden Eagle Books, Dublin, 1950.

LONERGAN, S.J., B.: *De constitutione Christi ontologica et psychologica*, altera editio, Pont. Univ. Gregoriana, Romae, 1958.

— *Divinarum personarum conceptionem analogicam*, Pont. Univ. Gregoriana, Romae, 1957.
— "*Insight:* Preface to a Discussion", in *Proceedings of the American Catholic Philosophical Association*, Cath. Univ., Wash., D.C., 1958, pp. 71–81.
— "The Concept of *Verbum* in the Writings of St. Thomas Aquinas", *Theological Studies*, VII, 1946, pp. 349–92; VIII, 1947, pp. 35–79 and 404–444; X, 1949, pp. 3–40 and 359–93.
LYTTKENS, H.: *The Analogy Between God and the World. An Investigation of its Background and Interpretation of its Use by Thomas of Aquino*, Universitets Arsskrift, Uppsala, 1953.
MANDONNET, O.P., P.: *Des écrits authentiques de S. Thomas d'Aquin*, 2e édit., revue et corrigée, St. Paul, Fribourg, 1910.
— *Siger de Brabant et l'averroïsme latin au XIIIe siècle*, 2. vols., Inst. Sup. de Phil., Louvain, 1911.
MANSION, A.: "Date de quelques commentaires de saint Thomas sur Aristote. (*De anima, Metaphysica*)", in *Studia Mediaevalia in honorem admodum Rev. Patris R. J. Martin, O.P.*, De Tempel, Brugis, 1948, pp. 271–87.
— *Introduction à la Physique Aristotélicienne*, 2e édit., revue et augmentée, Vrin, Louvain-Paris, 1945.
— "L'objet de la science philosophique suprême d'après Aristote, *Métaphysique E 1*", in *Mélanges de philosophie grecque offerts à Mgr. A. Diès*, Vrin, Paris, 1956, pp. 151–68.
— "L'origine du syllogisme et la théorie de la science chez Aristote", in *Aristote et les problèmes de méthode. Communications présentées au Symposium Aristotelicum tenu à Louvain du 24 août au 1er sept. 1960*, Nauwelaerts, Louvain-Paris, 1961, pp. 57–81.
— "Philosophie première, philosophie seconde, et métaphysique chez Aristote", *Revue Philosophique de Louvain*, LVI, 1958, pp. 165–221.
— "Pour l'histoire du commentaire de saint Thomas sur la Métaphysique d'Aristote", *Revue Neo-Scolastique de Philosophie*. XXVI, 1925, pp. 274–95.
— "Quelques travaux récents sur les versions latines des Ethiques et d'autres ouvrages d'Aristote", *Revue Neoscolastique de Louvain*, XXXIV, 1936, pp. 78–94.
— "Saint Thomas et le 'Liber de causis'. A propos d'une édition récente de son Commentaire", *Revue Philosophique de Louvain*, LIII, 1955, pp. 54–72.
— "'Universalis dubitatio de veritate'. S. Thomas in Metaphy., Lib. III, lect. I", *Revue Philosophique de Louvain*, LVII, 1959, pp. 513–42.
MANSION, S.: *Le jugement d'existence chez Aristote*, Desclée, Louvain-Paris, 1946.
— "Le rôle de l'exposé et de la critique des philosophies antérieures chez Aristote", in *Aristote et les problèmes de méthode. Communications présentées au Symposium Aristotelicum tenu à Louvain du 24 août au 1er sept. 1960*, Nauwelaerts, Louvain-Paris, 1961, pp. 35–56.
— "Les positions maîtresses de la philosophie d'Aristote", in *Aristote et S. Thomas d'Aquin*, Nauwelaerts, Louvain, 1955, pp. 43–91.
MARC, S.J., A.: *L'idée de l'être chez Saint Thomas et dans la scholastique postérieure*, Beauschene, Paris, 1933.
MARECHAL, S.J., J.: *Le point de départ de la métaphysique. Leçons sur le développement historique et théorique du problème de la connaissance*. Cahier V: *Le Thomisme devant la philosophie critique*, 2e édit., Desclée, Bruxelles-Paris, 1949.
MARITAIN, J.: *Distinguer pour unir ou les degrés du savoir*, 5e édit., Desclée, Paris, 1948.
— *Existence and the Existent*, trans. by L. Galantiere and G. Phelan, Panthéon, New York, 1948.

— *A Preface to Metaphysics. Seven Lectures on Being*, Mentor Omega Books, New York, 1962.

MASIELLO, R.: "The Analogy of Proportion According to the Metaphysics of St. Thomas", *The Modern Schoolman*, XXXV, Jan., 1958, pp. 91–105.

MAURER, C.S.B., A.: "Form and Essence in the Philosophy of St. Thomas", *Mediaeval Studies*, XIII, 1951, pp. 165–76.

— "The *De Quidditatibus Entium* of Dietrich of Freiburg and its Criticism of Thomistic Metaphysics", *Mediaeval Studies*, XVIII, 1956, pp. 173–88.

MCARTHUR, R.: "Universal *in praedicando*, universal *in causando*", *Laval Théologique et Philosophique*, XVIII, 1962, pp. 59–95.

MCINERNY, R.: "Notes on Being and Predication", *Laval Théologique et Philosophique*, XV, 1959, pp. 236–74.

— "Some Notes on Being and Predication", *The Thomist*, XXII, 1959, pp. 315–35.

— *The Logic of Analogy. An Interpretation of St. Thomas*, Nijhoff, The Hague, 1961.

— "The *Ratio Communis* of the Analogous Name", *Laval Théologique et Philosophique*, XVIII, 1962, pp. 9–34.

MCWILLIAMS, S.J., J.: "Notes and Discussion. Judgmental Knowledge", *The Modern Schoolman*, XXXIX, 1962, pp. 372–78.

MEERSSEMAN, O.P., G.: *Introductio in opera omnia B. Alberti Magni, O.P.*, (No publisher), Brugis, 1931.

MERLAN, P.: *From Platonism to Neoplatonism*, 2nd revised edit., Nijhoff, The Hague, 1960.

MILLER, R.: "An Aspect of Averroes' Influence on St. Albert", *Mediaeval Studies*, XVI, 1954, pp. 57–71.

MINIO-PALUELLO, L.: "Note sull'Aristotele latino medievale", *Rivista di Filosofia Neo-Scholastica*, XLII, 1950, pp. 222–37.

MONTAGNES, O.P., B.: *La doctrine de l'analogie de l'être d'après saint Thomas d'Aquin*, Philosophes Médiévaux, VI, Louvain, Pub. Univ.; Paris, Béatrice-Nauwelaerts, 1963.

MOREAU, J.: "Aristote et la vérité antéprédicative" in *Aristote et les problèmes de méthode. Communications présentées au Symposium Aristotelicum tenu à Louvain du 24 août au 1er sept. 1960*. Nauwelaerts, Louvain-Paris, 1961, pp. 21–33.

NARDI, B.: "L'aristotelismo della scholastica e i Francescani", in *Scholastica ratione historico-critica instaurando*, Acta congressus scholastici internationalis Romae anno sancto MCML celebrati, Pont. Athen. Antonianum, Romae, 1951, pp. 607–26.

NIELSEN, H.: "Discussion: Father Owens on Elucidation: A Comment", *The New Scholasticism*, XXXVI, 1962, pp. 233–36.

O'BRIEN, O.P., T.: *Metaphysics and the Existence of God. A Reflection on the Question of God's Existence in Contemporary Thomistic Metaphysics*, Thomist Press, Wash., D.C., 1960.

O'NEIL, M.: "Some Remarks on the Analogy of God and Creatures in St. Thomas Aquinas", *Mediaeval Studies*, XXIII, 1961, pp. 206–15.

O'SHAUGHNESSY, S.J., T.: "St. Thomas and Avicenna on the Nature of the One", *Gregorianum*, XLII, 1960, pp. 665–97.

— "St. Thomas' Changing Estimate of Avicenna's Teaching on Existence as an Accident", *The Modern Schoolman*, XXXVI, May 1959, pp. 245–60.

OWENS, C.Ss.R., J.: "Common Nature: A Point of Comparison Between Thomistic and Scotistic Metaphysics", *Mediaeval Studies*, XIX, 1957, pp. 1–14.

— "Review of: *Metaphysics and the Existence of God*. By Thomas C. O'Brien, O.P. ...", *The New Scholasticism*, XXXVI, 1962, pp. 250–53.

— "St. Thomas and Elucidation", *The New Scholasticism*, XXXV, 1961, pp. 421–44.
— *St. Thomas and The Future of Metaphysics*, Marquette Univ. Press, Milwaukee, 1957.
— "The Accidental and Essential Characteristics of Being in the Doctrine of St. Thomas Aquinas", *Mediaeval Studies*, XX, 1958, pp. 1–41.
— "The Causal Proposition – Principle or Conclusion?", *The Modern Schoolman*, XXXII, 1955, pp. 159–71, 257–70, and 323–39.
— *The Doctrine of Being in the Aristotelian Metaphysics*, preface by E. Gilson, Pont. Inst. of Med. Studies, Toronto, 1957.
— "Unity and Essence in St. Thomas Aquinas", *Mediaeval Studies*, XXIII, 1961, pp. 240–59.
PAULUS, J.: *Henri de Gand. Essai sur les tendances de sa métaphysique*, préface de M. E. Gilson, Vrin, Paris, 1938.
— "Le caractère métaphysique des preuves thomiste de l'existence de Dieu", *Archives d'histoire doctrinale et littéraire du Moyen Age*, IX, 1934, pp. 143–53.
PEGHAIRE, C.S.Sp., J.: *Intellectus et Ratio selon s. Thomas d'Aquin*, Vrin, Paris, 1936.
PEGIS, A.: "A Note on St. Thomas. *Summa Theologica*, I, 44, 1–2". *Mediaeval Studies*, VIII, 1946, pp. 159–68.
PELSTER, S.J., F.: "Die griechisch-lateinischen Metaphysik-übersetzungen des Mittelalters", in *Festgabe C. Baeumker zum 70. Geburtstag (16. sept. 1923)*, Abhandlungen zur Geschichte der Philosophie des Mittelalters, Aschendorf, Münster, 1923, pp. 39–118.
— "Die Uebersetzungen der aristotelischen Metaphysik in den Werken des hl. Thomas von Aquin", *Gregorianum*, XVI, 1935, pp. 325–48; 531–61; XVII, 1936, pp. 377–406.
— "Neure Forschungen über die Aristotelesübersetzungen des 12. und 13. Jahrhunderts. Eine kritische Ubersicht", *Gregorianum*, XXX, 1949, pp. 46–77.
PHELAN, F.: "Verum Sequitur Esse Rerum", *Mediaeval Studies*, I, 1939, pp. 11–22.
RABEAU, G.: *Species. Verbum. L'activité intellectuelle élémentaire selon S. Thomas d'Aquin*, Vrin, Paris, 1938.
REGIS, O.P., L.-M.: "Analyse et synthèse dans l'oeuvre de saint Thomas", in *Studia Mediaevalia in honorem Rev. Patris R. J. Martin O.P.*, De Tempel, Brugis, 1948, pp. 303–30.
— *L'Odyssée de la métaphysique*, Vrin, Montréal-Paris, 1949.
RENARD, S.J., H.: "What is St. Thomas' Approach to Metaphysics?", *The New Scholasticism*, XXX, 1956, pp. 64–83.
ROBERT, O.P., J.: "La métaphysique, science distincte de toute autre discipline philosophique selon saint Thomas d'Aquin", *Divus Thomas* (Piacenza), L, 1947, pp. 206–22.
ROHNER, O.P., P.: *Das Schöpfungsproblem bei Moses Maimonides, Albertus Magnus und Thomas von Aquin*, Beiträge zur Geschichte der Philosophie des Mittelalters, Band XI, Heft 5, Aschendorf, Münster, 1913.
ROLAND-GOSSELIN, O.P., M.-D.: "Introduction, Notes et Etudes historiques" in *Le "De ente et essentia" de St. Thomas d'Aquin*. Texte établi d'après les manuscrits parisiens. Introduction, Notes et Etudes historiques, Vrin, Paris, 1948.
ROSS, W.: *Aristotle*, 5th edit. revised, Methuen, London, 1949.
— "Introduction and Commentary" in *Aristotle's Metaphysics. A revised Text with Introduction, and Commentary*, 2 vols, Clarendon Press, Oxford, 1958.

— "Introduction and Commentary" in *Aristotle's Physics. A Revised Text with Introduction and Commentary*, Clarendon Press, Oxford, 1955.
SALIBA, D.: *Etude sur la Métaphysique d'Avicenne*, Press Univ. de France, Paris, 1926.
SALMAN, O.P., D.: "Albert le Grand et l'averroïsme latin", *Revue des Sciences Philosophiques et Théologiques*, XXIV, 1935, pp. 38–64.
— "Algazel et les latins", *Archives d'histoire doctrinale et littéraire du Moyen Age*, X, 1935–36, pp. 103–27.
— "Saint Thomas et les traductions latines des Métaphysiques d'Aristote", *Archives d'histoire doctrinale et littéraire du Moyen Age*, VII, 1932, pp. 85–120.
— "Versions latines et commentaires d'Aristote", *Bulletin Thomiste*, XIV, 1937, pp. 95–107.
— "Note sur la première influence d'Averroes", *Revue Neo-Scholastique de Philosophie*, XL, 1937, pp. 203–16.
SAUTIER, C.: *Avicennas Bearbeitung der Aristotelischen Metaphysik*, Herder, Freiburg, i. Br., 1912.
SCHMIDT, S.J., R.: "L'emploi de la séparation en métaphysique", *Revue Philosophique de Louvain*, LVIII, 1960, pp. 373–93.
SCHOOYANS, M.: "La distinction entre philosophie et théologie d'après les commentaires aristotéliciens de saint Albert le Grand", *Revista da Universidade Catolica de Sao Paulo*, XVIII, 1959, pp. 255–79.
SERTILLANGES, O.P., A.-D.: *La philosophie de S. Thomas d'Aquin*, 2 vols., Nouvelle édit., revue et augmentée, Aubier, Paris, 1940.
SESTILI, G.: "L'universale nella dottrina di S. Alberto Magno", *Angelicum*, IX, 1932, pp. 168–86.
SIEWERTH, G.: *Die Abstraktion und das Sein nach der Lehre des Thomas von Aquin*, Müller, Salzburg, 1958.
SIGER DE BRABANT: *Questions sur la métaphysique*. Texte inédit. Edité par C. A. Graiff, O.S.B., Inst. Sup. de Phil., Louvain, 1948.
SIMMONS, E.: "The Nature and Limits of Logic", *The Thomist*, XXIV, 1961, pp. 47–71.
SMITH, S.J., G.: "Avicenna and the Possible", *The New Scholasticism*, XVII, 1943, pp. 340–57.
TRICOT, J.: "Commentaire" in *Aristote. La Métaphysique*, nouvelle édit. entièrement refondue, avec commentaire, 2 vols., Vrin, Paris, 1953.
TURNER, W.: "St. Thomas' Exposition of Aristotle: A Rejoinder", *The New Scholasticism*, XXXV, 1961, pp. 210–24.
VAN DER BERGH, S.: "Introduction and Notes" in *Averroes' Tahafut Al-Tahafut* (*The Incoherence of the Incoherence*). Translated from the Arabic with Introduction and Notes, 2 vols., University Press, Oxford, 1954.
VAN RIET, G.: "La notion centrale du réalisme thomiste: l'abstraction", in *Problèmes d'Epistémologie*, Nauwelaerts, Louvain-Paris, 1960, pp. 1–45.
— "Philosophie et existence. A propos de 'L'être et l'essence' de M. Etienne Gilson", in *Problèmes d'Epistémologie*, Nauwelaerts, Louvain-Paris, 1960, pp. 144–69.
VAN STEENBERGHEN, F.: *Aristotle in the West. The Origins of Latin Aristotelianism*, Nauwelaerts, Louvain, 1955.
— "La composition constitutive de l'être fini", *Revue Philosophique de Louvain*, XII, 1938, pp. 489–518.
— "La démonstration de l'existence de Dieu par la finalité d'après les 'Quaestiones de veritate' de Saint Thomas d'Aquin", in *Medioevo e Rinascimento. Studi in onore di Bruno Nardi, 2 vols.*, Sansoni, Firenze, 1955, pp. 715–31.
— *La philosophie au XIIIe siècle*, (Philosophes médiévaux 9), Louvain, Pub. Universitaires; Paris, Béatrice-Nauwelaerts, 1966.

— "Le problème de l'existence de Dieu dans le 'De ente et essentia' de saint Thomas d'Aquin", in *Mélanges J. De Ghellinch, S.J.*, Vol. II, (no publisher), Gembloux, 1951, pp. 837–47.
— "Le problème de l'existence de Dieu dans le 'Scriptum super Sententiis' de saint Thomas", in *Studia mediaevalia in honorem Rev. Patris R. J. Martin, O.P.*, De Tempel, Brugis, 1948, pp. 331–49.
— "Le XIIIe siècle", in *Le mouvement doctrinal du XIe au XIVe siècle*, Vol. XIII of *Histoire de l'Eglise*, 2e édit., Bloud et Gay, Paris, 1956, pp. 177–328.
— *Ontologie*, 3e édit., Publ. Univ. de Louvain, Louvain, 1961.
— *Siger de Brabant d'après ses oeuvres inédits*, Vol. II: *Siger dans l'histoire de l'aristotélisme*, Editions de l'Inst. Sup. de Phil., Louvain, 1942.
— *The Philosophical Movement in the Thirteenth Century*, Nelson, Edinburgh, 1955.
Vansteenkiste, C.: "Il posto del tomismo nella storia del pensiero medioevale", *Aquinas*, III, 1960, pp. 307–27.
Verbeke, G.: "Authenticité et chronologie des écrits de saint Thomas d'Aquin", *Revue Philosophique de Louvain*, XLVIII, 1950, pp. 260–68.
— "Démarches de la réflexion métaphysique chez Aristote", in *Aristote et les problèmes de méthode. Communications présentées au Symposium Aristotelicum tenu à Louvain du 24 août au 1er sept. 1960.* Nauwelaerts, Louvain-Paris, 1961, pp. 107–29.
— "Note sur la date du commentaire du saint Thomas au De anima d'Aristote", *Revue Philosophique de Louvain*, L, 1952, pp. 56–63.
— "La date du commentaire de S. Thomas sur l'Ethique à Nicomaque", *Revue Philosophique de Louvain*, XLVII, 1949, pp. 203–20.
— "Le développement de la vie cognitive d'après S. Thomas", *Revue Philosophique de Louvain*, XLVII, 1949, pp. 437–57.
— "Philosophie et conceptions préphilosophiques chez Aristote", *Revue Philosophique de Louvain*, LIX, 1961, pp. 405–30.
Walz, A.: "Chronotaxis vitae et operum S. Thomas", *Angelicum*, XVI, 1934, pp. 467–83.
— *Saint Thomas d'Aquin*, adaptation française par P. Novarina, Nauwelaerts, Louvain-Paris, 1962.

INDEX OF TOPICS

(Because of the analytical and comparative nature of a large part of this book, an exhaustive index would first, contain numerous topics of secondary importance, and second, contain an excessive number of references for important topics. Such an index would do more to confuse the reader than help him. Accordingly, in the index that follows, references are given only to the basic treatments of the more important topics.

References given in parentheses are to footnotes: e.g. (106–3) refers to page 106, footnote 3. The abbreviations used are as follows: "Al" refers to Albert's position on a given topic; "Aq" to Aquinas'; "Ar" to Aristotle's; "Aver" to Averroes'; "Avic" to Avicenna's.)

INDEX OF TEXTS

I. ALBERT THE GREAT

Metaphysica

II. THOMAS AQUINAS

IV. AVERROES

Commentarium in Metaphysicorum